Antimalarial Medicinal Plants

Malaria continues to affect a large population of the world, especially in third world countries. The spread of drug-resistant parasites demonstrates the need for antimalarial agents with various modes of action. The search for remedies derived from medicinal plants for the treatment of malaria is reliant on accurate ethnobotanical and ethnopharmacological information obtained from traditional medical practitioners. *Antimalarial Medicinal Plants* provides information on bioactive compounds and therapeutic potentials of several antimalarial plant species found around the globe. This book evaluates these plant species with respect to their biology, diversity, distribution, and pharmacological values.

A volume in the *Exploring Medicinal Plants* series, this book highlights trends, technologies, processes, and services important to and necessary for efficient production, use, and understanding of medicinal qualities of antimalarial plants. It critically examines claims made by traditional medical practitioners with scientific validations for safe herbal drug formulation. It is a reference work for researchers of herbal medicine, traditional healers, pharmacists, and students associated with plant sciences and economic botany.

Exploring Medicinal Plants

Series Editor: Azamal Husen, Wolaita Sodo University, Wolaita, Ethiopia

Medicinal plants render a rich source of bioactive compounds used in drug formulation and development; they play a key role in traditional or indigenous health systems. As the demand for herbal medicines increases worldwide, supply is declining because most of the harvest is derived from naturally growing vegetation. Considering global interests and covering several important aspects associated with medicinal plants, the *Exploring Medicinal Plants* series comprises volumes valuable to academia, practitioners, and researchers interested in medicinal plants. Topics provide information on a range of subjects, including diversity, conservation, propagation, cultivation, physiology, molecular biology, growth response under extreme environment, handling, storage, bioactive compounds, secondary metabolites, extraction, therapeutics, mode of action, and healthcare practices.

Led by Azamal Husen, PhD, this series is directed to a broad range of researchers and professionals, consisting of topical books exploring information related to medicinal plants. It includes edited volumes, references, and textbooks available for individual print and electronic purchases.

Antidiabetic Medicinal Plants and Herbal Treatments
Edited by Azamal Husen

Ethnobotany and Ethnopharmacology of Medicinal and Aromatic Plants
Steps Towards Drugs Discovery
Edited by Mohd Adnan, Mitesh Patel, and Mejdi Snoussi

Wild Mushrooms and Health
Diversity, Phytochemistry, Medicinal Benefits, and Cultivation
Edited by Kamal Ch. Semwal, Steve L. Stephenson, and Azamal Husen

Medicinal Roots and Tubers for Pharmaceutical and Commercial Applications
Edited by Rakesh Kumar Bachheti and Archana Bachheti

Pharmacological Aspects of Essential Oils
Current and Future Trends
Edited by Sunita Singh, Pankaj Kumar Chaurasia, and Shashi Lata Bharati

Promising Antiviral Herbal and Medicinal Plants
Edited by Nadeem Akhtar, Azamal Husen, Vagish Dwibedi, and Santosh Kumar Rath

Antimalarial Medicinal Plants
Edited by Azamal Husen

For more information about this series, please visit: www.routledge.com/Exploring-Medicinal-Plants/book-series/CRCEMP

Antimalarial Medicinal Plants

Edited by Azamal Husen

CRC Press
Taylor & Francis Group
Boca Raton London New York

CRC Press is an imprint of the
Taylor & Francis Group, an **informa** business

Designed cover image: © Shutterstock

First edition published 2024
by CRC Press
2385 NW Executive Center Drive, Suite 320, Boca Raton FL 33431

and by CRC Press
4 Park Square, Milton Park, Abingdon, Oxon, OX14 4RN

CRC Press is an imprint of Taylor & Francis Group, LLC

Library of Congress Cataloging-in-Publication Data
Names: Husen, Azamal, editor.
Title: Antimalarial medicinal plants / Azamal Husen.
Other titles: Exploring medicinal plants ; 19.
Description: First edition. | Boca Raton : CRC Press, 2024. | Series: Exploring medicinal plants; 19 | Includes bibliographical references and index. | Summary: "Malaria continues to affect a large population of the world especially in third world countries, the spread of drug-resistant parasites demonstrates the need for antimalarial agents with various modes of action. The search for remedies derived from medicinal plants for the treatment of malaria is reliant on accurate ethnobotanical and ethnopharmacological information obtained from traditional medical practitioners. Antimalarial Medicinal Plants provides information on bioactive compounds and therapeutic potentials of several antimalarial plant species found around the globe. The book evaluates these plant species with respect to their biology, diversity, distribution, and pharmacological values. A volume in the Exploring Medicinal Plants series, the book highlights trends, technologies, processes, and services important to and necessary for efficient production, use, and understanding of medicinal qualities of antimalarial plants. It critically examines claims made by traditional medical practitioners with scientific validations for safe herbal drug formulation. It is a reference work for researchers of herbal medicine, traditional healers, pharmacists, and students associated with plant sciences and economic botany"— Provided by publisher.
Identifiers: LCCN 2023043583 (print) | LCCN 2023043584 (ebook) | ISBN 9781032457154 (hardback) | ISBN 9781032457253 (paperback) | ISBN 9781003378396 (ebook)
Subjects: LCSH: Antimalarials. | Medicinal plants. | Traditional medicine. | Malaria.
Classification: LCC RC159.A5 A586 2024 (print) | LCC RC159.A5 (ebook) | DDC 616.9/362—dc23/eng/20240209
LC record available at https://lccn.loc.gov/2023043583
LC ebook record available at https://lccn.loc.gov/2023043584

ISBN: 978-1-032-45715-4 (hbk)
ISBN: 978-1-032-45725-3 (pbk)
ISBN: 978-1-003-37839-6 (ebk)

DOI: 10.1201/9781003378396

Typeset in Times
by Apex CoVantage, LLC

Contents

Preface

Malaria is a serious and challenging, life-threatening infectious disease. This parasitic infectious disease is caused by protozoan parasites, namely, *Plasmodium falciparum*, *P. vivax*, *P. malariae*, *P. ovale*, *P. knowlesi*, and *P. cynomolgi*. The parasite is transmitted from human to human through the bite of an infective *Anopheles* mosquito. *Plasmodium* travels through the blood, matures and reproduces in the liver, and causes malaria disease. The common symptoms of malaria are fever and headache, and in severe cases, malaria leads to death. For its cure, normally, artemisinin-based medicines are used. However, antimalarial drug resistance has emerged as a great challenge to the malaria control programme. *P. falciparum* resistance to artemisinins has already been noticed. Thus, because malaria affects a large population of the world, the spread of drug-resistant parasites and the limited number of effective drugs for the treatment warrant discovery of new antimalarial drugs with new modes of action.

It has also been noticed that since the earliest times, mankind has used medicinal plants in order to cure various diseases and relieve suffering. Similarly, the search for new remedies from medicinal plant species used for the treatment of malaria depends on the accurate and specific ethnobotanical and ethnopharmacological information obtained from traditional medical practitioners. As we know, in these days, people are seeking an alternative to modern medical treatments, turning to phytomedicines for the treatment of malaria, and to cure various diseases and relieve suffering. Thus, in this connection, plant-based pharmaceuticals play a significant role. Numerous plant species belonging to the *Aristolochia*, *Aspidosperma*, *Cinchona*, *Croton*, *Momordica*, *Piper*, *Senna*, *Solanum*, *Stachytarpheta*, *Tabebuia*, *Swertia*, etc. genera have shown immense antimalarial activity and are reviewed in detail. Moreover, this book critically examines and reviews the claims made by traditional medical practitioners and scientific validations. Other information on aspects concerning diversity, biology, phytochemistry, harvest and conservation techniques, genetic improvement, *in vitro* practice for higher antimalarial agents, and so on have been presented.

Taken together, the aim of the book is to provide an overview of the most important and selected antimalarial plant species around the world and the natural products obtained from these plants. This book provides valuable information to scientists, researchers, and students working especially on medicinal plants or in areas of ethnomedicine, natural products, economic botany, plant biochemistry, biotechnology, pharmacognosy, industrial chemistry, and other allied subjects. This book would also help in testifying on the various reports on antimalarial plants and their other potential uses in new drug discovery based on their bioactive molecules, and may also attract the attention of pharmaceutical companies, industrialists, and health policymakers.

I am grateful to all contributors for readily accepting my invitation, sharing their knowledge in specialized areas of research, and readily adjusting the suggestions for improving the shape of their contributions. With great pleasure, I also extend my sincere thanks to Randy Brehm, Tom Connelly, and all associates at Taylor & Francis Group, LLC/CRC Press, for their active cooperation. Finally, my special thanks go to Shagufta, Zaara, Mehwish, and Huzaifa for providing their time, and overall support, to put everything together.

Azamal Husen

About the Editor

Azamal Husen is a Professor at Sankalchand Patel University, Visnagar, India; and Adjunct Professor at Graphic Era (Deemed to be University), Dehradun, Uttarakhand, India. He is also working as a Visiting Professor at University Putra Malaysia, Selangor, Malaysia; and acting as a Foreign Delegate at Wolaita Sodo University, Wolaita, Ethiopia. Previously, he served as Professor and Head of the Department of Biology, University of Gondar, Ethiopia. He also worked as a Visiting Faculty of the Forest Research Institute and the Doon College of Agriculture and Forest at Dehra Dun, India. His research and teaching experience of 25 years encompasses biogenic nanomaterial fabrication and application; plant responses to nanomaterials; plant adaptation to harsh environments at the physiological, biochemical, and molecular levels; herbal medicine; and clonal propagation for the improvement of tree species.

He has conducted research sponsored by the World Bank, the National Agricultural Technology Project, the Indian Council of Agriculture Research, the Indian Council of Forest Research Education, and the Japan Bank for International Cooperation. Husen has published extensively (over 250) and served on the Editorial Board and as a reviewer of reputed journals published by Elsevier, Frontiers Media, Taylor & Francis, Springer Nature, RSC, Oxford University Press, Sciendo, the Royal Society, CSIRO, PLOS, MDPI, John Wiley & Sons, and UPM Journals. He is on the advisory board of Cambridge Scholars Publishing, UK. He is a fellow of the Plantae group of the American Society of Plant Biologists, and a member of the International Society of Root Research, Asian Council of Science Editors, and the International Natural Product Sciences. He is Editor-in-Chief of the *American Journal of Plant Physiology*, and a Series Editor of *Exploring Medicinal Plants* (Taylor & Francis Group, USA); *Plant Biology, Sustainability, and Climate Change* (Elsevier, USA); and *Smart Nanomaterials Technology* (Springer Nature, Singapore). He has been awarded the distinguished honour of being recognized as one of the 'World's Top 2% Scientists' for the year 2022, and again for the year 2023 by Stanford University, USA. This recognition has also been prominently featured in the Elsevier Data Repository.

Contributors

Sahar Ahangari
Department of Agricultural Extension and Education
Faculty of Agriculture
Tarbiat Modares University
Tehran, Iran

Jamal Akhtar
Central Council for Research in Unani Medicine
Ministry of AYUSH
Government of India
New Delhi, India

Shah Alam
Central Council for Research in Unani Medicine
Ministry of AYUSH
Government of India
New Delhi, India

Ahmad Ali
Department of Life Sciences
University of Mumbai
Mumbai, India

Hina Alim
Department of Life Sciences
University of Mumbai
Mumbai, India

Arvind Arya
Department of Biotechnology
Noida Institute of Engineering
Technology
Greater Noida (UP), India

Atul Arya
Medicinal Plant Research Laboratory
Department of Botany, Ramjas College
University of Delhi
Delhi, India

Anoja Priyadarshani Attanayake
Department of Biochemistry
Faculty of Medicine
University of Ruhuna, Sri Lanka

Issakou Bakarnga-Via
Department of Biomedical and Pharmaceutical Sciences
Faculty of Human Health Sciences
University of Ndjamena
Ndjamena, Chad

Fouzia Bashir
Central Council for Research in Unani Medicine
Ministry of AYUSH
Government of India
New Delhi, India

Mangal Chandra Biswas
Molecular and Applied Mycology and Plant Pathology Laboratory
Department of Botany
University of Calcutta
Kolkata, India

Kolagani Chandramohan
Forest Survey of India, Central Zone
Seminary Hills, Nagpur
Maharashtra, India

Uttra Chauhan
Department of Microbiology
Graphic Era (Deemed-to-Be University)
Dehradun, Uttarakhand, India

Nayomi Deshani De Silva
Department of Biochemistry
Faculty of Medicine
University of Ruhuna
Sri Lanka

Abhijit Dey
Department of Life Sciences
Presidency University
Kolkata, India

Patrick Valere Tsouh Fokou
Department of Biochemistry
Faculty of Science
University of Bamenda
Bambili, Cameroon

Leila Hakimi
Department of Agriculture
Saveh Branch
Islamic Azad University
Saveh, Iran

Azamal Husen
Department of Biotechnology
Smt. S. S. Patel Nootan Science & Commerce College
Sankalchand Patel University
Visnagar-384315, Gujarat, India

Department of Biotechnology
Graphic Era (Deemed to be University)
Uttarakhand, India

Harshika Joshi
Department of Microbiology
Graphic Era (Deemed-to-Be University)
Dehradun, Uttarakhand, India

Raoul Kemzeu
Department of Biochemistry
Faculty of Science
University of Yaoundé 1
Yaoundé, Cameroon

Salman Khan
Forest Survey of India, Central Zone
Seminary Hills, Nagpur
Maharashtra, India

Tasneem Khan
Department of Life Sciences
University of Mumbai
Mumbai, India

Esmail Khosropour
Department of Forestry and Forest Economics
University of Tehran
Karaj, Iran

Sandeep Kumar
Department of Biotechnology
Shobhit Institute of Engineering and Technology (Deemed-to-Be University)
Meerut, UP, India

Suresh Kumar
Medicinal Plant Research Laboratory
Department of Botany
Ramjas College, University of Delhi,
Delhi, India

Ishaan Lakhani
Department of Life Sciences
J. C. Bose University of Science and Technology,
YMCA Faridabad
Haryana, India

Sachin Mahawar
Medicinal Plant Research Laboratory
Department of Botany Ramjas College,
University of Delhi
Delhi, India

Ahmad Mirjalili
Department of Horticultural Science and Agronomy
Science and Research Branch
Islamic Azad University
Tehran, Iran

Shuvadip Mondal
Department of Botany
Fakir Chand College,
Diamond Harbour
West Bengal, India

Shivansh Negi
Department of Biotechnology
Graphic Era (Deemed-to-Be University)
Dehradun, Uttarakhand, India

Vincent Ngouana
Department of Pharmaceutical Sciences
Faculty of Medicine and Pharmaceutical Sciences
University of Dschang
Dschang, Cameroon

Shakirdeen Mayowa Obidola
Department of Crop Production Technology
Federal College of Forestry
Jos, Nigeria

Olalekan Bukunmi Ogunro
Department of Biological Sciences
KolaDaisi University
Ibadan, Nigeria

Ahmed Olatunde
Department of Medical Biochemistry
Abubakar Tafawa Balewa University
Bauchi, Nigeria

Gaurav Pant
Department of Microbiology
Graphic Era (Deemed-to-Be University)
Dehradun, Uttarakhand, India

Manu Pant
Department of Biotechnology
Graphic Era (Deemed-to-Be University)
Dehradun, Uttarakhand, India

Apekshakumari Patel
Department of Life Sciences
University of Mumbai
Mumbai, India

Nimisha Patel
Department of Life Sciences
J.C. Bose University of Science and Technology, YMCA
Faridabad, Haryana, India

Yashashree Pradhan
Department of Life Sciences
University of Mumbai
Mumbai, Maharashtra, India

Shakeelur Rahman
Prakriti Bachao Foundation
Jharkhand, India

Manjula Rai
Department of Botany
St. Joseph's College
Darjeeling, West Bengal, India

Rudra Narayan Sahoo
Department of Botany
Jamia Hamdard
New Delhi, India

Rupa Sanyal
Department of Botany
Bhairab Gnaguly College
Kolkata, India

Noor Sayed
Department of Biotechnology
Graphic Era (Deemed-to-Be University)
Dehradun, Uttarakhand, India

Surjit Sen
Department of Botany
Fakir Chand College
West Bengal, India

Gunjan Sharma
Department of Life Sciences
J. C. Bose University of Science and Technology YMCA Faridabad
Haryana, India

Harsha Singh
Medicinal Plant Research Laboratory
Department of Botany
Ramjas College
University of Delhi
Delhi, India

Habibu Tijjani
Department of Biochemistry
Bauchi State University
Gadau, Nigeria

Pankaj Kumar Tyagi
Department of Biotechnology
Noida Institute of Engineering Technology
Greater Noida (UP), India

Keddagoda Gamage Piyumi Wasana
Department of Biochemistry
Faculty of Medicine
University of Ruhuna,
Sri Lanka

Aubin Kamche Youbi
Department of Biochemistry
Faculty of Science
University of Yaoundé
Yaoundé, Cameroon

1 Diversity and Geographic Distribution of Some Antimalarial Plant Species

Salman Khan, Kolagani Chandramohan, and Azamal Husen

1.1 INTRODUCTION

Worldwide, humans are exposed to various types of diseases, of which some are natural and some are anthropogenic-driven (Murray and Lopez, 1996). Malaria is such an important disease which has global impact on humans. It is widespread in tropical areas, such as Africa, China, India, and other Asian countries (O'Neill et al., 1985). Castro and Peterka (2023) reported the increasing trends of malaria cases specifically in indigenous and artisanal mining areas of Brazilian Amazon. It is due to increased deforestation and mining (Figure 1.1). In the year 1980, more than 5 million cases per annum were documented from Africa (O'Neill et al., 1985). Dash et al. (2008) mentioned that as per WHO (World Health Organization), India's contribution to malaria is about 70% in South-East Asian countries.

Anopheles spp. mosquitoes act as vector organisms for malaria-causing protozoan *Plasmodium* spp. The bite of the female *Anopheles* mosquito leads to the transfer of *Plasmodium* into human blood, causing malaria. In India, *Anopheles culicifacies*, *A. stephensi*, *A. dirus*, *A. fluviatilis*, *A. minius*, *A. sundaicus*, *A. annularis*, *A. varuna*, *A. jeyporiensis*, and *A. philippinensis* are reported to carry *Plasmodium* (Rao, 1984; Dash et al., 2008). The former six species are considered as primary vectors, and the rest are secondary.

Due to the global climate change, insects and other microorganisms are modifying their body and behaviour. Many insecticides are not effective against mosquitoes due to resistance development in them. Similarly, protozoans are being resistant to various chemotherapeutic potential agents (O'Neill et al., 1985).

There are so many factors which influence the dynamics of malarial cases, but the major reasons are population movement, climate change, deforestation, mining, agricultural activity, and floods (Figure 1.1). It is estimated that global warming tends to increase in global temperature, and this will change the growth rate of vector mosquitoes, increase blood digestion rate, and cause a surge in feeding rate (Dash et al., 2008).

Various programmes and projects for minimizing malarial cases have been formulated by many agencies, including WHO and the government of India. Programme such as PfCP (*Plasmodium falciparum* containment programmes), urban malarial schemes, global malaria eradication programmes, national malarial eradication programmes, and others are effective in minimizing the cases (Sharma, 1996).

1.2 GEOGRAPHIC DISTRIBUTION OF MALARIA IN THE WORLD

Dash et al. (2008) reported that 36% of the world population is affected by malaria. Around 107 countries and their nearby areas have shown malarial cases worldwide. India, being diverse in

DOI: 10.1201/9781003378396-1

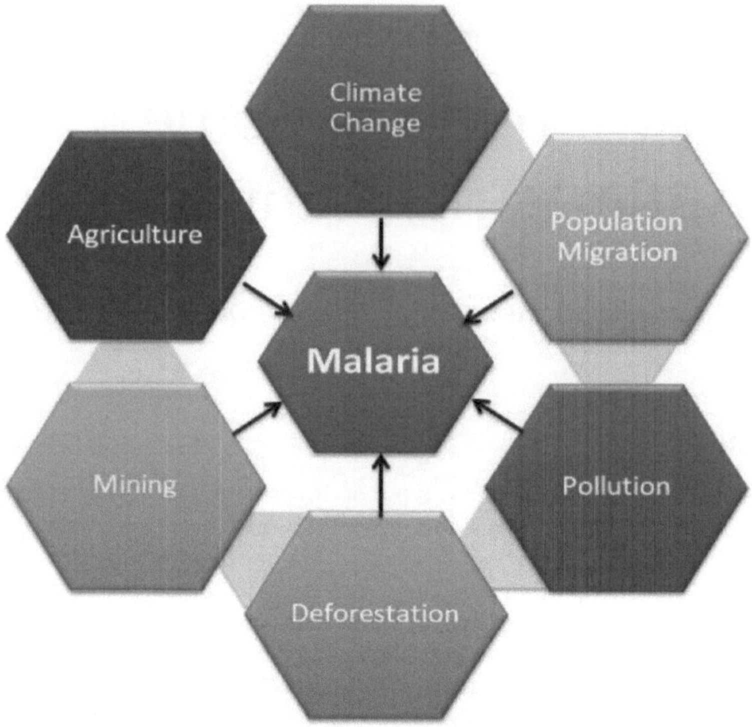

FIGURE 1.1 Factors driving dynamics of malarial cases in humans.

climate, topography, and biodiversity, provides the ideal environment for vector growth and its spread (Das et al., 2012).

- **In world other than India:** Africa, Mexico, South-East Asia, South America, Australia, Europe, Iran, etc. (Richards et al., 2001; Dash et al., 2008; Das et al., 2012; Narain and Nath, 2018).
- **In India:** Chhattisgarh, Gujarat, Haryana, Jharkhand, Madhya Pradesh, North-East India, Orissa, Uttarakhand, and West Bengal (Sharma et al., 1983; Misra, 1996; Bhattacharya et al., 2006; Kumar et al., 2007; Dash et al., 2008).

1.3 DESCRIPTION OF MAJOR ANTIMALARIAL PLANTS OF INDIA

Muñoz et al. (2000a, 2000b) identified the natural biological compound used by Mostene and Chacobo tribes against malaria derived from plants in Bolivia. They mentioned species such as *Apuleia leiocarpa*, *Bauhinia guianensis*, *Nectandra cuspidata*, *Parattanthelium amazonum*, *Tanaecium jaroba*, *Qualea paraensis*, *Derris amazonica*, and *Sclerolobium guianense*. A similar report on medicinal plants with antimalarial potential used by the Tacana Indians was also presented by Deharo et al. (2001) from Bolivia. Mahato and Sen (1997) and Matte (2023) mentioned that biochemical compound terpenoid exhibits tremendous antimalarial properties. O'Neill et al. (1985) reported antimalarial properties in *Artemisia annua* due to the presence of artemisinin. It was also reported that the most important families of plants which show great potential against malaria are Amaryllidaceae and Simaroubaceae. Spencer et al. (1947) surveyed and formulated a document mentioning a list of antimalarial plants belonging to 33 genera. Muregi et al. (2007) mentioned that *Vernonia lasiopus*, *Toddalia asiatica*, *Ficus sur*, *Rhamnus prinoides*, and *R. staddo* possess significant antimalarial properties. Mahagany (*Swietenia mahagoni*) seeds possess great antimalarial activity against *Plasmodium berghei* due to the presence of flavonoids and its derivates (Sari et al., 2023). Similarly, Chinchilla et al. (2012) listed 25 important *in vitro* antimalarial plant species from Costa Rica. These include *Cinnamomum chavarria*, *Hampea appendiculata*, *Nectandra membranacea*, *Persea povedae*, *Psidium guajava*, *Senna papillosa*, and *Siparuna thecaphora*. Mariath et al. (2009) formulated a detailed list of 476 antimalarial plants belonging to 103 plant families, such as Anacardiacea, Aracaceae, Bixaceae, Brassicaceae, Fabaceae, Lauraceae, Moraceae, Rhamnaceae, and Velloziaceae, from the American continent. Şener et al. (2003) provided details on important alkaloids extracted from the different species of Amaryllidaceae family. Similarly, Bray et al. (1990) mentioned the antimalarial potential of few important species of family Meliaceae.

Various researchers have also identified and reported important medicinal plants having higher antimalarial properties. Koudouvo et al. (2011) reported 52 plant species related to 29 families having antimalarial properties from the Togo Maritime Region. From West and Central Africa, alkaloids extracted from different antimalarial plants were described by Ancolio et al. (2002). It includes *Combretrum micranthum*, *Feretia apodanthera*, *Guiera senegalensis*, *Morinda*sp., *Pycnanthus angolensis*, and *Securidaca longepedunculata*. Abosi and Raseroka (2003) provided details on the antimalarial potential of the leaves and roots of *Vernonia amygdalina*. Achenbach (1992) mentioned the antimalarial property of *Hoslundia opposita*, which grows in East and West Africa. From Cameroon (Central Africa), 217 antimalarial plants' descriptions were given by Titanji et al. (2008). They also mentioned that about 100 bioactive compounds were isolated from 26 different species of plants from Cameroon.

Similarly, Tona et al. (1999) reported nine plants with antimalarial properties from the Democratic Republic of Congo. Likewise, Kaou et al. (2008) mentioned nine different antimalarial plant species from Africa. Azas et al. (2002) reported an experiment on the use of a combination of medicinal plants crude extract for antimalarial activity in Mali. Asase and Oppong-Mensah (2009) surveyed and identified about 29 antimalarial plants belonging to 24 different families from Ghana. Important medicinal plants (Figures 1.2 and 1.3) and various chemical features are shown in Figure 1.4.

Other than in Africa, many plants are being used as antimalarial/antiplasmodial agents. Researchers such as Mojab (2012) mentioned antimalarial plants from Iran, such as *Artemisia* spp., *Cichorium intybus*, *Glycyrrhiza glabra*, *Prosopis juliflora*, and *Solanum surattense*. Zhang et al. (2002) reported the novel antimalarial alkaloid (decursivine) extracted from the leaves and stem of *Rhaphidophora decursiva*. In India, Kamaraj et al. (2012) reported a list of 23 medicinal antimalarial plants which were used in the village area. They reported the following species with great antimalarial properties in different parts of the plant due to the activity of secondary metabolites: *Adhatoda vasica*, *Aegle marmelos*, *Andrographis lineata*, *Argemone Mexicana*, *Cassia nictitans*, *Coriandrum sativum*, *Cuminum cyminum*, *Datura metel*, *Diospyros melanoxylon*, *Eclipta prostrata*, *Lantana camara*, *Momordica charantia*, *Nelumbo nucifera*, *Piper nigrum*, and *Tagetes erecta*. Rahman et al. (1999) mentioned about the antimalarial properties of *Andrographis paniculata*, *Piper sarmentosum*, and *Tinospora crispa* from Malaysia. Garavito et al. (2006) reported the presence and activity of Colombian medicinal plants as antimalarial agents belonging to the families Annonaceae, Euphorbiaceae, Menispermaceae, Miosaceae, Piperaceae, and Solanaceae. Nguyen-Pouplin et al. (2007) identified and reported 49 different species of plants having antimalarial properties. Moreover, researchers from Indonesia have reported 72 species of plants belonging to 40 families as antimalarial potential species (Budiarti et al., 2020). Details of the major plant species, common name, their distribution, important chemical compound, from the world are discussed in Table 1.1.

1.4 BIOCHEMICAL COMPOUNDS

1.4.1 Alkaloids

They are an important biological compound which is extracted from various parts of the plant. They are capable of treating various diseases, including malaria. They have physiologically active

FIGURE 1.2 (a) *Vachellia nilotica* (L.) PJH Hurter and Mabb.; (b) *Ageratum conyzoides* L.; (c) *Argemone mexicana* L.; (d) *Annona squamosa* L.; (e) *Azadirachta indica* A. Juss.; (f) *Bixa orellana* L.; (g) *Cissampelos pareira* L.

nitrogen bases in a heterocyclic ring obtained from amino acids (Figure 1.4). The alkaloid groups that have antimalarial potency include terpenoidal and steroidal alkaloids, indole alkaloids, phenanthroindolizine alkaloids, isoquinoline, benzylisoquinoline and hasubanane alkaloids, naphthoisoquinoline alkaloids, aporphine and morphinandienone alkaloids, proterberine alkaloids,

FIGURE 1.3 (a) *Tridax procumbens* L.; (b) *Lippia javanica* (Burm.f.) Spreng.; (c) *Cassia fistula* L.; (d) *Glinus oppositifolius* (L.) Aug.DC.; (e) *Lantana camara* L.; (f) *Holarrhena pubescens* Wall. ex G. Don; (g) *Cestrum nocturnum* L.; (h) *Terminalia bellirica* (Gaertn.) Roxb.; (i) *Tinospora cordifolia* (Willd.) Hook.f. and Thomson.

amaryllidaceae alkaloids, cyclopeptide alkaloids, quinoline alkaloids, pyridocoumarin alkaloids, acridone alkaloids, and macrocyclic alkaloids (Uzor, 2020). Few examples of major alkaloids are anonaine, augustine, caesalminines A and B, carpaine, crinamine, lycorine, tubulosine, microthecaline, N-methyltelobine, obtusipetadione, simplicifolianine, virekine, etc. (Saxena et al., 2003).

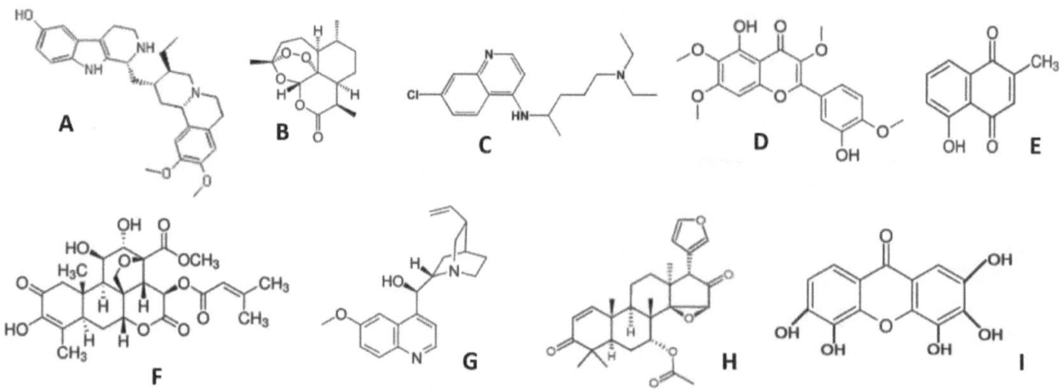

FIGURE 1.4 (a) Alkaloid, (b) artemisinin, (c) chloroquine, (d) flavonoid, (e) quinone, (f) quassinoid, (g) quinine, (h) triterpenoid, and (i) xanthone.

1.4.2 Chloroquine

It is a type of aminoquinoline which is primarily used for the treatment of malaria. It has many other uses, such as antiviral, extraintestinal amoebiasis treatment, antirheumatic, anti-parasitic, anti-inflammatory, etc. (Scholar, 2007). The structure of chloroquine is discussed in Figure 1.4. However, chloroquine resistance against malarial parasite is also reported from various countries. The chemical formula of chloroquine is $C_{18}H_{26}ClN_3$. Other names of chloroquine are Artichin, Chlorochine, Resochin, Sanoquin, Solprine, and others (Scholar, 2007).

1.4.3 Artemisinin

It is a type of sesquiterpenes lactone having an endoperoxide bridge derived from plants like *Artesimia annua* (Guo, 2016). The chemical formula of artemisinin is $C_{15}H_{22}O_5$. Its synonyms are artemisinine, artemisininum, and quinghaosu. Artemisinin is mostly used in malaria treatment worldwide. It is also used as an anti-infectious agent and has capabilities to treat cancer. A novel derivative of artemisinin based on the structural configuration is dihydroartemisinin (Guo, 2016).

1.4.4 Quinine

It is a medication that is used to treat malaria solely or in combination with other medicines (Achan et al., 2011). The chemical formula of quinine is $C_{20}H_{24}N_2O_2$ (Figure 1.4). Apart from malaria, it is also used to treat piroplasmosis. Like with other chemical compounds, resistance to quinine is also developed in malarial parasites. Its synonyms include chinine, chinin, and cinchonidine. It can be obtained from the bark of *Cinchona* spp. (Muñoz et al., 2000a).

1.4.5 Quinones

Quinones are a biochemical compound that has a complex structure and shows antimalarial activities. They are either 1,4-diketocyclohexa-2,5-dienoid or 1,2-diketocyclohexa-3,5-dienoid. Naphthoquinones are one of the important structures of quinones in which the asexual erythrocytic

TABLE 1.1
List of Important Medicinal Plants throughout the World Having Antimalarial Properties

Species	Common/Local Name	Family	Bioactive Compound	Plant Part Used	Distribution	Key References
Acacia nilotica	Babul	Fabaceae	Terpenoids	Leaves	India	Matte, 2023
Acalypha fruticosa	Birch-leaved cattail	Euphorbiaceae	Sesquiterpenoids	Leaves	Africa, India	Alshawsh et al., 2009
Acalyphawilkensiana	Copper plant	Euphorbiaceae	Alkaloids, flavonoids, saponins, tannins	Leaves	Pacific islands	Udobang et al., 2010
Acanthospermum hispidum	Bristly starbur	Asteraceae	Alkaloids, flavonoids	Aerial parts	Central and South America	Ganfon et al., 2012
Achillea millefolium	Yarrow	Asteraceae	Sesquiterpenes	Whole plant	North America, Europe, Asia	Saxena et al., 2003
Acokanthera schimperi	Arrow poison Tree	Apocynaceae	Glycosides	Leaves	Africa	Mohammed et al., 2014
Acridocarpus alternifolius	Papao	Malpighiaceae	Flavonoids	Leaves	Ghana	Laryea and Borquaye, 2019
Adhatoda spp.	Vasa	Acanthaceae	Alkaloids, flavonoids, tannins, saponins	Leaves	Malabar nut	Bora et al., 2007; Kamaraj et al., 2012
Aegle marmelos	Bel, golden apple	Rutaceae	Anthraquinone	Leave, bark	India	Kamaraj et al., 2012
Ageratum conyzoides	Billygoa weed	Asteraceae	Alkaloids, flavonoids, coumarins, essential oils	Aerial parts	Brazil, Kenya	Nour et al., 2010
Ajugaintegrifolia	Bugleweed	Lamiaceae	Terpenoids, iridoid glycosides, flavonoids, essential oils	Roots, root bark	Ethiopia, Asia	Asnake et al., 2015
Ajugaremota	Armagusa	Lamiaceae	Flavonoids, alkaloids, tannins, terpenoids	Aerial parts	Ethiopia, Africa	Njoroge and Bussmann, 2006; Gitua et al., 2012
Albiziazygia	West African walnut	Fabaceae	Flavonoids	Aerial parts	Africa	Abdalla and Laatsch, 2012
Allanblackia floribunda	Tallow tree	Clusiaceae	Flavonoids, xanthanoids, essential oil	Aerial parts	Africa	Azebaze et al., 2015
Allophylus africanus	False currant	Sapindaceae	Flavonoids, polyphenols, essential oils	Whole plant	Africa	Oladosu et al.,2013
Aloe spp.	Mugwanugu	Asphodelaceae	Alkaloids, flavonoids	Various parts	Kenya	Njoroge and Bussmann, 2006; Singh et al., 2022
Alstonia angustifolia	Pulai	Apocynaceae	Alkaloids	Leaves and bark	Asia	Saxena et al., 2003
Alstonia boonei	Ahun	Apocynaceae	Alkaloids, triterpenoids	Stem bark, leaves	Nigeria	Idowu et al., 2010
Alstoniamacrophylla	Hard milkwood	Apocynaceae	Alkaloids, flavonoids, terpenoids, saponins, tannins	Bark, stem bark, root bark	India, Malaysia, Thailand	Saxena et al., 2003; Cheenpracha et al., 2013

(*Continued*)

TABLE 1.1 (*Continued*)
List of Important Medicinal Plants throughout the World Having Antimalarial Properties

Species	Common/Local Name	Family	Bioactive Compound	Plant Part Used	Distribution	Key References
Amomum kravanh	Bai Dou Kou	Zingiberaceae	Myrtenol	Fruit	South-East Asia	Saxena et al., 2003
Ampelozizyphus amazonicus	Saracura mira	Rhamnaceae	Saponins, terpenoids, steroids	Stem, root	Amazon region	do Carmo et al., 2015
Anadenanthera colubrina	Cebil, willca	Fabaceae	Flavonoids	Bark	Africa	Muñoz et al., 2000b
Ancistrocladus spp.	Cameroonian vine	Ancistrocladaceae	Korupensamine, alkaloids	Leaf, tubers, root bark, stem	Africa	Saxena et al., 2003
Andira inermis	Brown heart	Fabaceae	Flavonoids, xanthones	Leaves, stem	Africa, America	Saxena et al., 2003
Andrographis lineata	Striped chiretta	Acanthaceae	Flavonoids	Whole plant	India	Kamaraj et al., 2012
Andrographis paniculata	Sambiroto	Acanthaceae	Andrographolide	Whole plant	Malaysia	Rahman et al., 1999
Angelica purpuraefolia	Bai jhi	Apiaceae	Khellactone, triterpenes	Rhizome	South America, Africa	Chung et al., 2010
Anisochilus harmandii	Kapurli	Lamiaceae	Terpenoids, flavonoids	Aerial parts	Asia	Lekphrom et al., 2010
Annona squamosa	Wild sweetsop	Annonaceae	Alkaloids, glycosides, flavonoids	Aerial parts	South America	Johns et al., 2011
Anthemis nobilis	Roman chamomile	Asteraceae	Sesquiterpenes	Aerial parts	North America, Western Europe	Saxena et al., 2003
Aphanamixis grandifolia	Rohitukka tree	Meliaceae	Terpenoids	Bark	Asia	Astulla et al., 2011
Apuleia leiocarpa	Garapa	Fabaceae	Quinone	Leaves	Brazil, Uruguay, Argentina, Paraguay	Muñoz et al., 2000a
Argemone mexicana	Mexican prickly poppy	Papaveraceae	Allocryptopine, protopine, berberine	Leaves, seeds	Nigeria, Brazil	Idowu et al., 2010; Kamaraj et al., 2012
Aristolochia griffithii	Birth wort	Aristolochiaceae	Alkaloids	Root	NE India	Das et al., 2016
Arrabidaea florida	Fridericia Mart	Bignoniaceae	Quinone	Leaves	America	Muñoz et al., 2000b
Artemisia abrotanum	Southernwood	Asteraceae	Sesquiterpenes	Leaves	Europe, Asia	Saxena et al., 2003
Artemisia absinthium	Common wormwood	Asteraceae	Sesquiterpenes	Leaves	Europe, Asia, Middle East	Saxena et al., 2003
Artemisia annua	Sweet wormwood	Asteraceae	Artemisinin, sesquiterpenes	Aerial part	Africa	O'Neill et al., 1985; Saxena et al., 2003; Youyou et al., 2015
Artemisia armeniaca	Sagebrush	Asteraceae	Essential oils, flavonoids	Aerial parts	Iran	Mojarrab et al., 2014

Artemisia indica	Wormwood	Asteraceae	Flavonoids, xanthones	Stem	India	Saxena et al., 2003
Asparagus africanus	African asparagus	Asparagaceae	Alkaloids, terpenoids, muzanzagenin	Whole plant	Africa	Saxena et al., 2003; Dikasso et al., 2006
Aspidosperma olivaceum	Copperpod	Apocynaceae	Alkaloids	Leaves, bark	Brazil	Chierrito et al., 2014
Aspilia pluriseta	Dwarf Aspilia	Asteraceae	Terpenoids	Aerial parts	Uganda	Sebisubi et al., 2010
Azadirachta indica	Neem	Meliaceae	Terpenoids, essential oils, triterpenoids	Whole plant	India, Nigeria	Saxena et al., 2003; Chianese et al., 2010; Idowu et al., 2010
Azanza garckeana	Snot apple	Malvaceae	Flavonoids, alkaloids	Whole plant	Africa	Njoroge and Bussmann, 2006
Bambusa vulgaris	Cana brava	Poaceae	Flavonoids, lactones	Aerial parts	Asia	Valdés et al., 2010
Bauhinia guianensis	Escalera de mono	Fabaceae	Flavonoids	Stem bark	Bolivia, South America	Muñoz et al., 2000a
Beilschmiedia spp.	Akolodo	Lauraceae	Alkaloids, flavonoids	Aerial parts, wood	Cameroon, Congo, Asia	Saxena et al., 2003; Lenta et al., 2009
Berberis aristata	Indian barberry	Berberidaceae	Alkaloids, tannins	Roots	India	Chandel et al., 2015
Bergenia ciliata	Hairy bergenia	Saxifragaceae	Alkaloids, terpenoids, phenols	Leaves	Africa	Walter et al., 2013
Bidens pilosa	Nyamaradza	Asteraceae	Flavonoids	Leaves and twigs	Ghana	Laryea and Borquaye, 2019
Bixaorellana	Achiote	Bixaceae	Essential oils	Whole plant	America	Valdés et al., 2011
Boswellia elongata	Frankincense tree	Burseraceae	Flavonoids, terpenoids	Bark	Yemen	Alshawsh et al., 2009
Brassica nigra	Indian mustard	Brassicaceae	Flavonoids, polyphenols, essential oils	Seeds	Asia	Muluye et al., 2015
Bridelia ferruginea	Kizni	Phyllanthaceae	Alkaloids, saponins, flavonoids	Stem bark	Africa, Nigeria	Mbah et al., 2012
Brucea javanica	Macassar kernels	Simaroubaceae	Quassinoids, triterpenoids	Stem, leaves	Asia	O'Neill, 1985; Saxena et al., 2003
Bruceamollis	Karbi	Simaroubaceae	Alkaloids, terpenoids, quassinoids	Root	India, Nepal, Malaysia	Prakash et al., 2013
Brunsvigia spp.	Kandelaarblom	Amaryllidaceae	Alkaloids	Bulb	South-Eastern Asia	Saxena et al., 2003
Caesalpinia bonducella	Fever nut	Fabaceae	Alkaloids, terpenoids, glycosides, saponins	Whole part	Tanzania	Nondo et al., 2016
Caesalpinia minax	Peacock flower	Fabaceae	Diterpene alkaloids	Seeds	South-East Asia	Ma et al., 2014
Caesalpinia sappan	Heartwood	Fabaceae	Alkaloids, terpenoids, glycosides, saponins	Seeds	Brazil	Ma et al., 2015

(*Continued*)

TABLE 1.1 (*Continued*)
List of Important Medicinal Plants throughout the World Having Antimalarial Properties

Species	Common/Local Name	Family	Bioactive Compound	Plant Part Used	Distribution	Key References
Caesalpinia volkensii	Kikuyu	Fabaceae	Flavonoids, tannins	Stem bark	Tanzania, Kenya	Njoroge and Bussmann, 2006; Ochieng et al., 2012
Cajanus cajan	Pigeon pea	Fabaceae	Cajachalcone	Leaves	Nigeria	Idowu et al., 2010
Canavalia beniensis	Jackbeans	Fabaceae	Alkaloids	Leaves, stem	Bolivia	Muñoz et al., 2000b
Canella winterana	Wild cinnamon	Canellaceae	Essential oils, sesquiterpenoids	Leaves	West Indies	Grace et al., 2010
Canthium multiflorum	Laager	Rubiaceae	Alkaloids, terpenoids, tannins	Whole plant	Cameroon	Akomo et al., 2009
Carapa guianensis	Crabwood	Meliaceae	Essential oil	Seeds	Amazon region, America	Pereira et al., 2014
Carica papaya	Papaw	Caricaceae	Alkaloids, flavonoids, glucosides	Leaves	India, Africa, America	Ofori-Attah et al., 2012
Cassia edulis	Mukawa	Fabaceae	Flavonoids	Root bark	Africa, South Asia, Australia	Njoroge and Bussmann, 2006
Cassia fistula	Golden shower	Fabaceae	Flavonoids, anthraquinones	Leaves, fruit, bark	India, Amazon, Sri Lanka	Grace et al., 2012
Cassia nictitans	Sensitive partridge pea	Fabaceae	Emodin, luteolin	Leaves	India	Kamaraj et al., 2012; Osunga et al., 2023
Cassia siamea	Kasood tree	Fabaceae	Alkaloids	Leaves	SE Asia	Deguchi et al., 2012
Cassia sieberiana	Drumstick tree	Fabaceae	Flavonoids, alkaloids, stilbenes	Root, stem	Africa	Abdulrazak et al., 2015
Cassipourea malosana	Muthathi	Rhizophoraceae	Alkaloids	Stem bark	Africa	Njoroge and Bussmann, 2006
Cedrelopsis grevei	Katrafay	Rutaceae	Essential oils	Leaves	Madagascar	Afoulous et al., 2013
Cestrum sp.	Jessamines	Solanaceae	Alkaloids, quassinoids, sesquiterpenes, triterpenoids, flavonoids	Leaves	Bolivia	Muñoz et al., 2000b
Chamaedorea elegans	Parlor palm	Arecaceae	Triterpenes	Leaves, stem, bark	Bolivia	Muñoz et al., 2000b
Christia vespertilionis	Red butterfly wing	Fabaceae	Triterpenes, alkaloids, phenols	Roots, leaves, stem	South-East Asia	Upadhyay et al., 2013
Chukrasia tabularis	White cedar	Meliaceae	Limonoids, tetranorriterpenoids	Stem bark	India, China, Bangladesh	Ogbole et al., 2016
Cichorium intybus	Chicory	Asteraceae	Lactucin, lactucopicrin	Root	America, Australia, India	Mojab, 2012
Cinchona spp.	Cinchona	Rubiaceae	Alkaloids	Bark	Sri Lanka, India, America, Africa	O'Neill et al., 1985; Muñoz et al., 2000a
Cinnamomum chavarria	Cinnamon	Lauraceae	Alkaloids	Bark, fruits, leaves	North America	Chinchilla et al., 2012

Cissampelos pareira	Velvet leaf	Menispermaceae	Alkaloids, terpenoids, tannins	Roots	Asia, Africa	Singh and Banyal, 2011
Cissus quadrangularis	Veldt grape	Vitaceae	Terpenoids	Leaves	Africa, tropical Asia	Matte, 2023
Cissus rotundifolia	Venezuelan treebine	Vitaceae	Flavonoids	Leaves	Yemen	Alshawsh et al., 2009
Citropsis articulata	African cherry orange	Rutaceae	Alkaloids, terpenoids, saponins, tannins	Root bark	Africa	Lacroix et al., 2011
Citrus limetta	Sweet lime	Rutaceae	Flavonoids, glycosides, phenols, essential oils	Fruit peel	South-East Asia	Mohanty et al., 2015
Citrus medica	Citron	Rutaceae	Essential oil	Aerial parts	South-East Asia	Idowu et al., 2010
Clappertonia ficifolia	Sahomia	Malvaceae	Flavonoids	Leaves	Ghana	Laryea and Borquaye, 2019
Clausena harmandiana	Prong faa	Rutaceae	Alkaloids, coumarins, essential oils	Aerial parts, roots	Thailand	Saxena et al., 2003; Thongthoom et al., 2010
Clausena anisata	Horsewood	Rutaceae	Alkaloids, essential oils	Leaves	Africa	Okokon et al., 2012
Clerodendrum rotundifolium	Bagflower	Lamiaceae	Alkaloids	Aerial parts	Asia, Africa, America	Adia et al., 2016
Clerodendrum Viscosum=C. infortunatum	Thuner	Lamiaceae	Alkaloids, sesquiterpene lactones	Whole plant	Asia, Africa, America	Moshi et al., 2012
Clutia abyssinica	Muthima mburi	Peraceae	Alkaloids	Various parts	Africa	Njoroge and Bussmann, 2006
Combretum micranthum	Kinkeliba	Combretaceae	Alkaloids	Leaves	Central and West Africa	Ancolio et al., 2002
Commiphora eminii	Mukungugu	Burseraceae	Flavonoids	Whole plant	Africa	Mutie et al., 2023
Coriandrum sativum	Cilantro	Apiaceae	Decenal, decanal, geranyl vinyl ether	Leaves	India	Kamaraj et al., 2012
Corydalis dubia	Crested lark	Papaveraceae	Alkaloids	Whole plant	Bhutan, India	Wangchuk et al., 2012
Corymbia watsoniana	Yellow bloodwood	Myrtaceae	Triketones	Flower	Australia	Carroll et al., 2013
Corynanthe pachyceras	Mende paa	Rubiaceae	Alkaloids	Stem bark	Africa	Saxena et al., 2003
Crinum amabile	Giant spider lily	Amaryllidaceae	Alkaloids	Bulb	South-East Asia	Saxena et al., 2003
Croton gratissimus	Lavender croton	Euphorbiaceae	Cembranolides	Leaves	Africa	Langat et al., 2011
Croton macrostachyus	Broad-leaved cotton	Euphorbiaceae	Alkaloids, flavonoids, terpenes, saponins	Leaves	Kenya, Ethiopia, Nigeria	Mohammed et al., 2014
Croton penduliflorus	Nyamaradza	Euphorbiaceae	Tannins, flavonoids	Leaves, twigs	Ghana	Laryea and Borquaye, 2019
Cryptocarya rigidifolia	Peumo	Lauraceae	Tetrahydropyrone derivatives	Root wood	Africa, Indonesia	Liu et al., 2015

(*Continued*)

TABLE 1.1 (*Continued*)
List of Important Medicinal Plants throughout the World Having Antimalarial Properties

Species	Common/Local Name	Family	Bioactive Compound	Plant Part Used	Distribution	Key References
Cryptocarya nigra	Medang	Lauraceae	Alkaloids	Stem bark	Indonesia	Nasrullah et al., 2013
Cryptolepis sanguinolenta	Nibima, Kadze	Apocynaceae	Alkaloids	Roots	Africa	Saxena et al., 2003; Samy et al., 2011
Cucumis aculeatus	Gakungui	Cucurbitaceae	Nitidine	Fruits	Africa	Njoroge and Bussmann, 2006
Cuminum cyminum	Cumin	Apiaceae	Essential oils, alkaloids, flavonoids, saponins, coumarins	Seeds	India, Pakistan, Iran	Kamaraj et al., 2012; Zheljazkov et al., 2015
Curcuma longa	Haldi, Turmmeric	Zingiberaceae	Flavonoids, curcumin	Rhizomes	Asia, Africa	Okello and Kang, 2019
Cyclea barbata	Green grass jelly	Menispermaceae	Alkaloids	Roots	India, Java	Saxena et al., 2003
Cyperus rotundus	Nut grass	Cyperaceae	Sesquiterpenes	Twigs	Europe, South-East Asia	Saxena et al., 2003
Dasymaschalon obtusipetalum	Jing wang	Annonaceae	Alkaloids	Twigs	Thailand	Jaidee et al., 2015
Datisca glomerata	Durango	Datiscaceae	Triterpene, cucurbitacin	Whole plant	North America	Graziose et al., 2013
Datura metel	Indian thornapple	Solanaceae	Alkaloids, flavonoids	Leaf	India	Kamaraj et al., 2012
Datura stramonium	Apple of Peru	Solanaceae	Flavonoids	Leaves and twigs	Ghana	Laryea and Borquaye, 2019
Dehaasia incrassata	Margapali, Paitan	Lauraceae	Alkaloids	Leaves, bark	Asia, Africa	Saxena et al., 2003
Dendrobium venustum	The lovely Dendrobium	Orchidaceae	Phenolic compounds	Whole plant	Thailand, Cambodia	Sukphan et al., 2014
Derris amazonica	Timbo Branco	Fabaceae	Flavonoids	Roots	Brazil	Muñoz et al., 2000a
Dichroa febrifuga	Chinese quinine	Hydrangeaceae	Febriguine	Roots	China	O'Neill et al., 1985; Saxena et al., 2003
Dicoma tomentosa	Hookiah bel	Asteraceae	Sesquiterpene lactone	Whole plant	Africa, Asia	Jansen et al., 2012
Dillenia andamanica	Andaman Dillenia	Dilleniaceae	Flavonoids, triterpenoids, saponins, phenolics	Whole plant	Australia, India	Chander et al., 2016
Diospyros melanoxylon	East Indian ebony	Ebenaceae	Lupeol, betulin, β-sitosterol	Roots, bark, wood	India	Kamaraj et al., 2012
Dovyalis spp.	Kau apple	Salicaceae	Catechin, flavonoids	Fruits	Africa	Njoroge and Bussmann, 2006; Qanash et al., 2022
Duguetia hadrantha	Pindaiva	Annonaceae	Sampangine Alkaloids	Stem bark	South America	Saxena et al., 2003

Duranta repens	Golden dewdrop	Verbenaceae	Iridoid glycosides, flavonoids, alkaloids	Whole plant	Mexico, S. America	Ijaz et al., 2010
Echinops kebericho	Amhar	Asteraceae	Essential oils	Roots	Ethiopia	Toma et al., 2015
Eclipta prostrata	False daisy	Asteraceae	Flavonoids	Leaves	India	Kamaraj et al., 2012
Ekebergia capensis	Mukuriahungu	Meliaceae	Alkaloids, flavonoids	Stem bark	Africa	Njoroge and Bussmann, 2006
Eleusine indica	Yard grass	Poaceae	Phenolic compounds, flavonoids	Whole plant	Warmer areas of the world	Adoho et al., 2021
Eleutherine citriodora	Jharpo peyaz	Iridaceae	Flavonoids	Entire plant	South America, South Africa	Muñoz et al., 2000b
Enantia polycarpa	Osopupa	Annonaceae	Alkaloids	Stem bark	Nigeria	Anosa et al., 2014
Enkleia siamensis	Chamar jasmine	Thymelaeaceae	Flavonoids	Roots	Australia	Rajachan et al., 2014
Eremostachys macrophylla	Desert spike	Lamiaceae	Essential oils	Aerial parts, rhizome	Iran	Asnaashari et al., 2015
Erythrina abyssinica	Muhuti	Fabaceae	Flavonoids	Various parts	Kenya	Njoroge and Bussmann, 2006
Erythrina burttii	Rattlepod	Fabaceae	Flavonoids	Root, stem bark	Kenya, Ethiopia	Yenesew et al., 2012
Erythrina fusca	Lucky bean tree	Fabaceae	Flavonoids, pterocarpans	Stem bark	Asia, Africa	Innok et al., 2010
Erythrina indica	Indian coral	Fabaceae	Alkaloids, flavonoids, steroids	Stem	Asia	Govindarajan and Sivakumar, 2014
Erythrina schliebenii	Mlindimila	Fabaceae	Flavonoids, alkaloids, terpenoids	Stem, bark	Tanzania	Nondo et al., 2017
Eucalyptus globulus	Mubau, blue gum	Myrtaceae	Pinene, cineole	Leaves	Africa, India	Njoroge and Bussmann, 2006; Sheikh et al., 2021
Euphorbia tirucalli	Pencil cactus	Euphorbiaceae	Sesquiterpenes	Various parts	Africa	Njoroge and Bussmann, 2006
Eurycoma longifolia	Longjack	Simaroubaceae	Alkaloids, quassinoids	Root	Asia	Saxena et al., 2003
Fagaropsis angolensis	Mukaragati	Rutaceae	Chloroquine	Stem bark, leaves	Kenya	Njoroge and Bussmann, 2006
Faidherbia albida	Winter thorn	Mimosoideae	Alkaloids, tannins, saponins	Stem bark	Africa, Nigeria	Bruno et al., 2020
Faurea speciosa=F. rochetiana	Setingo sebari	Proteaceae	Tannins, alkaloids	Leaves, twigs	Ghana	Laryea and Borquaye, 2019
Feretia apodanthera	Tiôbi, Tukulor	Rubiaceae	Proanthocyanidins, iridoids	Leaves	Tropical Africa	Ancolio et al., 2002

(Continued)

TABLE 1.1 (*Continued*)
List of Important Medicinal Plants throughout the World Having Antimalarial Properties

Species	Common/Local Name	Family	Bioactive Compound	Plant Part Used	Distribution	Key References
Ficus insipida	Fig	Moraceae	Alkaloids	Bark	Bolivia	Muñoz et al., 2000b; Vásquez-Ocmín et al., 2018
Ficus saussureana	Muwo	Moraceae	Alkaloids	Bark	Uganda	Okello and Kang, 2019
Ficus sur	Bush fig	Moraceae	Phenolic compounds	Leaves, stem bark, root bark	South Africa, Somalia, Namibia	Muregi et al., 2007
Ficus thonningii	Wild fig	Moraceae	Flavonoids, alkaloids, terpenoids, tannins, essential oils	Whole plant	Africa, Nigeria	Falade et al., 2014
Flacourtia indica	Batoku palm	Salicaceae	Poliothrysoside	Aerial parts	Asia, Africa	Sashidhara et al., 2013
Gallesia integrifolia	Ajo Ajo	Phytolaccaceae	Linolenic acid, phytol, methyl palmitate	Leaves	South America	Muñoz et al., 2000b; De-Campos-Bortolucci et al., 2021
Garcinia xanthochymus	Yellow mangosteen	Clusiaceae	Benzophenones	Aerial parts	Asia, Africa, Australia	Lyles et al., 2014
Garcinia cowa	Cowa mangosteen	Clusiaceae	Flavonoids and xanthones	Bark	Asia	Saxena et al., 2003
Garcinia mangostana	Purple mangosteen	Clusiaceae	Xanthones	Aerial parts	Asia, Africa, Australia	Thiengsusuk et al., 2013
Geissospermum vellosii	Bergibita	Apocynaceae	Alkaloids	Stem bark	Brazil	Mbeunkui et al., 2012
Geissospermum laeve	Pao pereira	Apocynaceae	Alkaloids	Bark	Bolivia, South America	Muñoz et al., 2000a; Vital et al., 2013
Glinus oppositifolius	Papait	Molluginaceae	Triterpenoids	All parts	Africa	Saxena et al., 2003
Glycyrrhiza glabra	Liquorice	Fabaceae	Flavonoids, saponin glycosides	Roots	Asia, Europe	Mojab, 2012; Cheema et al., 2014
Goniothalamus elegant	Kao nang nee	Annonaceae	Tetrahydropyran derivatives	Bark	Africa, Asia	Suchaichit et al., 2015
Goniothalamus australis	China pine	Annonaceae	Alkaloids, lactones	Aerial parts	Australia	Levrier et al., 2013
Guarea spp.	Muskwood	Meliaceae	Triterpenoids	Bark	Africa, America	Muñoz et al., 2000b

Guiera senegalensis	Sabara	Combretaceae	Alkaloids	Roots	Central and West Africa	Ancolio et al., 2002
Hampea appendiculata	Majagua	Malvaceae	Flavonoids	Fruits	Mexico, Central America	Chinchilla et al., 2012
Heterothalamus psiadioides	Heterothalamus	Asteraceae	Sesquiterpenes	Leaves	South America	Saxena et al., 2003
Himatanthus articulates	Sucuba	Apocynaceae	Triterpenoids, iridoids	Stem bark	Brazil	Vale et al., 2015
Holarrhena antidysenterica	Coral swirl	Apocynaceae	Alkaloids	Bark	India	Verma et al., 2011
Holarrhena pubescens	Tellicherry Bark	Apocynaceae	Triterpenoids, steroids, alkaloids	Roots	Africa, India	Ma et al., 2015
Hoslundia opposita	Bird gooseberry	Lamiaceae	3-O-benzoylholoppone	Root bark	Africa	Achenbach et al., 1992; Saxena et al., 2003
Hunteria zeylanica	Lahoi	Apocynaceae	Alkaloids	Bark	Asia, Africa	Nugroho et al., 2011
Hura crepitans	Sandbox tree	Euphorbiaceae	Alkaloids	Latex	Tropical areas of America	Muñoz et al., 2000b
Hydnora abyssinica	Muthigira	Aristolochiaceae	Alkaloids	Seeds, leaves	Africa	Njoroge and Bussmann, 2006
Hymenachne spp.	Marsh grass	Poaceae	Sterol	Entire plant	South America	Muñoz et al., 2000b
Hypericum spp.	St John's Wort	Hypericaceae	Terpenoids, xanthones, flavonoids	Stem bark, apical part	Africa	Saxena et al., 2003; Bahekar and Kale, 2013
Icacina senegalensis	False yam	Icacinaceae	Alkaloids, saponins, tannins, terpenoids	Leaves	Africa	Akuodor et al., 2017
Indigofera oblongifolia	Jhil	Papilionaceae	Alkaloids, tannins	Leaves	Africa	Dkhil et al., 2015
Isolona guesquiereina	Isolona	Annonaceae	Curine, Isochondodendrine	Stem bark	Africa	Saxena et al., 2003
Jacaratia digitata	Papaya casp	Caricaceae	Alkaloids	Bark	Columbia, Bolivia	Muñoz et al., 2000b
Jasminum Syringifolium= J. sambac	Jasmine	Oleaceae	Essential oils	Leaves	Andaman and Nicobar Islands	Ijaz et al., 2010
Jatropha curcas	Biodiesel plant	Euphorbiaceae	Alkaloids	Leaves	Uganda, Asia	Okello and Kang, 2019
Jatropha ribifolia	Pohl	Euphorbiaceae	Terpenoids, coumarins, essential oils	Whole plant	America	Pinto et al., 2015
Justicia betonica	Nalongo	Acanthaceae	Alkaloids, triterpenoids	Leaves, whole plant	Uganda	Okello and Kang 2019
Kalanchoe spp.	Mahuithia	Crassulaceae	Flavonoids	Entire plant	Africa	Muñoz et al., 2000b; Njoroge and Bussmann, 2006

(*Continued*)

TABLE 1.1 (*Continued*)
List of Important Medicinal Plants throughout the World Having Antimalarial Properties

Species	Common/Local Name	Family	Bioactive Compound	Plant Part Used	Distribution	Key References
Keetia leucantha	Buje	Rubiaceae	Essential oils, tripenic acids	Twigs	Africa	Beaufay et al., 2019
Khaya grandifoliola	African mahogany	Meliaceae	Triterpenoids	Stem bark, bark, seeds	Africa	Saxena et al., 2003; Idowu et al., 2010
Kniphofia foliosa	Torch lily	Asphodelaceae	Quinones	Rhizome	Africa	Induli et al., 2013
Kuloa usambarensis	Muthaiti	Lauraceae	Bisabolol, Sesquiterpenes	Bark	Africa	Njoroge and Bussmann, 2006; Dharani et al., 2010
Lantana camara	Lantana, Ruithiki	Verbenaceae	Triterpenes	Leaves	India, Africa	Njoroge and Bussmann, 2006; Kamaraj et al., 2012
Lettowianthus stellatus	Green star flower	Annonaceae	Geranylbenzoquinoids, aporphinoidalkaloids	Fruits	Tanzania, Kenya	Makangara et al., 2010
Lippia javanica	Lemon bark	Verbenaceae	Lippialactone	Aerial parts	Africa	Malahlela et al., 2019
Liriodendron tulipifera	Lily tree	Magnoliaceae	Alkaloids, sesquiterpenoids	Bark, leaves	North America	Kaur and Kaur, 2017
Lophira alata	Red ironwood	Ochnaceae	Chalcones, biflavonoids	Aerial parts	Africa	Falade et al., 2014
Lophira lanceolata	Dwarf red ironwood	Ochnaceae	Artemisinin	Stem bark, flowers, seeds, leaves	Cameroon	Azizi et al., 2023
Lycoris radiata	Red magic lily	Amaryllidaceae	Alkaloids	Bulbs	China, Korea	Hao et al., 2013
Magnifera indica	Mango	Anacardiaceae	Flavonoids, polyphenols, glycosides	Aerial parts, stem bark, leaves	Asia, Nigeria	Awe et al., 1998; Idowu et al., 2010
Mallotus oppositifolius	Geisel	Euphorbiaceae	Phloroglucinols	Leaves	Africa	Adetunji et al., 2022
Mammea africana	African apple	Calophyllaceae	Alkaloids, flavonoids, diterpenoids	Aerial parts	Africa	Singh and Banyal, 2011
Maquira coriacea	Envireira	Moraceae	Alkaloids	Stem bark	Africa	Muñoz et al., 2000a
Margaritaria discoidea	Pheasant-berry	Phyllanthaceae	Securinine	Root bark	Africa	Saxena et al., 2003
Markhamia tomentosa	Fula pulaar	Bignoniaceae	Triterpenoid saponins	Stem bark	Africa	Bankole and Abiodun, 2020
Mascagnia macrophylla	Butterfly vine	Malpighiaceae	Alkaloids	Bark	Bolivia, South America	Muñoz et al., 2000a
Maytenus mekongensis	Spike thorn	Celastraceae	Sesquiterpene, alkaloids	Roots	Asia, Africa	Kaur and Kaur, 2017

Maytenus senegalensis	Muthuthi	Celastraceae	Alkaloids, flavonoids	Root bark	Africa	Njoroge and Bussmann, 2006; Umar et al., 2019
Meconopsis simplicifolia	Blue poppy	Papaveraceae	Alkaloids	Aerial parts	Nepal	Wangchuk and Samten, 2015
Melia azedarach	Elira, Chinaberry, Mahaneem	Meliaceae	Caryophyllene, benzaldehyde	Leaves	India, Uganda	Okello and Kang, 2019
Mikania spp.	American rope	Asteraceae	Coumarin	Aerial part	Africa, America, Europe	Muñoz et al., 2000b
Mirabilis jalapa	Marvel-of-peru	Nyctaginaceae	Sitosterol, stigmasterol	Entire plant	Mexico, Central and North America	Muñoz et al., 2000b
Mitragyna ciliata	Subaha akori	Rubiaceae	Tannins	Leaves, twigs	Ghana	Laryea and Borquaye, 2019
Mitragyna inermis	Mitragyna	Rubiaceae	Alkaloids	Leaves	Tropical Africa, Sudan	Azas et al., 2002
Mitrephora diversifolia	Flying duck flower	Annonaceae	Azafluorenone alkaloids	Roots	Australia	Mueller et al., 2009
Momordica charantia	Karela, bitter gourd	Cucurbitaceae	Alkaloids, flavonoids, cardiac glycosides, triterpenes	Whole plant	Asia, Africa, Caribbean, South America	Muñoz et al., 2000b; Kamaraj et al., 2012
Momordica foetida	Wild cucumber	Cucurbitaceae	Alkaloids, flavonoids, terpenoids, steroids	Leaves	Africa	Adia et al., 2016
Monanthotaxis caffra	Ntetekon	Annonaceae	Tannins	Leaves, twigs	Ghana	Laryea and Borquaye, 2019
Morinda lucida	Oruwo	Rubiaceae	Quinones	Stem bark, leaf	Nigeria	Saxena et al., 2003; Idowu et al., 2010
Morinda morindoides	Ponju Owiwi	Rubiaceae	Quinones	Aerial part, root bark	Nigeria	Idowu et al., 2010
Moringa oleifera	Drumstick tree	Moringaceae	Triterpenoids, flavonoids, alkaloids	Seeds	Indian subcontinent, Central and South America	Okorie et al., 2023
Muntafara sessilifolia= *Tabernaemontana sessilifolia*	Baker	Apocynaceae	Indole alkaloids	Stem bark	Madagascar	Girardot et al., 2012
Myrtus communis	True myrtale	Myrtaceae	Coumarins, flavonoids, essential oils	Aerial parts	Iran	Sangian et al., 2013
Nardostachys chinensis= *N. jatamansi*	Spikenard	Caprifoliaceae	Sesquiterpenoids	Whole plant	South-East Asia	Saxena et al., 2003; Rehman and Ahmad, 2019

(Continued)

TABLE 1.1 (*Continued*)
List of Important Medicinal Plants throughout the World Having Antimalarial Properties

Species	Common/Local Name	Family	Bioactive Compound	Plant Part Used	Distribution	Key References
Nauclea latifolia	African peach	Rubiaceae	Alkaloids, tannins, saponins	Stem bark	Africa	Boucherle et al., 2016
Nectandra cuspidata	Nectandra	Lauraceae	Alkaloids	Leaves	South America	Muñoz et al., 2000a
Nectandra membranacea	Sweetwood	Lauraceae	Alkaloids, flavonoids, triterpenes	Fruits, leaves	Mexico, South America	Chinchilla et al., 2012
Nelumbo nucifera	Kamal, lotus	Nelumbonaceae	Alkaloids	Leaves	Asia, Australia	Kamaraj et al., 2012
Neoboutonia macrocalyx	Lace leaf	Euphorbiaceae	Triterpenoids	Stem bark	Africa	Namukobe et al., 2015
Neonauclea purpurea	Purple neo-cheesewood	Rubiaceae	Alkaloids	Stem bark	Asia, Australia	Kaur and Kaur, 2017
Nepenthes thorelii	Pitcher plant	Nepenthaceae	Quinones	Roots	Indochina region	Saxena et al., 2003
Neurolaena lobata	Jackass bitter	Asteraceae	Sesquiterpenes	Whole plant	Mexico, America	Saxena et al., 2003
Newbouldia laevis	Balanta, akoko	Bignoniaceae	Flavonoids, alkaloids, saponins	Aerial parts	Africa	Okagu et al., 2021
Ocimum basilicum	Basil, Tulsi	Lamiaceae	Flavonoids, essential oils	Leaves	Asia, Africa	Ileke and Adesina, 2019
Ocimum campechianum	Haiti	Lamiaceae	Flavonoids	Entire plant	America, Mexico, West Indies	Muñoz et al., 2000b
Ocimum gratissimum	African basil	Lamiaceae	Eugenol, essential oil	Leaves	Africa	Ileke and Adesina, 2019
Ocimum sanctum	Holy basil	Lamiaceae	Phenols, terpenoids	Leaves	India	Pattanayak et al., 2010
Ocimum suave	Wild basil	Lamiaceae	Essential oils	Leaves	Africa	Nguta et al., 2010
Olea europaea	Mutamaiyu	Oleaceae	Chloroquine	Leaves	Kenya	Njoroge and Bussmann, 2006
Ormocarpum kirkii	Curled caterpillar bush	Fabaceae	Flavonoids, coumarins	Roots	Tanzania, Zimbabwe	Dhooghe et al., 2010
Osyris lanceolata	African sandalwood	Santalaceae	Sesquiterpenes	Whole plant	Africa	Njoroge and Bussmann, 2006
Otostegia integrifolia	Abyssinia rose	Lamiaceae	Diterpenoids	Leaves	Ethiopia	Endale et al., 2013
Panicum maximum	Guinea grass	Poaceae	Essential oil, flavonoids	Leaves	Africa, Palestine	Okokon et al., 2016
Pappea capensis	Jacket plum	Sapindaceae	Triterpenoids	Twigs	Africa	Njoroge and Bussmann, 2006; Mabuza et al., 2022
Parinari congensis	Krenku	Chrysobalanaceae	Flavonoids	Stem bark	Ghana	Laryea and Borquaye, 2019
Paspalum scrobiculatum	Kodo millet	Poaceae	Tannins	Spinkelets	Africa	Laryea and Borquaye, 2019

Passiflora ligularis	Sweet granadilla	Passifloraceae	Flavonoids, alkaloids	Fruits	Africa, Mexico	Njoroge and Bussmann, 2006; Sebastian and Jose, 2018
Pedilanthus tithymaloides= *Euphorbia tithymaloides*	Red bird flower	Euphorbiaceae	Phytosterols, flavonoids	Aerial parts	India, North America	Ijaz et al., 2010
Pentasbussei	Egyptian starcluster	Rubiaceae	Naphthohydroquinones	Roots	Tanzania, Kenya	Endale et al., 2012
Pentas longiflora	Muhuha	Rubiaceae	Naphthoquinones	Roots, leaves	Kenya, Africa	Njoroge and Bussmann, 2006
Periploca linearifolia	Muimbathunu	Apocynaceae	Lupeol, artemisinin	Stem bark	Africa	Njoroge and Bussmann, 2006; Belay et al., 2022
Persea povedae	Avocado	Lauraceae	Alkaloids, α-pinene, linalool	Bark, fruit, leaves, roots	South and Central America, Mexico	Chinchilla et al., 2012
Peschiera (Tabernaemontana) fuchsiaefolia	Pinwheel flower	Apocynaceae	Affinisine, alkaloids, vocamines	Stem bark, root bark, seeds	Africa, Asia	Saxena et al., 2003
Petiveria alliacea	Awopa	Phytolaccaceae	Flavonoids	Aerial part, stem bark	Nigeria	Muñoz et al., 2000b; Idowu et al., 2010
Philodendron megalophyllum	Arosma Raf	Araceae	–	Sap, root, leaves	South America	Muñoz et al., 2000b
Phyllanthus acuminatus	Jamaican gooseberry tree	Phyllanthaceae	Alkaloids, flavonoids	Leaves	Central and South America	Muñoz et al., 2000b
Phyllanthus amarus	Black catnip	Phyllanthaceae	Alkaloids, flavonoids, tannins, terpenoids	Aerial parts	America, Asia	Appiah-Opong et al., 2011
Phyllanthus urinaria	Gripeweed	Phyllanthaceae	Alkaloids, flavonoids, tannins, terpenoids	Whole plant	Asia	Geethangili and Ding, 2018
Physalis angulata	Balloon cherry	Solanaceae	Flavonoids, alkaloids, steroids	Whole plant	America	Sharma et al., 2015
Phytolacca rivinoides	Venezuelan pokeweed	Phytolaccaceae	β-carboxymethoxy-oleanolic acid	Aerial part, fruit	South America	Muñoz et al., 2000b
Picramnia sellowii	Bitterbush	Picramniaceae	Flavonoids	Leaves	South America	O'Neill et al., 1985; Muñoz et al., 2000b
Picrorhiza kurroa	Kutki	Plantaginaceae	Kutkin	Whole plant	India	Masood et al., 2021
Picrorhiza scrophulariiflora	Figwort	Scrophulariaceae	Secoiridoid and caffeoyl glycosides	Whole plant	India, Nepal, China	Wang et al., 2013
Piper guineense	Kale	Piperaceae	Essential oil, saponins	Leaves and stems	Africa	Isikhuemen et al., 2020
Piper nigrum	Black pepper	Piperaceae	Alkaloids, essential oils, flavonoids	Fruit	Asia	Kamaraj et al., 2012; Mgbeahuruike et al., 2017

(Continued)

TABLE 1.1 (*Continued*)
List of Important Medicinal Plants throughout the World Having Antimalarial Properties

Species	Common/Local Name	Family	Bioactive Compound	Plant Part Used	Distribution	Key References
Piper peltatum	Monkeys hand	Piperaceae	Essential oils	Aerial parts	North America	Rocha e Silva et al., 2015
Piper sarmentosum	Kaduk	Piperaceae	Alkaloids	Leaves	Malaysia	Rahman et al., 1999
Piper umbellatum	Cow foot leaf	Piperaceae	4-nerolidylcathecol, saponin	Aerial parts, leaves, stem	South America, Africa	Bagatela et al., 2013; Isikhuemen et al., 2020
Piptocoma antillana	Velvet shrub	Asteraceae	Terpenoids, sesquiterpene lactone	Leaves, twigs	Latin America	Liu et al., 2014
Pittosporum lanatum	Munyamati	Pittosporaceae	Flavonoids	Stem bark	South Africa, tropical Africa	Madikizela and McGaw, 2017
Plectranthus barbatus	Forskohlii	Lamiaceae	Flavonoids, diterpenoids	Root bark	S. America, Asia	Liu et al., 2014
Pleurospermum amabile	Mirik	Apiaceae	Phenylpropanoids, furanocoumarins	Whole plant	Bhutan	Wangchuk et al., 2014
Plumbago zeylanica	Doctor bush	Plumbaginaceae	Alkaloids, steroids, tannins, triterpenoids, flavonoids, saponins	Roots	America, Mexico	Simonsen et al., 2001
Polyalthia longifolia	False ashoka	Annonaceae	Alkaloids, steroids, tannins, flavonoids	Stem	India, Sri Lanka	Kwansa-Bentum et al., 2019
Polybotrya caudata	Wood fern	Dryopteridaceae	Volatile organic compound	Aerial part	South America	Muñoz et al., 2000b
Prosopis juliflora	Mesquite	Fabaceae	Alkaloids	Leaves, pods	Africa, Asia, Australia	Mojab, 2012
Prunus spp.	African cherry	Rosaceae	Alkaloids, terpenes	Leaves	Africa	Muñoz et al., 2000a; Njoroge and Bussmann, 2006
Psidium acutangulum	Para guava	Myrtaceae	Triterpenoids	Aerial parts	French Guinea	Houël et al., 2015
Psidium guajava	Amrood, Guava	Myrtaceae	Terpenoids	Leaves, fruits	India, Costa Rica, Central America, Mexico, Argentina	Chinchilla et al., 2012; Matte, 2023
Psychotria camponutans	Brimstone tree	Rubiaceae	Quinones	Wood	Africa, South America	Saxena et al., 2003
Pterocarpus amazonum	Padauk	Fabaceae	Flavonoids	Stem bark	South America	Muñoz et al., 2000a

Punica granatum	Pomegranate	Lythraceae	Flavonoids, anthocyanins	Whole plant	Africa, Asia	Dell'Agli et al., 2009
Pycnanthus angolensis	Ilomba	Myristicaceae	Lignans, Pycnantolol, Heliobuphthalmin	Stem bark	West and Central Africa	Ancolio et al., 2002
Quassia amara	Bitter ash wood	Simaroubaceae	Quassinoids	Leaves	Africa	Ajaiyeoba et al., 1999
Quassia indica	Niepa bark tree	Simaroubaceae	Quassinoids	Stem	Africa, South-East Asia, Indian subcontinent	Saxena et al., 2003
Quassia undulata	Mjoho	Simaroubaceae	Quassinoids, samaderines	Leaves	Africa	Ajaiyeoba et al., 1999
Renealmia sp.	Mardi gras	Zingiberaceae	Terpenes	Aerial part, rhizome	Mexico, Peru, South America	Muñoz et al., 2000b
Rhamnus prinoides	Mukarakinga	Rhamnaceae	Chloroquine	Leaves	Kenya	Njoroge and Bussmann, 2006
Rhamnus spp.	Common buckthorn	Rhamnaceae	Alkaloids, flavonoids	Roots, root bark, leaves	Africa	Njoroge and Bussmann, 2006; Muregi et al., 2007
Rhaphidophora decursiva	Decursiva plant	Araceae	Serotobenine, decursivine	Leaves, stem	China, Indian subcontinent	Zhang et al., 2002
Rheedia gardneriana= Garcinia gardneriana	Bacuripari	Guttiferae	Xanthones	Aerial part, bark	South America	Muñoz et al., 2000b
Rhoicissus tridentata	Ndurutua	Vitaceae	Alkaloids	Leaf	Africa	Njoroge and Bussmann, 2006
Rhus vulgaris	Muthura	Anacardiaceae	Biflavone, flavonoids, terpenoids, alkaloids	Stem bark	Kenya	Njoroge and Bussmann, 2006
Ritchiea capparoides	Mtunguru	Capparaceae	Alkaloids, saponins, tannins, sesquiterpenes	Leaves	Africa, Nigeria	Taiwo et al., 2013
Rosa rugosa	Beach rose	Rosaceae	Sesquiterpenes	Various parts	Asia	Saxena et al., 2003
Rumex crispus	Curly dock	Polygonaceae	Flavonoids, anthraquinones, naphthalenes	Aerial parts	Asia, America	Uzor et al., 2020
Salacia kraussii	Ibhonsi	Celastraceae	Triterpenoids	Root	Africa	Saxena et al., 2003
Salacia longipes	Oliver	Celastraceae	Sesquiterpenoids	Seeds	Africa	Mba'ning et al., 2013
Salix humboldtiana	Humboldt's willow	Salicaceae	Phenolic compounds	Leaves	America	Muñoz et al., 2000b
Salvia rhytidea	Persian sage	Lamiaceae	Diterpenoids	Roots	Iran, Afghanistan	Jassbi et al., 2017
Sarcocephalus latifolius	African peach	Rubiaceae	Alkaloids, flavonoids	Stem bark, leaves	Nigeria	Idowu et al., 2010
Savignya parviflora	Gilgilan	Brassicaceae	Alkaloids	Twigs and leaves	South America	Muñoz et al., 2000b
Schima wallichii	Needlewood tree	Theaceae	Quinones, glycosides, coumarins, flavonic glycosides	Leaves	India, Nepal, Bhutan	Barliana et al., 2014

(Continued)

TABLE 1.1 (*Continued*)
List of Important Medicinal Plants throughout the World Having Antimalarial Properties

Species	Common/Local Name	Family	Bioactive Compound	Plant Part Used	Distribution	Key References
Schizanthus tricolor	Poor man's orchid	Solanaceae	Alkaloids	Aerial parts	Chile, Argentina	Cretton et al., 2010
Schkuhria pinnata	Gakuinini	Asteraceae	Flavonoids, alkaloids	Whole plant	Africa	Njoroge and Bussmann, 2006; Nuwagira et al., 2021
Scutia myrtina	Mwikunya, cat-thorn	Rhamnaceae	Alkaloids	Root bark	Africa	Njoroge and Bussmann, 2006
Securidaca longepedunculata	Violet tree	Polygalaceae	Alkaloids, flavonoids, triterpenes	Roots	Mali, Sao Tome	Ancolio et al., 2002
Senna didymobotrya	Mwinu	Fabaceae	Chloroquine	Leaves	Kenya	Njoroge and Bussmann, 2006
Senna occidentalis	Mogdad coffee	Fabaceae	Anthraquinone glycoside, fatty oils, glycosides, flavonoids, tannins	Leaves	America, Asia	Daskum et al., 2019
Senna papillosa	Senna	Fabaceae	Alkaloids	Fruits, leaves, roots	Mexico, Venezuela	Chinchilla et al., 2012
Serjania deltoidea	Soapberry	Sapindaceae	Flavonoids	Aerial part, root bark, leaf, stem bark	Columbia	Muñoz et al., 2000b
Sida rhombifolia	Arrowleaf sida	Malvaceae	Alkaloids, flavonoids	Entire plant	India, Sri Lanka	Muñoz et al., 2000b
Simaba spp.	Cedron	Simaroubaceae	Futolactone, cedronin, quassinoids	Bark	Columbia, Central America	O'Neill et al., 1985; Saxena et al., 2003
Siparuna andina	Siparuna	Siparunaceae	Sipandinolide, sesquiterpenes, flavonoids, xanthones	Leaves	Central and South America, Mexico	Saxena et al., 2003
Siparuna thecaphora	Siparuna	Siparunaceae	Sipandinolide	Fruits, leaves	Mexico, Bolivia	Chinchilla et al., 2012
Socratea exorrhiza	Cashapona	Arecaceae	Flavonoids	Roots	Central and South America	Muñoz et al., 2000b
Solanum spp.	Soda apple	Solanaceae	Alkaloids	Entire plant	Africa	Mojab, 2012; Schultz et al., 2021
Sophora mollis	Low mountain laurel	Fabaceae	Flavonoids	Roots	Asia, Australia	Zhang et al., 2009
Sparattanthelium amazonum	Bejuco de Quina	Hernandiaceae	Roemrefidine	Stem bark	South America, Bolivia	Muñoz et al., 2000a; Saxena et al., 2003
Spathodea campanulata	African tulip tree	Bignoniaceae	Ursolic acid, tomentosolic acid, other triterpenoids	Stem, bark	Africa	Saxena et al., 2003

Spirospermum sp.	Parul	Menispermaceae	Alkaloids	Seeds, roots	India, Bangladesh	Saxena et al., 2003
Spondias venosa	Mombins	Anacardiaceae	–	Leaves	Brazil	Muñoz et al., 2000b
Stephania abyssinica	Beiyanzi	Menispermaceae	Alkaloids	Leaves	Kenya, Zimbabwe	Muregi et al., 2004
Stephania erecta	Potato plant	Menispermaceae	Alkaloids	Tuber	Thailand, India	Saxena et al., 2003
Stephania pierrei	Binh voi	Menispermaceae	Alkaloids	Tuber	India, China, Vietnam	Saxena et al., 2003
Strychnos spp.	Boomer-rid	Loganiaceae	Alkaloids	Stem bark, leaves, roots	Africa, Asia	Saxena et al., 2003; Njoroge and Bussmann, 2006; Tepa et al., 2022
Swertia chirayita	Chiretta	Gentianaceae	Gentianine and other alkaloids, xanthones, flavonoids	Leaves	South-East Asia	Swati et al., 2023
Swietenia macrophylla	Mahogany	Meliaceae	Triterpenes	Bark	Mexico, South America	Muñoz et al., 2000b; Telrandhe et al., 2022
Swietenia mahagoni	American mahogany	Meliaceae	Flavonoids	Seeds	America, West Indies, Cuba, Jamaica	Sari et al., 2023
Symphonia globulifera	Chewstick	Clusiaceae	Benzophenones, xanthones	Leaves	America, Africa	Ateba et al., 2018
Syzygium guineense	Senza	Myrtaceae	Tannins, flavonoids	Leaves	Ghana	Laryea and Borquaye, 2019
Tabebuia ochracea	Corteza amarillia	Bignoniaceae	Quinones	Inner stem bark	South America	Saxena et al., 2003
Tachia grandiflora	Mahot noir	Gentianaceae	Xanthonoids, monoterpenoids	Roots, leaves	Brazil	Kffuri et al., 2016
Tagetes erecta	Marigold	Asteraceae	Bithienyl, flavonoids	Roots,	America, India, Mexico, Costa Rica, China, Australia, South Africa	Kamaraj et al., 2012; Chaniad et al., 2021
Tagetes minuta	Southern cone marigold	Asteraceae	Sesquiterpenes	Whole plant, leaves	Africa, Europe, Australia, America, Asia	Okello and Kang, 2019
Talinum triangulare	Waterleaf, Nela basale	Talinaceae	Alkaloids, flavonoids	Leaves	India, Mexico	Muñoz et al., 2000b; Adekanmi et al., 2020
Tamarindus indica	Imli	Fabaceae	Phenols	Root, bark	Indian subcontinent, Africa, Pakistan, Bangladesh, Nigeria, America	Okello and Kang, 2019

(Continued)

TABLE 1.1 (*Continued*)
List of Important Medicinal Plants throughout the World Having Antimalarial Properties

Species	Common/Local Name	Family	Bioactive Compound	Plant Part Used	Distribution	Key References
Tanaecium jaroba	Liana	Bignoniaceae	Terpenoids	Leaves	Central and South America	Muñoz et al., 2000a; Carvalho et al., 2021
Telfairia occidentalis	Fluted gourd	Cucurbitaceae	Alkaloids, flavonoids, saponins, tannins	Leaves, seeds	Africa, Nigeria	Okokon et al., 2009
Terminalia bellirica	Baheda	Combretaceae	Flavonoids, xanthones	Fruits	South-East Asia	Saxena et al., 2003
Tessaria integrifolia	Aliso del río	Asteraceae	Sesquiterpenes	Leaves, roots	Peru	Muñoz et al., 2000b
Tetradenia riparia	Kyewamala	Lamiaceae	Terpineol	Leaves	Uganda	Okello and Kang, 2019
Teucrium ramosissimum	Germanders	Lamiaceae	Sesquiterpenoids, flavonoids, essential oils	Aerial parts	North America, Tunisia	Tjitraresmi et al., 2020
Thalictrum spp.	Meadow rue	Ranunculaceae	Alkaloids, phenols, triterpenoids, saponins, phytosterols	Aerial parts, roots	India	Saxena et al., 2003; Das et al., 2016
Tinospora cordifolia	Guduchi	Menispermaceae	Alkaloids, glycosides, sesquiterpenoids	Aerial parts	Asia	Singh and Banyal, 2011
Tinospora crispa	Patawali	Menispermaceae	Alkaloids	Stem	Malaysia	Rahman et al., 1999
Tithonia diversifolia	Tree marigold	Asteraceae	Sesquiterpenes, triterpenes	Leaves	Africa, Mexico, Central America	Njoroge and Bussmann, 2006
Toddalia asiatica	Orange cucumber	Rutaceae	Flavonoids, alkaloids, steroids, saponins, coumarins	Root bark	Africa, Asia	Njoroge and Bussmann, 2006; Orwa et al., 2013
Trema micrantha	Jamaican nettletree	Cannabaceae	Terpenes	Bark	Mexico, Central and South America	Muñoz et al., 2000b; Leme et al., 2020
Trichilia megalantha	African nuttree	Meliaceae	Quassinoids, alkaloids, flavonoids	Leaves, roots	Africa	Fadare and Ajaiyeoba, 2020
Trichilia welwitschii	Bakoie	Meliaceae	Alkaloids, flavonoids	Leaves, roots	Africa	Fadare and Ajaiyeoba, 2020
Tridax procumbens	Tridex daisy	Asteraceae	Flavonoids, steroids, triterpenoids, alkyl esters	Aerial parts	America, Asia	Dattaray, 2022
Trimeria grandifolia	Muhindahindi	Salicaceae	Chloroquine	Roots	Kenya	Njoroge and Bussmann, 2006
Triplaris americana	Ant tree, palo santo	Polygonaceae	Triterpenes, flavonoids	Bark	South America	Muñoz et al., 2000b; Rodrigues et al., 2018

Triumfetta spp.	Burweed	Malvaceae	Flavonoids	Leaves	Central and South America	Muñoz et al., 2000b; Akintimehin et al., 2022
Uapaca nitida	Uapaca	Phyllanthaceae	Triterpenoids	Root, bark	Africa	Saxena et al., 2003
Urtica massaica	Maasai stinging nettle, Thabai	Urticaceae	Alkaloids	Aerial parts, roots, stem, leaves	Africa	Njoroge and Bussmann, 2006
Vepris simplicifolia	Munderendu	Rutaceae	Alkaloids	Root bark	Africa	Njoroge and Bussmann, 2006
Vernonia amygdalina	Bitterleaf	Asteraceae	Tannins, saponins, flavonoids, phenols, alkaloids, steroids, sesquiterpenes	Leaves, root bark	Africa	Abosi and Raseroka, 2003; Idowu et al., 2010; Bihonegn et al., 2019
Vernonia brasiliana	Ironweed	Asteraceae	Triterpenoids	Leaves	Africa	Saxena et al., 2003
Vernonia guineensis	Ewuro-olopaa-kan	Asteraceae	Tannins, saponins, flavonoids, alkaloids, steroids, phenols	Whole plant	Africa	Toyang et al., 2013
Vernonia lasiopus	Bitterleaf	Asteraceae	Sesquiterpenes, triterpenes	Root bark	Africa	Muregi et al., 2007
Viola canescens	Himalayan white violet	Violaceae	Chloroquine	Whole plant	Asia	Verma et al., 2011
Viola websteri	Wood violet	Violaceae	Alkaloids, quassinoids, sesquiterpenes	Whole plant	Asia	Lee et al., 2009
Warburgia ugandensis	Muthiga	Canellaceae	Chloroquine	Stem bark	Kenya	Njoroge and Bussmann, 2006
Werauhia sanguinolenta	Vriesea	Bromeliaceae	Flavonoids, xanthones	Leaves	Central and South America, Mexico	Saxena et al., 2003
Withania somnifera	Winter cherry	Solanaceae	Alkaloids, steroidal lactones	Aerial parts	India, Africa	Njoroge and Bussmann, 2006; Dame et al., 2013
Xanthium strumarium	Rough cocklebur	Asteraceae	Tomentosun, zanthumin	Fruit, aerial parts	South Europe, Asia	Saxena et al., 2003
Xylocarpus granatum	Cannonball mangrove	Meliaceae	Flavonoids, alkaloids, tannins, triterpenes, steroids	Fruits	Asia, Africa, Australia	Lakshmi et al., 2012
Zanthoxylum asiaticum	Orange climber	Rutaceae	Coumarin	Roots	Africa, Asia	Saxena et al., 2003
Zanthoxylum chalybeum	Knob wood	Rutaceae	Coumarins, alkaloids, essential oils	Stem bark	Africa	Njoroge and Bussmann, 2006; Adia et al., 2016
Zanthoxylum gilletii	East African satinwood	Rutaceae	Alkaloids	Roots	Africa	Saxena et al., 2003
Zanthoxylum monophyllum	Yellow prickle	Rutaceae	Coumarins, alkaloids, essential oils	Leaves, bark	Africa	Rodríguez-Guzmán et al., 2011

stage of the malarial parasite is targeted. Some of the examples of quinones include plumbagin, 2-methylnaphthazarin, etc. (Saxena et al., 2003).

1.4.6 Flavonoids

They are a group of compounds extracted from higher plants. They show a positive effect against *Plasmodium falciparum* when combined with artemisinin (Saxena et al., 2003). Examples of some antimalarial flavonoids are Artemetin, Casticin, Chrysosplenetin, and Luteolin-7-O-glufuranoside (Rudrapal and Chetia, 2017). The chemical structure of flavonoid is depicted in figure 1.4.

1.4.7 Xanthones

Similarly, xanthones are extracted from *Garcinia dulcis* (from wood, pericarp, and fruit), which represents an antiplasmodial activity (Ignatushchenko et al., 2000). They are basically phenolic compounds and act as an antioxidant agent. Examples of xanthone are 2,3,4,5,6-pentahydroxyxanthone (Riscoe et al., 2005).

1.4.8 Sesquiterpenes

It is a type of the chemical compound terpene which shows an equally active antimalarial activity. Examples of sesquiterpenes include artemisinin and isonardoperoxide.

1.4.9 Triterpenoids

They are a group of biocompound extracted from a large number of plant materials. Saxena et al. (2003) highlighted that the antimalarial potential of triterpenoids such as gedunin is triple of chloroquine but around 1/20th of quinine. The structure of a triterpenoid is shown in Figure 1.4.

1.4.10 Quassinoids

They are mostly 20-carbon oxygenated lactones. They are important in biological activity due to the presence of oxygen and methyl group attached to different carbons (Figure 1.4). Triterpenes such as quassinoids exhibit antimalarial properties. Some examples of quassinoids are cedronin, brusatol, bruceantin, chaparrin, glucarubol, and chloroquine diphosphate (O'Neill et al., 1986). Saxena et al. (2003) mentioned that if the alkyl group is added to the quassinoid structure, the potential of antimalarial activity increases.

1.5 FUTURE DEVELOPMENTS/RESEARCH AND RECOMMENDATIONS

Most of the malarial vectors *Anopheles* are complex species and identifiable using various techniques, such as polymerase chain reaction, RFLP (restriction fragment length polymorphism), paracentric inversion of polytene chromosome, electrophoretic variation of enzyme, and mitotic chromosome variation (Dash et al., 2007, 2008). Subbarao (1998) reported the distribution maps using GIS (geographic information system) for *A. culicifacies* and *A. fluviatilis* in India and emphasized its importance in understanding the disease (malaria) potential and formulating effective control measures. Detection of malarial cases is made using rapid test based on histidine-rich protein (HRP-2) for *Plasmodium falciparum* presence (Beadle et al., 1994).

Chemical control: DDT, HCH, malathion, pirimiphos-methyl, fenitrothion, carbamates such as propoxur and bendiocarb, synthetic pyrethroids, temephos, Malariol, and Baytex (Dash et al., 2008).

Biological control: Larvivorous fish, *Bacillus thuringiensis*, *B. sphaericus* (Kumar et al., 1998).

In rural and tribal area, villagers often use smoke by burning the wood/dry cowdung as preventive measure to avoid mosquitoes (Sharma et al., 2007). Integrated approach is advised for effective control of malaria (WHO, 2004). It includes the use of various resources, such as chemical, biological agents, and plants. In India, malarial drug programme has a set rule for its treatment, that is, chloroquine and artesunate-sulfadoxine pyrimethamine (Dash et al., 2008).

The major that problem arises in mosquito control is insecticide resistance. Most of the chemicals, importantly DDT, were used as insecticide against *Anopheles* species. Subsequently, with time, they developed resistance against DDT and other chemicals. Some of the chemicals, such as HCH and organochlorine, were banned due to their negative impact on humans and food (Dash et al., 2008). Similarly, *Plasmodium falciparum* was found to develop resistance against chloroquines (Sehgal et al., 1973). Narain and Nath (2018) reported that *Plasmodium* has shown resistance to the drug ACT (artemisinin-based combination therapy) in North-East India near Myanmar. According to IPCC (Intergovernmental Panel on Climate Change), the global temperature will rise from 1.4° to 5.8°C by 2100 (Bhattacharya, 2006). It is projected using mathematical modelling by Bhattacharya (2006) that climate change by 2050 will have more impact on malarial cases, especially in Kerala, Karnataka, and Himachal Pradesh, and states which are already affected, such as Madhya Pradesh, Jharkhand, Odisha, West Bengal, and some part of North-East India, will get worse. Das et al. (2012) mentioned that due to its large effect and distribution, malaria is a major disease in South-East Asia. They also reported that the major problems to tackle with malaria are insecticide resistance, antimalarial drug resistance, and implementation of control programmes. Rahi and Sharma (2020) emphasized the integration of all types of data and surveillance related to malaria on a single platform to eliminate it completely. Medicinal plants throughout the world are suitable for a variety of uses, such as antidiuretic, antiallergic, antipyretic, anti-inflammatory, antianalgesic, antiviral, antibacterial, antifungal, antifeedant, and others. Following are the recommendations for antimalarial plants uses in the world:

1. Proper identification of the plant upto species level needs to be done to disseminate correct information about the plant.
2. More social surveys are to be conducted for better understanding of the current and past use of antimalarial plants throughout the world.
3. Phytochemical constituents of different parts of any trees need to be carried out to identify the chemical compound responsible for any medicinal property.
4. Long-term study should be conducted to assess the effect of any antimalarial plant uses on humans.
5. Traditional knowledge (TKs) of any community or tribe related to medicinal plants should be conserved scientifically to support the medical systems.

REFERENCES

Abdalla, M. A. & Laatsch, H. 2012. Flavonoids from Sudanese *Albizia zygia* (Leguminosae, subfamily Mimosoideae), a plant with antimalarial potency. *African Journal of Traditional, Complementary and Alternative Medicines* 9(1): 56–58.

Abdulrazak, N., Asiya, U. I., Usman, N. S., Unata, I. M. & Farida, A. 2015. Anti-plasmodial activity of ethanolic extract of root and stem back of *Cassia sieberiana* DC on mice. *Journal of Intercultural Ethnopharmacology* 4(2): 96–101.

Abosi, A. O. & Raseroka, B. H. 2003. In vivo antimalarial activity of *Vernonia amygdalina*. *British Journal of Biomedical Science* 60(2): 89–91.

Achan, J., Talisuna, A. O., Erhart, A., Yeka, A., Tibenderana, J. K., Baliraine, F. N., Rosenthal, P. J. & D'Alessandro, U. 2011. Quinine, an old anti-malarial drug in a modern world: Role in the treatment of malaria. *Malaria Journal* 10(1): 1–12.

Achenbach, H., Waibel, R., Nkunya, M. H. & Weenen, H. 1992. Antimalarial compounds from *Hoslundia opposita. Phytochemistry* 31(11): 3781–3784.

Adekanmi, A. A., Adekanmi, U. T., Adekanmi, A. S. & Oyekanmi, H. A. 2020. Assessment of proximate composition and phytochemical properties of bitter leaf (*Vernonia Amygdalina*) and water leaf (*Talinum Triangular*). *United International Journal for Research & Technology* 1(9): 13–21.

Adetunji, T. L., Adetunji, A. E., Odebunmi, C. A., Van der Kooy, F. & Siebert, F. 2022. *Mallotus oppositifolius* (Geiseler) Müll. Arg.: The first review of its botany, ethnomedicinal uses, phytochemistry and biological activities. *South African Journal of Botany* 147: 245–262.

Adia, M. M., Emami, S. N., Byamukama, R., Faye, I. & Borg-Karlson, A. K. 2016. Antiplasmodial activity and phytochemical analysis of extracts from selected Ugandan medicinal plants. *Journal of Ethnopharmacology* 186: 14–19.

Adoho, A. C. C., Zinsou, F. T., Olounlade, P. A., Azando, E. V. B., Hounzangbe-Adote, M. S. & Gbangboche, A. B. 2021. Review of the literature of *Eleusine indica*: Phytochemical, toxicity, pharmacological and zootechnical studies. *Journal of Pharmacognosy and Phytochemistry* 10(3): 29–33.

Afoulous, S., Ferhout, H., Raoelison, E. G., Valentin, A., Moukarzel, B., Couderc, F. & Bouajila, J. 2013. Chemical composition and anticancer, antiinflammatory, antioxidant and antimalarial activities of leaves essential oil of *Cedrelopsis grevei. Food and Chemical Toxicology* 56: 352–362.

Ajaiyeoba, E. O., Abalogu, U. I., Krebs, H. C. & Oduola, A. M. J. 1999. In vivo antimalarial activities of *Quassia amara* and *Quassia undulata* plant extracts in mice. *Journal of Ethnopharmacology* 67(3): 321–325.

Akintimehin, E., Onoagbe, I. & Abu, O. 2022. Proximate composition, qualitative phytochemical screening and in-vitro antioxidant potential of *Triumfettar homboidea* leaves extracts. *Trends in Sciences* 19(24): 3033–3033.

Akomo, E. F. O., Zongo, C., Karou, S. D., Obame, L. C., Savadogo, A., Atteke, C. & Traore, A. S. 2009. In vitro antiplasmodial and antibacterial activities of *Canthium multiflorum* Schum and Thonn (Rubiacea) extracts. *Pakistan Journal of Biological Sciences* 12(12): 919–923.

Akuodor, G. C., Essien, D. O., Nkorroh, J. A., Essien, A. D., Nkanor, E. E., Ezeunala, M. N. & Chilaka, K. C. 2017. Antiplasmodial activity of the ethanolic root bark extract of *Icacina senegalensis* in mice infected by *Plasmodium berghei. Journal of Basic and Clinical Physiology and Pharmacology* 28(2): 181–184.

Alshawsh, M. A., Mothana, R. A., Al-Shamahy, H. A., Alsllami, S. F. & Lindequist, U. 2009. Assessment of antimalarial activity against *Plasmodium falciparum* and phytochemical screening of some Yemeni medicinal plants. *Evidence-based Complementary and Alternative Medicine* 6(4): 453–456.

Ancolio, C., Azas, N., Mahiou, V., Ollivier, E., Di Giorgio, C., Keita, A., Timon-David, P. & Balansard, G. 2002. Antimalarial activity of extracts and alkaloids isolated from six plants used in traditional medicine in Mali and Sao Tome. *Phytotherapy Research* 16(7): 646–649.

Anosa, G. N., Udegbunam, R. I., Okoro, J. O. & Okoroafor, O. N. 2014. In vivo antimalarial activities of Enantia polycarpa stem bark against *Plasmodium berghei* berghei in mice. *Journal of Ethnopharmacology* 153(2): 531–534.

Appiah-Opong, R., Nyarko, A. K., Dodoo, D., Gyang, F. N., Koram, K. A. & Ayisi, N. K. 2011. Antiplasmodial activity of extracts of *Tridax procumbens* and *Phyllanthus amarus* in in vitro *Plasmodium falciparum* culture systems. *Ghana Medical Journal* 45(4): 143–150.

Asase, A. & Oppong-Mensah, G. 2009. Traditional antimalarial phytotherapy remedies in herbal markets in southern Ghana. *Journal of Ethnopharmacology* 126(3): 492–499.

Asnaashari, S., Afshar, F. H., Ebrahimi, A., Moghadam, S. B. & Delazar, A. 2015. In vitro antimalarial activity of different extracts of *Eremostachys macrophylla* Montbr. & Auch. *Bio Impacts* 5(3): 135.

Asnake, S., Teklehaymanot, T., Hymete, A., Erko, B. & Giday, M. 2015. Evaluation of the antiplasmodial properties of selected plants in southern Ethiopia. *BMC Complementary and Alternative Medicine* 15(1): 1–12.

Astulla, A., Hirasawa, Y., Rahman, A., Kusumawati, I., Ekasari, W., Widyawaruyanti, A., Zaini, N. C. & Morita, H. 2011. Melidianolic acid A and B, new antimalarial acyclic diterpenes from *Aphanamixisgrandifolia. Natural Product Communications* 6(3): 323–326.

Ateba, J. E., Toghueo, R. M., Awantu, A. F., Mba'ning, B. M., Gohlke, S., Sahal, D., Rodrigues-Filho, E., Tsamo, E., Boyom, F. F., Sewald, N. & Lenta, B. N. 2018. Antiplasmodial properties and cytotoxicity of endophytic fungi from *Symphonia globulifera* (Clusiaceae). *Journal of Fungi* 4(2): 1–9.

Awe, S. O., Olajide, O. A., Oladiran, O. O. & Makinde, J. M. 1998. Antiplasmodial and antipyretic screening of *Mangifera indica* extract. *Phytotherapy Research Natural* 12(6): 437–438.

Azas, N., Laurencin, N., Delmas, F., Giorgio, C. D., Gasquet, M., Laget, M. & Timon-Daid, P. 2002. Synergistic in vitro antimalarial activity of plant extracts used as traditional herbal remedies in Mali. *Parasitology Research* 88: 165–171.

Azebaze, A. G. B., Teinkela, J. E. M., Nguemfo, E. L., Valentin, A., Dongmo, A. B. & Vardamides, J. C. 2015. Antiplasmodial activity of some phenolic compounds from *Cameroonians Allanblackia. African Health Sciences* 15(3): 835–840.

Azizi, M. A., Nadia, N. A. C., Cedric, Y., Guy-Armand, G. N., Sidiki, N. N. A., Sandra, T. N. J., Kevin, T. D. A. & Payne, V. C. 2023. In vitro antiplasmodial, cytotoxicity, and antioxidant activities of *Lophira lanceolata* (Ochnaceae): A cameroonian plant commonly used to treat malaria. *Journal of Tropical Medicine* 4061592: 1–9.

Bagatela, B. S., Lopes, A. P., Fonseca, F. L. A., Andreo, M. A., Nanayakkara, D. N., Bastos, J. K. & Perazzo, F. F. 2013. Evaluation of antimicrobial and antimalarial activities of crude extract, fractions and 4-nerolidylcathecol from the aerial parts of *Piper umbellata* L. (Piperaceae). *Natural Product Research* 27(23): 2202–2209.

Bahekar, S. & Kale, R. 2013. Herbal plants used for the treatment of malaria-A literature review. *Journal of Pharmacognosy and Phytochemistry*1(6): 141–146.

Bankole, A. E. & Abiodun, O. O. 2020. *Markhamia tomentosa* enhanced the antimalarial activity of chloroquine and amodiaquine in chloroquine resistant *P. berghei* infected mice. *The Journal of Phytopharmacology* 9(2): 83–88.

Barliana, M. I., Suradji, E. W., Abdulah, R., Diantini, A., Hatabu, T., Nakajima-Shimada, J., Subarnas, A. & Koyama, H. 2014. Antiplasmodial properties of kaempferol-3-O-rhamnoside isolated from the leaves of *Schima wallichii* against chloroquine-resistant *Plasmodium falciparum. Biomedical Reports* 2(4): 579–583.

Beadle, C., Long, G. W., Weiss, W. R., McElroy, P. D., Maret, S. M., Oloo, A. J. & Hoffman, S. L. 1994. Diagnosis of malaria by detection of *Plasmodium falciparum* HRP-2 antigen with a rapid dipstick antigen-capture assay. *Lancet* 343: 564–568.

Beaufay, C., Henry, G., Streel, C., Bony, E., Hérent, M. F., Bero, J. & Quetin-Leclercq, J. 2019. Optimization and validation of extraction and quantification methods of antimalarial triterpenic esters in *Keetia leucantha* plant and plasma. *Journal of Chromatography B*1 104: 109–118.

Belay, W. Y., Gurmu, A. E. & Wubneh, Z. B. 2022. Antimalarial activity of solvent fractions of the stem bark of *Periploca linearifolia* Quart.-Dill. & A. Rich. (Asclepiadaceae) against *Plasmodium berghei* in Mice. *Journal of Experimental Pharmacology* 4: 7–15.

Bhattacharya, S., Sharma, C., Dhiman, R. C. & Mitra, A. P. 2006. Climate change and malaria in India. *Current Science* 90(3): 369–375.

Bihonegn, T., Giday, M., Yimer, G., Animut, A. & Sisay, M. 2019. Antimalarial activity of hydromethanolic extract and its solvent fractions of *Vernonia amygdalina* leaves in mice infected with *Plasmodium berghei. SAGE Open Medicine* 7: 1–10.

Bora, U., Sahu, A., Saikia, A. P., Ryakala, V. K. & Goswami, P. 2007. Medicinal plants used by the people of Northeast India for curing malaria. *Phytotherapy Research* 21(8): 800–804.

Boucherle, B., Haudecoeur, R., Queiroz, E. F., De Waard, M., Wolfender, J. L., Robins, R. J. & Boumendjel, A. 2016. *Nauclea latifolia:* Biological activity and alkaloid phytochemistry of a West African tree. *Natural Product Reports* 33(9): 1034–1043.

Bray, D. H., Warhurst, D. C., Connolly, J. D., O'Neill, M. J. & Phillipson, J. D. 1990. Plants as sources of antimalarial drugs. Part 7. Activity of some species of Meliaceae plants and their constituent limonoids. *Phytotherapy Research* 4(1): 29–35.

Bruno, T., Soh, D. & Ernestine, N. 2020. Phytochemical composition and biological activity of *Faidherbia albida* (Mimosaceae) roots and leaves. *International Journal of Pharmaceutical Sciences Review and Research* 65(1): 124–130.

Budiarti, M., Maruzy, A., Mujahid, R., Sari, A. N., Jokopriyambodo, W., Widayat, T. & Wahyono, S. 2020. The use of antimalarial plants as traditional treatment in Papua Island, Indonesia. *Heliyon* 6(12): 1–10.

Carroll, A. R., Avery, V. M., Duffy, S., Forster, P. I. & Guymer, G. P. 2013. Watsonianone A – C, anti-plasmodial β-triketones from the Australian tree, *Corymbia watsoniana. Organic & Biomolecular Chemistry* 11(3): 453–458.

Carvalho, J. C., Frazão, A., Lohmann, L. G. & Ferreira, M. J. 2021. Leaf cuticular waxes of *Tanaecium* (Bignonieae, Bignoniaceae): Chemical composition and taxonomic implications. *Biochemical Systematics and Ecology* 98: 1–9.

Castro, M. C. & Peterka, C. 2023. Malaria is increasing in Indigenous and artisanal mining areas in the Brazilian Amazon. *Nature Medicine*: 1–3. https://doi.org/10.1038/s41591-023-02280-0.

Chandel, S., Bagai, U., Semwal, R. B. & Semwal, D. K. 2015. Antiplasmodial activity of aqueous extract of *Berberis aristata* roots against *Plasmodiumberghei*-infected BALB/c mice. *Pharmaceutical Biology* 53(12): 1735–1740.

Chander, M. P., Pillai, C. R., Sunish, I. P. & Vijayachari, P. 2016. Antimicrobial and antimalarial properties of medicinal plants used by the indigenous tribes of Andaman and Nicobar Islands, India. *Microbial Pathogenesis* 96: 85–88.

Chaniad, P., Techarang, T., Phuwajaroanpong, A., Na-Ek, P., Viriyavejakul, P. & Punsawad, C. 2021. In vivo antimalarial activity and toxicity study of extracts of *Tagetes* erlecta L. and *Synedrella nodiflora* (L.) Gaertn. from the asteraceae family. *Evidence-Based Complementary and Alternative Medicine* 2021: 1–9.

Cheema, H. S., Prakash, O., Pal, A., Khan, F., Bawankule, D. U. & Darokar, M. P. 2014. Glabridin induces oxidative stress mediated apoptosis like cell death of malaria parasite *Plasmodium falciparum*. *Parasitology International* 63(2): 349–358.

Cheenpracha, S., Ritthiwigrom, T. & Laphookhieo, S. 2013. Alstoniaphyllines A – C, unusual nitrogenous derivatives from the bark of *Alstonia macrophylla*. *Journal of Natural Products* 76(4): 723–726.

Chianese, G., Yerbanga, S. R., Lucantoni, L., Habluetzel, A., Basilico, N., Taramelli, D., Fattorusso, E. & Taglialatela-Scafati, O. 2010. Antiplasmodial triterpenoids from the fruits of neem, *Azadirachta indica*. *Journal of Natural Products* 73(8): 1448–1452.

Chierrito, T. P., Aguiar, A. C., de Andrade, I. M., Ceravolo, I. P., Gonçalves, R. A., de Oliveira, A. J. & Krettli, A. U. 2014. Anti-malarial activity of indole alkaloids isolated from *Aspidosperma olivaceum*. *Malaria Journal* 13: 1–10.

Chinchilla, M., Valerio, I., Sánchez, R., Mora, V., Bagnarello, V., Martínez, L., Gonzalez, A., Vanegas, J. C. & Apestegui, Á. 2012. In vitro antimalarial activity of extracts of some plants from a biological reserve in Costa Rica. *Revista de Biología Tropical* 60(2): 881–891.

Chung, I. M., Ghimire, B. K., Kang, E. Y. & Moon, H. I. 2010. Antiplasmodial and cytotoxic activity of khellactone derivatives from *Angelica purpuraefolia* Chung. *Phytotherapy Research* 24(3): 469–471.

Cretton, S., Glauser, G., Humam, M., Jeannerat, D., Muñoz, O., Maes, L., Christen, P. & Hostettmann, K. 2010. Isomeric tropane alkaloids from the aerial parts of *Schizanthus tricolor*. *Journal of Natural Products* 73(5): 844–847.

Dame, Z. T., Petros, B. & Mekonnen, Y. 2013. Evaluation of anti-*Plasmodium berghei* activity of crude and column fractions of extracts from *Withania somnifera*. *Turkish Journal of Biology* 37(2): 147–150.

Das, A., Anvikar, A. R., Cator, L. J., Dhiman, R. C., Eapen, A., Mishra, N., Nagpal, B. N., Nanda, N., Raghavendra, K., Read, A. F., Sharma, S. K., Singh, O. P., Singh, V., Sinnis, P., Srivastava, H. C., Sullivan, S. A. Sutton, P. L., Thomas, M. B., Carlton, J. M. & Valecha, N. 2012. Malaria in India: The center for the study of complex malaria in India. *Acta Tropica*1 21(3): 267–273.

Das, N. G., Rabha, B., Talukdar, P. K., Goswami, D. & Dhiman, S. 2016. Preliminary in vitro antiplasmodial activity of *Aristolochia griffithii* and *Thalictrum foliolosum* DC extracts against malaria parasite *Plasmodium falciparum*. *BMC Research Notes* 9: 1–5.

Dash, A. P., Adak, T., Raghavendra, K. & Singh, O. P. 2007. The biology and control of malaria vectors in India. *Current Science* 92: 1571–1578.

Dash, A. P., Valecha, N., Anvirkar, A. R. & Kumar, A. 2008. Malaria in India: Challenges and opportunities. *Journal of Bioscience* 33(4): 583–592.

Daskum, A. M., Godly, C., Qadeer, M. A. & Ling, L. Y. 2019. Effect of *Senna occidentalis* (Fabaceae) leaves extract on the formation of β-hematin and evaluation of in vitro antimalarial activity. *International Journal of Herbal Medicine* 7(3): 46–51.

Dattaray, D. 2022. Traditional uses and pharmacology of plant *Tridax procumbens*: A review. *Systematic Review Pharmacy* 13(5): 511–517.

De-Campos-Bortolucci, W., Marko-de-Oliveira, H. L., Roque-Oliva, L., Gonçalves, J. E., Piau-Júnior, R., Mariano-Fernandez, C. M., Barros-Colauto, N., Linde, G. A. & Gazim, Z. C. 2021. Crude extract of the tropical tree *Gallesia integrifolia* (Phytolaccaceae) for the control of *Aedes aegypti* (Diptera: Culicidae) larvae. *Revista de Biología Tropical* 69(1): 153–169.

Deguchi, J., Hirahara, T., Hirasawa, Y., Ekasari, W., Widyawaruyanti, A., Shirota, O., Shiro, M. & Morita, H. 2012. New tricyclic alkaloids, cassiarins G, H, J, and K from leaves of *Cassia siamea*. *Chemical and Pharmaceutical Bulletin* 60(2): 219–222.

Deharo, E., Bourdy, G., Quenevo, C., Muñoz, V., Ruiz, G. & Sauvain, M. 2001. A search for natural bioactive compounds in Bolivia through a multidisciplinary approach. Part V. Evaluation of the antimalarial activity of plants used by the Tacana Indians. *Journal of Ethnopharmacology* 77(1): 91–98.

Dell'Agli, M., Galli, G. V., Corbett, Y., Taramelli, D., Lucantoni, L., Habluetzel, A., Maschi, O., Caruso, D., Giavarini, F., Romeo, S. & Bhattacharya, D. 2009. Antiplasmodial activity of *Punica granatum* L. fruit rind. *Journal of Ethnopharmacology* 125(2): 279–285.

Dharani, N., Rukunga, G., Yenesew, A., Mbora, A., Mwaura, L., Dawson, I. & Jamnadass, R. 2010. *Common Antimalarial Trees and Shrubs of East Africa: A Description of Species and a Guide to Cultivation and Conservation through Use*. Edited by Dawson, I. The World Agroforestry Centre (ICRAF), pp. 1–100.

Dhooghe, L., Maregesi, S., Mincheva, I., Ferreira, D., Marais, J. P., Lemière, F., Matheeussen, A., Cos, P., Maes, L., Vlietinck, A. & Apers, S. 2010. Antiplasmodial activity of (I-3, II-3)-biflavonoids and other constituents from *Ormocarpum kirkii*. *Phytochemistry* 71(7): 785–791.

Dikasso, D., Makonnen, E., Debella, A., Abebe, D., Urga, K., Makonnen, W., Melaku, D., Assefa, A. & Makonnen, Y. 2006. In vivo anti-malarial activity of hydroalcoholic extracts from *Asparagus africanus* Lam. in mice infected with *Plasmodium berghei*. *Ethiopian Journal of Health Development* 20(2): 112–118.

Dkhil, M. A., Lubbad, M. Y., Al-Shaebi, E. M., Delic, D. & Al-Quraishy, S. 2015. The antiplasmodial and spleen protective role of crude *Indigofera oblongifolia* leaf extract traditionally used in the treatment of malaria in Saudi Arabia. *Drug Design, Development and Therapy* 9: 6235–6246.

do Carmo, D. F. M., Amaral, A. C. F., Machado, M., Lopes, D., Echevarria, A., Rosário, V. E. & Silva, J. R. D. A. 2015. Evaluation of Antiplasmodial activity of extracts and constituents from *Ampelozizyphus amazonicus*. *Pharmacognosy Magazine* 11(44): 244–250.

Endale, A., Bisrat, D., Animut, A., Bucar, F. & Asres, K. 2013. In vivo antimalarial activity of a labdane diterpenoid from the leaves of *Otostegia integrifolia* Benth. *Phytotherapy Research* 27(12): 1805–1809.

Endale, M., Ekberg, A., Akala, H. M., Alao, J. P., Sunnerhagen, P., Yenesew, A. & Erdélyi, M. 2012. Busseihydroquinones A – D from the roots of *Pentas bussei*. *Journal of Natural Products* 75(7): 1299–1304.

Fadare, D. A. & Ajaiyeoba, E. O. 2020. Activity of *Trichilia megalantha* harms and *Trichilia welwitschii* CDC extracts and fractions on *Anopheles gambiae* larvae. *Anchor University Journal of Science and Technology* 1(2): 35–43.

Falade, M. O., Akinboye, D. O., Gbotosho, G. O., Ajaiyeoba, E. O., Happi, T. C., Abiodun, O. O. & Oduola, A. M. J. 2014. In vitro and in vivo antimalarial activity of *Ficus thonningii* Blume (Moraceae) and *Lophira alata* Banks (Ochnaceae), identified from the ethnomedicine of the Nigerian Middle Belt. *Journal of Parasitology Research*: 1–7.

Ganfon, H., Bero, J., Tchinda, A. T., Gbaguidi, F., Gbenou, J., Moudachirou, M., Frédérich, M. & Quetin-Leclercq, J. 2012. Antiparasitic activities of two sesquiterpenic lactones isolated from *Acanthospermum hispidum* DC. *Journal of Ethnopharmacology* 141(1): 411–417.

Garavito, G., Rincón, J., Arteaga, L., Hata, Y., Bourdy, G., Gimenez, A., Pinzón, R. & Deharo, E. 2006. Antimalarial activity of some Colombian medicinal plants. *Journal of Ethnopharmacology* 107(3): 460–462.

Geethangili, M. & Ding, S. T. 2018. A review of the phytochemistry and pharmacology of *Phyllanthus urinaria* L. *Frontiers in Pharmacology* 9: 1–20.

Girardot, M., Deregnaucourt, C., Deville, A., Dubost, L., Joyeau, R., Allorge, L., Rasoanaivo, P. & Mambu, L. 2012. Indole alkaloids from *Muntafara sessilifolia* with antiplasmodial and cytotoxic activities. *Phytochemistry* 73: 65–73.

Gitua, J. N., Muchiri, D. R. & Nguyen, X. A. 2012. In vivo antimalarial activity of *Ajuga remota* water extracts against *Plasmodium berghei* in mice. *Southeast Asian Journal of Tropical Medicine and Public Health* 43(3): 545–548.

Govindarajan, M. & Sivakumar, R. 2014. Larvicidal, ovicidal, and adulticidal efficacy of *Erythrina indica* (Lam.)(Family: Fabaceae) against *Anopheles stephensi, Aedes aegypti*, and *Culex quinquefasciatus* (Diptera: Culicidae). *Parasitology Research* 113: 777–791.

Grace, M. H., Lategan, C., Graziose, R., Smith, P. J., Raskin, I. & Lila, M. A. 2012. Antiplasmodial activity of the ethnobotanical plant *Cassia fistula*. *Natural Product Communications* 7(10): 1263–1266.

Grace, M. H., Lategan, C., Mbeunkui, F., Graziose, R., Smith, P. J., Raskin, I. & Lila, M. A. 2010. Antiplasmodial and cytotoxic activities of drimane sesquiterpenes from *Canella winterana*. *Natural Product Communications* 5(12): 1869–1872.

Graziose, R., Grace, M. H., Rathinasabapathy, T., Rojas-Silva, P., Dekock, C., Poulev, A., Lila, M. A., Smith, P. & Raskin, I. 2013. Antiplasmodial activity of cucurbitacin glycosides from *Datisca glomerata* (C. Presl) Baill. *Phytochemistry* 87: 78–85.

Guo, Z. 2016. Artemisinin anti-malarial drugs in China. *Acta Pharmaceutica Sinica B* 6(2): 115–124.

Hao, B., Shen, S. F. & Zhao, Q. J. 2013. Cytotoxic and antimalarial amaryllidaceae alkaloids from the bulbs of *Lycoris radiata*. *Molecules* 18(3): 2458–2468.

Houël, E., Fleury, M., Odonne, G., Nardella, F., Bourdy, G., Vonthron-Sénécheau, C., Villa, P., Obrecht, A., Eparvier, V., Deharo, E. & Stien, D. 2015. Antiplasmodial and anti-inflammatory effects of an antimalarial remedy from the Wayana Amerindians, French Guiana: Takamalaimë (*Psidium acutangulum* Mart. ex DC., Myrtaceae). *Journal of Ethnopharmacology* 166: 279–285.

Idowu, O. A., Soniran, O. T., Ajana, O. & Aworinde, D. O. 2010. Ethnobotanical survey of antimalarial plants used in Ogun State, Southwest Nigeria. *African Journal of Pharmacy and Pharmacology* 4(2): 55–60.

Ignatushchenko, M. V., Winter, R. W. & Riscoe, M. 2000. Xanthones as antimalarial agents: Stage specificity. *The American Journal of Tropical Medicine and Hygiene* 62(1): 77–81.

Ijaz, F., Ahmad, N., Ahmad, I., ul Haq, A. & Wang, F. 2010. Two new anti-plasmodial flavonoid glycosides from *Duranta repens*. *Journal of Enzyme Inhibition and Medicinal Chemistry* 25(6): 773–778.

Ileke, K. D. & Adesina, J. M. 2019. Toxicity of *Ocimum basilicum* and *Ocimum gratissimum* extracts against main malaria vector, *Anopheles gambiae* (Diptera: Culicidae) in Nigeria. *Journal of Arthropod-Borne Diseases* 13(4): 362–368.

Induli, M., Gebru, M., Abdissa, N., Akala, H., Wekesa, I., Byamukama, R., Heydenreich, M., Murunga, S., Dagne, E. & Yenesew, A. 2013. Antiplasmodial quinones from the rhizomes of *Kniphofia foliosa*. *Natural Product Communications* 8(9): 1261–1264.

Innok, P., Rukachaisirikul, T., Phongpaichit, S. & Suksamrarn, A. 2010. Fuscacarpans A – C, new pterocarpans from the stems of *Erythrina fusca*. *Fitoterapia* 81(6): 518–523.

Isikhuemen, E. M., Ogbomwan, B. O. & Efenudu, I. U. 2020. Evaluation of phytochemical and mineral constituents of *Piper guineense* Schum. & Thonn. and *Piper Umbellatum* Linn: Implications for ethnomedicine. *European Journal of Medicinal Plants* 31(1): 84–97.

Jaidee, A., Promchai, T., Trisuwan, K., Laphookhieo, S., Rattanajak, R., Kamchonwongpaisan, S., Pyne, S. G. & Ritthiwigrom, T. 2015. Cytotoxic and antimalarial alkaloids from the twigs of *Dasymaschalon obtusipetalum*. *Natural Product Communications* 10(7): 1175–1177.

Jansen, O., Tits, M., Angenot, L., Nicolas, J. P., De Mol, P., Nikiema, J. B. & Frédérich, M. 2012. Antiplasmodial activity of *Dicoma tomentosa* (Asteraceae) and identification of urospermal A-15-O-acetate as the main active compound. *Malaria Journal* 11(1): 1–9.

Jassbi, A. R., Eghtesadi, F., Hazeri, N., Ma'sumi, H., Valizadeh, J., Chandran, J. N., Schneider, B. & Baldwin, I. T. 2017. The roots of *Salvia rhytidea*: A rich source of biologically active diterpenoids. *Natural Product Research* 31(4): 477–481.

Johns, T., Windust, A., Jurgens, T. & Mansor, S. M. 2011. Antimalarial alkaloids isolated from *Annona squamosa*. *Phytopharmacology* 1(3): 49–53.

Kamaraj, C., Kaushik, N. K., Rahuman, A. A., Mohanakrishnan, D., Bagavan, A., Elango, G., Zahir, A. A., Santhoshkumar, T., Marimuthu, S., Jayaseelan, C. & Kirthi, A. V. 2012. Antimalarial activities of medicinal plants traditionally used in the villages of Dharmapuri regions of South India. *Journal of Ethnopharmacology* 141(3): 796–802.

Kaou, A. M., Mahiou-Leddet, V., Hutter, S., Aïnouddine, S., Hassan, S., Yahaya, I., Azas, N. & Ollivier, E. 2008. Antimalarial activity of crude extracts from nine African medicinal plants. *Journal of Ethnopharmacology* 116: 74–83.

Kaur, R. & Kaur, H. 2017. Plant derived antimalarial agents. *Journal of Medicinal Plants Studies* 5(1): 346–363.

Kffuri, C. W., Lopes, M. A., Ming, L. C., Odonne, G. & Kinupp, V. F. 2016. Antimalarial plants used by indigenous people of the Upper Rio Negro in Amazonas, Brazil. *Journal of Ethnopharmacology* 178: 188–198.

Koudouvo, K., Karou, D. S., Kokou, K., Essien, K., Aklikokou, K., Glitho, I. A., Simpore, J., Sanogo, R., De Souza, C. & Gbeassor, M. 2011. An ethnobotanical study of antimalarial plants in Togo Maritime Region. *Journal of Ethnopharmacology* 134(1): 183–190.

Kumar, A., Sharma, V. P., Sumodan, P. K. & Thavaselvam, D. 1998. Field trials of biolarvicide *Bacillus thuringiensis* var. israelensis strain 164 and the larvivorous fish *Aplocheilus blocki* against *Anopheles stephensi* for malaria control in Goa, India. *Journal of the American Mosquito Control Association* 14(4): 457–462.

Kumar, A., Valecha, N., Jain, T. & Dash, A. P. 2007. Burden of malaria in India: Retrospective and prospective view. *American Journal of Tropical Medicine and Hygiene* 77(6): 69–78.

Kwansa-Bentum, B., Agyeman, K., Larbi-Akor, J., Anyigba, C. & Appiah-Opong, R. 2019. In vitro assessment of antiplasmodial activity and cytotoxicity of *Polyalthia longifolia* leaf extracts on *Plasmodium falciparum strain NF54*. *Malaria Research and Treatment*: 1–9.

Lacroix, D., Prado, S., Kamoga, D., Kasenene, J. & Bodo, B. 2011. Structure and in vitro antiparasitic activity of constituents of *Citropsis articulata* root bark. *Journal of Natural Products* 74(10): 2286–2289.

Lakshmi, V., Srivastava, S., Mishra, S. K., Srivastava, M. N., Srivastava, K. & Puri, S. K. 2012. Antimalarial activity in *Xylocarpus granatum* (Koen). *Natural Product Research* 26(11): 1012–1015.

Langat, M. K., Crouch, N. R., Smith, P. J. & Mulholland, D. A. 2011. Cembranolides from the leaves of *Croton gratissimus*. *Journal of Natural Products* 74(11): 2349–2355.

Laryea, M. K. & Borquaye, L. S. 2019. Antimalarial efficacy and toxicological assessment of extracts of some Ghanaian medicinal plants. *Journal of Parasitology Research* 1630405: 1–9.

Lee, S. J., Park, W. H. & Moon, H. I. 2009. Bioassay-guided isolation of antiplasmodial anacardic acids derivatives from the whole plants of *Viola websteri* Hemsl. *Parasitology Research* 104: 463–466.

Lekphrom, R., Kanokmedhakul, S. & Kanokmedhakul, K. 2010. Bioactive diterpenes from the aerial parts of *Anisochilus harmandii*. *Planta Medica* 76(7): 726–728.

Leme, F. M., Borella, P. H., Marinho, C. R. & Teixeira, S. P. 2020. Expanding the laticifer knowledge in Cannabaceae: Distribution, morphology, origin, and latex composition. *Protoplasma* 257: 1183–1199.

Lenta, B. N., Tantangmo, F., Devkota, K. P., Wansi, J. D., Chouna, J. R., Soh, R. C. F., Neumann, B., Stammler, H. G., Tsamo, E. & Sewald, N. 2009. Bioactive constituents of the stem bark of *Beilschmiedia zenkeri*. *Journal of Natural Products* 72(12): 2130–2134.

Levrier, C., Balastrier, M., Beattie, K. D., Carroll, A. R., Martin, F., Choomuenwai, V. & Davis, R. A. 2013. Pyridocoumarin, aristolactam and aporphine alkaloids from the Australian rainforest plant *Goniothalamus australis*. *Phytochemistry* 86: 121–126.

Liu, Y., Rakotondraibe, L. H., Brodie, P. J., Wiley, J. D., Cassera, M. B., Goetz, M. & Kingston, D. G. 2014. Antiproliferative and antimalarial sesquiterpene lactones from *Piptocoma antillana* from Puerto Rico [1]. *Natural Product Communications* 9(10): 1403–1406.

Liu, Y., Rakotondraibe, L. H., Brodie, P. J., Wiley, J. D., Cassera, M. B., Miller, J. S., Ratovoson, F., Rakotobe, E., Rasamison, V. E. & Kingston, D. G. 2015. Antimalarial 5, 6-dihydro-α-pyrones from *Cryptocarya rigidifolia*: Related bicyclic tetrahydro-α-pyrones are artifacts. *Journal of Natural Products* 78(6): 1330–1338.

Lyles, J. T., Negrin, A., Khan, S. I., He, K. & Kennelly, E. J. 2014. In vitro antiplasmodial activity of benzophenones and xanthones from edible fruits of *Garcinia* species. *Planta Medica* 80(8/9): 676–681.

Ma, G., Sun, Z., Sun, Z., Yuan, J., Wei, H., Yang, J., Wu, H. & Xu, X. 2014. Antimalarial diterpene alkaloids from the seeds of *Caesalpinia minax*. *Fitoterapia* 95: 234–239.

Ma, G., Wu, H., Chen, D., Zhu, N., Zhu, Y., Sun, Z., Li, P., Yang, J., Yuan, J. & Xu, X. 2015. Antimalarial and antiproliferative cassane diterpenes of *Caesalpiniasappan*. *Journal of Natural Products* 78(10): 2364–2371.

Mabuza, J. M., Kaiser, M. & Bapela, M. J. 2022. In vitro antiplasmodial activity and cytotoxicity of extracts and chromatographic fractions of twigs from *Pappea capensis* EckI & Zeyh. (Sapindaceae). *Journal of Ethnopharmacology* 298: 115659.

Madikizela, B. & McGaw, L. J. 2017. *Pittosporum viridiflorum* Sims (Pittosporaceae): A review on a useful medicinal plant native to South Africa and tropical Africa. *Journal of Ethnopharmacology* 205: 217–230.

Mahato, S. B. & Sen, S. 1997. Advances in triterpenoid research, 1990–1994. *Phytochemistry* 44: 1185–1236.

Makangara, J. J., Nkunya, M. H. & Jonker, S. S. 2010. Geranylbenzoquinonoids from *Lettowianthus stellatus* fruits. *Natural Product Research* 24(8): 710–717.

Malahlela, O. E., Adjorlolo, C. & Olwoch, J. M. 2019. Mapping the spatial distribution of *Lippia javanica* (Burm. f.) Spreng using Sentinel-2 and SRTM-derived topographic data in malaria endemic environment. *Ecological Modelling* 392: 147–158.

Mariath, I. R., Falcão, H. D. S., Barbosa-Filho, J. M., de Sousa, L. C., Tomaz, A. C. D. A., Batista, L. M., Diniz, M. D. F. F., Athayde-Filho, P. F., Tavares, J. F., Silva, M. S. & Cunha, E. V. L. D. 2009. Plants of the American continent with antimalarial activity. *Revista Brasileira de Farmacognosia* 19: 158–191.

Masood, M., Arshad, M., Rahmatullh, Q., Sabir, S., Shoaib, M., Amjad, H. Q. & Tahir, Z. 2021. Picrorhiza kurroa: An ethnopharmacologically important plant species of Himalayan region. *Pure and Applied Biology* 4(3): 407–417.

Matte, R. S. 2023. Phytochemical analysis of medicinal plants occurring in local area of Zari tahsil Dist. Yavatmal Maharashtra. *International Journal for Multidisciplinary Research* 5(1): 1–6.

Mbah, C. C., Akuodor, G. C., Anyalewechi, N. A., Iwuanyanwu, T. C. & Osunkwo, U. A. 2012. In vivo antiplasmodial activities of aqueous extract of *Bridelia ferruginea* stem bark against *Plasmodium berghei* berghei in mice. *Pharmaceutical Biology* 50(2): 188–194.

Mba'ning, B. M., Lenta, B. N., Noungoué, D. T., Antheaume, C., Fongang, Y. F., Ngouela, S. A., Boyom, F. F., Rosenthal, P. J., Tsamo, E., Sewald, N. & Laatsch, H. 2013. Antiplasmodial sesquiterpenes from the seeds of *Salacia longipes* var. camerunensis. *Phytochemistry* 96: 347–352.

Mbeunkui, F., Grace, M. H., Lategan, C., Smith, P. J., Raskin, I. & Lila, M. A. 2012. In vitro antiplasmodial activity of indole alkaloids from the stem bark of *Geissospermum vellosii*. *Journal of Ethnopharmacology* 139(2): 471–477.

Mgbeahuruike, E. E., Yrjönen, T., Vuorela, H. & Holm, Y. 2017. Bioactive compounds from medicinal plants: Focus on *Piper* species. *South African Journal of Botany* 112: 54–69.

Misra, S. P. 1996. In-vivo resistance to chloroquine and sulfa-pyrimethamine combination in *Plasmodium falciparum* in India. *Proceedings of National Academy of Science India* 66: 123–128.

Mohammed, T., Erko, B. & Giday, M. 2014. Evaluation of antimalarial activity of leaves of *Acokantheraschimperi* and *Croton macrostachyus* against *Plasmodium berghei* in Swiss albino mice. *BMC Complementary and Alternative Medicine* 14: 1–10.

Mohanty, S., K Maurya, A., Saxena, A., Shanker, K., Pal, A. & U Bawankule, D. 2015. Flavonoids rich fraction of *Citrus limetta* fruit peels reduces proinflammatory cytokine production and attenuates malaria pathogenesis. *Current Pharmaceutical Biotechnology* 16(6): 544–552.

Mojab, F. 2012. Antimalarial natural products: A review. *Avicenna Journal of Phytomedicine* 2(2): 52.

Mojarrab, M., Shiravand, A., Delazar, A. & Heshmati Afshar, F. 2014. Evaluation of in vitro antimalarial activity of different extracts of *Artemisia aucheri* Boiss. and *A. armeniaca* Lam. and fractions of the most potent extracts. *The Scientific World Journal*: 1–6.

Moshi, M. J., Otieno, D. F. & Weisheit, A. 2012. Ethnomedicine of the Kagera region, north western Tanzania. Part 3: Plants used in traditional medicine in Kikuku village, Muleba District. *Journal of Ethnobiology and Ethnomedicine* 8(1): 1–11.

Mueller, D., Davis, R. A., Duffy, S., Avery, V. M., Camp, D. & Quinn, R. J. 2009. Antimalarial activity of azafluorenone alkaloids from the Australian tree *Mitrephora diversifolia*. *Journal of Natural Products* 72(8): 1538–1540.

Muluye, A. B., Melese, E. & Adinew, G. M. 2015. Antimalarial activity of 80% methanolic extract of *Brassica nigra* (L.) Koch. (Brassicaceae) seeds against *Plasmodium berghei* infection in mice. *BMC Complementary and Alternative Medicine* 15(1): 1–8.

Muñoz, V., Sauvain, M., Bourdy, G., Callapa, J., Bergeron, S., Rojas, I., Bravo, J. A., Balderrama, L., Ortiz, B., Gimenez, A. & Deharo, E. 2000a. A search for natural bioactive compounds in Bolivia through a multidisciplinary approach: Part I. Evaluation of the antimalarial activity of plants used by the Chacobo Indians. *Journal of Ethnopharmacology* 69(2): 127–137.

Muñoz, V., Sauvain, M., Bourdy, G., Callapa, J., Rojas, I., Vargas, L., Tae, A. & Deharo, E. 2000b. The search for natural bioactive compounds through a multidisciplinary approach in Bolivia. Part II. Antimalarial activity of some plants used by Mosetene indians. *Journal of Ethnopharmacology* 69(2): 139–155.

Muregi, F. W., Chhabra, S. C., Njagi, E. N. M., Lang'at-Thoruwa, C. C., Njue, W. M., Orago, A. S. S., Omar, S. A. & Ndiege, I. O. 2004. Anti-plasmodial activity of some Kenyan medicinal plant extracts singly and in combination with chloroquine. *Phytotherapy Research* 18(5): 379–384.

Muregi, F. W., Ishih, A., Miyase, T., Suzuki, T., Kino, H., Amado, T., Mkoji, G. M. & Terada, M. 2007. Antimalarial activity of methonolic extracts from plants used in Kenyan ethnomedicine and their interaction with chloroquine (CQ) against a CQ-tolerant rodent parasite, in mice. *Journal of Ethnopharmacology* 111(1): 190–195.

Murray, C. J. L. & Lopez, A. D. 1996. Evidence-based health policy – Lessons from the Global Burden of Disease study. *Science* 274: 740–743.

Mutie, F. M., Mbuni, Y. M., Rono, P. C., Mkala, E. M., Nzei, J. M., Phumthum, M., Hu, G. W. & Wang, Q. F. 2023. Important medicinal and food taxa (Orders and Families) in Kenya, based on three quantitative approaches. *Plants* 12(1145): 1–143.

Namukobe, J., Kiremire, B. T., Byamukama, R., Kasenene, J. M., Akala, H. M., Kamau, E. & Dumontet, V. 2015. Antiplasmodial compounds from the stem bark of *Neoboutonia macrocalyx* pax. *Journal of Ethnopharmacology* 162: 317–322.

Narain, J. P. & Nath, L. M. 2018. Eliminating malaria in India by 2027: The countdown begins! *The Indian Journal of Medical Research* 148(2): 123.

Nasrullah, A. A., Zahari, A., Mohamad, J. & Awang, K. 2013. Antiplasmodial alkaloids from the bark of *Cryptocarya nigra* (Lauraceae). *Molecules* 18(7): 8009–8017.

Nguta, J. M., Mbaria, J. M., Gakuya, D. W., Gathumbi, P. K. & Kiama, S. G. 2010. Antimalarial herbal remedies of Msambweni, Kenya. *Journal of Ethnopharmacology* 128(2): 424–432.

Nguyen-Pouplin, J., Tran, H., Tran, H., Phan, T. A., Dolecek, C., Farrar, J., Tran, T. H., Caron, P., Bodo, B. & Grellier, P. 2007. Antimalarial and cytotoxic activities of ethnopharmacologically selected medicinal plants from South Vietnam. *Journal of Ethnopharmacology* 109(3): 417–427.

Njoroge, G. N. & Bussmann, R. W. 2006. Diversity and utilization of antimalarial ethnophytotherapeutic remedies among the Kikuyus (Central Kenya). *Journal of Ethnobiology and Ethnomedicine* 2: 1–7.

Nondo, R. S. O., Erasto, P., Moshi, M. J., Zacharia, A., Masimba, P. J. & Kidukuli, A. W. 2016. In vivo antimalarial activity of extracts of Tanzanian medicinal plants used for the treatment of malaria. *Journal of Advanced Pharmaceutical Technology & Research* 7(2): 59–63.

Nondo, R. S. O., Moshi, M. J., Erasto, P., Masimba, P. J., Machumi, F., Kidukuli, A. W., Heydenreich, M. & Zofou, D. 2017. Anti-plasmodial activity of Norcaesalpin D and extracts of four medicinal plants used traditionally for treatment of malaria. *BMC Complementary and Alternative Medicine* 17: 1–8.

Nour, A. M., Khalid, S. A., Kaiser, M., Brun, R., Wai'l, E. A. & Schmidt, T. J. 2010. The antiprotozoal activity of methylated flavonoids from *Ageratum conyzoides* L. *Journal of Ethnopharmacology* 129(1): 127–130.

Nugroho, A. E., Sugai, M., Hirasawa, Y., Hosoya, T., Awang, K., Hadi, A. H. A., Ekasari, W., Widyawaruyanti, A. & Morita, H. 2011. New antiplasmodial indole alkaloids from *Hunteria zeylanica*. *Bioorganic & Medicinal Chemistry Letters* 21(11): 3417–3419.

Nuwagira, C., Peter, E. L., Ajayi, C. O., Adriko, J., Kagoro, G. R., Olet, E. A., Ogwang, P. E. & Tolo, C. U. 2021. Developmental stages influence in vivo antimalarial activity of aerial part extracts of *Schkuhria pinnata*. *Journal of Ethnopharmacology* 279: 1–8.

Ochieng, C. O., Owuor, P. O., Mang'uro, L. A., Akala, H. & Ishola, I. O. 2012. Antinociceptive and antiplasmodial activities of *Cassane furanoditerpenes* from *Caesalpinia volkensii* H. root bark. *Fitoterapia* 83(1): 74–80.

Ofori-Attah, K., Oseni, L. A., Quasie, O., Antwi, S. & Tandoh, M. 2012. A comparative evaluation of in vivo antiplasmodial activity of aqueous leaf extracts of *Carica papaya, Azadirachta indica, Magnifera indica* and the combination thereof using plasmodium infected balb/c mice. *International Journal of Applied Biology and Pharmaceutical Technology* 3(3): 372–378.

Ogbole, O. O., Saka, Y. A., Fasinu, P. S., Fadare, A. A. & Ajaiyeoba, E. O. 2016. Antimalarial and cytotoxic properties of *Chukrasia tabularis* A. Juss and *Turraeavogelii* Hook F. Ex. Benth. *Parasitology Research* 115: 1667–1674.

Okagu, I. U., Ndefo, J. C. & Agbo, M. O. 2021. Trado-medical uses, chemical constituents and biological activities of *Newbouldia laevis* (Bignoniaceae): A review. *Pharmaceutical Sciences* 28(1): 51–75.

Okello, D. & Kang, Y. 2019. Exploring antimalarial herbal plants across communities in Uganda based on electronic data. *Evidence-Based Complementary and Alternative Medicine* 3057180: 1–27.

Okokon, J. E., Ekpo, A. J. & Eseyin, O. A. 2009. Evaluation of in vivo antimalarial activities of ethanolic leaf and seed extracts of *Telfairia occidentalis*. *Journal of Medicinal Food* 12(3): 649–653.

Okokon, J. E., Etebong, E. O., Udobang, J. A. & Essien, G. E. 2012. Antiplasmodial and analgesic activities of *Clausena anisata*. *Asian Pacific Journal of Tropical Medicine* 5(3): 214–219.

Okokon, J. E., Koofreh, D., Azare, B. A. & Okokon, P. J. 2016. Antimalarial and analgesic activities of root extract of *Panicum maximum*. *African Journal of Pharmacology and Therapeutics* 5(3): 128–135.

Okorie, C. C., Elechi-Amadi, K. N., Briggs, O. N., Davies-Nwalele, O., Giami, L. K. & Ebifa, J. O. 2023. Antiplasmodial potential of ethanolic extract of *Moringa oleifera* Lam seeds in *Plamodium Berghei*-infected mice. *Ghana Alternative Medicine Journal* 4(1): 17–23.

Oladosu, I. A., Balogun, S. O. & Ademowo, G. O. 2013. Phytochemical screening, antimalarial and histopathological studies of *Allophylus africanus* and *Tragiabenthamii*. *Chinese Journal of Natural Medicines* 11(4): 371–376.

O'Neill, M. J., Bray, D. H., Boardman, P., Phillipson, J. D. & Warhurst, D. C. 1985. Plants as sources of antimalarial drugs part. 1. In vitro test method for the evaluation of crude extracts from plants. *Planta Medica* 51(5): 394–398.

O'Neill, M. J., Bray, D. H., Boardman, P., Phillipson, J. D., Warhurst, D. C., Peters, W. & Suffness, M. 1986. Plants as sources of antimalarial drugs: In vitro antimalarial activities of some quassinoids. *Antimicrobial Agents and Chemotherapy* 30(1): 101–104.

Orwa, J. A., Ngeny, L., Mwikwabe, N. M., Ondicho, J. & Jondiko, I. J. O. 2013. Antimalarial and safety evaluation of extracts from *Toddalia asiatica* (L) Lam. (Rutaceae). *Journal of Ethnopharmacology* 145(2): 587–590.

Osunga, S., Amuka, O., Machocho, A., Getabu, A. & Onani, M. 2023. Some members of genus *Cassia* (Senna): Their ethnobotany, potency and prospects of drug discovery. *Indian Journal of Pharmacy & Drugs Studies* 4(2): 33–57.

Pattanayak, P., Behera, P., Das, D. & Panda, S. K. 2010. *Ocimum sanctum* Linn. A reservoir plant for therapeutic applications: An overview. *Pharmacognosy Reviews* 4(7): 95–105.

Pereira, T. B., Rocha e Silva, L. F., Amorim, R. C., Melo, M. R., Zacardi de Souza, R. C., Eberlin, M. N., Lima, E. S., Vasconcellos, M. C. & Pohlit, A. M. 2014. In vitro and in vivo anti-malarial activity of limonoids isolated from the residual seed biomass from *Carapa guianensis* (andiroba) oil production. *Malaria Journal* 13: 1–8.

Pinto, M. E. F., Batista Jr, J. M., Koehbach, J., Gaur, P., Sharma, A., Nakabashi, M., Cilli, E. M., Giesel, G. M., Verli, H., Gruber, C. W. & Blanch, E. W. 2015. Ribifolin, an orbitide from *Jatropha ribifolia*, and its potential antimalarial activity. *Journal of Natural Products* 78(3): 374–380.

Prakash, A., Sharma, S. K., Mohapatra, P. K., Bhattacharjee, K., Gogoi, K., Gogoi, P., Mahanta, J. & Bhattacharyya, D. R. 2013. In vitro and in vivo antiplasmodial activity of the root extracts of *Brucea mollis* Wall. ex Kurz. *Parasitology Research* 112: 637–642.

Qanash, H., Yahya, R., Bakri, M. M., Bazaid, A. S., Qanash, S., Shater, A. F. & TM, A. 2022. Anticancer, antioxidant, antiviral and antimicrobial activities of Kei Apple (*Dovyalis caffra*) fruit. *Scientific Reports* 12(1): 1–15.

Rahi, M. & Sharma, A. 2020. For malaria elimination India needs a platform for data integration. *BMJ Global Health* 5: 1–5. https://doi.org/10.1136/bmjgh-2020-004198.

Rahman, N. N. N. A., Furuta, T., Takane, K. & Mohd, M. A. 1999. Antimalarial activity of extracts of Malaysian medicinal plants. *Journal of Ethnopharmacology* 64(3): 249–254.

Rajachan, O. A., Kanokmedhakul, S., Nasomjai, P. & Kanokmedhakul, K. 2014. Chemical constituents and biological activities from roots of *Enkleia siamensis*. *Natural Product Research* 28(4): 268–270.

Rao, T. R. 1984. *The Anopheles of India Delhi: Malaria Research Centre*. Indian Council of Medical Research, p. 518.

Rehman, T. & Ahmad, S. 2019. *Nardostachys chinensis* Batalin: A review of traditional uses, phytochemistry, and pharmacology. *Phytotherapy Research* 33(10): 2622–2648.

Richards, Jr. F. O., Boatin, B., Sauerbrey, M. & Sékétéli, A. 2001. Control of onchocerciasis today: Status and challenges. *Trends in Parasitology* 17(12): 558–563.

Riscoe, M., Kelly, J. X. & Winter, R. 2005. Xanthones as antimalarial agents: Discovery, mode of action, and optimization. *Current Medicinal Chemistry* 12(21): 2539–2549.

Rocha e Silva, L. F., Nogueira, K. L., Pinto, A. C. D. S., Katzin, A. M., Sussmann, R. A., Muniz, M. P., Neto, V. F. D. A., Chaves, F. C. M., Coutinho, J. P., Lima, E. S. & Krettli, A. U. 2015. In vivo antimalarial activity and mechanisms of action of 4-nerolidylcatechol derivatives. *Antimicrobial Agents and Chemotherapy* 59(6): 3271–3280.

Rodrigues, R. A. F., Figueira, G. M., Sartoratto, A., Yamane, L. T. & de Freitas-Blanco, V. S. 2018. Chemical diversity and ethnopharmacological survey of South American medicinal and aromatic plant species. In: Medicinal and Aromatic Plants of South America: Brazil. *Medicinal and Aromatic Plants of the World* 5, Springer. pp. 17–44. https://doi.org/10.1007/978-94-024-1552-0_2.

Rodríguez-Guzmán, R., Fulks, L. C. J., Radwan, M. M., Burandt, C. L. & Ross, S. A. 2011. Chemical constituents, antimicrobial and antimalarial activities of *Zanthoxylum monophyllum*. *Planta Medica* 77(13): 1542–1544.

Rudrapal, M. & Chetia, D. 2017. Plant flavonoids as potential source of future antimalarial leads. *Systematic Reviews in Pharmacy* 8(1): 13–18.

Samy, C. K. T., Dankwa, K., Stephen, Y. G. & Archibald, S. 2011. Antimalarial activity of *Cryptolepisanguinolenta* based herbal capsules in *Plasmodium berghei* infected mice. *International Research Journal of Pharmacy* 2(5): 127–131.

Sangian, H., Faramarzi, H., Yazdinezhad, A., Mousavi, S. J., Zamani, Z., Noubarani, M. & Ramazani, A. 2013. Antiplasmodial activity of ethanolic extracts of some selected medicinal plants from the northwest of Iran. *Parasitology Research* 112: 3697–3701.

Sari, N. K. Y., Arwati, H. & Tantular, I. S. 2023. Antimalarial activity of mahagony seed ethanolic extract in BALB/c mice infected with *Plasmodium berghei* ANKA and the correlation of parasitemia and plasma level of IFN-γ. *Malaysian Journal of Medicine and Health Sciences* 19(3): 27–32.

Sashidhara, K. V., Singh, S. P., Singh, S. V., Srivastava, R. K., Srivastava, K., Saxena, J. K. & Puri, S. K. 2013. Isolation and identification of β-hematin inhibitors from *Flacourtia indica* as promising antiplasmodial agents. *European Journal of Medicinal Chemistry* 60: 497–502.

Saxena, S., Pant, N., Jain, D. C. & Bhakuni, R. S. 2003. Antimalarial agents from plant sources. *Current Science* 85(9): 1314–1329.

Scholar, E. 2007. Chloroquine. In: *xPharm: The Comprehensive Pharmacology Reference*. Edited by Enna, S. J., & Bylund, D. B. Elsevier, pp. 1–7.

Schultz, F., Osuji, O. F., Nguyen, A., Anywar, G., Scheel, J. R., Caljon, G., Pieters, L. & Garbe, L. A. 2021. Pharmacological assessment of the antiprotozoal activity, cytotoxicity and genotoxicity of medicinal plants used in the treatment of Malaria in the greater Mpigi Region in Uganda. *Frontiers in Pharmacology* 12: 1–20.

Sebastian, J. & Jose, S. 2018. Antidiabetic and antioxidant profile of three varieties of passion fruit. *FoodSci: Indian Journal of Research in Food Science and Nutrition* 5(1): 16–21.

Sebisubi, F. M., Odyek, O., Anokbonggo, W. W., Ogwal-Okeng, J., Carcache-Blanco, E. J., Ma, C., Orjala, J. & Tan, G. T. 2010. Antimalarial activity of *Aspilia pruliseta*, a medicinal plant from Uganda. *Planta Medica* 76(16): 1870–1873.

Sehgal, P. N., Sharma, M. I. D., Sharma, S. L. & Gogai, S. 1973. Resistance to chloroquine in falciparum malaria in Assam State, India. *Journal of Communicable Diseases* 5(4): 175–180.

Şener, B., Orhan, I. & Satayavivad, J. 2003. Antimalarial activity screening of some alkaloids and the plant extracts from Amaryllidaceae. *Phytotherapy Research* 17(10): 1220–1223.

Sharma, A. K., Bhasin, S. & Chaturvedi, S. 2007. Predictors of knowledge about malaria in India. *Journal of Vector Borne Disease* 44: 189–197.

Sharma, N., Bano, A., Dhaliwal, H. S. & Sharma, V. 2015. A pharmacological comprehensive review on "Rassbhary" *Physalis angulata* (L.). *International Journal of Pharmacy and Pharmaceutical Sciences* 7(8): 34–38.

Sharma, V. P. 1996. Re-emergence of malaria in India. *Indian Journal of Medical Research* 103: 26–45.

Sharma, V. P., Choudhury, D. S., Ansari, M. A., Malhotra, M. S., Menon, P. K. B., Razdan, R. K. & Batra, C. P. 1983. Studies on the true incidence of malaria in Kharkhoda (district Sonepat, Haryana) and Kichha (district Nainital, UP) Primary Health Centres. *Indian Journal of Malariol* 20(1): 21–34.

Sheikh, Z., Amani, A., Basseri, H. R., Kazemi, S. H. M., Sedaghat, M. M., Azam, K., Azizi, M. & Amirmohammadi, F. 2021. Repellent efficacy of *Eucalyptus globulus* and *Syzygium aromaticum* essential oils against malaria vector, *Anopheles stephensi* (diptera: Culicidae). *Iranian Journal of Public Health* 50(8): 1668–1677.

Simonsen, H. T., Nordskjold, J. B., Smitt, U. W., Nyman, U., Palpu, P., Joshi, P. & Varughese, G. 2001. In vitro screening of Indian medicinal plants for antiplasmodial activity. *Journal of Ethnopharmacology* 74(2): 195–204.

Singh, K., Ajao, A. A. N. & Sabiu, S. 2022. Ethnobotanical, phytochemistry, toxicological and pharmacological significance of the underutilized indigenous *Aloe* species of West Africa. *South African Journal of Botany* 147: 1007–1015.

Singh, V. & Banyal, H. S. 2011. Antimalarial effect of *Tinospora cordifolia* (Willd.) Hook. f. & Thoms and *Cissampelos pareira* L. on *Plasmodium berghei*. *Current Science* 101(10): 1356–1358.

Spencer, C. F., Koniuszy, F. R., Rogers, E. F., Shavel, J., Easton, N. R., Kaczka, E. A., Kuehl, F. A., Phillips, R. F., Walti, A., Folkers, K. & Malanga, C. 1947. Survey of plants for antimalarial activity. *Lloydia* 10(3): 145–174.

Subbarao, S. K. 1998. *Anopheles Species Complexes in South East Asian Region (World Health Organization, South East Asia Regional Office, New Delhi*. Technical Publication, SEARO 18, pp. 1–82.

Suchaichit, N., Kanokmedhakul, K., Panthama, N., Poopasit, K., Moosophon, P. & Kanokmedhakul, S. 2015. A 2H-tetrahydropyran derivative and bioactive constituents from the bark of *Goniothalamus elegants* Ast. *Fitoterapia* 103: 206–212.

Sukphan, P., Sritularak, B., Mekboonsonglarp, W., Lipipun, V. & Likhitwitayawuid, K. 2014. Chemical constituents of *Dendrobium venustum* and their antimalarial and anti-herpetic properties. *Natural Product Communications* 9(6): 825–827.

Swati, K., Bhatt, V., Sendri, N., Bhatt, P. & Bhandari, P. 2023. *Swertia chirayita*: A comprehensive review on traditional uses, phytochemistry, quality assessment and pharmacology. *Journal of Ethnopharmacology* 300: 115714.

Taiwo, B. J., Akinkunmi, E. O. & Omisore, N. O. 2013. Antimicrobial and antiplasmodial activities of a quaternary compound from *Ritchiea capparoides* var. longipedicellata. *African Journal of Traditional, Complementary and Alternative Medicines* 10(6): 528–531.

Telrandhe, U. B., Kosalge, S. B., Parihar, S., Sharma, D. & Lade, S. N. 2022. Phytochemistry and pharmacological activities of *Swietenia macrophylla* King (Meliaceae). *Scholars Academic Journal of Pharmacy* 11(1): 6–12.

Tepa, A. G. N., Ambassa, P., Ayong, L. S., Biapa Nya, P. C. & Pieme, C. A. 2022. The antiplasmodial potential of medicinal plants used in the Cameroonian pharmacopoeia: An updated systematic review and meta-analysis. *Evidence-Based Complementary and Alternative Medicine*: 1–18.

Thiengsusuk, A., Chaijaroenkul, W. & Na-Bangchang, K. 2013. Antimalarial activities of medicinal plants and herbal formulations used in Thai traditional medicine. *Parasitology Research* 112: 1475–1481.

Thongthoom, T., Songsiang, U., Phaosiri, C. & Yenjai, C. 2010. Biological activity of chemical constituents from *Clausena harmandiana*. *Archives of Pharmacal Research* 33: 675–680.

Titanji, V. P., Zofou, D. & Ngemenya, M. N. 2008. The antimalarial potential of medicinal plants used for the treatment of malaria in Cameroonian folk medicine. *African Journal of Traditional, Complementary, and Alternative Medicines* 5(3): 302–321.

Tjitraresmi, A., Moektiwardoyo, M., Susilawati, Y. & Shiono, Y. 2020. Antimalarial activity of lamiaceae family plants. *Systematic Reviews in Pharmacy* 11(7): 324–333.

Toma, A., Deyno, S., Fikru, A., Eyado, A. & Beale, A. 2015. In vivo antiplasmodial and toxicological effect of crude ethanol extract of *Echinops kebericho* traditionally used in treatment of malaria in Ethiopia. *Malaria Journal* 14(1): 1–5.

Tona, L., Ngimbi, N. P., Tsakala, M., Mesia, K., Cimanga, K., Apers, S., Bruyne, T. D., Pieters, L., Totte, J. & Vlietinck. 1999. Antimalarial activity of 20 crude extracts from nine African medicinal plants used in Kinshasa, Congo. *Journal of Ethnopharmacology* 68: 193–203.

Toyang, N. J., Krause, M. A., Fairhurst, R. M., Tane, P., Bryant, J. & Verpoorte, R. 2013. Antiplasmodial activity of sesquiterpene lactones and a sucrose ester from *Vernoniaguineensis* Benth. (Asteraceae). *Journal of Ethnopharmacology* 147(3): 618–621.

Udobang, J. A., Nwafor, P. A. & Okokon, J. E. 2010. Analgesic and antimalarial activities of crude leaf extract and fractions of *Acalypha wilkensiana*. *Journal of Ethnopharmacology* 127(2): 373–378.

Umar, S. I., Ndako, M., Jigam, A. A., Adefolalu, S. F., Ibikunle, G. F. & Lawal, B. 2019. Anti-plasmodial, Anti-inflammatory, anti-nociceptive and safety profile of *Maytenus senegalensis* root bark extract on hepato-renal integrity in experimental animals. *Comparative Clinical Pathology* 28: 1571–1579.

Upadhyay, H. C., Sisodia, B. S., Cheema, H. S., Agrawal, J., Pal, A., Darokar, M. P. & Srivastava, S. K. 2013. Novel antiplasmodial agents from *Christia vespertilionis*. *Natural Product Communications* 8(11): 1591–1594.

Uzor, P. F. 2020. Alkaloids from plants with antimalarial activity: A review of recent studies. *Evidence-Based Complementary and Alternative Medicine* 8749083: 1–17.

Uzor, P. F., Prasasty, V. D. & Agubata, C. O. 2020. Natural products as sources of antimalarial drugs. *Evidence-Based Complementary and Alternative Medicine*: 1–2.

Valdés, A. F. C., Martínez, J. M., Lizama, R. S., Gaitén, Y. G., Rodríguez, D. A. & Payrol, J. A. 2010. In vitro antimalarial activity and cytotoxicity of some selected Cuban medicinal plants. *Revista do Instituto de Medicina Tropical de Sao Paulo* 52: 197–201.

Valdés, A. F. C., Martínez, J. M., Rodríguez, D. A., Lizama, R. S. & Gaitén, Y. G. 2011. Antimalarial activity of hydroalcoholic extract from *Bixa orellana* L. *Revista Cubana de Medicina Tropical* 63(2): 181–185.

Vale, V. V., Vilhena, T. C., Trindade, R. C. S., Ferreira, M. R. C., Percário, S., Soares, L. F., Pereira, W. L. A., Brandão, G. C., Oliveira, A. B., Dolabela, M. F. & De Vasconcelos, F. 2015. Anti-malarial activity and toxicity assessment of *Himatanthus articulatus*, a plant used to treat malaria in the Brazilian Amazon. *Malaria Journal* 14(1): 1–10.

Vásquez-Ocmín, P., Cojean, S., Rengifo, E., Suyyagh-Albouz, S., Guerra, C. A. A., Pomel, S., Cabanillas, B., Mejía, K., Loiseau, P. M., Figadère, B. & Maciuk, A. 2018. Antiprotozoal activity of medicinal plants used by Iquitos-Nauta road communities in Loreto (Peru). *Journal of Ethnopharmacology* 210: 372–385.

Verma, G., Dua, V. K., Agarwal, D. D. & Atul, P. K. 2011. Anti-malarial activity of *Holarrhena antidysenterica* and *Viola canescens*, plants traditionally used against malaria in the Garhwal region of north-west Himalaya. *Malaria Journal* 10: 1–5.

Vital, M. J. S., Carneiro, A. L. B., Rocha, L. F., das Neves Amorim, R. C., Camargo, M. R. M. & Pohlit, A. M. 2013. Chemical composition, ethnopharmacology and biological activity of *Geissospermum Allemão* species (Apocynaceae Juss.). *Revista Fitos* 8(2): 137–146.

Walter, N. S., Bagai, U. & Kalia, S. 2013. Antimalarial activity of *Bergenia ciliata* (Haw.) Sternb. against *Plasmodium berghei*. *Parasitology Research* 112: 3123–3128.

Wang, H., Zhao, W., Choomuenwai, V., Andrews, K. T., Quinn, R. J. & Feng, Y. 2013. Chemical investigation of an antimalarial Chinese medicinal herb *Picrorhiza scrophulariiflora*. *Bioorganic & Medicinal Chemistry Letters* 23(21): 5915–5918.

Wangchuk, P., Keller, P. A., Pyne, S. G., Willis, A. C. & Kamchonwongpaisan, S. 2012. Antimalarial alkaloids from a Bhutanese traditional medicinal plant *Corydalis dubia*. *Journal of Ethnopharmacology* 143(1): 310–313.

Wangchuk, P., Pyne, S. G., Keller, P. A., Taweechotipatr, M. & Kamchonwongpaisan, S. 2014. Phenylpropanoids and furanocoumarins as antibacterial and antimalarial constituents of the Bhutanese medicinal plant *Pleurospermum amabile*. *Natural Product Communications* 9(7): 957–960.

Wangchuk, P. & Samten. 2015. GC-FID coupled GC-MS analysis of the essential oil and the recorded biological activities of *Meconopsis simplicifolia*. *Journal of Biologically Active Products from Nature* 5(6): 365–372.

WHO. 2004. *World Health Organization: Global Strategic Framework for Integrated Vector Management* (No. WHO/CDS/CPE/PVC/2004.10). World Health Organization (WHO).

Yenesew, A., Akala, H. M., Twinomuhwezi, H., Chepkirui, C., Irungu, B. N., Eyase, F. L., Kamatenesi-Mugisha, M., Kiremire, B. T., Johnson, J. D. & Waters, N. C. 2012. The antiplasmodial and radical scavenging activities of flavonoids of *Erythrinaburttii*. *Acta Tropica* 123(2): 123–127.

Youyou, T., Muyun, N., Yurong, Z., Lanna, L., Shulian, G., Muqun, Z., Xiuzhen, W. & Xiaotian, L. 2015. Studies on the constituents of *artemisia annua* L. *Acta Pharmaceutica Sinica* 50(10): 366–370.

Zhang, G. P., Xiao, Z. Y., Rafique, J., Arfan, M., Smith, P. J., Lategan, C. A. & Hu, L. H. 2009. Antiplasmodial isoflavanones from the roots of *Sophora mollis*. *Journal of Natural Products* 72(7): 1265–1268.

Zhang, H., Qiu, S., Tamez, P., Tan, G. T., Aydogmus, Z., Hung, N. V., Cuong, N. M., Angerhofer, C., Doel Soejarto, D., Pezzuto, J. M. & Fong, H. H. 2002. Antimalarial agents from plants II. Decursivine, a new antimalarial indole alkaloid from *Rhaphidophora decursiva*. *Pharmaceutical Biology* 40(3): 221–224.

Zheljazkov, V. D., Gawde, A., Cantrell, C. L., Astatkie, T. & Schlegel, V. 2015. Distillation time as tool for improved antimalarial activity and differential oil composition of cumin seed oil. *PLoS ONE* 10(12): 1–12.

2 Biology and Genetic Improvement of Antimalarial Plant Species

Apekshakumari Patel, Tasneem Khan, Nimisha Patel, Ahmad Ali, and Hina Alim

ABBREVIATIONS

ACTs	antimalarial effective medications
CDCs	Centers for Disease Control
CSP	Circumsporozoite protein
DHFR	dihydrofolate reductase
DHPS	dihydropteroate synthase
DOXP	1-deoxy-d-xylulose 5-phosphate
FDA	Food and Drug Administration
IC_{50}	half-maximal inhibitory concentration
NADH	nicotinamide adenine dinucleotide (NAD) + hydrogen (H)
NIAID	National Institute of Allergy and Infectious Diseases
PDCs	plant-derived compounds
PfAMA1	Plasmodium falciparumapical membrane antigen 1
PfCRT	*Plasmodium falciparum* chloroquine resistance transporter
QSAR	quantitative structure–activity relationship
RBCs	red blood cells
RE	reticulum endoplasmic
WHO	World Health Organization

2.1 INTRODUCTION

Malaria is unquestionably the most virulent and lethal infectious agent in the developing world of all parasite diseases (Greenwood et al., 2005; Winter et al., 2006). Since it causes more energy loss, more debilitation, more loss of job capacity, and more economic harm than any other human parasitic diseases, this vector-borne infectious disease is a perfect example of one that affects the productivity of people, families, and the entire society (Sachs & Malaney, 2002). Although malaria is frequently linked to poverty, it also contributes to it and is a significant barrier to economic growth. In 2006, there were 3.3 billion individuals at risk of contracting malaria, which resulted in an estimated 247 million cases and about a million fatalities, largely in children.

One of the most lethal infectious diseases brought on by *Plasmodium* parasites is malaria. Female *Anopheles* mosquitoes carrying the malaria parasite spread it throughout people. There are five parasite species known to cause malaria in humans, including *P. knowlesi*, *P. ovale*, *P. malariae*, *P. vivax*, and *P. falciparum* (Gamo et al., 2010). The two last parasite species pose the greatest threat, because *P. vivax* causes the most morbidity and *P. falciparum* causes the most mortality, mostly in children under 5 age group (Lopez et al., 2014). Almost half of contemporary civilization has been

DOI: 10.1201/9781003378396-2

impacted by malaria, which is endemic in more than 100 nations (Qayum et al., 2016). There are a number of reasons that malaria-related fatalities are disproportionately higher in underdeveloped nations than in technologically developed nations. Regrettably, access to antimalarial medications has been constrained in poor nations due to the socioeconomic position of the population. In Africa, where 85% of the world's malaria cases and 94% of deaths have been reported, there have been 228 million reported cases of malaria, leading to 435,000 deaths, according to the world malaria report published in December 2019 (WHO, 2019). In light of this, current projections indicate that a majority of people on the planet have a high risk of contracting malaria. Malaria infections can cause significant economic losses in addition to rising mortality rates, with estimated healthcare-related costs of US$2.7 billion in 2018 (WHO, 2019); as a result, resource-poor nations' economies tend to suffer.

In many nations, malaria is still a scourge that contributes significantly to population mortality. The World Health Organization (WHO, 2019) reports that 216 million clinical cases of malaria were reported globally in 2016, resulting in 445,000 fatalities; 80% of the world's malaria burden was carried by 15 nations, all of which were in sub-Saharan Africa and India, out of the 91 countries that reported indigenous malaria cases. Five *Plasmodium* species, namely, *P. malariae*, *P. falciparum*, *P. vivax*, *P. ovale*, and *P. knowlesi*, are responsible for the development of human malaria (Autino et al., 2012; Amir et al., 2018). While *P. vivax* predominates outside of Africa, *P. falciparum* is the species that is most common within sub-Saharan Africa, accounting for 99% of estimated malaria cases in 2016 (WHO, 2017). Although human malaria transmitted by female *Anopheles* mosquitoes has four *Plasmodium* species as its aetiological agents – *P. falciparum*, *P. vivax*, *P. ovale*, and *P. malariae* – the most widespread and severe disease is caused by *P. falciparum*, which transiently infects the liver before invading the red blood cells of the mammalian host (Figure 2.1). The erythrocytic stage of the disease is characterized by clinical signs, such as fever, chills, prostration, anaemia, delirium, metabolic acidosis, cerebral malaria, and multi–organ system failure, which may be followed by coma and death (Fidock et al., 2004; Deprez-Poulain & Melnyk, 2005; Jones & Good, 2006). *Anopheles* mosquito females are the only mosquito species that can spread malaria through bites. The infected *Anopheles* injects sporozoites into the human host during a blood meal. The sporozoites mature and multiply after they enter the parenchymal cells of the liver. They grow into schizonts that contain a large number of merozoites (Prudêncio et al., 2006; Krishnan & Williamson, 2018). *Pre-erythrocytic schizogony* is the word for the phenomenon. This phase is unique to each species and takes 6 days on average for *P. falciparum* and 16 days on average for *P. malariae* to mature (Antinori et al., 2012). Red blood cells are recognized and invaded by the merozoites that the liver releases. The parasite population in the red cells grows swiftly. Each merozoite either undergoes a sexual or an erythrocytic schizogony to grow into a schizont. Merozoites, which are present in the adult erythrocytic-stage schizont, are ruptured and released into the blood, where they infect other erythrocytes. The time required for the erythrocytic schizogony varies with the *Plasmodium* species and is responsible for the traditional periodicity of fever in malaria. The mosquitoes that consume the gametocytes while feeding on an infected human are contagious (Crutcher & Hoffman, 1996; Wells & Andrew, 2015). Gametocytes in the *Anopheles* grow into male and female gametes, which mate in the mosquito's gut to create zygotes. The zygotes then grow into elongated, motile ookinetes, which infiltrate the mosquito's midgut wall and develop into oocysts. Each oocyst grows and divides into hundreds of active haploid forms called sporozoites, which travel to the salivary glands of the mosquito. The malaria life cycle is continued by injecting the sporozoites into a fresh human host.

Years of research since the groundbreaking work by Nussenzweig et al. (Nussenzweig et al., 1967) on the development of protective immunity in mice by vaccination with X-irradiated sporozoites of *P. berghei* have shown that it is challenging to develop an effective vaccine due to the high complexity of the *Plasmodium* life cycle. The greatest method for controlling the disease and averting fatalities is early diagnosis and chemotherapy, despite the fact that a number of candidate vaccines are currently through various stages of testing (Birkett, 2016; Draper et al., 2018; Tham et al., 2017; WHO, 2019). This method aids in reducing the spread of the malaria parasite (Birkett, 2016).

The bark of the *Cinchona* species (Rubiaceae) was used to make the first antimalarial medication, quinine, which was discovered in 1820 by two French chemists, Joseph Pelletier and Joseph Caventou (Pai-Dhungat & Caventou, 2015). It is the earliest antimalarial medicine that is still in use today (Achan et al., 2011). Thereafter, synthetic quinine substitutes have been created. Chloroquine was the most effective of them, but widespread resistance against *P. falciparum* started to emerge in the 1970s. An altogether new medicine was needed because all structural variations of chloroquine had the same mode of action. Thankfully, Chinese researchers discovered artemisinin from the sweet wormwood plant (*A. annua*) in the 1970s. This compound has proven to be highly efficient against malaria parasites that are resistant to chloroquine (Butler et al., 2010). As of now, ACT (artemisinin-based combination therapy) is advised. The primary benefit of ACT is that the companion medication eliminates any leftover parasites, while artemisinin rapidly decreases the majority of malaria parasites (WHO, 2017). Various combination therapy treatments have been suggested; however, the evolution of resistance also poses a threat to their efficacy (WHO, 2017; Bloland, 2001).

Finding novel medications that target the blood stage to reduce symptoms, the liver stage to prevent relapses, and the transmission stage to prevent more human contaminations is urgently needed (Biamonte et al., 2013). It has been suggested to quickly discover novel antimalarial medicines by repurposing medications (Yuan et al., 2011; Devillers, 2018). For the purpose of finding new active compounds, the quantitative structure–activity relationship (QSAR) methodology has also been extensively applied (Karcher & Devillers, 1990). Plants are a great source for discovering new antimalarial medications because the earliest antimalarial treatments were derived from them and because >80% of the world's population uses plants as their primary source of therapeutic substances (Ekor, 2013). In this context, attempts have been made to find and select plant-derived antimalarial drugs from chemoinformatics techniques. The approach typically entails the following actions: (1) (PDCs) plant-derived compounds with *in vitro* antiplasmodial activities are gathered from a variety of sources; (2) PDCs are categorized based on their IC_{50} values obtained on *P. falciparum*–sensitive and/or *P. falciparum*–resistant strains; (3) structure and activity value curations are made; (4) a more or less thorough in silico pharmacokinetic profiling of the PDCs is carried out; and (5) toxicity of the PDCs is evaluated. Sadly, this final step is frequently made briefly or even skipped (Onguene et al., 2014; Egieyeh et al., 2016).

2.2 THE LIFE CYCLE OF MALARIA

The human and the insect are both hosts in different stages of the malarial parasite's life cycle. A single bite from an infected female *Anopheles* mosquito causes the injection of uninucleated sporozoites into the bloodstream, beginning the life cycle. After less than a minute, sporozoites move from blood to the liver tissue. After five days, they transform into uninucleated merozoites with a population of 10,000–40,000, which can then invade a healthy erythrocyte (Hoffman et al., 2015). Merozoites can be discharged upon liver maturation, continuing the cycle of intraerythrocytic-stage development, rupture, and reinvasion, as well as developing into gametocytes, the sexual-stage parasites. Gametocytes detach from erythrocytes in the stomach of the mosquito and create gametes to finish the life cycle by merging to form a zygote (ookinetes), which then changes into an oocyst, in which additional sporozoites are produced. Ultimately, while the mosquito bites, the freshly created sporozoites move to the salivary glands and eventually infect people (Cox, 2010). Figure 2.1 depicts the complex life cycle of the human malaria parasite. When an insect bites the skin, it explores for a blood vessel to feed from, releases various vasodilators to boost its probability of locating a vessel, and salivates into the blood to prevent clotting. This process transmits the motile infectious form of *Plasmodium*, called a sporozoite. The vascular system transports the thread-like sporozoites to the liver within 30 to 60 minutes of inoculation (NIAID, 2007; Krettli & Miller, 2001).

The sporozoites evolve into schizonts over a period of 7–12 days, and up to 30,000 merozoites can form, which rupture the hepatocytes (WHO, 2013; Ricardo et al., 2014). However, certain sporozoites from the *Plasmodium vivax* and other species change into hypnozoites, which can remain dormant in the liver for months or years and result in relapses in infected individuals (Walker et al., 2010). Fascinatingly, *falciparum* malaria recurrence has been documented in patients who left an endemic area some years prior. It indicates that *falciparum* has a dormant state at least periodically (Szmitko et al., 2008; Poilane et al., 2009). The asexual cycle then starts with the merozoites encroaching on RBCs to feed on haemoglobin and develop. Depending on the parasite type, the parasite develops inside the host RBC from the early ring stage to the late trophozoite stage, followed by mitotic divisions to the schizont stage, which comprises 6 to 32 merozoites (Jiraprapa et al., 2002). The discharged merozoites continue the life cycle by invading other RBCs after the erythrocytic schizont ruptures. Cyclical fevers frequently occur just prior to or simultaneously with RBC lysis as schizonts burst to release fresh infectious merozoites. Infection with tertian malaria happens every 48 hours, while with quartan malaria happens every 72 hours. Some merozoites undergo sexual differentiation during this cycle, resulting in erythrocytic gametocytes, which have a single nucleus and are ready to mate when a bloodsucking female *Anopheles* mosquito appears (NIAID, 2007; Jiraprapa et al., 2002). The mosquito's consumption of gametocytes causes gametogenesis to occur. The macrogametes that produce zygotes are penetrated or fertilized by the exflagellated forms of microgametes. The zygotes develop into ookinetes before becoming a spherical oocyst. The oocyst's nucleus repeatedly divides, resulting in the growth of numerous sporozoites and an enlarged oocyst (WHO, 2013). The sporozoites are released into the haemocoel (the mosquito's bodily cavity) when the oocyst bursts after they are fully grown. The life cycle is finished when the sporozoites move on to the salivary glands (Figure 2.1). The malaria life cycle is continued by the sporozoites from the salivary glands of the mosquito entering a new human host (CDC, 2016; WHO, 2013).

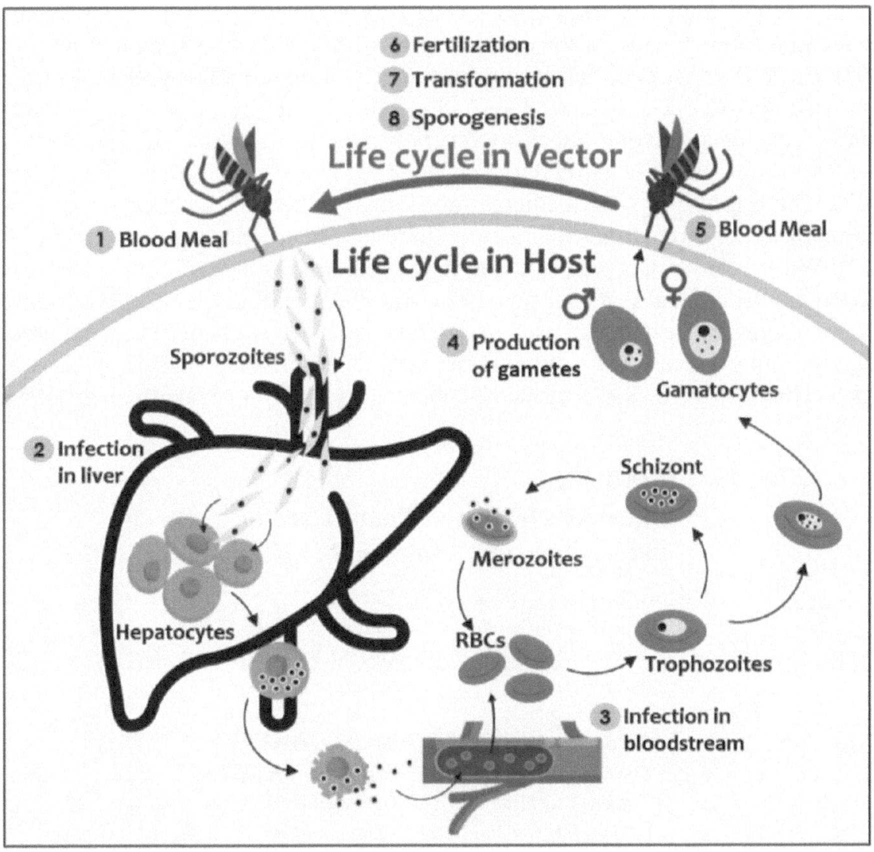

FIGURE 2.1 Malaria parasite life cycle involves two hosts.

2.3 ANTIMALARIAL DRUG HISTORY

Quinine, an aminoquinoline alkaloid discovered in the bark of the *Cinchona* species (Rubiaceae) by Pelletier and Caventou in 1820, is one of the first and most significant antimalarial medications still in use today (Greenwood et al., 2005). After the Second World War, this alkaloid was thought to be the driving force behind the creation of synthetic antimalarial medications belonging to the classes of 4- and 8-aminoquinolines, including chloroquine (Winter et al., 2006), primaquine (Sachs & Malaney, 2002), and others. For nearly three centuries, this alkaloid was the only active principle effective against *P. falciparum.* The only medication used to treat malaria up until recently was chloroquine (Saxena et al., 2003; Winter, 2006; Viegas et al., 2006).

Since 1960, drug-resistant *P. falciparum* strains have emerged, particularly those that are resistant to chloroquine, making it more difficult to treat malaria in almost all malaria-endemic areas of the world (Winter et al., 2006). As an alternative to chloroquine, several researchers have worked hard to produce novel active molecules, mainly from artemisinin (Winter et al., 2006; WHO, 2008). Although no single medicine is currently effective for treating multidrug-resistant malaria, effective combination therapies include artemisinin derivatives like artesunate (Fidock et al., 2004) or combinations with older medications like Malarone, which contains the pharmaceuticals atovaquone (Deprez-Poulain & Melnyk, 2005) and proguanil (Jones & Good, 2006). Regrettably, there have already been initial reports on drug resistance to artemisinin derivatives (Jambou et al., 2005) and pharmacological combination therapy (Wichmann et al., 2004). There are various commonly used antimalarial plants that have been used as therapeutics to treat malaria, which are listed in Table 2.1. However, attempts to develop novel antimalarial medications are still urgently needed in the absence of a malaria vaccine that is effective, safe, and widely available.

The scientific community as a whole agrees that natural products have been playing a significant role in the identification of leads for the creation of medications to treat human ailments (Newman et al., 2003). Since the vast majority of currently used antimalarial chemotherapeutic agents are derived from natural sources, this fact foreshadows the possibility of new leads emerging from sources of tropical plants, as biological chemodiversity continues to be a crucial source of molecular templates for the development of antimalarial drugs (Ziegler et al., 2002; Kalauni et al., 2006; Portet et al., 2007). The use of drugs for medicinal purposes has been in practice since ancient times, and the development of drugs from traditional to modern science has been a process of innovation (Figure 2.2). While many drugs we use today have their origins in traditional medicine practices from different cultures globally, modern medicine and the scientific method have transformed the way drugs are discovered, tested, and used. At present, drugs go through a stringent process of scientific research, clinical trials, and regulatory approval to ensure they are safe and effective before being made available to the public. Advanced technology such as molecular biology techniques and

TABLE 2.1
Common Medicinal Plants Used for Prevention and Treatment of Malaria

Artemisia annua	*Ficus sur*
Aspidosperma rigidum	*Lantana camara*
Azadirachta indica	*Lippiauckambensis*
Bertholletia excels	*Ocinum suave*
Cassia alata	*Plumbago zeylanica*
Cassia occidentalis	*Simaba cedron*
Cinchona calisaya	*Tithonia diversifolia*
Cinchona succirubra	*Turraeanilotica*
Cochlospermum planchonii	*Turraearobusta*
Eucalyptus camaldulensis	*Vernonia amigdalina*

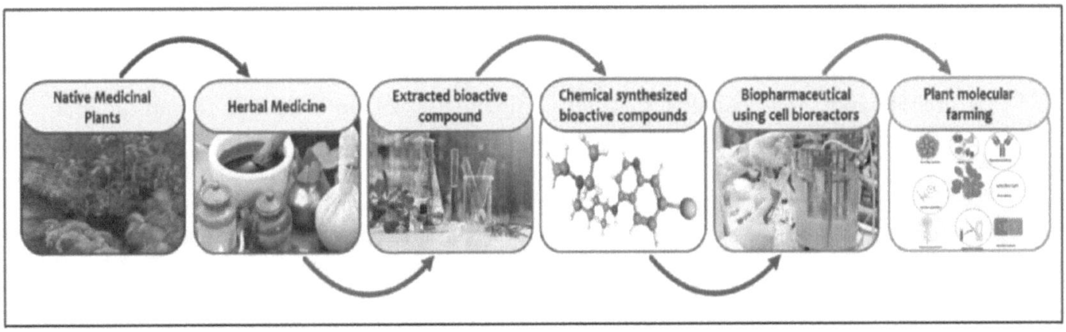

FIGURE 2.2 History of remedies and therapeutics used for malaria.

imaging are utilized by scientists to understand the mechanisms of diseases and to develop specific, effective drugs to treat various conditions. This exceptional innovation has enhanced the quality of life for millions of people across the world and has saved countless lives.

There haven't been many studies on natural compounds with antiplasmodial effects described in recent literature. A review study by Saxena and colleagues, published in 2003, provided a critical evaluation of crude extracts, essential oils, and antiplasmodial secondary metabolites from higher plants, collected from the years 1993 to 2003. In their study, they highlighted 25 more substances in addition to 127 alkaloids, 18 quassinoids, 23 sesquiterpenes, 27 triterpenoids, 21 flavonoids/xanthones, and 9 quinones (Saxena et al., 2003). In their 2008 review, Frederich and colleagues covered 31 indole alkaloids that had been isolated from natural sources and had high antiplasmodial activity (both *in vitro* and in vivo). The majority of these compounds had IC_{50} values that were in the micromolar range and had high selectivity indices (Frederich et al., 2008). A review of antimalarial chemicals found between 1998 and 2008 from all natural sources, including unprocessed plant and marine extracts, was published in 2009 by Kaur and co-authors; 37 interesting semi-synthetic antimalarials were also identified in this compilation, which included 266 antiplasmodial natural products from the groups of alkaloids, terpenes, quassinoids, flavonoids, limonoids, chalcones, peptides, xanthones, quinones, and other substances (Kaur et al., 2009).

Luckily, many publications reporting plant-based antiplasmodial natural products have recently been published. Since a review by Oliveira and colleagues in 2009 only covered plant-derived alkaloids that have antiplasmodial and antimalarial properties, this review is meant to be a continuation of that work. It will cover the majority of recently published reports of non-alkaloidal natural compounds with antiplasmodial and antimalarial properties from the years 2008–2009 that have not yet been reviewed, as well as recently published active crude extracts. Although many plant species employed in traditional remedies for the treatment of malaria include a variety of classes of natural chemicals, alkaloids, quassinoids, and sesquiterpene lactones have been found to have the most significant and diversified biopotency (Saxena et al., 2003). The majority of antiplasmodial non-alkaloidal natural products published between January 2008 and May 2009 that belong to the classes of terpenes, limonoids, flavonoids, chromones, xanthones, anthraquinones, miscellaneous, and related compounds will be highlighted in this review because alkaloids have recently been discussed in the literature (Frederich et al., 2008; Oliveira et al., 2009).

2.4 IMPROVED BIOLOGICAL CHARACTERISTICS OF ANTIMALARIAL PLANT SPECIES

2.4.1 Optimising Treatment Using Currently Available Drugs

Optimising therapy with currently available drugs is one initial strategy. A new dosage schedule or formulation could improve activity. Combination therapy is a highly drug-resistant treatment that has been developed for use in first-line therapy in Africa and many other regions. It includes both new medications, such as atovaquone and artemisinin derivatives, and new combinations of older agents, such as amodiaquine/pyrimethamine and chlorproguanil/sulfadoxine/dapsone.

There are two potential benefits for the amalgamation of antimalarial therapies. First and foremost, the combination should increase the effectiveness of antimalarial medications by generating additional/synergistic antiparasitic activity. Artemisinin derivatives and atovaquone alone have unacceptable failure rates for *falciparum* malaria, but when combined with other well-known antimalarials, they have proved quite successful. The second, and arguably most crucial, benefit of combination therapy is that it should halt the development of parasite resistance to the novel medications. As one works to create new treatments that will remain active for a long time, this last aspect is crucial to take into account.

To avoid low blood levels due to monotherapy, combination therapy regimens that avoid the development of resistance should include at least two drugs that have not yet developed parasite resistance and have similar pharmacokinetics. There is not presently a perfect treatment plan, but chlorproguanil/dapsone/artesunate might meet the need. Alternatively, combining a short-acting, extremely potent substance with a longer-acting drug may be successful, provided the initial reduction in parasite load is significant enough to prevent the development of a resistance to the long-acting treatment (such as artesunate/mefloquine). Another option is to combine two medications with similar pharmacokinetics, even if the community possesses resistance to each treatment individually (e.g. amodiaquine/sulfadoxine/pyrimethamine). When individual medications with established efficacy are taken into account as components of combinations late in the drug development process, considerations of combination therapy typically have an impact on drug development. Combination therapy, however, may need to be taken into account earlier in the drug research process. In contrast to appearances, relatively slow-acting antimalarials (such as antibiotics) may be effective therapeutic agents when used in combination regimens (e.g. quinine and doxycycline).

Even if drug resistance has greatly reduced the effectiveness of some older medications, combinations with other medications may still work. Resistance to any of the drugs was considered unlikely, as resistance to each treatment was considered sufficiently low. Sulfadoxine/pyrimethamine, although a combination treatment, loses rapidly once resistance is found; in this case, the effectiveness depends on synergy, so once resistance to one of the drugs develops, the effectiveness will disappear. Sulfadoxine/pyrimethamine and amodiaquine, a chloroquine analogue that is still potent against a number of chloroquine-resistant parasites, work together to provide two powerful long-acting medications. Notwithstanding the fact that both of the individual components are already constrained by drug resistance due to media resistance and security concerns, it is significant that this combination regimen delivers two readily available and affordable medications (with long-lasting preventative use but with concerned toxicity). The amalgamation of amodiaquine and sulfadoxine/pyrimethamine, in spite of the concerns, which has relatively high resistance to each drug, has shown significant antimalarial effects in parts of East Africa (Staedke et al., 2001; Dorsey et al., 2002; Schellenberg et al., 2002). Use of chloroquine in continuation, especially in combination with different drugs, is another attractive option in areas where chloroquine has not been used for a long time. Recently, it was discovered that chloroquine sensitivity has returned to Malawi after its use had been reduced for roughly ten years (Kublin et al., 2003).

The fast-acting and strong antimalarial properties of artemisinin analogues, particularly artesunate and artemether, have recently shown great results; however, due to the short half-lives of these compounds and the frequent late recurrences they cause after therapy, it is likely that combination therapies will be required to fully utilize the potency of this class. Although artesunate partners having effective resistance lead to many years of late regeneration, such as in Uganda sulfadoxine/artesunate/pyrimethamine, artesunate combined with sulfadoxine/pyrimethamine and amodiaquine has been studied in Africa with good results (Dorsey et al., 2002). The most effective antimalarial medications may turn out to be those that combine artemisinins with longer-acting medications that lack underlying resistance. The combination of artesunate and mefloquine has shown to be quite successful in Thailand, where drug resistance is particularly severe, even where resistance to mefloquine was previously quite common (Price et al., 1997). Lumefantrine, a new drug related to halofantrine, has been coupled with artemethrin to provide a highly effective treatment (Lefevre et al., 2001). Proguanil, a dihydrofolate reductase inhibitor originally used to cure pneumocystis pneumonia, has been used in amalgamation with atovaquone to give antimalarial synergistic activity (Canfield et al., 1995) and even compete with parasites resistant to particular drugs in combination (Vaidya, 2001). Atovaquone/proguanil, artesunate/mefloquine, and artemether/benzanthra quinone are very cost-effective, lack ingredients with comparable pharmacokinetics, may pose toxicity issues (especially mefloquine), and in some cases, while treating, do not have the best optimal dosage regimen (for artemether/lumefantrine can be taken with a fatty meal, having two doses in a day). So while these amalgamations of drugs show effective indications, they may not be the best choice for general use in many places, especially in Africa.

Chlorproguanil/dapsone is a novel antimalarial drug development strategy that combines a proguanil analogue with dapsone, a dihydropteroate synthase (DHPS) inhibitor with a long history of usage in the treatment of leprosy. In Africa, for the treatment of malaria, resistance to chloroquine is widespread, and also, because sulfadoxine/pyrimethamine resistance is higher, chlorproguanil/dapsone was developed particularly. The new treatment plan has the same effects of sulfadoxine/pyrimethamine, but it is generally effective against sulfadoxine/pyrimethamine-resistant parasites because the common mutations in DHFR (dihydrofolate reductase) and DHPS that mediate this resistance do not result in clinical resistance to chlorproguanil/dapsone, and because higher-level antifolate mutations that result in chlorproguanil/dapsone resistance are not so common in Africa (Nzila et al., 2000; Mutabingwa et al., 2001; Kublin et al., 2002). Chlorproguanil/dapsone has a comparatively short half-life, which is enough to enable effective therapy with three-day daily doses and not so long as to rapidly select for resistance. Very soon, chlorproguanil/dapsone for the treatment of malaria should be accessible. Yet a three-drug combination may be its best application. Chlorproguanil/dapsone and artesunate may be the best combination since it combines the high influence of artesunate with the lower curative efficiency of chlorproguanil/dapsone, but further research is required.

2.4.2 Creation of Analogues for Current Agents

By chemically altering current antimalarials, existing antimalarials can be made even more effective as a form of antimalarial chemotherapy. Understanding the parent compound's biological target or mechanism of action is not necessary for this method. In fact, this technology has led to the invention of many existing antimalarial drugs. For instance, chemical methods to enhance quinine led to the discovery of chloroquine, primaquine, and mefloquine (Stocks et al., 2001). In recent years, 4-aminoquinolines, which are closely related to chloroquine, have shown to have the same antimalarial effectiveness as the parent medication, even with chloroquine-resistant parasites (Kaschula et al., 2002; Raynes et al., 1999). China has invented a similar substance, pyrrolidine, which in several areas has undergone clinical trial (Ringwald et al., 1996). Tafenoquine, an 8-aminoquinoline having greater effect against liver-stage parasites than primaquine, its parent drug (Walsh et al., 1999), can be used in chemoprevention with antimalarials (Lell et al., 2000). The development of the analogue benzothraquinone, which is currently part of the new artemether conjoint compound (artemether/benzothraquinone; Van Vugt et al., 2002), was made possible because the toxicity of halofantrine limits its use. Novel folate antagonists and artemisinin-related endoperoxides are both now being researched (Posner et al., 2003; Vennerstrom et al., 2000).

2.4.3 Natural Products

Using substances produced from plants is a third way to treat cancer. A crucial point is that this strategy can profit from the residents of malarial areas' knowledge of medicinal plants, where enthusiasm for the practice of using botanicals to treat feverish illnesses has accumulated over many generations. Plant products with specific clinical activity might thus serve as the beginning point for a medicinal chemistry endeavour, which is a significant improvement over random screening. The two main medications used for the treatment of severe *falciparum* malaria, quinine and artemisinin derivatives, are both derived from natural materials. With regard to artemisinin, relatively straightforward chemical changes of the natural product's parent molecule have produced a number of extremely effective antimalarials that are becoming more and more crucial in the malarial disease management (Meshnick, 2001). However, these compounds may be prohibitive because of price range, so it is crucial to make an attempt to build entirely artificial endoperoxides that are not that expensive to produce (Posner et al., 2003; Vennerstrom et al., 2000). Many studies are being conducted to assess natural products as potential novel treatments for a variety of human ailments (Tagboto &Townson, 2001). It is essential to check the evaluation of several trials of the antimalarial efficiency of compounds extracted from plants and prospective medications of these extracts. Antimalarial natural products are expected to serve as the parent molecules for the manufacture of new medications that are semi- or fully synthetic, just like with quinolines and artemisinins.

2.4.4 Active Ingredients against Other Disorders

Finding drugs that are created or sold as therapies for other diseases is a fourth strategy for antimalarial chemotherapy. These substances may exert their effects on malaria parasites through various pathways or against orthologs of their targets in other systems. Whatever their mechanism of action, they have been previously identified for human indications, and this is the advantage of these compounds and will be fairly affordable to develop as antimalarials. This is significant, given the challenges associated with funding antimalarial drug research. Even while they don't require significant development costs, some novel compounds, particularly those created for diseases common in affluent countries, like cancer, may be prohibitively expensive to produce as antimalarials. But frequently, manufacturing of these drugs can be at a low cost, and they can be sold as affordable antimalarials, especially after patents expire, as is the case with some antibiotics. Tetracyclines, folic acid antagonists, and other antibiotics originally created for their antimicrobial properties were later also found to be effective against the malaria parasite (Clough &Wilson, 2001). Antimalarial atovaquone's development was accelerated after the discovery of its effect on pneumatocystis. Its antimalarial efficiency was recently re-evaluated (which is a compound involved in the Malarone drug combination), and it shows a significant antimalarial synergistic effect with proguanil (Canfield et al., 1995). Malarone is now approved for both the treatment and chemoprophylaxis of malaria because it was later demonstrated to be effective in these two areas (Hogh et al., 2000). Iron chelators have proven antimalarial effectiveness and are for the treatment of iron overload disorders (Loyevsky & Gordeuk, 2001). These results show that it is suitable to investigate existing compounds and new antibacterial agents for potential as antimalarial medicines. The availability of high-throughput testing for putative antimalarials facilitates this strategy. The recent development of Malarone provides evidence that compounds with action against other, more lucrative microbial targets should be taken into account as a potential source of future antimalarials. Although efforts to find effective anticancer treatments for protein farnesyl transferases have not yet succeeded, they have accelerated the idea of using these targets for antimalarial chemotherapy (Gelb et al., 2003).

2.4.5 Anti–Drug Resistance Agents

Another method of chemotherapy involves combining previously efficient substances with substances that counteract parasite resistance to those substances. Several medications, in particular, used to treat high blood pressure (Martin et al., 1987) and the anxiolytic desipramine, have been shown to reverse chloroquine resistance in *P. falciparum in vitro* (Bitonti et al., 1988). Studies with verapamil, desipramine, and trifluoperazine suggest that amalgamation of two or more of these drugs may provide clinically relevant reverse resistance at pharmacological concentration in cases where resistance reversers are required for their effects in unacceptably high concentration (Van Schalkwyk et al., 2001). Chlorpheniramine, a widely available and affordable antihistamine, reversed resistance at acceptable dosage levels, but its common adverse effect of sleepiness may limit its acceptance as a treatment (Sowunmi et al., 1997). Development of new approaches to counteract chloroquine resistance has taken place (Alibert et al., 2002; Batra et al., 2000). Therefore, even though chloroquine has been disapproved as a primary-line antimalarial in most people of the sector, it is able to be revived by combining with strong resistance reversers.

Active compounds against novel objectives, the development of new goals and subsequent compounds that function on these objectives, are arguably the most avant-garde technique of chemotherapy. The discovery of technologies to disrupt plasmodial genes, albeit this method is still time-consuming and ineffective, as well as the sequencing and annotation of the *P. falciparum* genome have sped up progress in characterizing the biology of malaria parasites. The widely available genome sequence improves genomic approaches to drug discovery, but progress is often slowed by the need for more complex and dangerous biochemical and parasitological confirmation of potential therapeutic targets. Based on where they are located inside the malaria parasite, new targets for antimalarial therapy will be taken into consideration (Table 2.2).

2.4.6 Cytosolic Targets

Many metabolic pathways and hundreds of likely important enzymes can be found in the cytosol, making them prospective therapeutic targets. It may be challenging to find drugs that specifically inhibit parasite enzymes, because many of these pathways have undergone good evolutionary conservation, making parasite and host targets extremely similar. The folate metabolism route has demonstrated to be a useful target, as was mentioned previously (Plowe, 2001). In spite of having identical targets, substantial research has found antifolates that efficiently and safely cure both bacterial and protozoan diseases. The availability of substances created to treat other ailments, in this case bacterial infections, has also assisted this strategy. Sadly, resistance to specific DHFR and DHPS inhibitors, such as pyrimethamine, proguanil, and sulfas, causes even combination treatments to significantly lose their effectiveness (Plowe, 2001). Sulfadoxine/pyrimethamine is a cheap

TABLE 2.2
Drugs Used for the Treatment of Malaria

Cellular Location	Target	Drugs	Key References
Cytoplasmic matrix	Lactate dehydrogenase	Sulfadoxine, dapsone	Razakantoanina et al., 2000
	Thymidylate synthase	–	Rathod et al., 1992
	Dihydrofolate reductase	Pyrimethamine, proguanil	Nzila et al., 2000; Mutabingwa et al., 2001
Parasite membrane	Unique channels	–	Gero et al., 2003
	Choline transporter	–	Wengelnik et al., 2002
Food vacuole	Unknown	Artemisinins	Vennerstrom et al., 2000; Borstnik et al., 2002
	Falcipains	Quinolines	Rosenthal, 2001; Shenai et al., 2003
	Hemozoin	–	De et al., 1998
Mitochondrion	Cyt. *c* oxidoreductase	Atovaquone	–
Apicoplast	DNA gyrase	Quinolones	–
	Apicoplast ribosome	Antibiotics	–
	RNA polymerase	Rifampin	–
	Farnesyl transferase	–	Ohkanda et al., 2001; Chakrabarti et al., 2002
	FabI	–	Surolia & Surolia, 2001
	DOXP reductoisomerase	–	Jomaa et al., 1999

alternative to chloroquine as the first-line treatment for malaria in a number of African nations. In spite of this, resistance to this agent is already giant in lots of regions, which include sections of Africa, and it appears to spread unexpectedly, at the least in a few instances, with huge use. Due to variations in *P. falciparum* DHFR and DHPS mutation patterns around the world, new chlorproguanil/dapsone combinations will likely be very effective in Africa but less so in some other regions. However, it is far uncertain how quickly this new drug will pick up resistance if it is used by one to deal with malaria, as it will probably take place soon. Improved DHFR inhibitor antimalarials, such as biguanides related to proguanil, are now being developed (Kinyanjui et al., 1999). Moreover, other folate pathway enzyme inhibitors may work well as antimalarials. For instance, 5-fluoroorotate inhibits thymidylate synthase to produce antimalarial effects (Rathod et al., 1992).

Another cytosolic route of importance is glycolysis. This pathway is necessary for the synthesis of energy in malaria parasites. The structural characterization of *P. falciparum* lactate dehydrogenase has provided potential for the development of selective inhibitors due to its different location of attachment for the NADH cofactor (Dunn et al., 1996). The discovery of specific inhibitors of *P. falciparum* lactate dehydrogenase took place (Deck et al., 1998), and *in vitro* antimalarial efficiency has been shown by some drugs (Razakantoanina et al., 2000). Potential therapeutic targets include the mechanisms for purine salvage and pyrimidine synthesis. Because they are unable to generate purines, malaria parasites must obtain host purines for the production of nucleic acids. Hypoxanthine appears to be the main purine source in *P. falciparum*, and hypoxanthine-guanine phosphoribosyl transferase has been put forward as an effective therapeutic target (Keough et al., 1999). Given the distinctive characteristics of parasite enzymes and key parasite purine scavenging functions, but not for humans, other enzymes involved in purine salvage may provide useful targets. Malaria parasites are dependent on pyrimidine production because, unlike purines, they cannot recycle pyrimidines. It may be possible to use the differences between human and parasitic pyrimidine synthesising enzymes as therapeutic targets. In order to connect these targets to those in the mitochondrion, pyrimidine synthesis also depends on mitochondrial electron transport.

2.4.7 Targets for Membrane Parasites

One possible target is the phospholipid metabolism of the parasite. The membranes required to encapsulate the parasitophorous vacuole, cytosol, and several subcellular compartments are produced by intraerythrocytic malaria parasites by substantial phospholipid synthesis. Phosphatidylcholine is the most prevalent lipid in plasmodial membranes. Synthesis of phosphatidylcholine does require a host choline, and blocking choline transport has been shown to be a viable therapeutic approach (Vial &Calas, 2001). Compounds that have strong antimalarial properties have been developed through extensive medicinal chemistry studies, likely by inhibiting choline transport. *In vitro P. falciparum* parasite growth was suppressed by the lead compound G25 at doses 1,000 times lower than those which are not good to mammalian cells (Calas et al., 2000; Wengelnik et al., 2002). Also, the effectiveness of G25 in vivo against primates and mice infected with *P. falciparum* and *P. cynomolgi* (a model for *P. vivax*). However, efficacy with oral administration was significantly lower than that with parenteral dosing. This is significant because the drug was found to be very effective in primate models and showed activity at doses well below 1 mg/kg/day.

Transport networks that are exclusive to malaria parasites are other membrane targets. Erythrocyte transport routes are significantly altered by intraerythrocytic parasites. Although one still does not fully understand how parasites move, it is possible that there are distinctions between host and parasite pathways that could lead to the development of selective antimalarial medications (Haldar & Akompong, 2001; Kirk, 2001). Using the preferential transport of cytotoxic substances into erythrocytes infected with *P. falciparum* is one option. To increase selective access to parasite targets, one method under investigation is to utilize nucleoside phosphate dimers attached to antimalarial drugs (Gero et al., 2003).

2.4.8 Targets for Food Vacuoles

Erythrocyte haemoglobin is degraded inside the acidic feeding vacuoles that are present in malaria parasites. *P. falciparum* trophozoites contain a solitary, sizable feeding vacuole. Numerous currently available antimalarials appear to function at the food vacuole, and it also presents prospects for therapy focused on novel targets (Banerjee & Goldberg, 2001). Heme, which is polymerized into the insoluble hemozoin pigment and globin, and haemoglobin are all digested into individual amino acids in the food vacuole. Antimalarial medications seem to work by inhibiting the synthesis of hemozoin, generating the food vacuole free radicals or, in experimental substances case, inhibiting the hydrolysis of globin.

As heme molecules are released from haemoglobin, 4-aminoquinoline chloroquine appears to work by preventing the formation of hemozoin (Sullivan, 2002). It is likely that the toxicity of unbound heme, possibly through membrane rupture, causes antiparasitic effects. It is unknown whether antimalarials with similar chemical properties operate similarly. The drug has been used successfully for over 50 years, with only modest development of resistance, which can be attributed to its non-enzymatic mode of action, even if drug resistance has severely restricted its usage currently. Several experimental and commercial antimalarials inhibit particular enzymes, but this method is expected to frequently suffer from the quick selection of parasites with mutations, as with other antimalarials, the enzymes responsible for drug resistance are targeted. Chloroquine resistance is slowly starting to emerge. It is now understood that resistance is predominantly caused by mutations in the putative transporter PfCRT, and that many mutations were required to select for clinical resistance even if a single mutation mediated resistance *in vitro* (Fidock et al., 2000). Although parasites that are resistant to chloroquine are already widespread in practically all malarious regions, it makes sense to explore additional drugs that target heme polymerization. In this regard, significant attempts are being made to create enhanced quinolines or similar compounds as antimalarials (De et al., 1998; Stocks et al., 2001, 2002).

Moreover, it appears that artemisinin antimalarials target the food vacuole. As previously mentioned, this novel class of chemicals has extremely strong efficacy with no known cases of drug resistance to date. The antimalarial activity of artemisinins depends on an endoperoxide bridge, which appears to decompose into free radicals under the influence of iron (Meshnick, 2001). Although the precise pharmacological targets are unknown, the compounds appear to have free radical damage, which causes antimalarial effects, effectively by alkylating plasmodial proteins (Asawamahasakda et al., 1994; Bhisutthibhan et al., 1998). A factor in the selective antimalarial toxicity of artemisinins, as it is with other substances active in the feeding vacuole, may be the drug's unique concentration in this parasite organelle. Analogues of artemisinin are already effective antimalarials, but production of them is cost-effective, because the plant products are not naturally produced. There are significant attempts being made to create antimalarial entirely from synthetic peroxides and some other chemicals (Vennerstrom et al., 2000; Borstnik et al., 2002).

The hydrolysis of globin seems to be mediated by a variety of protease classes, including metalloproteases (Falcilysin; Eggleson et al., 1999), cytosolic metallo-aminopeptidases, and the aspartic food vacuole (plasmepsins; Banerjee et al., 2002) and cysteine (Shenai et al., 2000; Sijwali (Gavigan et al., 2001). All these enzymes provide prospective chemotherapeutic targets (Rosenthal, 2001). The repertoire of proteases known to mediate haemoglobin hydrolysis in the case of plasma proteases and *P. falciparum* proteases is now more critical than primarily thought, as biochemical studies and the full sequence of the *P. falciparum* genome have identified as the involved members of these families. Haemoglobin hydrolysis by four plasmepsins is thought to occur in the food vacuole (Banerjee et al., 2002). The precise functions of these or any other proteases in the process are yet unknown; however, it has been suggested that plasmepsin II and maybe other plasmepsins mediate the early cleavages of haemoglobin, permitting further processing by other proteases. Antimalarial activity has been shown by inhibitors of plasmepsin (Francis et al., 1994; Silva et al., 1996; Moon et al., 1998; Haque et al., 1999; Jiang et al., 2001; Nezami et al., 2002; Noteberg et al., 2003), but these restraints' effects have not been convincingly linked to inhibition of haemoglobin hydrolysis. The primary cysteine protease mediators of haemoglobin hydrolysis, falcipain-2 and falcipain-3, appear to be the targets of cysteine protease inhibitors (Shenai et al., 2000; Sijwali et al., 2001). Demonstration of stoppage of the formation of parasites, stoppage of haemoglobin hydrolysis, and treatment of mouse malaria has been done by falcipain inhibitors (Batra et al., 2003; Rosenthal et al., 1991, 1993, 1996, 2002; Shenai et al., 2003; Singh &Rosenthal, 2001; Rosenthal, 2001). An important note is that cysteine antimalarial acts on then stops the hydrolysis of haemoglobin by the parasite, resulting in a build-up of intact haemoglobin in the feeding vacuole. This morphological anomaly supports the chemicals' targeted activity on falcipain targets. In note, both in vivo and *in vitro* antimalarial actions of cysteine and aspartic protease inhibitors are synergistic (Semenov et al., 1998). These findings imply that both kinds of protease inhibitors may be present in an ideal protease inhibitor antimalarial.

One recent antimalarial drug targets the mitochondria. Atovaquone interferes with electron transport, collapses mitochondrial membrane potential, and inhibits ubiquinol-cytochrome c oxidoreductase (complex III), all of which are necessary for various parasite metabolic processes (Vaidya, 2001). Although the medicine has effective antimalarial efficacy, a quick selection of parasites with mutations in the target enzyme makes it unsuitable for use as monotherapy. For both chemotherapeutic (Radloff et al., 1996; Hogh et al., 2000) *falciparum* malaria therapy, Malarone is a high-cost/price medicine but effective; it was developed after the discovery of the incredible success of atovaquone when used in combination with the antifolate proguanil. Combination more effective than expected in regions with high levels of antifolate resistance because it may benefit from the combined effect of atovaquone and proguanil inspite of antifolate activity of the metabolite cyclidine (Vaidya, 2001).

2.4.9 Targets of the Apicoplast

Recent research has revealed the apicoplast, an organelle similar to chloroplasts found in apicomplexan parasites (Kohler et al., 1997). It appears that endosymbiosis produced the apicoplast, an organelle that maintains a number of distinct roles; this can include fatty acid, heme, as well as amino acid metabolism (Roos et al., 2002). The fact that the apical body, like the mitochondria, has a unique prokaryotic-like genome may explain why some antibacterial substances that do not normally affect eukaryotes have antimalarial effects. Yet a unique mechanism combining two amino-terminal targeting sequences is used to transport the majority of apicoplast proteins, which are encoded in the nucleus (Waller et al., 1998; Foth et al., 2003).

Many antibacterial substances are efficient, albeit slow-acting, antimalarials, as was mentioned earlier (Clough &Wilson, 2001). These substances probably exert their effects by concentrating on mitochondrial and/or apicoplast mechanisms that are comparable to those in bacteria (Ralph et al., 2001). Several processes involved in prokaryote-like protein synthesis are inhibited by tetracyclines, clindamycin, macrolides, and chloramphenicol. Both rifampin and quinolone antibiotics specifically suppress prokaryote-like activity in RNA polymerase and DNA gyrase, respectively. Why each of these chemicals appears to have just minimal antimalarial action is unclear. Based in part on research with the related protozoan toxoplasma, most likely, the toxicity of the apicoplast mainly affects its life cycle after the initial incubation with this drug (Fichera & Roos, 1997). Mitochondrial toxicity is also sometimes important. Despite the sluggish action of antibacterial chemicals as antimalarials, several, such as tetracycline and clindamycin, have been widely validated as effective antimalarial drugs, and more research on related drugs is needed.

Apicoplast biology has metabolic pathways that are lacking in many human hosts but are abundant in bacteria, plants, and apicomplexan parasites, offering clear results for chemotherapy (Ralph et al., 2001; Roos et al., 2002). Humans do not have the biosynthetic pathway of type II fatty acid; however, the *P. falciparum* genome contains genes that encode homologs of bacterial enzymes from this route and may contain putative apicoplast coding signals (Waller et al., 1998).

The biosynthesis of type II fatty acid subunit known as ketoacyl-acyl-carrier protein synthase (FabH) was targeted by thiolactomycin antibiotic, and the cultured malaria parasites got affected by this thiolactomycin antibiotic (Waller et al., 1998). *P. falciparum* encodes the enoyl-acyl-carrier protein reductase (FabI) subunit, which is targeted by triclosan antimicrobial (Surolia & Surolia, 2001). Triclosan inhibited the target enzyme and prevented the synthesis of fatty acids in the parasites, validating this target by showing efficacy against cultured *P. falciparum* parasites and against murine malaria.

In bacteria and chloroplasts, a different mechanism called the non-mevalonate pathway or 1-deoxy-d-xylulose 5-phosphate (DOXP) pathway exists in addition to the mevalonate pathway, which is how plants and animals do the biosynthesis of isopentenyl diphosphate, which is an isoprenoid precursor. *P. falciparum* encodes the genes for DOXP reductoisomerase and DOXP synthase, two of the enzymes involved in this process, and they have potential signals of apicoplast targeting (Jomaa et al., 1999). *P. falciparum* inhibits the activity of recombinant DOXP reductoisomerase and *P. falciparum* parasite development in culture, and curing mouse malaria is an effect of the antibiotic fosmidomycin (Jomaa et al., 1999). As fosmidomycin had already been created as an antibiotic, it could be tested on humans for malaria very quickly. Fosmidomycin was well tolerated and showed 100% initial cure rates in a first trial of safety and efficacy for uncomplicated malaria (Lell et al., 2000). However, the apparent necessity for frequent and prolonged dosing as well as the frequent occurrence of recrudescence after therapy will limit the use of fosmidomycin as monotherapy. It may be appropriate to include this substance or other apicoplast process inhibitors in combination antimalarial treatments.

Protein farnesylation precursors are made available by the DOXP process. Protein farnesyltransferase inhibitors have been investigated as potential cancer treatments. According to Chakrabarti et al. (2002), the biochemical properties of *Plasmodium* farnesyltransferase activity are unique, and inhibitors of this *Plasmodium* farnesyltransferase activity exhibit antimalarial effects in invitro condition (Ohkanda et al., 2001; Chakrabarti et al., 2002).

2.4.10 Malaria Vaccine Made from Plants

The strategy for developing vaccines is based on the idea that ongoing medication resistance can reduce the effectiveness of both traditional (using insecticide-treated bed nets and indoor residual spraying) and modern antimalarial drugs (Organization, 2015). Despite extensive research towards the creation of a malaria vaccine, there is still no vaccine that can deliver long-lasting immunity. The vaccine has been shown to be an effective and affordable tool for managing and controlling other infectious diseases, making it a promising choice for managing and preventing malaria. The liver, blood, and mosquito phases of the parasite life cycle have been studied in an effort to create an effective malaria vaccine. The creation of a malaria vaccine at any of these stages has a great chance of producing an immunological response (Riley & Stewart, 2013).

In order to develop a potential malaria vaccine, numerous factors have been studied. Some of these have moved into preclinical and advanced clinical studies (Laurens, 2018). The most advanced malaria vaccine that potentially hit the market is GSK's Mosquirix (RTS'S), which is based on *P. falciparum* circumsporozoite protein (CSP) (Clemente & Corigliano, 2012). According to Figure 2.3, the process of developing a plant-based product can take many years to receive regulatory approval, and at the moment, it is difficult to advance any plant-based products in Europe and the USA due to stringent rules. However, the recent COVID-19 outbreak has forced regulators to search for alternative production methods, such as plants, for scalable and affordable ways to reduce infections.

Many prospective antigens against various stages of the parasite life cycle have been generated via transient expression or stably transformed plants, yet there is currently no report on the creation of an effective vaccine that confers protection against malaria (Farrance et al., 2011). Production of neutralizing antibodies has been observed in plant-made vaccine candidates following oral

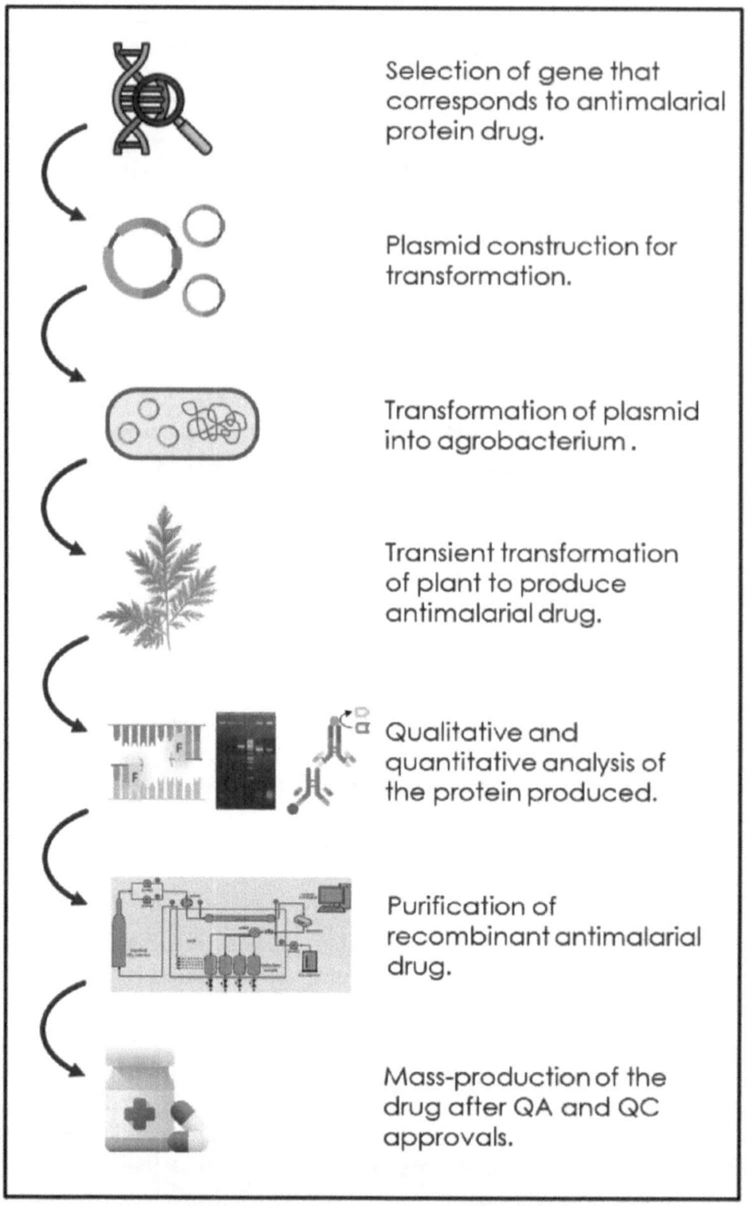

FIGURE 2.3 Flowchart of developing plant-based antimalarial drugs.

administration for immunization. Long-lasting TBA (thiobarbituric acid) and improved immunogenicity produced a protective immune response in an animal model (Farrance et al., 2011; Clemente & Corigliano, 2012). The malaria recombinant vaccine's subcellular targeting in the reticulum endoplasmic (RE) revealed that N-glycosylation was only present in high-mannose glycoforms, which are free of potentially allergic plant-specific glycan alterations (Bardor et al., 2003; Sriraman et al., 2004; Sriraman et al., 2004). Recently, PfAMA1 (>510 g/g fresh leaf weight) was produced at high levels in *Nicotiana benthamiana* plants using this idea. It was discovered that non-glycosylated antigens could adopt a natural shape and trigger potent growth inhibitory antibody responses in rabbits (Boes et al., 2015).

2.4.11 Transient Expression Systems to Treat Malaria

Transient expression systems are preferred for the creation of recombinant proteins because they enable cost-effective large-scale manufacture in a matter of weeks without the need for genetic alterations. Transient expression methods are promising alternatives to traditional cell culture–based and transgenic plants–based technologies that call for lengthy development cycles in order to produce antimalarial vaccines in a comparatively shorter amount of time. By intentionally infecting a plant with a replicating virus that contains the desired recombinant protein's sequence, temporary expression can be produced (Glick et al., 2010). Injecting tobacco or the closely related *N. benthamiana* with an *Agrobacterium tumefaciens*–carrying vector that carries the appropriate gene sequence is another way to achieve temporary expression.

Systemic and agro-infections were regarded as independent and competitive ways to create recombinant proteins in the early stages of molecular farming. Evidence over the past few decades, however, has shown that agro-infection is a more efficient strategy for reproducing viruses that contain genes of interest. The rise of deconstructed vectors and the most significant recent development in molecular farming both rely on readily accessible plant viruses (Gleba et al., 2013; Peyret & Lomonossoff, 2015). Unwanted viral genome components have been removed in this new generation of vectors, allowing for the construction of a bigger trans gene while preserving viral transcription and replication. Moreover, agro-inoculation can circumvent restrictions associated with mechanically injecting nucleic acids via virus vectors into plants and transfer specific genes into plant cells (Peyret & Lomonossoff, 2015). MagnICON system as a deconstructed viral vector has been demonstrated for rapid, high-level accumulation of numerous pharmaceuticals, up to 1 mg/g (FW) in shorter duration (7–10 days after agroinfiltration), some of which are currently undergoing clinical evaluation (Phoolcharoen et al., 2011; Peyret & Lomonossoff, 2015). PyMSP4/5(1–2mg/g/Fwt) and MSP119(3–4mg/g/Fwt), two plant-made proteins that are highly effective malaria asexual blood-stage subunit vaccines, have also been shown to be produced in large quantities using thissy stems in *N. benthamiana* (Webster et al., 2009). The yield variability across batches, however, is one of the difficulties with transient expression systems (O'Neill et al., 2008), which highlights the issues with process repeatability (Fischer et al., 2012). Improvement of process parameter controls, such as incubation time and temperature using VFUs facility, can solve this problem and maintain consistency (Buyela, 2018). Fischer and colleagues have examined all available data on VFUs (Fischer et al., 2012). Government approval in the USA was given to this strategy when it was successfully used for the transient expression of recombinant proteins involved in the 2014 influenza pandemic (Stoger et al., 2014). Spiegel and associates looked at the scalable transient expression platform for malaria vaccine development (Spiegel et al., 2015). Several targets representing different *P. falciparum* life cycle stages were looked into as potential vaccine candidates. Their study's findings showed that temporary expression is necessary for creating multi-domain malaria vaccine candidates that are both scalable and practical.

2.5 CONCLUSION

Due to socioeconomic factors, climate change, migration, the absence of a cogent health policy to stop the spread of the illness, and resistance to antimalarial medications, particularly single-drug delivery in developing nations, the incidence of malaria infection is projected to rise. The need for antimalarial drugs will rise as malaria infections spread like wildfire in the developing world's growing population. Even while biotechnological processes have advanced to the point where there is a sufficient supply of these agents, the production method for those is based on microbe or mammalian cell cultures with a restricted capacity, which is unlikely to satisfy rising future market demands. Plants may therefore naturally produce antimalarial substances like quinine, lapinone, and artemisinin, as well as the potential to meet the problem of creating efficient and economical protein medicines, making them an attractive option. The scientific community will be encouraged to submit more of their product to clinical studies using the same regulatory pathway in light of the FDA's (Food and Drug Administration) recent regulatory approval of orally administered Palforzia. Moreover, development in stable, transient-based technologies is needed to produce malaria-related chemicals that can lower costs, do away with the need for a cold chain, bypass oral delivery's downstream processing, and boost bioavailability through passing degradation. Also, the scalability of these transitory production systems is crucial and needs further study in order to fulfil future demands and save millions of lives globally.

REFERENCES

Achan, A.O., Talisuna, A., Erhart, A., Yeka, A., Tibenderana, J.K., Baliraine, F.N., Rosenthal, P.J. & Alessandro, U.D. 2011. Quinine, an old anti-malarial drug in a modern world: Role in the treatment of malaria. *Malar. J.* 10: 144. https://doi.org/10.1186/1475-2875-10-144.

Alibert, S., Santelli-Rouvier, C., Pradines, B., Houdoin, C., Parzy, D., Karolak-Wojciechowska, J. & Barbe, J. 2002. Synthesis and effects on chloroquine susceptibility in *Plasmodium falciparum* of a series of new dihydroanthracene derivatives. *J. Med. Chem.* 45: 3195–3209. https://doi.org/10.1021/jm011046l.

Amir, A., Cheong, F.W., de Silva, J.R., Liew, J.W.K. & Lau, Y.L. 2018. Plasmodium knowlesi malaria: Current research perspectives. *Infect. Drug. Resist.* 11: 1145–1155. https://doi.org/10.2147/IDR.S148664.

Antinori, S., Galimberti, L., Milazzo, L. & Corbellino, M. 2012. Biology of human malaria plasmodia including Plasmodium knowlesi. *Mediterr. J. Hematol. Infect. Dis.* 4(1): e2012013. https://doi.org/10.4084/MJHID.2012.013.

Asawamahasakda, W., Ittarat, I., Pu, Y.M., Ziffer, H. & Meshnick, S.R. 1994. Reaction of antimalarial endoperoxides with specific parasite proteins. *Antimicrob. Agents Chemother.* 38: 1854–1858.

Autino, B., Noris, A., Russo, R. & Castelli, F. 2012. Epidemiology of malaria in endemic areas. *Mediterr. J. Hematol. Infect. Dis.* 4(1): e2012060. https://doi.org/10.4084/MJHID.2012.060.

Banerjee, R. & Goldberg, D.E. 2001. The Plasmodium food vacuole. In *Antimalarial Chemotherapy: Mechanisms of Action, Resistance, and New Directions in Drug Discovery* (ed. P. J. Rosenthal), pp. 43–63. Totawa, NJ: Humana Press. https://doi.org/10.1007/978-1-59259-111-4_4.

Banerjee, R., Liu, J., Beatty, W., Pelosof, L., Klemba, M. & Goldberg, D.E. 2002. Four plasmepsins are active in the *Plasmodium falciparum* food vacuole, including a protease with an active-site histidine. *Proc. Natl. Acad. Sci. USA.* 99: 990–995. https://doi.org/10.1073/pnas.022630099.

Bardor, M., Faveeuw, C., Fitchette, A.C., Gilbert, D., Galas, L., Trottein, F., Faye, L. & Lerouge, P. 2003. Immunoreactivity in mammals of two typical plant glyco-epitopes, core α(1,3)-fucose and core xylose. *Glycobiology.* 13: 427–434.

Batra, S., Sabnis, Y.A., Rosenthal, P.J. & Avery, M.A. 2003. Structure-based approach to falcipain-2 inhibitors: Synthesis and biological evaluation of 1,6,7-trisubstituted dihydroisoquinolines and isoquinolines. *Bioorg. Med. Chem.* 11: 2293–2299. https://doi.org/10.1016/s0968-0896(03)00117-2.

Batra, S., Srivastava, P., Roy, K., Pandey, V.C. & Bhaduri, A.P. 2000. A new class of potential chloroquine-resistance reversal agents for Plasmodia: Syntheses and biological evaluation of 1-(3′-diethylaminopropyl)-3-(substituted phenylmethylene) pyrrolidines. *J. Med. Chem.* 43: 3428–3433. https://doi.org/10.1021/jm000083u.

Bhisutthibhan, J., Pan, X.Q., Hossler, P.A., Walker, D.J., Yowell, C.A., arlton, J., Dame, J.B. & Meshnick, S.R. 1998. The *Plasmodium falciparum* translationally controlled tumor protein homolog and its reaction with the antimalarial drug artemisinin. *J. Biol. Chem.* 273: 16192–16198. https://doi.org/10.1074/jbc.273.26.16192.

Biamonte, M.A., Wanner, J. & LeRoch, K.G. 2013. Recent advances in malaria drug discovery. *Bioorg. Med. Chem. Lett.* 23: 2829–2843. https://doi.org/10.1016/j.bmcl.2013.03.067.

Birkett, A.J. 2016. Status of vaccine research and development of vaccines for malaria. *Vaccine.* 34: 2915–2920. https://doi.org/10.1016/j.vaccine.2015.12.074.

Bitonti, A.J., Sjoerdsma, A., McCann, P.P., Kyle, D.E., Oduola, A.M., Rossan, R.N., Milhous, W.K. & Davidson, D.E. Jr. 1988. Reversal of chloroquine resistance in malaria parasite Plasmodium falciparum by desipramine. *Science.* 242: 1301–1303. https://doi.org/10.1126/science.3057629.

Bloland, P.B. 2001. *Drug Resistance in Malaria.* World Health Organization, Switzerland, WHO/CDS/CSR/DRS/2001.4.

Boes, A., Spiegel, H., Edgue, G., Kapelski, S., Scheuermayer, M., Fendel, R., Remarque, E., Altmann, F., Maresch, D. & Reimann, A. 2015. Detailed functional characterization of glycosylated and nonglycosylated variants of malaria vaccine candidate Pf AMA I produced in *Nicotiana benthamiana* and analysis of growth inhibitory responses in rabbits. *Plant Biotechnol. J.* 13: 222–234.

Borstnik, K., Paik, I.H. & Posner, G.H. 2002. Malaria: New chemotherapeutic peroxide drugs. *Mini Rev. Med. Chem.* 2: 573–583. https://doi.org/10.2174/1389557023405620.

Butler, A.R., Khan, S. & Ferguson, E. 2010. A brief history of malaria chemotherapy. *J. R. Coll. Physicians Edinb.* 40: 172–177. https://doi.org/10.4997/jrcpe.2010.216.

Buyela, J.F. 2018. Plants as sources of natural and recombinant anti-cancer agents. *Biotechnol. Adv.* 36: 506–520.

Calas, M., Ancelin, M.L., Cordina, G., Portefaix, P., Piquet, G., Vidal-Sailhan, V. & Vial, H. 2000. Antimalarial activity of compounds interfering with Plasmodium falciparum phospholipid metabolism: Comparison between mono- and bisquaternary ammonium salts. *J. Med. Chem.* 43: 505–516. https://doi.org/10.1021/jm9911027.

Canfield, C.J., Pudney, M. & Gutteridge, W.E. 1995. Interactions of atovaquone with other antimalarial drugs against Plasmodium falciparum in vitro. *Exp. Parasitol.* 80: 373–381. https://doi.org/10.1006/expr.1995.1049.

CDC. 2016. *The history of malaria, an ancient disease*. https://www.cdc.gov/malaria/about/history/index.html. Accessed 17 Aug 2022.

Chakrabarti, D., Da Silva, T., Barger, J., Paquette, S., Patel, H., Patterson, S. & Allen, C.M. 2002. Protein farnesyltransferase and protein prenylation in Plasmodium falciparum. *J. Biol. Chem.* 277: 42066–42073. https://doi.org/10.1074/jbc.M202860200.

Clemente, M. & Corigliano, M.G. 2012. Overview of plant-made vaccine antigens against malaria. *J. Biomed. Biotechnol.* https://doi.org/10.1155/2012/206918.

Clough, B. & Wilson, R.J.M. 2001. Antibiotics and the plasmodial plastid organelle. In *Antimalarial Chemotherapy: Mechanisms of Action, Resistance, and New Directions in Drug Discovery* (ed. P.J. Rosenthal), pp. 265–286. Totawa, NJ: Humana Press.

Cox, F.E. 2010. History of the discovery of the malaria parasites and their vectors. *Parasites Vectors.* 3(5).

Crutcher, J.M. & Hoffman, S.L. 1996. Malaria. In *Medical Microbiology* (ed. S. Baron), 4th ed. Galveston: University of Texas Medical Branch at Galveston, Chapter 83. Available at: www.ncbi.nlm.nih.gov/books/NBK8584/.

De, D., Krogstad, F.M., Byers, L.D. & Krogstad, D.J. 1998. Structure-activity relationships for antiplasmodial activity among 7-substituted 4-aminoquinolines. *J. Med. Chem.* 41: 4918–4926. https://doi.org/10.1021/jm980146x.

Deck, L.M., Royer, R.E., Chamblee, B.B., Hernandez, V.M., Malone, R.R., Torres, J.E., Hunsaker, L.A., Piper, R.C., Makler, M.T. & Vander Jagt, D.L. 1998. Selective inhibitors of human lactate dehydrogenases and lactate dehydrogenase from the malarial parasite *Plasmodium falciparum*. *J. Med. Chem.* 41: 3879–3887. https://doi.org/10.1021/jm980334n.

Deprez-Poulain, R. & Melnyk, P. 2005. 1,4-Bis(3-aminopropyl)piperazine libraries: From the discovery of classical chloroquine-like antimalarials to the identification of new targets. *Comb. Chem. High Throughput Screen.* 8: 39–48. https://doi.org/10.2174/1386207053328165.

Devillers, J. 2018. Repurposing insecticides and drugs for the control of mosquitoes and their diseases. In *Computational Design of Chemicals for the Control of Mosquitoes and Their Diseases* (ed. J. Devillers), pp. 1–37. Boca Raton: CRC Press.

Dorsey, G., Njama, D., Kamya, M.R., Cattamanchi, A., Kyabayinze, D., Staedke, S.G., Gasasira, A. & Rosenthal, P.J. 2002. Sulfadoxine/pyrimethamine alone or with amodiaquine or artesunate for treatment of uncomplicated malaria: A longitudinal randomised trial. *Lancet.* 360: 2031–2038. https://doi.org/10.1016/S0140-6736(02)12021-6.

Draper, S.J., Sack, B.K., King, C.R., Nielsen, C.M., Rayner, J.C., Higgins, M.K., Long, C.A. & Seder, R.A. 2018. Malaria vaccines: Recent advances and new horizons. *Cell Host Microbe.* 24: 43–56. https://doi.org/10.1016/j.chom.2018.06.008.

Dunn, C.R., Banfield, M.J., Barker, J.J., Higham, C.W., Moreton, K.M., Turgut-Balik, D., Brady, R.L. & Holbrook, J.J. 1996. The structure of lactate dehydrogenase from Plasmodium falciparum reveals a new target for anti-malarial design. *Nat. Struct. Biol.* 3: 912–915. https://doi.org/10.1038/nsb1196-912.

Eggleson, K.K., Duffin, K.L. & Goldberg, D.E. 1999. Identification and characterization of falcilysin, a metallopeptidase involved in hemoglobin catabolism within the malaria parasite *Plasmodium falciparum. J. Biol. Chem.* 274: 32411–32417. https://doi.org/10.1074/jbc.274.45.32411.

Egieyeh, S.A., Syce, J., Malan, S.F. & Christoffels, A. 2016. Prioritization of anti-malarial hits from nature: Chemo-informatic profiling of natural products with in vitro antiplasmodial activities and currently registered anti-malarial drugs. *Malar. J.* 15: 50. https://doi.org/10.1186/s12936-016-1087-y.

Ekor, M. 2013. The growing use of herbal medicines: Issues relating to adverse reactions and challenges in monitoring safety. *Front. Pharmacol.* 4: 177. https://doi.org/10.3389/fphar.2013.00177.

Farrance, C.E., Chichester, J.A., Musiychuk, K., Shamloul, M., Rhee, A. & Manceva, S.D. 2011. Antibodies to plant-produced Plasmodium falciparum sexual stage protein Pfs25 exhibit transmission blocking activity. *Human Vaccines* 7: 191–198.

Fichera, M.E. & Roos, D.S. 1997. A plastid organelle as a drug target in apicomplexan parasites. *Nature* 390: 407–409. https://doi.org/10.1038/37132.

Fidock, D.A., Nomura, T., Talley, A.K., Cooper, R.A., Dzekunov, S.M., Ferdig, M.T., Ursos, L.M., Sidhu, A.B., Naude, B. & Deitsch, K.W. 2000. Mutations in the P. falciparum digestive vacuole transmembrane protein PfCRT and evidence for their role in chloroquine resistance. *Mol. Cell.* 6: 861–871. https://doi.org/10.1016/s1097-2765(05)00077-8.

Fidock, D.A., Rosenthal, P.J., Croft, S.L., Brun, R. & Nwaka, S. 2004. Antimalarial drug discovery: Efficacy models for compound screening. *Nat. Rev. Drug Discov*. 3: 509–520. https://doi.org/10.1038/nrd1416.

Fischer, R., Schillberg, S., Hellwig, S., Twyman, R.M. & Drossard, J. 2012. GMP issues for recombinant plant-derived pharmaceutical proteins. *Biotechnol. Adv*. 30: 434–439.

Foth, B.J., Ralph, S.A., Tonkin, C.J., Struck, N.S., Fraunholz, M., Roos, D.S., Cowman, A.F. & McFadden, G.I. 2003. Dissecting apicoplast targeting in the malaria parasite *Plasmodium falciparum*. *Science*. 299: 705–708. https://doi.org/10.1126/science.1078599.

Francis, S.E., Gluzman, I.Y., Oksman, A., Knickerbocker, A., Mueller, R., Bryant, M.L., Sherman, D.R., Russell, D.G. & Goldberg, D.E. 1994. Molecular characterization and inhibition of a Plasmodium falciparum aspartic hemoglobinase. *EMBO J*. 13: 306–317. https://doi.org/10.1002/j.1460-2075.1994.tb06263.x.

Frederich, M., Tits, M. & Angenot, L. 2008. Potential antimalarial activity of indole alkaloids. *Trans. R. Soc. Trop. Med. Hyg*. 102: 11–19. https://doi.org/10.1016/j.trstmh.2007.10.002.

Gamo, F.J., Sanz, L.M., Vidal, J., de Cozar, C., Alvarez, E., Lavandera, J.L., Vanderwall, D.E., Green, D.V., Kumar, V. & Hasan, S. 2010. Thousands of chemical starting points for antimalarial lead identification. *Nature*. 465: 305. https://doi.org/10.1038/nature09107.

Gavigan, C.S., Dalton, J.P. & Bell, A. 2001. The role of aminopeptidases in haemoglobin degradation in Plasmodium falciparum-infected erythrocytes. *Mol. Biochem. Parasitol*. 117: 37–48. https://doi.org/10.1016/s0166-6851(01)00327-9.

Gelb, M.H., Van Voorhis, W.C., Buckner, F.S., Yokoyama, K., Eastman, R., Carpenter, E.P., Panethymitaki, C., Brown, K.A. & Smith, D.F. 2003. Protein farnesyl and N-myristoyl transferases: Piggy-back medicinal chemistry targets for the development of antitrypanosomatid and antimalarial therapeutics. *Mol. Biochem. Parasitol*. 126: 155–163. https://doi.org/10.1016/s0166-6851(02)00282-7.

Gero, A.M., Dunn, C.G., Brown, D.M., Pulenthiran, K., Gorovits, E.L., Bakos, T. & Weis, A.L. 2003. New malaria chemotherapy developed by utilization of a unique parasite transport system. *Curr. Pharm. Des*. 9: 867–877. https://doi.org/10.2174/1381612033455233.

Gleba, Y.Y., Tusé, D. & Giritch, A. 2013. Plant viral vectors for delivery by Agrobacterium. In *Plant Viral Vectors* (eds. K. Palmer & Y. Gleba), pp. 155–192. Berlin and Heidelberg: Springer.

Glick, B.R., Patten, J.P.J. & Li, C. 2010. *Molecular Bio-Technology: Principles and Applications of Recombinant DNA* (4th ed.). American Society for Microbiology.

Greenwood, B.M., Bojang, K., Whitty, C.J. & Targett, G.A. 2005. Malaria. *Lancet*. 365: 1487–1498. https://doi.org/10.1016/S0140-6736(05)66420-3.

Haldar, K. & Akompong, T. 2001. Transport and trafficking in Plasmodium-infected red cells. In *Antimalarial Chemotherapy: Mechanisms of Action, Resistance, and New Directions in Drug Discovery* (ed. P.J. Rosenthal), pp. 27–41. Totowa, NJ: Humana Press.

Haque, T.S., Skillman, A.G., Lee, C.E., Habashita, H., Gluzman, I.Y.T.J., Goldberg, D.E., Kuntz, I.D. & Ellman, J.A. 1999. Potent, low-molecular-weight non-peptide inhibitors of malarial aspartyl protease plasmepsin II. *J. Med. Chem*. 42(8): 1428–1440. https://doi.org/10.1021/jm980641t.

Hoffman, S.L., Vekemans, J., Richie, T.L. & Duffy, P.E. 2015. The march toward malaria vaccines. *Vaccine*. 33: D13–D23. https://doi.org/10.1016/j.amepre.2015.09.011.

Hogh, B., Clarke, P.D., Camus, D., Nothdurft, H.D., Overbosch, D., Gunther, M., Joubert, I., Kain, K.C., Shaw, D. & Roskell, N.S. 2000. Atovaquone-proguanil versus chloroquine-proguanil for malaria prophylaxis in non-immune travellers: A randomised, double-blind study. Malarone International Study Team. *Lancet*. 356: 1888–1894. https://doi.org/10.1016/s0140-6736(00)03260-8.

Jambou, R., Legrand, E., Niang, M., Khim, N., Lim, P., Volney, B., Ekala, M.T., Bouchier, C., Esterre, P., Fandeur, T. & Mercereau-Puijalon, O. 2005. Resistance of *Plasmodium falciparum* field isolates to in vitro artemether and point mutations of the SERCA-type PfATPase6. *Lancet*. 366: 1960–1963. https://doi.org/10.1016/S0140-6736(05)67787-2.

Jiang, S., Prigge, S.T., Wei, L., Gao, Y., Hudson, T.H., Gerena, L., Dame, J.B. & Kyle, D.E. 2001. New class of small nonpeptidyl compounds blocks *Plasmodium falciparum* development in vitro by inhibiting plasmepsins. *Antimicrob. Agents Chemother*. 45: 2577–2584. https://doi.org/10.1128/AAC.45.9.2577-2584.2001.

Jiraprapa, W., Salenna, E., Huji, X. & Good, M.F. 2002. Immunity to asexual blood stage malaria and vaccine approaches. *Immunol. Cell Biol*. 80: 401–414. https://doi.org/10.1046/j.1440-1711.2002.01107.x.

Jomaa, H., Wiesner, J., Sanderbrand, S., Altincicek, B., Weidemeyer, C., Hintz, M., Turbachova, I., Eberl, M., Zeidler, J. & Lichtenthaler, H.K. 1999. Inhibitors of the nonmevalonate pathway of isoprenoid biosynthesis as antimalarial drugs. *Science* 285: 1573–1576. https://doi.org/10.1126/science.285.5433.1573.

Jones, M.K. & Good, M.F. 2006. Malaria parasites up close. *Nat. Med*. 12: 170–171. https://doi.org/10.1038/nm0206-170.

Kalauni, S.K., Awale, S., Tezuka, Y., Banskota, A.H., Linn, T.Z., Asih, P.B., Syafruddin, D. & Kadota, S. 2006. Antimalarial activity of cassane- and norcassane-type diterpenes from Caesalpinia crista and their structure-activity relationship. *Biol. Pharm. Bull*. 29: 1050–1052. https://doi.org/10.1248/bpb.29.1050.

Karcher, W.& Devillers, J. 1990. *Practical Applications of Quantitative Structure-Activity Relationships (QSAR) in Environmental Chemistry and Toxicology*. Dordrecht, The Netherlands: Kluwer Academic Publishers.

Kaschula, C.H., Egan, T.J., Hunter, R., Basilico, N., Parapini, S., Taramelli, D., Pasini, E. & Monti, D. 2002. Structure-activity relationships in 4-aminoquinoline antiplasmodials. The role of the group at the 7-position. *J. Med. Chem*. 45: 3531–3539.

Kaur, K., Jain, M., Kaur, T. & Jain, R. 2009. Antimalarials from nature. *Bioorg. Med. Chem*. 17: 3229–3256. https://doi.org/10.1016/j.bmc.2009.02.050.

Keough, D.T., Ng, A.L., Winzor, D.J., Emmerson, B.T. & de Jersey, J. 1999. Purification and characterization of *Plasmodium falciparum* hypoxanthine-guanine-xanthine phosphoribosyl transferase and comparison with the human enzyme. *Mol. Biochem. Parasitol*. 98: 29–41. https://doi.org/10.1016/s0166-6851(98)00139-x.

Kinyanjui, S.M., Mberu, E.K., Winstanley, P.A., Jacobus, D.P. & Watkins, W.M. 1999. The antimalarial triazine WR99210 and the prodrug PS-15: Folate reversal of in vitro activity against *Plasmodium falciparum* and a non-antifolate mode of action of the prodrug. *Am. J. Trop. Med. Hyg*. 60: 943–947. https://doi.org/10.4269/ajtmh.1999.60.943.

Kirk, K. 2001. Membrane transport in the malaria-infected erythrocyte. *Physiol. Rev*. 81: 495–537. https://doi.org/10.1152/physrev.2001.81.2.495.

Kohler, S., Delwiche, C.F., Denny, P.W., Tilney, L.G., Webster, P., Wilson, R.J., Palmer, J.D. & Roos, D.S. 1997. A plastid of probable green algal origin in apicomplexan parasites. *Science*. 275: 1485–1489. https://doi.org/10.1126/science.275.5305.1485.

Krettli, A.U. & Miller, L.H. 2001. Malaria: A sporozoite runs through it. *Curr. Biol*. 11(10): R409–R412. https://doi.org/10.1016/s0960-9822(01)00221-4.

Krishnan, K.M. &Williamson, K.C. 2018. The proteasome as a target to combat malaria: Hits and misses. *Transl. Res*. 198: 40–47. https://doi.org/10.1016/j.trsl.2018.04.007.

Kublin, J.G., Cortese, J.F., Njunju, E.M., Mukadam, R.A.G., Wirima, J.J., Kazembe, P.N., Djimde, A.A., Kouriba, B., Taylor, T.E. & Plowe, C.V. 2003. Reemergence of chloroquine-sensitive *Plasmodium falciparum* malaria after cessation of chloroquine use in Malawi. *J. Infect. Dis*. 187: 1870–1875. https://doi.org/10.1086/375419.

Kublin, J.G., Dzinjalamala, F.K., Kamwendo, D.D., Malkin, E.M., Cortese, J.F., Martino, L.M., Mukadam, R.A., Rogerson, S.J., Lescano, A.G. & Molyneux, M.E. 2002. Molecular markers for failure of sulfadoxine-pyrimethamine and chlorproguanil-dapsone treatment of *Plasmodium falciparum* malaria. *J. Infect. Dis*. 185: 380–388. https://doi.org/10.1086/338566.

Laurens, M.B. 2018. The promise of a malaria vaccine-Are we closer? *Annu. Rev. Microbiol*. 72: 273–292.

Lefevre, G., Looareesuwan, S., Treeprasertsuk, S., Krudsood, S., Silachamroon, U., Gathmann, I., Mull, R. & Bakshi, R. 2001. A clinical and pharmacokinetic trial of six doses of artemether-lumefantrine for multidrug-resistant *Plasmodium falciparum* malaria in Thailand. *Am. J. Trop. Med. Hyg*. 64: 247–256. https://doi.org/10.4269/ajtmh.2001.64.247.

Lell, B., Faucher, J.F., Missinou, M.A., Borrmann, S., Dangelmaier, O., Horton, J. & Kremsner, P.G. 2000. Malaria chemoprophylaxis with tafenoquine: A randomised study. *Lancet*. 355: 2041–2045.

Lopez, G.R., Hernan, C., Moreno, L., Fernandez, E.V., Munoz, M.F., Delgado, A., Polo, M.J. & Andres, I. 2014. Malaria in developing countries. *J. Infect. Dev. Ctries*. 8: 1–4. https://doi.org/10.3855/jidc.4610.

Loyevsky, M. & Gordeuk, V.R. 2001. Iron chelators. In *Antimalarial Chemotherapy: Mechanisms of Action, Resistance, and New Directions in Drug Discovery* (ed. P.J. Rosenthal), pp. 307–324. Totowa, NJ: Humana Press.

Martin, S.K., Oduola, A.M. & Milhous, W.K. 1987. Reversal of chloroquine resistance in *Plasmodium falciparum* by verapamil. *Science*. 235: 899–901. https://doi.org/10.1126/science.3544220.

Meshnick, S.R. 2001. Artemisinin and its derivatives. In *Antimalarial Chemotherapy: Mechanisms of Action, Resistance, and New Directions in Drug Discovery* (ed. P.J. Rosenthal), pp. 191–201. Totowa, NJ: Humana Press.

Moon, R.P., Bur, D., Loetscher, H., D'Arcy, A., Tyas, L., Oefner, C., Grueninger-Leitch, F., Mona, D., Rupp, K. & Dorn, A. 1998. Studies on plasmepsins I and II from the malarial parasite *Plasmodium falciparum* and their exploitation as drug targets. *Adv. Exp. Med. Biol.* 436: 397–406. https://doi.org/10.1007/978-1-4615-5373-1_56.

Mutabingwa, T., Nzila, A., Mberu, E., Nduati, E., Winstanley, P., Hills, E. & Watkins, W. 2001. Chlorproguanil-dapsone for treatment of drug-resistant falciparum malaria in Tanzania. *Lancet*. 358: 1218–1223. https://doi.org/10.1016/S0140-6736(01)06344-9.

Newman, D.J., Cragg, G.M. & Snader, K.M. 2003. Natural products as sources of new drugs over the period 1981–2002. *J. Nat. Prod.* 66: 1022–1037. https://doi.org/10.1021/np030096l.

Nezami, A., Luque, I., Kimura, T., Kiso, Y. & Freire, E. 2002. Identification and characterization of allophenylnorstatine-based inhibitors of plasmepsin II, an antimalarial target. *Biochemistry*. 41: 2273–2280. https://doi.org/10.1021/bi0117549.

NIAID. 2007. Understanding malaria. *Fighting an Ancient Scourge*, No. 07–7139, U.S. Dept. of Health and Human Services, National Institutes of Health, National Institute of Allergy and Infectious Diseases, Bethesda, MD.

Noteberg, D., Hamelink, E., Hulten, J., Wahlgren, M., Vrang, L., Samuelsson, B. & Hallberg, A. 2003. Design and synthesis of plasmepsin I and plasmepsin II inhibitors with activity in *Plasmodium falciparum*-infected cultured human erythrocytes. *J. Med. Chem.* 46: 734–746. https://doi.org/10.1021/jm020951i.

Nussenzweig, R.S., Vanderberg, J., Most, H. & Orton, C. 1967. *Protective immunity produced by the injection of x-irradiated sporozoites of* Plasmodium berghei. *Nature*. 216: 160–162. https://doi.org/10.1038/216160a0.

Nzila, A.M., Nduati, E., Mberu, E.K., Hopkins Sibley, C., Monks, S.A., Winstanley, P.A. & Watkins, W.M. 2000. Molecular evidence of greater selective pressure for drug resistance exerted by the long-acting antifolate Pyrimethamine/Sulfadoxine compared with the shorter-acting chlorproguanil/dapsone on Kenyan *Plasmodium falciparum*. *J. Infect. Dis.* 181: 2023–2028. https://doi.org/10.1086/315520.

Ohkanda, J., Lockman, J.W., Yokoyama, K., Gelb, M.H., Croft, S.L., Kendrick, H., Harrell, M.I., Feagin, J.E., Blaskovich, M.A. & Sebti, S.M. 2001. Peptidomimetic inhibitors of protein farnesyltransferase show potent antimalarial activity. *Bioorg. Med. Chem. Lett.* 11: 761–764. https://doi.org/10.1016/s0960-894x(01)00055-5.

Oliveira, A.B., Dolabela, M.F., Braga, F.C., Jácome, R.L.R.P., Varotti, F.P. & Póvoa, M.M. 2009. Plant-derived antimalarial agents: New leads and efficient phytomedicines. Part I. Alkaloids. *Ann. Braz. Acad. Sci.* 81(4). https://doi.org/10.1590/s0001-37652009000400011.

O'Neill, K.M., Larsen, J.S. & Curtis, W.R. 2008. Scale-up of Agrobacterium-mediated transient protein expression in bioreactor-grown Nicotiana glutinous a plant cell suspension culture. *Biotechnol. Prog.* 2: 372–376.

Onguéné, P.A., NtieKang, F., Mbah, J., Lifongo, A.L.L., Ndom, J.C., Sippl, W. & Mbaze, L.M. 2014. The potential of anti-malarial compounds derived from African medicinal plants, part III: An in silico evaluation of drug metabolism and pharmacokinetics profiling. *Org. Med. Chem. Lett.* 4: 6. https://doi.org/10.1186/s13588-014-0006-x.

Pai-Dhungat, J.V. & Caventou, Pelletier. 2015. History of quinine. *J. Assoc. Phys. India*. 63: 58.

Peyret, H. & Lomonossoff, G.P. 2015. When plant virology met Agrobacterium: The rise of the deconstructed clones. *Plant Biotechnol. J.* 13: 1121–1135.

Phoolcharoen, W., Bhoo, S.H., Lai, H., Ma, J., Arntzen, C.J., Chen, Q. & Mason, H.S. 2011. Expression of an immuno-genic Ebola immune complex in Nicotiana benthamiana. *Plant Biotechnol. J.* 9: 807–816.

Plowe, C.V. 2001. Folate antagonists and mechanisms of resistance. In *Antimalarial Chemotherapy: Mechanisms of Action, Resistance, and New Directions in Drug Discovery* (ed. P.J. Rosenthal), pp. 173–190. Totowa, NJ: Humana Press.

Poilane, I., Jeantils, V. & Carbillon, L. 2009. Découvertefortuite de paludisme à Plasmodium falciparum au cours de la grossesse: à propos de deux cas. *Gynécol. Obstét. Fertil.* 37(10): 824–826.

Portet, B., Fabre, N., Roumy, V., Gornitzka, H., Bourdy, G., Chevalley, S., Sauvain, M., Valentin, A. & Moulis, C. 2007. Activity-guided isolation of antiplasmodial dihydrochalcones and flavanones from *Piperhostmannianum* var. *berbicense*. *Phytochemistry*. 68: 1312–1320. https://doi.org/10.1016/j.phytochem.2007.02.006.

Posner, G.H., Paik, I.H., Sur, S., McRiner, A.J., Borstnik, K., Xie, S. & Shapiro, T.A. 2003. Orally active, antimalarial, anticancer, artemisinin-derived trioxane dimers with high stability and efficacy. *J. Med. Chem.* 46: 1060–1065. https://doi.org/10.1021/jm020461q.

Price, R.N., Nosten, F., Luxemburger, C., Van Vugt, M., Phaipun, L., Chongsuphajaisiddhi, T. & White, N.J. 1997. Artesunate/mefloquine treatment of multi-drug resistant falciparum malaria. *Trans. R. Soc. Trop. Med. Hyg.* 91: 574–577. https://doi.org/10.1016/s0035-9203(97)90032-8.

Prudencio, M., Rodriguez, A. & Mota, M.M. 2006. The silent path to thousands of merozoites: The Plasmodium liver stage. *Nat. Rev. Microbiol.* 4: 849–856. https://doi.org/10.1038/nrmicro1529.

Qayum, A., Arya, R. & Lynn, A.M. 2016. Ethnobotanical perspective of antimalarial plants: Traditional knowledge based study. *BMC Res. Notes.* 9: 67. https://doi.org/10.1186/s13104-015-1827-z.

Radloff, P.D., Philipps, J., Nkeyi, M., Hutchinson, D. & Kremsner, P.G. 1996. Atovaquone and proguanil for *Plasmodium falciparum* malaria. *Lancet.* 347: 1511–1514. https://doi.org/10.1016/s0140-6736(96)90671-6.

Ralph, S.A., D'Ombrain, M.C. & McFadden, G.I. 2001. The apicoplast as an antimalarial drug target. *Drug Resist. Update.* 4: 145–151. https://doi.org/10.1054/drup.2001.0205.

Rathod, P.K., Leffers, N.P. & Young, R.D. 1992. Molecular targets of 5-fluoroorotate in the human malaria parasite, *Plasmodium falciparum. Antimicrob. Agents Chemother.* 36: 704–711. https://doi.org/10.1128/AAC.36.4.704.

Raynes, K.J., Stocks, P.A., O'Neill, P.M., Park, B.K. &Ward, S.A. 1999. New 4-aminoquinoline Mannich base antimalarials. 1. Effect of an alkyl substituent in the 5′-position of the 4′-hydroxyanilino side chain. *J. Med. Chem.* 42: 2747–2751.

Razakantoanina, V., Nguyen Kim, P.P. & Jaureguiberry, G. 2000. Antimalarial activity of new gossypol derivatives. *Parasitol. Res.* 86: 665–668. https://doi.org/10.1007/pl00008549.

Ricardo, T., Parisa, K., Katherine, A. & Douglas, G. 2014. Plasmodium life cycle and the pathogenesis of malaria. From innate sensing of malaria parasites. *Nat. Rev. Immunol.* 14: 744–757. https://doi.org/10.1038/nri3742.

Riley, E.M. & Stewart, V.A. 2013. Immune mechanisms in malaria: New insights in vaccine development. *Nat. Med.* 19: 168.

Ringwald, P., Bickii, J. & Basco, L. 1996. Randomised trial of pyronaridine versus chloroquine for acute uncomplicated falciparum malaria in Africa. *Lancet.* 347: 24–28.

Roos, D.S., Crawford, M.J., Donald, R.G., Fraunholz, M., Harb, O.S., He, C.Y., Kissinger, J.C., Shaw, M.K. & Striepen, B. 2002. Mining the Plasmodium genome database to define organellar function: What does the apicoplast do? *Philos. Trans. R. Soc. Lond. B. Biol. Sci.* 357: 35–46. https://doi.org/10.1098/rstb.2001.1047.

Rosenthal, P.J. 2001. Protease inhibitors. In *Antimalarial Chemotherapy: Mechanisms of Action, Resistance, and New Directions in Drug Discovery* (ed. P.J. Rosenthal), pp. 325–345. Totowa, NJ: Humana Press.

Rosenthal, P.J., Lee, G.K. & Smith, R.E. 1993. Inhibition of a *Plasmodium vinckei* cysteine proteinase cures murine malaria. *J. Clin. Invest.* 91: 1052–1056. https://doi.org/10.1172/JCI116262.

Rosenthal, P.J., Olson, J.E., Lee, G.K., Palmer, J.T., Klaus, J.L. & Rasnick, D. 1996. Antimalarial effects of vinyl sulfone cysteine proteinase inhibitors. *Antimicrob. Agents Chemother.* 40: 1600–1603. https://doi.org/10.1128/AAC.40.7.1600.

Rosenthal, P.J., Sijwali, P.S., Singh, A. & Shenai, B.R. 2002. Cysteine proteases of malaria parasites: Targets for chemotherapy. *Curr. Pharm. Des.* 8: 1659–1672. https://doi.org/10.2174/1381612023394197.

Rosenthal, P.J., Wollish, W.S., Palmer, J.T. & Rasnick, D. 1991. Antimalarial effects of peptide inhibitors of a *Plasmodium falciparum* cysteine proteinase. *J. Clin. Invest.* 88: 1467–1472. https://doi.org/10.1172/JCI115456.

Sachs, J. & Malaney, P. 2002. The economic and social burden of malaria. *Nature.* 415: 680–685. https://doi.org/10.1038/415680a.

Saxena, S., Pant, N., Jain, D.C. & Bhakuni, R.S. 2003. Antimalarial agents from plant sources. *Curr. Sci.* 85: 1314–1329.

Schellenberg, D., Kahigwa, E., Drakeley, C., Malende, A., Wigayi, J., Msokame, C., Aponte, J.J., Tanner, M., Mshinda, H. & Menendez, C. 2002. The safety and efficacy of sulfadoxine-pyrimethamine, amodiaquine, and their combination in the treatment of uncomplicated *Plasmodium falciparum* malaria. *Am. J. Trop. Med. Hyg.* 67: 17–23. https://doi.org/10.4269/ajtmh.2002.67.17.

Semenov, A., Olson, J.E. & Rosenthal, P.J. 1998. Antimalarial synergy of cysteine and aspartic protease inhibitors. *Antimicrob. Agents Chemother.* 42: 2254–2258. https://doi.org/10.1128/AAC.42.9.2254.

Shenai, B.R., Lee, B.J., Alvarez-Hernandez, A., Chong, P.Y., Emal, C.D., Neitz, R.J., Roush, W.R. & Rosenthal, P.J. 2003. Structure-activity relationships for inhibition of cysteine protease activity and development of *Plasmodium falciparum* by peptidyl vinyl sulfones. *Antimicrob Agents Chemother.* 47(1). https://doi.org/10.1128/AAC.47.1.154-160.

Shenai, B.R., Sijwali, P.S., Singh, A. & Rosenthal, P.J. 2000. Characterization of native and recombinant falcipain-2, a principal trophozoite cysteine protease and essential hemoglobinase of *Plasmodium falciparum*. *J. Biol. Chem.* 275: 29000–29010. https://doi.org/10.1074/jbc.M004459200.

Sijwali, P.S., Shenai, B.R., Gut, J., Singh, A. & Rosenthal, P.J. 2001. Expression and characterization of the *Plasmodium falciparum* haemoglobinase falcipain-3. *Biochem. J.* 360: 481–489. https://doi.org/10.1042/0264-6021:3600481.

Silva, A.M., Lee, A.Y., Gulnik, S.V., Maier, P., Collins, J., Bhat, T.N., Collins, P.J., Cachau, R.E., Luker, K.E. & Gluzman, I.Y. 1996. Structure and inhibition of plasmepsin II, a hemoglobin-degrading enzyme from *Plasmodium falciparum*. *Proc. Natl. Acad. Sci. USA.* 93: 10034–10039. https://doi.org/10.1073/pnas.93.19.10034.

Singh, A. & Rosenthal, P.J. 2001. Comparison of efficacies of cysteine protease inhibitors against five strains of *Plasmodium falciparum*. *Antimicrob. Agents Chemother.* 45: 949–951. https://doi.org/10.1128/AAC.45.3.949-951.2001.

Sowunmi, A., Oduola, A.M., Ogundahunsi, O.A., Falade, C.O., Gbotosho, G.O. & Salako, L.A. 1997. Enhanced efficacy of chloroquine-chlorpheniramine combination in acute uncomplicated falciparum malaria in children. *Trans. R. Soc. Trop. Med. Hyg.* 91: 63–67. https://doi.org/10.1016/s0035-9203(97)90399-0.

Spiegel, H., Boes, A., Voepel, N., Beiss, V., Edgue, G., Rademacher, T., Sack, M., Schillberg, S., Reimann, A. & Fischer, R. 2015. Application of a scalable plant transient gene expression platform for malaria vaccine development. *Front. Plant Sci.* 6: 1169.

Sriraman, R., Bardor, M., Sack, M., Vaquero, C., Faye, L., Fischer, R., Finnern, R. & Lerouge, P. (2004). Recombinant anti-hCG antibodies retained in the endoplasmic reticulum of transformed plants lack core-xylose and core-α (1, 3)-fucose residues. *Plant Biotechnol. J.* 2: 279–287.

Staedke, S.G., Kamya, M.R., Dorsey, G., Gasasira, A., Ndeezi, G., Charlebois, E.D. & Rosenthal, P.J. 2001. Amodiaquine, sulfadoxine/pyrimethamine, and combination therapy for treatment of uncomplicated falciparum malaria in Kampala, Uganda: A randomised trial. *Lancet.* 358: 368–374. https://doi.org/10.1016/S0140-6736(01)05557-X.

Stocks, P.A., Raynes, K.J., Bray, P.G., Park, B.K., O'Neill, P.M. & Ward, S.A. 2002. Novel short chain chloroquine analogues retain activity against chloroquine resistant K1 *Plasmodium falciparum*. *J. Med. Chem.* 45: 4975–4983.

Stocks, P.A., Raynes, K.J. & Ward, S.A. 2001. Novel quinoline antimalarials. In *Antimalarial Chemotherapy: Mechanisms of Action, Resistance, and New Directions in Drug Discovery* (ed. P.J. Rosenthal), pp. 235–253. Totowa, NJ: Humana Press.

Stoger, E., Fischer, R., Moloney, M. & Ma, J.K.C. 2014. Plant molecular pharming for the treatment of chronic and infectious diseases. *Ann. Rev. Plant Biol.* 65: 743–768.

Sullivan, D.J. 2002 Theories on malarial pigment formation and quinoline action. *Int. J. Parasitol.* 32(13): 1645–1653. https://doi.org/10.1016/s0020-7519(02)00193-5.

Surolia, N. & Surolia, A. 2001. Triclosan offers protection against blood stages of malaria by inhibiting enoyl-ACP reductase of *Plasmodium falciparum*. *Nat. Med.* 7: 167–173. https://doi.org/10.1038/84612.

Szmitko, P., Kohn, M.L. & Simor, A.E. 2008. Plasmodium falciparum malaria occurring eight years after leaving an endemic area. *Diagn. Microbiol. Infect. Dis.* 61(1): 105–107. https://doi.org/10.1016/j.diagmicrobio.2008.08.017.

Tagboto, S. & Townson, S. 2001. Antiparasitic properties of medicinal plants and other naturally occurring products. *Adv. Parasitol.* 50: 199–295. https://doi.org/10.1016/s0065-308x(01)50032-9.

Tham, W.H., Beeson, J.G. &Rayner, J.C. 2017. Plasmodium vivax vaccine research – We've only just begun. *Int. J. Parasitol.* 47: 111–118. https://doi.org/10.1016/j.ijpara.2016.09.006.

Vaidya, A.B. 2001. Atovaquone-proguanil combination. In *Antimalarial Chemotherapy: Mechanisms of Action, Resistance, and New Directions in Drug Discovery* (ed. P.J. Rosenthal), pp. 203–218. Totowa, NJ: Humana Press.

Van Schalkwyk, D.A., Walden, J.C. & Smith, P.J. 2001. Reversal of chloroquine resistance in *Plasmodium falciparum* using combinations of chemosensitizers. *Antimicrob. Agents Chemother.* 45: 3171–3174. https://doi.org/10.1128/AAC.45.11.3171-3174.2001.

Van Vugt, M., Leonardi, E., Phaipun, L., Slight, T., Thway, K.L., McGready, R., Brockman, A., Villegas, L., Looareesuwan, S., White, N.J. & Nosten, F. 2002. Treatment of uncomplicated multidrug-resistant falciparum malaria with artesunate-atovaquone-proguanil. Clin. Infect. *Dis.* 35: 1498–1504.

Vennerstrom, J.L., Dong, Y., Andersen, S.L., Ager, A.L., Jr, Fu, H., Miller, R.E., Wesche, D.L., Kyle, D.E., Gerena, L. & Walters, S.M. 2000. Synthesis and antimalarial activity of sixteen dispiro-1,2,4, 5-tetraoxanes: alkyl-substituted 7,8,15,16-tetraoxadispiro[5.2.5. 2]hexadecanes. *J. Med. Chem.* 43: 2753–2758.

Vial, H.J. & Calas, M. 2001. Inhibitors of phospholipid metabolism. In *Antimalarial Chemotherapy: Mechanisms of Action, Resistance, and New Directions in Drug Discovery* (ed. P.J. Rosenthal), pp. 347–365. Totowa, NJ: Humana Press.

Viegas, Júnior C., Bolzani, V.S. & Barreiro, E.J. 2006. Osprodutosnaturais e a química medicinal moderna. *Quím. Nova.* 29: 326–337.https://doi.org/10.1590/S0100-40422006000200025.

Walker, N.F., Nadjm, B. & Whitty, C.J.M. 2010. Malaria. *Medicine* 38(1): 41–46.

Waller, R.F., Keeling, P.J., Donald, R.G., Striepen, B., Handman, E., Lang-Unnasch, N., Cowman, A.F., Besra, G.S., Roos, D.S. & McFadden, G.I. 1998. Nuclear-encoded proteins target to the plastid in *Toxoplasma gondii* and *Plasmodium falciparum*. *Proc. Natl. Acad. Sci. USA*. 95: 12352–12357. https://doi.org/10.1073/pnas.95.21.12352.

Walsh, D.S., Looareesuwan, S., Wilairatana, P., Heppner, D.G., Jr, Tang, D.B., Brewer, T.G., Chokejindachai, W., Viriyavejakul, P., Kyle, D.E. & Milhous, W.K. 1999. Randomized dose-ranging study of the safety and efficacy of WR 238605 (Tafenoquine) in the prevention of relapse of Plasmodium vivax malaria in Thailand. *J. Infect. Dis.* 180: 1282–1287.

Webster, D.E., Wang, L., Mulcair, M., Ma, C., Santi, L., Mason, H.S., Wesselingh, S.L. & Coppel, R.L. 2009. Production and characterization of an orally immunogenic Plasmodium antigen in plants using a virus-based expression system. *Plant Biotechnol. J.* 7: 846–855.

Wells, M.B. & Andrew, D.J. 2015. Salivary gland cellular architecture in the Asian malaria vector mosquito Anopheles stephensi. *Parasit. Vect.* 8: 617. https://doi.org/10.1186/s13071-015-1229-z.

Wengelnik, K., Vidal, V., Ancelin, M.L., Cathiard, A.M., Morgat, J.L., Kocken, C.H., Calas, M., Herrera, S., Thomas, A.W. & Vial, H.J. 2002. A class of potent antimalarials and their specific accumulation in infected erythrocytes. *Science*. 295: 1311–1314. https://doi.org/10.1126/science.1067236.

WHO. 2008. *World Malaria Report*. Geneva, Switzerland: World Health Organization. Available at: http://apps.who.int/malaria/wmr2008/.

WHO. 2013. *Training Module on Malaria Control Malaria Entomology and Vector Control. Guide for Participants*. Geneva, Switzerland: WHO.

WHO. 2015. *Guidelines for the Treatment of Malaria*, 3rd ed.; World Health Organization: Geneva, Switzerland.

WHO. 2017. *World Malaria Report 2017*. Geneva, Switzerland: WHO.

WHO. 2019. *Malaria vaccine pilot launched in Malawi*. Available at: www.who.int/news-room/detail/23-04-2019-malaria-vaccine-pilot-launched-in-malawi.

Wichmann, O., Muhlen, M., Grub, H., Mockenhaupt, F.P., Suttorp, N. & Jelinek, T. 2004. Malarone treatment failure not associated with previously described mutations in the cytochrome *b* gene. *Malaria J.* 3: 1–3. https://doi.org/10.1186/1475-2875-3-14.

Winter, R.W., Kelly, J.X., Smilkstein, M.J., Dodean, R., Bagby, G.C., Rathbun, R.K., Levin, J.I., Hinrichs, D. & Riscoe, M.K. 2006. Evaluation and lead optimization of anti-malarial acridones. *Exp. Parasitol.* 114(1): 47–56. https://doi.org/10.1016/j.exppara.2006.03.014.

Yuan, J., Cheng, K.C.C., Johnson, R.L., Huang, R., Pattaradilokrat, S., Liu, A., Guha, R., Fidock, D., Inglese, J., Wellems, T.E., Austin, C.P. & Su, X.Z. 2011. Chemical genomic profiling for antimalarial therapies, response signatures and molecular targets. *Science*. 333: 724–729. https://doi.org/10.1126/science.1205216.

Ziegler, H.L., Staerk, D., Christensen, J., Hviid, L., Hagerstr, H. & Jaroszewski, J.W. 2002. *In vitro Plasmodium falciparum* drug sensitivity assay: Inhibition of parasite growth by incorporation of stomatocytogenic amphiphiles into the erythrocyte membrane. *Antimicrob. Agents Chemother.* 46: 1441–1446. https://doi.org/10.1128/AAC.46.5.1441-1446.2002.

3 Plant Tissue Culture Practices for Antimalarial Agents

Arvind Arya, Sandeep Kumar, Pankaj Kumar Tyagi, and Azamal Husen

3.1 INTRODUCTION

Plant tissue culture is a laboratory technique used for growing and manipulating plant cells, tissues, or organs in an artificial environment under sterile conditions. Plant tissue culture is used for a variety of purposes, including the mass propagation of plants, the study of plant growth and development, and the production of genetically modified plants (Ammirato et al., 1984). Plant tissue culture is primarily used in agriculture, horticulture, and forestry to propagate plants (Brown & Thorpe, 1995; Husen, 2022; Arya & Husen, 2022; Arya et al., 2022a, 2022b). However, it has also been found to have potential applications in the field of medicine. Specifically, plant tissue culture has been explored for its ability to produce secondary metabolites and proteins that have potential use in treating human diseases (Garcia, 2020; Roy, 2020; Ramakrishna et al., 2021).

One example of this is the use of plant tissue culture to produce human insulin. In addition to insulin, plant tissue culture has also been explored for its potential in producing other therapeutic proteins, such as antibodies, vaccines, and growth factors (Fischer & Emans, 2000; Schillberg, Fischer & Emans, 2003; Knödler & Buyel, 2021). Some of these proteins have shown promise in the treatment of cancer, autoimmune diseases, and other conditions (Garcia, 2020; Ramakrishna et al., 2021). Over the years, the plant tissue culture has shown potential to revolutionize the way we treat human diseases. In a similar way, plant tissue culture has been used to find novel antimalarial drugs.

The present discourse entails an in-depth analysis of the malaria ailment, encompassing its pathogenicity, morbidity, and socioeconomic impact. Further, the discussion delves into the pharmacological interventions offered by antimalarial compounds and plant-based drugs and highlights the utilization of plant tissue culture technology in the synthesis of antimalarial agents.

3.2 MALARIA

Malaria is a parasitic disease caused by the *Plasmodium* parasite and transmitted to humans through the bite of infected *Anopheles* mosquitoes. The following are some key points about malaria. Malaria symptoms include fever, chills, headache, muscle pain, and anaemia. In severe cases, it can lead to life-threatening complications, such as cerebral malaria, respiratory distress, and organ failure. Malaria is a major global health problem, particularly in sub-Saharan Africa, where most of the cases and deaths occur. According to the World Health Organization (WHO), there were 229 million reported cases of malaria and 409,000 deaths in 2019. Risk factors for malaria include living or traveling to endemic areas, poor housing conditions, and a weakened immune system. Children under the age of 5 and pregnant women are particularly vulnerable. Malaria is diagnosed through blood tests and can be treated with antimalarial drugs. Prompt and effective treatment is crucial to prevent severe disease and death.

 DOI: 10.1201/9781003378396-3

There are several ways to prevent malaria, including using insecticide-treated bed nets, wearing protective clothing, and taking preventive antimalarial drugs for high-risk individuals (Hamre et al., 2020; Shah et al., 2020). The global effort to control and eliminate malaria involves a combination of measures, including vector control, prompt diagnosis and treatment, and efforts to increase access to malaria prevention and care (Ashley & Poespoprodjo, 2020; Mohammadi et al., 2020).

Malaria remains a significant global health problem, particularly in sub-Saharan Africa, where the burden of disease is highest (Stratton et al., 2008; Chico et al., 2015). Despite progress in reducing malaria cases and deaths in recent years, the disease continues to pose a serious threat to public health and development.

According to WHO's latest world malaria report, there were an estimated 241 million malaria cases and 627,000 malaria deaths worldwide in 2020. This represents about 14 million more cases in 2020 compared to 2019, and 69,000 more deaths (World Malaria Report, 2021). Approximately two-thirds of these additional deaths (47,000) were linked to disruptions in the provision of malaria prevention, diagnosis, and treatment during the pandemic (Weiss et al., 2021). Most cases and deaths occur in sub-Saharan Africa, where young children and pregnant women are particularly vulnerable. Malaria can also have a significant economic impact, as it can reduce productivity and increase healthcare costs (Sarma et al., 2019). In addition, the disease can contribute to poverty and perpetuate the cycle of poverty and illness.

3.2.1 Economic Effect of Malaria

The economic effect of malaria is substantial and affects both individual families and entire countries. Malaria treatment and management can be expensive, particularly in low- and middle-income countries, where access to healthcare is limited. The cost of treating severe cases of malaria can be particularly high and can strain already-limited health budgets (Ogutu, 2009; Nahrendorf, Ivens & Spence, 2021). Malaria can cause individuals to miss work or school, leading to lost productivity and income. This can have a significant impact on families and communities, especially in areas where most of the population relies on manual labour for their livelihood (Booth & MacLean, 2020; Jochum et al., 2021). Malaria can also slow economic growth by reducing the productivity of workers and deterring foreign investment. For example, the disease can reduce agricultural productivity, limit access to education, and increase healthcare costs. In areas where malaria is endemic, the disease can also negatively impact tourism, as travellers may avoid visiting due to fear of contracting the disease. The COVID-19 pandemic has also had an impact on malaria control efforts, as health systems have been stretched and access to preventive measures and treatment has been disrupted in some areas (Chiang & Sony, 2020; Colson, Rolain & Raoult, 2020; Zawawi et al., 2020). The COVID-19 pandemic has also led to a reduction in funding for malaria control, as resources are diverted to respond to the pandemic. The disruption of malaria control activities and increased movement of people due to the pandemic could lead to increased malaria transmission in some areas.

3.2.2 Malaria in Developing vs Developed Countries

The burden of malaria differs between developing and developed countries. In developing countries, particularly in sub-Saharan Africa, malaria is a major public health problem. Most malaria cases and deaths occur in these countries, where access to preventive measures, diagnosis, and treatment is often limited. In developed countries, malaria is rare and can often be effectively treated. However, travellers to malaria-endemic areas can still be at risk of contracting the disease, and imported cases can occur. In developing countries, access to preventive measures such as insecticide-treated bed nets and preventive antimalarial drugs is often limited (Tatem et al., 2010). In addition, access to prompt and effective treatment for malaria can be limited, leading to higher rates of severe disease and death.

The economic impact of malaria is often more severe in developing countries, where it can reduce productivity, increase healthcare costs, and perpetuate the cycle of poverty and illness (Talisuna et al., 2002; Millet et al., 2009).

3.2.3 About Antimalarial Agents

Antimalarial agents are drugs used to treat malaria, a parasitic disease that affects millions of people worldwide (Christensen, 2021). Antimalarial agents work by either killing the parasite or inhibiting its growth and replication (Sahu, Sahu & Kohli, 2008; Belete, 2020). There are several types of antimalarial drugs, including blood schizonticides, which kill the parasite in the blood, and gametocides, which target the sexual stage of the parasite's life cycle. The efficacy of antimalarial drugs depends on factors, such as the type of parasite, the patient's age, and the presence of other infections (Rosenthal, 2003). Malaria parasites can develop resistance to antimalarial drugs, reducing their effectiveness. This is why it is important to use antimalarial drugs in combination and to rotate the use of different drugs over time. Antimalarial drugs are often not available or affordable in endemic regions, especially in low- and middle-income countries. This can limit access to treatment and contribute to the spread of drug-resistant malaria.

3.2.4 Drug-Resistant Malaria

Drug-resistant malaria refers to the phenomenon where the parasites that cause malaria have developed resistance to antimalarial drugs that were previously effective in treating the disease (Hyde, 2005). This can make it more difficult to effectively treat and control malaria, leading to increased rates of illness and death. When antimalarial drugs are overused or misused, such as when they are not taken in the correct dose or for the full course of treatment, this can lead to the selection of drug-resistant parasites. Inadequate quality or counterfeit antimalarial drugs can also contribute to the development of drug-resistant malaria, as they may not contain the active ingredients needed to effectively treat the disease. In areas where access to effective treatment is limited, people may turn to less-effective antimalarial drugs, which can also contribute to the development of drug-resistant malaria.

3.2.5 Antimalarial Plants

In 2018, Suleman and colleagues conducted a review of 124 medicinal plant species from 55 families and 99 genera in Ethiopia to evaluate their effectiveness against malaria from 1991 to 2016 (Suleman et al., 2018). Additionally, the study analyzed the antiplasmodial activity of 25 plant extracts, along with the structures, and 28 medicinal plant species from 24 families commonly used for treating malaria. The review identified *Allium sativum* L., *Croton macrostachyus* Del., *Carica papaya* L., *Lepidium sativum* L., and *Vernonia amygdalina* Del. as notable plant species based on ethnomedicinal surveys, with *A. sativum* L. and *C. papaya* L. specifically mentioned by traditional healers as effective treatments for malaria. Leaves were found to be the most used plant part, while alkaloids and terpenoids were the most frequently identified chemical classes for their antiplasmodial properties.

A review conducted by Kaur and Stud (2017) has summarized 356 active chemical constituents from antimalarial plants, highlighting 26 specific active components with their respective chemical structures. The compounds identified in this review belong to various classes, including but not limited to alkaloids, flavonoids, terpenoids, saponins, tannins, essential oils, sesquiterpene lactones, xanthanoids, polyphenol, iridoid glycosides, coumarins, anthraquinones, limonoids, tetranortriterpenoids, triketones, cembranolides, tetrahydropyrone derivatives, triterpene hydrocarbon derivatives, cucurbitacin glycosides, poliothrysoside, benzophenones, chalcones, and phloroglucinols.

3.2.6 Applications of Plant Tissue Culture in Antimalarial Agents

Plant tissue culture has been utilized in the search for new antimalarial agents. Some of the practices used in this regard include:

Callus induction and regeneration. This technique is used to produce large quantities of callus, a type of plant tissue, which can then be screened for antimalarial activity. Screening of callus culture for antimalarial drugs involves testing callus tissue, which is a type of plant tissue culture, for its ability to produce compounds with antimalarial activity. This can be done in a laboratory setting using *in vitro* cultures of plant cells.

The process typically involves the following steps, as shown in Figure 3.1.

Screening of callus culture for antimalarial drugs is a useful approach for discovering new antimalarial compounds and advancing the development of new treatments for malaria. Some examples of plant secondary metabolites with antimalarial activity include the following.

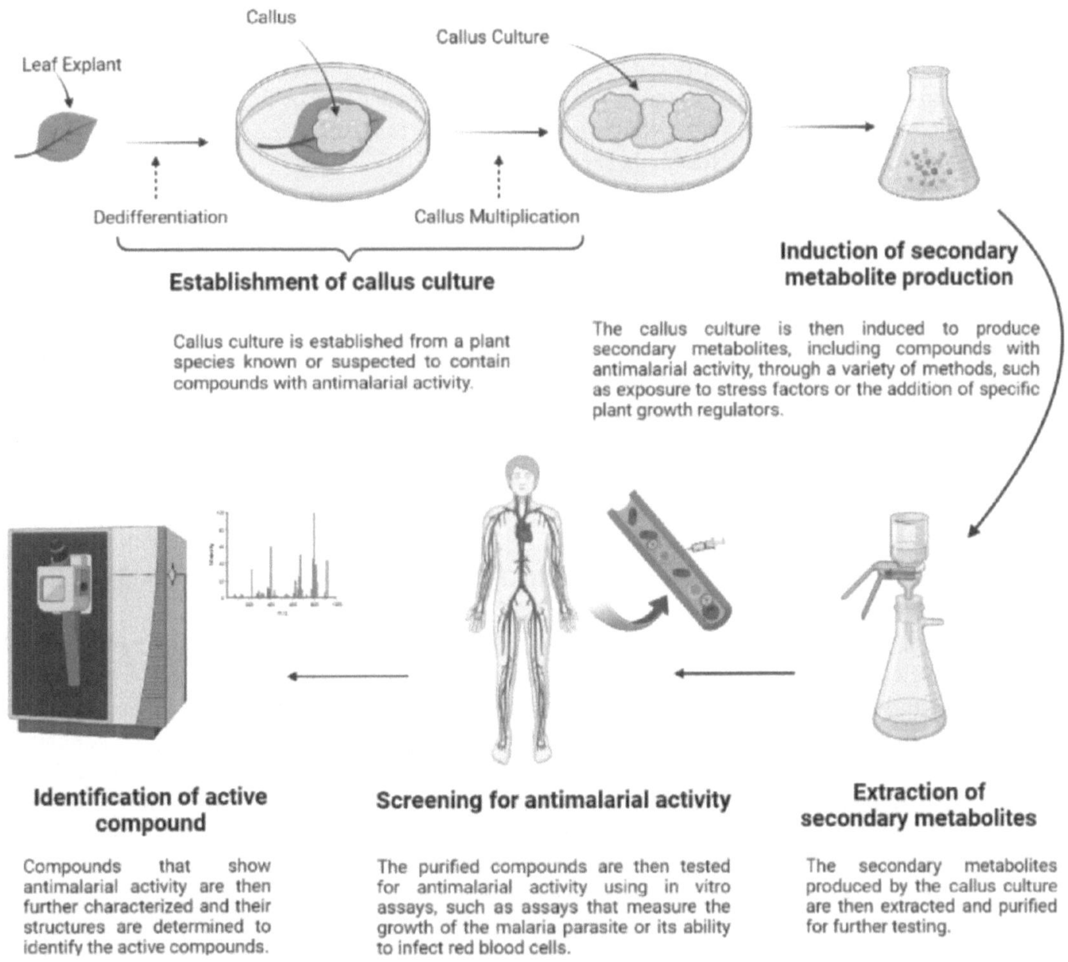

FIGURE 3.1 Steps involved in the production of drug compounds from plants using tissue culture methods.

3.2.7 Artemisinin

This compound, which is derived from the sweet wormwood plant, is used as a first-line treatment for malaria in many countries. It has been shown to be highly effective against the malaria parasite and is used in combination with other antimalarial drugs to reduce the risk of drug resistance (Klayman, 1985). Artemisinin is a natural compound derived from the sweet wormwood plant (*Artemisia annua*) that is used as a treatment for malaria.

Artemisinin works by attacking the malaria parasite at a critical stage in its life cycle, rapidly reducing the parasite burden in the patient and leading to a rapid improvement in their symptoms (O'Neill, Barton & Ward, 2010). It is typically used in combination with other antimalarial drugs to reduce the risk of the development of drug resistance (Balint, 2001).

Artemisinin is also being studied for its potential in the treatment of other diseases, including cancer and autoimmune disorders (Lu et al., 2019).

3.2.7.1 Production of Artemisinin through Plant Tissue Culture

Plant cell cultures have the potential to produce valuable pharmaceuticals and other biologically active compounds, but many cultures do not synthesize secondary compounds in amounts comparable to those found in the whole plant. Manipulation of culture conditions and phytohormone levels has not yet allowed for commercial production of useful phytochemicals. However, microbial molecules called 'elicitors' can stimulate synthesis of secondary products in cultured cells, leading to immense biotechnological potential for large-scale production of these compounds. In one research, DiCosmo and Misawa (1985) highlighted the importance of 'elicitor' interactions in plant cell culture and suggested areas for further research.

Various techniques have been explored for artemisinin production in *A. annua*, including plant tissue culture using hairy root or cell suspension cultures (Nair et al., 1986; Baldi & Dixit, 2008). Artemisinin can be produced through plant tissue culture techniques, such as callus culture and micropropagation. These techniques involve growing plant cells *in vitro* in a controlled environment, where they can be manipulated to produce high levels of secondary metabolites, including artemisinin. Elhag et al. (1992) developed an *in vitro* micropropagation method to clone *Artemisia annua* L. plants from axenic seedlings or field-grown plants, and clones were maintained in culture through continuous transfer. The clones were compared to field-grown plants for their performance and contents of artemisinin and related sesquiterpenes using an HPLC method. Artemisinic acid was found to be the major sesquiterpene for all clones, followed by arteannuin B and then artemisinin. However, there was a high degree of variability between various clones for all sesquiterpenes tested. Unorganized callus cultures initiated from the leaves of the corresponding clones did not indicate the formation of artemisinin or related sesquiterpenes in these cultures.

Growth conditions such as different sugar supply, light irradiation, UV-B radiation, and chilling treatment have been manipulated to increase artemisinin production (Pras et al., 1991; Liu, Zhao & Wang, 2006; Baldi & Dixit, 2008). Somaclonal variants have also been generated through gamma rays irradiation, resulting in a stable salt-tolerant line with a higher expression profile of artemisinin key genes and artemisinin content compared to the wild type (Liu et al., 2002; Baldi & Dixit, 2008; Han et al., 2016). Additionally, treatments with elicitors such as methyl jasmonate, chitosan, gibberellic acid, and salicylic acid have shown to increase artemisinin production and GT index (Banyai, Kirdmanee et al., 2010; Banyai, Sangthong et al., 2010; Guo et al., 2010).

The artemisinin content in plants grown in the open is too low for industrial purposes. Pras et al. (1991) developed a new method for laboratory selection of high-yielding individual plants using thin-layer chromatography and high-performance liquid chromatography for analysis. Selected plantlets were transferred to the open, and high-yielding plants in the laboratory were observed to be so in the field. The artemisinin content of the selected plants was three to four times higher than in plants sown directly in the field.

Mahendran and Rahman (2023) highlighted the significance of artemisinin as the most potent antimalarial drug and the challenges of its insufficient production due to low amounts found in natural sources. Mahendran and Rahman (2023) discussed the selection of high-yielding cell lines, culture conditions, plant growth regulators, and elicitation as methods to boost artemisinin levels in plant cells. It also considers the importance of developing continuous and reliable manufacturing systems to meet market demand and prospects to produce artemisinin.

Ikram and Simonsen (2017) described the global threat of malaria and the use of artemisinin as a treatment in their research. They mentioned that natural artemisinin production is limited, and it is difficult to synthesize chemically. Therefore, they highlighted current efforts in genetic and metabolic engineering to improve in planta production of artemisinin, both in *A. annua* and in other plants, and discussed the progress made in bioengineering to obtain higher production of artemisinin in *A. annua* and stable heterologous in planta systems.

Basile et al. (1993) studied the effect of plant growth regulators on the production of artemisinin through tissue culture. The study aimed to determine the suitable conditions for producing friable callus from *Artemisia annua* L. through tissue culture, as a source of artemisinin. Fourteen multifactorial experiments were conducted using three different nutrient formulations and different combinations of auxins and cytokinins. Both Gamborg B5 and Whetmore and Rier nutrients supported friable callus formation from leaf explants. The remaining experiments were conducted solely with Gamborg B5, and it was found that friable callus formed when combinations of NAA with kinetin or 2,4-D and BA were used with B5 medium. Ranges of 2,4-D from 0.5 to 2.0 with BA between 0.025 and 0.1, or NAA between 0.5 and 2.0 with kinetin between 0.5 and 1.0 mg/liter, produced acceptable results.

Artemisinin is certainly an effective antimalarial drug recommended by the WHO. However, due to its low concentration in the plant, transgenic approaches are being explored to increase production and aid in global eradication efforts (Kundu, Mukherjee & Ghosh, 2016). Recent trends have also focused on bioengineering the artemisinin biosynthetic pathway in heterologous expression platforms, with successful in planta production in tobacco and moss and achievements in heterologous microorganisms like *Escherichia coli* and *Saccharomyces cerevisiae*. The latter is favoured due to its compatibility with plant-origin genes and the presence of the MVA biosynthetic pathway for production of universal sesquiterpenes precursor, farnesyl pyrophosphate (Beyraghdar Kashkooli, Farmanpour-Kalalagh & Babaei, 2022).

3.2.8 Quinine

This alkaloid, which is derived from the bark of the cinchona tree (native to the Andes Mountains in South America), has been used for centuries to treat malaria (Soon, Kim & Lee, 2007; Panda et al., 2013). The bark of cinchona tree contains several alkaloids, including quinine, quinidine, cinchonine, and cinchonidine, but quinine is the most important and widely used. Quinine is one of the first antimalarial drugs discovered, and although it is now less commonly used due to the development of resistance to the drug, it remains an important treatment option in some areas (White, 1998; Taylor & White, 2004). Quinine works by interfering with the replication of the malaria parasite, preventing it from reproducing and spreading within the body. Quinine is also used to treat leg cramps and has been used in traditional medicine to treat a variety of other conditions, including fever, headache, and arthritis (Achan et al., 2011). The demand for quinine has increased substantially over the past few decades due to the growing resistance of the malaria parasite to synthetic drugs. However, the natural sources of quinine are limited, and traditional extraction methods are expensive and laborious.

The traditional method of extracting quinine from the bark involves stripping the outer layers and soaking the inner layers in water, followed by extraction with an organic solvent. This method is time-consuming and requires large quantities of bark, which can lead to deforestation and loss of biodiversity (Cross, 1924; Kaufman et al., 2005).

3.2.8.1 Production of Quinine through Plant Tissue Culture

The production of quinine can be achieved through plant tissue culture techniques, such as callus culture and micropropagation. Several studies have reported successful quinine production through plant tissue culture. Ratnadewi and Sumaryono (2010) worked on the development of cell culture to produce *Cinchona* alkaloids. Different substances were added to the growth media to promote alkaloid production in *Cinchona ledgeriana* cell cultures. After seven weeks of culture, the addition of paclobutrazol and abscisic acid resulted in suppressed quinine and cinchonine content, while cinchonidine content was enhanced by tryptophan. An addition of 5 mg/L paclobutrazol to the culture media produced the highest level of total quinine/quinidine after seven weeks. The study suggests that further research is needed to improve the production of quinine in cell cultures to match or exceed that produced by conventional sources. Further, Ratnadewi, Satriawan, and Sumaryono (2013) tested an improved method for culturing cell suspensions of *C. ledgeriana* to investigate the possibility of increasing quinine production. The results showed that plant tissue culture techniques can be used to increase the production of quinine and other alkaloids in cinchona plants, with paclobutrazol producing the highest level of quinine after seven weeks of culture. Jean et al. (2018) used plant cell culture to increase quinine production in *C. ledgeriana*. Stress-inducing agents like abscisic acid and paclobutrazol were used to stimulate the cells to produce quinine, and sorbitol was found to be more effective than mannitol in inducing stress. The combination of ABA 3 mg/L or PBZ 7 mg/L with sorbitol produced the highest concentration of quinine. According to the findings, elongated cells predominated in the mature cells culture, and alkaloids were discovered to be present in the mature cells.

An improved method for culturing cell suspensions of *C. ledgeriana* was tested to investigate the possibility of increasing production of quinine. The clone QRC 315 was treated with either a growth retardant (abscisic acid or paclobutrazol) or precursor feeding of L-tryptophan1. Cell suspension cultures of *C. ledgeriana* clone QRC 315 exhibited a high percentage of quinine, 7 to 11% (Ratnadewi & Satriawan, 2013). The cells were treated with plant growth retardants abscisic acid (ABA) or paclobutrazol (PBZ), combined with sucrose as a source of carbohydrate or with mannitol as a partial substitution of sucrose3.

Previous studies have reported on the production of cinchona alkaloids using cell suspension cultures and various methods, such as stress, precursors, elicitors, and the hairy roots system (Wijnsma et al., 1986; Hamill, Robins & Rhodes, 1989). In *C. ledgeriana*, treating leaf, shoot, and organ cultures with benzyladenin in Murashige-Skoog (MS) media resulted in an increase in alkaloid content with the age of the cultures.

The potential of plant tissue culture for quinine production goes beyond just being a sustainable and cost-effective method (Staba & Chung, 1981a; Anderson, Keene & Phillipson, 1982a; Hunter, 1988). Plant tissue culture can also be used for genetic modification of the plant cells to improve the yield and quality of quinine. Genetic engineering can be used to introduce genes involved in the biosynthesis of quinine or to silence genes that limit the production of quinine (Pratiwi et al., 2018; Trenti et al., 2021).

These are just a few examples of plant secondary metabolites with antimalarial activity. The discovery and development of new antimalarial compounds from plants is an ongoing area of research, and the hope is that these compounds will provide new and effective treatments for malaria in the future.

Overall, plant tissue culture techniques provide a controlled environment for the study and cultivation of plants with potential antimalarial activity, which can aid in the discovery and development of new antimalarial drugs.

3.3 CONCLUSION

In conclusion, this book chapter has provided an overview of the applications of plant tissue culture in the production of antimalarial drugs. The use of plant tissue culture techniques for the production of artemisinin, quinine, and other antimalarial agents has been discussed and the findings summarized in Table 3.1. The potential benefits of using plant tissue culture in the production of antimalarial drugs include increased efficiency, reduced costs, and a more sustainable production process. It is evident that plant tissue culture offers promising opportunities for the future of antimalarial drug production. Further research and development in this area may lead to the discovery of new antimalarial compounds and the creation of a more effective and efficient production process.

TABLE 3.1
Plant Species and Their Corresponding Antimalarial Drugs Produced via Tissue Culture along with the Relevant Findings from Research Studies

Name of Plant	Antimalarial Drug	Tissue Culture Findings	Key References
Artemisia annua	Artemisinin	Production of artemisinin in *A. annua* hairy root or cell suspension cultures	(Smith, Weathers & Cheetham, 1997; Wang et al., 2001, 2009; Ahlawat et al., 2014; Patra & Srivastava, 2014)
Artemisia annua	Artemisinin	Generation of somaclonal variants tolerant against salt stress, resulting in a stable salt-tolerant line with a higher expression profile of artemisinin key genes (ADS, CYP71AV1, DBR2, and ALDH1) and a higher artemisinin content	(Baldi & Dixit, 2008)
Artemisia annua	Artemisinin	Elicitation with methyl jasmonate, resulting in increased artemisinin production and upregulated expression of artemisinin biosynthesis genes as well as increased GT index	(Baldi & Dixit, 2008; Dangash et al., 2014)
Artemisia annua	Artemisinin	Elicitation with chitosan, gibberellic acid, and salicylic acid, resulting in increased accumulation of artemisinin	(Guo et al., 2010; Banyai, Mii & Supaibulwatana, 2011)
Artemisia annua	Artemisinin	Combination of various cultivation and elicitation methods, resulting in mass production of artemisinin in *A. annua* hairy roots via bioreactors	(Patra & Srivastava, 2015a, 2015b)
Catharanthus roseus	Vincristine and Vinblastine	Production of vincristine and vinblastine in *C. roseus* cell cultures	(Zhao & Verpoorte, 2007; Zhou, Shao & Tang, 2009)
Cinchona ledgeriana	Quinine	Production of quinine in *C. ledgeriana* cell cultures	(Staba & Chung, 1981b; Anderson, Keene & Phillipson, 1982b; Rhodes, Payne & Robins, 1986; Payne, Rhodes & Robins, 1987)
Taxus baccata	Paclitaxel	Production of paclitaxel in *T. baccata* cell cultures	(Yukimune et al., 1996; Cusidó et al., 2002; Bruňáková et al., 2004; Bentebibel et al., 2005; Bonfill et al., 2006)
Curcuma longa	Curcumin	Production of curcumin in *C. longa* cell cultures	(Chaturvedi et al., 2014; Jie et al., 2019)
Andrographis paniculata	Andrographolide	Production of andrographolide in *A. paniculata* cell cultures	(Gandi et al., 2012; Zaheer & Giri, 2017; Das & Bandyopadhyay, 2020)

REFERENCES

Achan, J., Talisuna, A.O., Erhart, A., Yeka, A., Tibenderana, J.K., Baliraine, F.N., Rosenthal, P.J. & D'Alessandro, U. 2011. Quinine, an old anti-malarial drug in a modern world: Role in the treatment of malaria. *Malaria Journal*. 10. https://doi.org/10.1186/1475-2875-10-144.

Ahlawat, S., Saxena, P., Alam, P., Wajid, S. & Abdin, M.Z. 2014. Modulation of artemisinin biosynthesis by elicitors, inhibitor, and precursor in hairy root cultures of Artemisia Annua L. *Journal of Plant Interactions*. 9(1):811–824. https://doi.org/10.1080/17429145.2014.949885.

Ammirato, P.V., Evans, D.A., Flick, C.E., Whitaker, R.J. & Sharp, W.R. 1984. Biotechnology and agricultural improvement. *Trends in Biotechnology*. 2(3):53–58. https://doi.org/10.1016/0167-7799(84)90010-6.

Anderson, L.A., Keene, A.T. & Phillipson, J.D. 1982a. Production of quinine and quinidine by plant cell cultures of *Cinchona ledgeriana*. *Journal of Pharmacy and Pharmacology*. 34(12 S):45P–45P. https://doi.org/10.1111/J.2042-7158.1982.TB00876.X.

Anderson, L.A., Keene, A.T. & Phillipson, J.D. 1982b. Alkaloid production by leaf organ, root organ and cell suspension cultures of *Cinchona ledgeriana*. *Planta Medica*. 46(1):25–27. https://doi.org/10.1055/S-2007-970011/BIB.

Arya, A., Gola, D., Tyagi, P.K. & Husen, A. 2022b. Molecular control of adventitious root formation. In: *Environmental, Physiological and Chemical Controls of Adventitious Rooting in Cuttings* (Ed. Husen, A.). Elsevier Inc., Cambridge, MA, pp. 25–46.

Arya, A. & Husen, A. 2022. Role of various auxins in adventitious root formation. In: *Environmental, Physiological and Chemical Controls of Adventitious Rooting in Cuttings* (Ed. Husen, A.). Elsevier Inc., Cambridge, MA, pp. 213–238.

Arya, A., Sharma, V., Tyagi, P.K., Gola, D. & Husen, A. 2022a. Role of cytokinins in adventitious root formation. In: *Environmental, Physiological and Chemical Controls of Adventitious Rooting in Cuttings* (Ed. Husen, A.). Elsevier Inc., Cambridge, MA, pp. 239–249.

Ashley, E.A. & Poespoprodjo, J.R. 2020. Treatment and prevention of malaria in children. *The Lancet Child & Adolescent Health*. 4(10):775–789. https://doi.org/10.1016/S2352-4642(20)30127-9.

Baldi, A. & Dixit, V.K. 2008. Yield enhancement strategies for artemisinin production by suspension cultures of Artemisia annua. *Bioresource Technology*. 99(11):4609–4614. https://doi.org/10.1016/J.BIORTECH.2007.06.061.

Balint, G.A. 2001. Artemisinin and its derivatives: An important new class of antimalarial agents. *Pharmacology & Therapeutics*. 90(2–3):261–265. https://doi.org/10.1016/S0163-7258(01)00140-1.

Banyai, W., Kirdmanee, C., Mii, M. & Supaibulwatana, K. 2010. Overexpression of farnesyl pyrophosphate synthase (FPS) gene affected Artemisinin content and growth of Artemisia annua L. *Plant Cell, Tissue and Organ Culture*. 103(2):255–265. https://doi.org/10.1007/S11240-010-9775-8.

Banyai, W., Mii, M. & Supaibulwatana, K. 2011. Enhancement of artemisinin content and biomass in Artemisia annua by exogenous GA3 treatment. *Plant Growth Regulation*. 63(1):45–54. https://doi.org/10.1007/S10725-010-9510-9.

Banyai, W., Sangthong, R., Karaket, N., Inthima, P., Mii, M. & Supaibulwatana, K. 2010. Overproduction of artemisinin in tetraploid Artemisia annua L. *Plant Biotechnology*. 27(5):427–433. https://doi.org/10.5511/PLANTBIOTECHNOLOGY.10.0726A.

Basile, D.V., Akhtari, N., Durand, Y. & Nair, M.S.R. 1993. Toward the production of artemisinin through tissue culture: Determining nutrient-hormone combinations suitable for cell suspension cultures. *In Vitro Cellular & Developmental Biology – Plant*. 29(3):143–147. https://doi.org/10.1007/BF02632286.

Belete, T.M. 2020. Recent progress in the development of new antimalarial drugs with novel targets. *Drug Design, Development and Therapy*. 14:3875–3889. https://doi.org/10.2147/DDDT.S265602.

Bentebibel, S., Moyano, E., Palazón, J., Cusidó, R.M., Bonfill, M., Eibl, R. & Piñol, M.T. 2005. Effects of immobilization by entrapment in alginate and scale-up on paclitaxel and baccatin III production in cell suspension cultures of Taxus baccata. *Biotechnology and Bioengineering*. 89(6):647–655. https://doi.org/10.1002/BIT.20321.

Beyraghdar Kashkooli, A., Farmanpour-Kalalagh, K. & Babaei, A. 2022. Yeast synthetic biology for production of artemisinin as an antimalarial drug. *Synthetic Biology of Yeasts*: 157–180. https://doi.org/10.1007/978-3-030-89680-5_6.

Bonfill, M., Expósito, O., Moyano, E., Cusidó, R.M., Palazón, J. & Piñol, M.T. 2006. Manipulation by culture mixing and elicitation of paclitaxel and baccatin III production in Taxus baccata suspension cultures. *In Vitro Cellular and Developmental Biology – Plant*. 42(5):422–426. https://doi.org/10.1079/IVP2006761/METRICS.

Booth, C.M. & MacLean, J.D. 2020. Knowledge, treatment-seeking, and socioeconomic impact of malaria on the Essequibo Coast of Guyana. *McGill Journal of Medicine*. 6(1):17–25. https://doi.org/10.26443/mjm.v6i1.765.

Brown, D.C.W. & Thorpe, T.A. 1995. Crop improvement through tissue culture. *World Journal of Microbiology & Biotechnology*. 11(4):409–415. https://doi.org/10.1007/BF00364616/METRICS.

Bruňáková, K., Babincová, Z., Takáč, M. & Čellárová, E. 2004. Selection of callus cultures of taxus baccata L. as a potential source of paclitaxel production. *Engineering in Life Sciences*. 4(5):465–469. https://doi.org/10.1002/ELSC.200420050.

Chaturvedi, P., Soundar, S., Parekh, K., Lokhande, S. & Chowdhary, A. 2014. Media optimization in immobilized culture to enhance the content of Kaempferol in Tylophora indica (Asclepeadaceae) and Curcumin in Curcuma longa (Zingiberaceae). *IOSR Journal of Pharmacy and Biological Sciences (IOSR-JPBS)*. 9(2):86–90. Available: www.iosrjournals.org [2023, April 22].

Chiang, C.Y. & Sony, A.E. 2020. Tackling the threat of COVID-19 in Africa: An urgent need for practical planning. *International Journal of Tuberculosis and Lung Disease*. 24(5):541–542. https://doi.org/10.5588/IJTLD.20.0192.

Chico, R.M., Dellicour, S., Roman, E., Mangiaterra, V., Coleman, J., Menendez, C., Majeres-Lugand, M., Webster, J., et al. 2015. Global call to action: Maximize the public health impact of intermittent preventive treatment of malaria in pregnancy in sub-Saharan Africa. *Malaria Journal*. 14(1):1–6. https://doi.org/10.1186/S12936-015-0728-X/TABLES/1.

Christensen, S.B. 2021. Natural products that changed society. *Biomedicines*. 9(5):472. https://doi.org/10.3390/BIOMEDICINES9050472.

Colson, P., Rolain, J.M. & Raoult, D. 2020. Chloroquine for the 2019 novel coronavirus SARS-CoV-2. *International Journal of Antimicrobial Agents*. 55(3). https://doi.org/10.1016/j.ijantimicag.2020.105923.

Cross, S. 1924. *Quinine: Production and Marketing*. Available: https://books.google.com/books?hl=en&lr=&id=A5v6FMan4zsC&oi=fnd&pg=PA1&dq=quinine+production+from+bark+is+time+consuming&ots=vtqpQjary_&sig=axcNG73lDi6XQYEq4AhK7f2Inqs [2023, April 03].

Cusidó, R.M., Palazón, J., Bonfill, M., Navia-Osorio, A., Morales, C. & Piñol, M.T. 2002. Improved paclitaxel and baccatin III production in suspension cultures of taxusmedia. *Biotechnology Progress*. 18(3):418–423. https://doi.org/10.1021/BP0101583.

Dangash, A.J., Pandya, N., Bharillya, A., Jhala, A. & Jain, D. 2014. Impact of exogenous elicitors on artemisinin production and trichome density in *Artemisia annua* L. under subtropical conditions. *Notulae Scientia Biologicae*. 6(3). https://doi.org/10.15835/NSB.6.3.9109.

Das, D. & Bandyopadhyay, M. 2020. Novel approaches towards over-production of andrographolide in in vitro seedling cultures of Andrographis paniculata. *South African Journal of Botany*. 128:77–86. https://doi.org/10.1016/J.SAJB.2019.10.015.

DiCosmo, F. & Misawa, M. 1985. Eliciting secondary metabolism in plant cell cultures. *Trends in Biotechnology*. 3(12):318–322. https://doi.org/10.1016/0167-7799(85)90036-8.

Elhag, H.M., El-Domiaty, M.M., El-Feraly, F.S., Mossa, J.S. & El-Olemy, M.M. 1992. Selection and micropropagation of high artemisinin producing clones of Artemisia annua L. *Phytotherapy Research*. 6(1):20–24. https://doi.org/10.1002/PTR.2650060106.

Fischer, R. & Emans, N. 2000. Molecular farming of pharmaceutical proteins. *Transgenic Research*. 9(4–5):279–299. https://doi.org/10.1023/A:1008975123362/METRICS.

Gandi, S., Rao, K., Chodisetti, B. & Giri, A. 2012. Elicitation of andrographolide in the suspension cultures of andrographis paniculata. *Applied Biochemistry and Biotechnology*. 168(7):1729–1738. https://doi.org/10.1007/S12010-012-9892-4/METRICS.

Garcia, S. 2020. Pandemics and traditional plant-based remedies. A historical-botanical review in the era of COVID19. *Frontiers in Plant Science*. 11:1353. https://doi.org/10.3389/FPLS.2020.571042/BIBTEX.

Guo, X.X., Yang, X.Q., Yang, R.Y. & Zeng, Q.P. 2010. Salicylic acid and methyl jasmonate but not Rose Bengal enhance artemisinin production through invoking burst of endogenous singlet oxygen. *Plant Science*. 178(4):390–397. https://doi.org/10.1016/J.PLANTSCI.2010.01.014.

Hamill, J., Robins, R. & Rhodes, M. 1989. Alkaloid production by transformed root cultures of cinchona ledgeriana. *Planta Medica*. 55(4):354–357. https://doi.org/10.1055/S-2006-962027.

Hamre, K.E.S., Ayodo, G., Hodges, J.S. & John, C.C. 2020. A mass insecticide-treated bed net distribution campaign reduced malaria risk on an individual but not population level in a highland epidemic-prone area of Kenya. *The American Journal of Tropical Medicine and Hygiene*. 103(6):2183–2188. https://doi.org/10.4269/AJTMH.19-0306.

Han, J., Wang, H., Kanagarajan, S., Hao, M., Lundgren, A. & Brodelius, P.E.E. 2016. Promoting artemisinin biosynthesis in artemisia annua plants by substrate channeling. *Molecular Plant*. 9(6):946–948. https://doi.org/10.1016/J.MOLP.2016.03.004.

Hunter, C.S. 1988. Cinchona spp.: Micropropagation, and the in vitro production of quinine and quinidine. In: *Medicinal and Aromatic Plants I*, pp. 367–387. https://doi.org/10.1007/978-3-642-73026-9_20.

Husen, A. 2022. *Environmental, Physiological and Chemical Controls of Adventitious Rooting in Cuttings*. Elsevier Inc., Cambridge, MA.

Hyde, J.E. 2005. Drug-resistant malaria. *Trends in Parasitology*. 21(11):494–498. https://doi.org/10.1016/j.pt.2005.08.020.

Ikram, N.K.B.K. & Simonsen, H.T. 2017. A review of biotechnological artemisinin production in plants. *Frontiers in Plant Science*. 8:1966. https://doi.org/10.3389/FPLS.2017.01966/BIBTEX.

Jean, E., Delage, M.-È., Beaulieu, L., Pik, M., Piech, R., Paczosa-Bator, B., Pratiwi, D.R., Sari, P.T., et al. 2018. Cinchona cells performance in in vitro culture: Quinine alkaloid production with application of different elicitors. *IOP Conference Series: Earth and Environmental Science*. 185(1):012029. https://doi.org/10.1088/1755-1315/185/1/012029.

Jie, E.Y., Ahn, M.S., Lee, J., Cheon, Y.I., Kim, C.Y. & Kim, S.W. 2019. Establishment of a high-frequency plant regeneration system from rhizome-derived embryogenic cell-suspension cultures of Curcuma longa L. *Plant Biotechnology Reports*. 13(2):123–129. https://doi.org/10.1007/S11816-019-00519-2/METRICS.

Jochum, J., Kreuels, B., Tannich, E., Huber, S., Schulze Zur Wiesch, J., Schmiedel, S., Ramharter, M. & Addo, M.M. 2021. Malaria in the time of COVID-19: Do not miss the real cause of illness. *Tropical Medicine and Infectious Disease*. 6(2):40. https://doi.org/10.3390/TROPICALMED6020040.

Kaufman, T.S., Rfflveda, E.A., Kaufman, T.S. & Rfflveda, E.A. 2005. The quest for quinine: Those who won the battles and those who won the war. *Wiley Online Library*. 44(6):854–885. https://doi.org/10.1002/anie.200400663.

Kaur, R. & Stud, H.K. 2017. Plant derived antimalarial agents. *Researchgate.net*. 5(1):346–363. Available: www.researchgate.net/profile/Rajandeep-Kaur-2/publication/347944225_Plant_Derived_Antimalarial_Agents/links/5fe92fb9a6fdccdcb80cb054/Plant-Derived-Antimalarial-Agents.pdf [2023, April 15].

Klayman, D.L. 1985. Qinghaosu (artemisinin): An antimalarial drug from China. *Science*. 228(4703): 1049–1055. https://doi.org/10.1126/SCIENCE.3887571.

Knödler, M. & Buyel, J.F. 2021. Plant-made immunotoxin building blocks: A roadmap for producing therapeutic antibody-toxin fusions. *Biotechnology Advances*. 47:107683. https://doi.org/10.1016/J.BIOTECHADV.2020.107683.

Kundu, S., Mukherjee, S. & Ghosh, B. 2016. Antimalarial compound synthesis from transgenic cultures. *Transgenesis and Secondary Metabolism*: 1–30. https://doi.org/10.1007/978-3-319-27490-4_22-1.

Liu, C.Z., Guo, C., Wang, Y.C. & Ouyang, F. 2002. Effect of light irradiation on hairy root growth and artemisinin biosynthesis of Artemisia annua L. *Process Biochemistry*. 38(4):581–585. https://doi.org/10.1016/S0032-9592(02)00165-6.

Liu, C.Z., Zhao, Y. & Wang, Y. 2006. Artemisinin: Current state and perspectives for biotechnological production of an antimalarial drug. *Applied Microbiology and Biotechnology*. 72(1):11–20. https://doi.org/10.1007/S00253-006-0452-0/METRICS.

Lu, B.W., Baum, L., So, K.F., Chiu, K. & Xie, L.K. 2019. More than anti-malarial agents: Therapeutic potential of artemisinins in neurodegeneration. *Neural Regeneration Research*. 14(9):1494–1498. https://doi.org/10.4103/1673-5374.255960.

Mahendran, G. & Rahman, L.U. 2023. Recent biotechnological approaches for the enhancement of artemisinin production from cell culture system of artemisia annua L.: Medicinal plant with potent antimalarial properties. In: *Natural Product Experiments in Drug Discovery*, pp. 361–376. https://doi.org/10.1007/978-1-0716-2683-2_22.

Millet, J.P., De Olalla, P.G., Gascón, J., Prat, J.G.I., Treviño, B., Pinazo, M.J., Cabezos, J., Muñoz, J., et al. 2009. Imported malaria among African immigrants: Is there still a relationship between developed countries and their ex-colonies? *Malaria Journal*. 8(1):1–8. https://doi.org/10.1186/1475-2875-8-111/TABLES/2.

Mohammadi, S., Jafari, B., Asgharian, P., Martorell, M. & Sharifi-Rad, J. 2020. Medicinal plants used in the treatment of Malaria: A key emphasis to Artemisia, Cinchona, Cryptolepis, and Tabebuia genera. *Phytotherapy Research*. 34(7):1556–1569. https://doi.org/10.1002/PTR.6628.

Nahrendorf, W., Ivens, A. & Spence, P.J. 2021. Inducible mechanisms of disease tolerance provide an alternative strategy of acquired immunity to malaria. *eLife*. 10. https://doi.org/10.7554/ELIFE.63838.

Nair, M.S.R., Acton, N., Klayman, D.L., Kendrick, K., Basile, D.V. & Mante, S. 1986. Production of artemisinin in tissue cultures of artemisia annua. *Journal of Natural Products*. 49(3):504–507. https://doi.org/10.1021/NP50045A021.

Ogutu, B.R. 2009. Malaria management – Progress made and challenges still to face. *Malaria Journal*. 8(Suppl. 1): 1–3. https://doi.org/10.1186/1475-2875-8-S1-S1/METRICS.

O'Neill, P.M., Barton, V.E. & Ward, S.A. 2010. The molecular mechanism of action of artemisinin – The debate continues. *Molecules*. 15(3):1705–1721. https://doi.org/10.3390/MOLECULES15031705.

Panda, S.S., Ibrahim, M.A., Küçükbay, H., Meyers, M.J., Sverdrup, F.M., El-Feky, S.A. & Katritzky, A.R. 2013. Synthesis and antimalarial bioassay of quinine – peptide conjugates. *Chemical Biology & Drug Design*. 82(4):361–366. https://doi.org/10.1111/CBDD.12134.

Patra, N. & Srivastava, A.K. 2014. Enhanced production of artemisinin by hairy root cultivation of artemisia annua in a modified stirred tank reactor. *Applied Biochemistry and Biotechnology*. 174(6):2209–2222. https://doi.org/10.1007/S12010-014-1176-8.

Patra, N. & Srivastava, A.K. 2015a. Use of model-based nutrient feeding for improved production of artemisinin by hairy roots of artemisia annua in a modified stirred tank bioreactor. *Applied Biochemistry and Biotechnology*. 177(2):373–388. https://doi.org/10.1007/S12010-015-1750-8/METRICS.

Patra, N. & Srivastava, A.K. 2015b. Artemisinin production by plant hairy root cultures in gas- and liquid-phase bioreactors. *Plant Cell Reports*. 35(1):143–153. https://doi.org/10.1007/S00299-015-1875-9.

Payne, J., Rhodes, M.J.C. & Robins, R.J. 1987. Quinoline alkaloid production by transformed cultures of Cinchona ledgeriana. *Planta Medica*. 53(4):367–372. https://doi.org/10.1055/S-2006–962740/BIB.

Pras, N., Visser, J.F., Batterman, S., Woerdenbag, H.J., Malingré, T.M. & Lugt, C.B. 1991. Laboratory selection of Artemisia annua L. for high artemisinin yielding types. *Phytochemical Analysis*. 2(2):80–83. https://doi.org/10.1002/PCA.2800020208.

Pratiwi, D.R., Sumaryono, Sari, P.T. & Ratnadewi, D. 2018. Cinchona cells performance in in vitro culture: Quinine alkaloid production with application of different elicitors. *IOP Conference Series: Earth and Environmental Science*. 185(1):012029. https://doi.org/10.1088/1755-1315/185/1/012029.

Ramakrishna, W., Kumari, A., Rahman, N. & Mandave, P. 2021. Anticancer activities of plant secondary metabolites: Rice callus suspension culture as a new paradigm. *Rice Science*. 28(1):13–30. https://doi.org/10.1016/J.RSCI.2020.11.004.

Ratnadewi, D. & Satriawan, D. 2013. Enhanced production level of quinine in cell suspension culture of cinchona ledgeriana moens by paclobutrazol. *BIOTROPIA – The Southeast Asian Journal of Tropical Biology*. 20(1):10–18. https://doi.org/10.11598/BTB.2013.20.1.291.

Ratnadewi, D., Satriawan, D. & Sumaryono. 2013. Enhanced production level of quinine in cell suspension culture of cinchona ledgeriana moens by paclobutrazol. *Biotropia*. 20(1):10–18. https://doi.org/10.11598/BTB.2013.20.1.10.

Ratnadewi, D. & Sumaryono. 2010. Quinoline alkaloids in suspension cultures of cinchona ledgeriana treated with various substances. *HAYATI Journal of Biosciences*. 17(4):179–182. https://doi.org/10.4308/HJB.17.4.179.

Rhodes, M.J.C., Payne, J. & Robins, R.J. 1986. Cell suspension cultures of cinchona ledgeriana; II. The effect of a range of auxins and cytokinins on the production of quinoline alkaloids. *Planta Medica*. 3(3): 226–229. https://doi.org/10.1055/S-2007-969129/BIB.

Robins, R.J., Payne, J. & Rhodes, M.J.C. 1986. Cell suspension cultures of Cinchona ledgeriana; I. Growth and quinoline alkaloid production. *Planta Medica*. 3(3):220–226. https://doi.org/10.1055/S-2007-969128/BIB.

Rosenthal, P.J. 2003. Antimalarial drug discovery: Old and new approaches. *Journal of Experimental Biology*. 206(21):3735–3744. https://doi.org/10.1242/JEB.00589.

Roy, A. 2020. Hairy root culture an alternative for bioactive compound production from medicinal plants. *Current Pharmaceutical Biotechnology*. 22(1):136–149. https://doi.org/10.2174/13892010216662012 29110625.

Sahu, N.K., Sahu, S. & Kohli, D.V. 2008. Novel molecular targets for antimalarial drug development. *Chemical Biology and Drug Design*. 71(4):287–297. https://doi.org/10.1111/J.1747-0285.2008.00640.X.

Sarma, N., Patouillard, E., Cibulskis, R.E. & Arcand, J.L. 2019. The economic burden of malaria: Revisiting the evidence. *The American Journal of Tropical Medicine and Hygiene*. 101(6):1405–1415. https://doi.org/10.4269/AJTMH.19-0386.

Schillberg, S., Fischer, R. & Emans, N. 2003. "Molecular farming" of antibodies in plants. *Naturwissenschaften*. 90(4):145–155. https://doi.org/10.1007/S00114-002-0400-5/METRICS.

Shah, M.P., Steinhardt, L.C., Mwandama, D., Mzilahowa, T., Gimnig, J.E., Bauleni, A., Wong, J., Wiegand, R., et al. 2020. The effectiveness of older insecticide-treated bed nets (ITNs) to prevent malaria infection in an area of moderate pyrethroid resistance: Results from a cohort study in Malawi. *Malaria Journal*. 19(1):1–12. https://doi.org/10.1186/S12936-020-3106-2/TABLES/4.

Smith, T.C., Weathers, P.J. & Cheetham, R.D. 1997. Effects of gibberellic acid on hairy root cultures of Artemisia annua: Growth and artemisinin production. *In Vitro Cellular and Developmental Biology – Plant*. 33(1):75–79. https://doi.org/10.1007/S11627-997-0044-4.

Soon, S.L., Kim, H.S. & Lee, D.U. 2007. In vitro antimalarial activity of flavonoids and chalcones. *Bulletin of the Korean Chemical Society*. 28(12):2495–2497. https://doi.org/10.5012/BKCS.2007.28.12.2495.

Staba, E.J. & Chung, A.C. 1981a. Quinine and quinidine production by cinchona leaf, root and unorganized cultures. *Phytochemistry*. 20(11):2495–2498. https://doi.org/10.1016/0031-9422(81)83079-8.

Staba, E.J. & Chung, A.C. 1981b. Quinine and quinidine production by cinchona leaf, root and unorganized cultures. *Phytochemistry*. 20(11):2495–2498. https://doi.org/10.1016/0031-9422(81)83079-8.

Stratton, L., O'Neill, M.S., Kruk, M.E. & Bell, M.L. 2008. The persistent problem of malaria: Addressing the fundamental causes of a global killer. *Social Science & Medicine*. 67(5):854–862. https://doi.org/10.1016/J.SOCSCIMED.2008.05.013.

Suleman, S., Tufa, T., Kebebe, D., S Belew, Mekonnen, Y., Gashe, F., Mussa, S., Wynendaele, E., et al. 2018. Treatment of malaria and related symptoms using traditional herbal medicine in Ethiopia. *Journal Ethnopharmacol*. 262–279. Available: www.sciencedirect.com/science/article/pii/S0378874117322997 [2023, April 15].

Talisuna, A.O., Langi, P., Bakyaita, N., Egwang, T., Mutabingwa, T.K., Watkins, W., Van Marck, E. & D'Alessandro, U. 2002. Intensity of malaria transmission, antimalarial-drug use and resistance in Uganda: What is the relationship between these three factors? *Transactions of the Royal Society of Tropical Medicine and Hygiene*. 96(3):310–317. https://doi.org/10.1016/S0035-9203(02)90108-2.

Tatem, A.J., Smith, D.L., Gething, P.W., Kabaria, C.W., Snow, R.W. & Hay, S.I. 2010. Ranking of elimination feasibility between malaria-endemic countries. *The Lancet*. 376(9752):1579–1591. https://doi.org/10.1016/S0140-6736(10)61301-3.

Taylor, W.R.J. & White, N.J. 2004. Antimalarial drug toxicity: A review. *Drug Safety*. 27(1):25–61. https://doi.org/10.2165/00002018-200427010-00003.

Trenti, F., Yamamoto, K., Hong, B., Paetz, C., Nakamura, Y. & O'Connor, S.E. 2021. Early and late steps of quinine biosynthesis. *Organic Letters*. 23(5):1793–1797. https://doi.org/10.1021/ACS.ORGLETT.1C00206/SUPPL_FILE/OL1C00206_SI_001.PDF.

Wang, J.W., Zheng, L.P., Zhang, B. & Zou, T. 2009. Stimulation of artemisinin synthesis by combined cerebroside and nitric oxide elicitation in Artemisia annua hairy roots. *Applied Microbiology and Biotechnology*. 85(2):285–292. https://doi.org/10.1007/S00253-009-2090-9.

Wang, Y., Zhang, H., Zhao, B. & Yuan, X. 2001. Improved growth of artemisia annua L hairy roots and artemisinin production under red light conditions. *Biotechnology Letters*. 23(23):1971–1973. https://doi.org/10.1023/A:1013786332363.

Weiss, D.J., Bertozzi-Villa, A., Rumisha, S.F., Amratia, P., Arambepola, R., Battle, K.E., Cameron, E., Chestnutt, E., et al. 2021. Indirect effects of the COVID-19 pandemic on malaria intervention coverage, morbidity, and mortality in Africa: A geospatial modelling analysis. *The Lancet Infectious Diseases*. 21(1):59–69. https://doi.org/10.1016/S1473-3099(20)30700-3.

White, N.J. 1998. Why is it that antimalarial drug treatments do not always work? *Annals of Tropical Medicine & Parasitology*. 92(4):449–458. https://doi.org/10.1080/00034983.1998.11813303.

Wijnsma, R., Verpoorte, R., Harkes, P.A.A., Van Vliet, T.B., Ten Hoopen, H.J.G. & Svendsen, A.B. 1986. The influence of initial sucrose and nitrate concentrations on the growth of Cinchona ledgeriana cell suspension cultures and the production of alkloids and anthraquinones. *Plant Cell, Tissue and Organ Culture*. 7(1): 21–29. https://doi.org/10.1007/BF00043917.

World Malaria Report 2021. n.d. Available: www.who.int/teams/global-malaria-programme/reports/world-malaria-report-2021 [2023, March 25].

Yukimune, Y., Tabata, H., Higashi, Y. & Hara, Y. 1996. Methyl jasmonate-induced overproduction of paclitaxel and baccatin III in Taxus cell suspension cultures. *Nature Biotechnology*. 14(9):1129–1132. https://doi.org/10.1038/nbt0996-1129.

Zaheer, M. & Giri, C.C. 2017. Enhanced diterpene lactone (andrographolide) production from elicited adventitious root cultures of Andrographis paniculata. *Research on Chemical Intermediates*. 43(4):2433–2444. https://doi.org/10.1007/S11164-016-2771-9/METRICS.

Zawawi, A., Alghanmi, M., Alsaady, I., Gattan, H., Zakai, H. & Couper, K. 2020. The impact of COVID-19 pandemic on malaria elimination. *Parasite Epidemiology and Control*. 11:e00187. https://doi.org/10.1016/J.PAREPI.2020.E00187.

Zhao, J. & Verpoorte, R. 2007. Manipulating indole alkaloid production by Catharanthus roseus cell cultures in bioreactors: From biochemical processing to metabolic engineering. *Phytochemistry Reviews*. 6(2–3): 435–457. https://doi.org/10.1007/S11101-006-9050-0/METRICS.

Zhou, M.-L., Shao, J.-R. & Tang, Y.-X. 2009. Production and metabolic engineering of terpenoid indole alkaloids in cell cultures of the medicinal plant Catharanthus roseus (L.) G. Don (Madagascar periwinkle). *Biotechnology and Applied Biochemistry*. 52(4):313–323. https://doi.org/10.1042/BA20080239.

4 Sustainable Techniques for the Harvest and Conservation of Antimalarial Plant Species

Harsha Singh, Atul Arya, and Suresh Kumar

4.1 INTRODUCTION

The use of herbal medicinal plants is a tradition that has been followed for ages across the globe (Husen, 2021, 2022a, 2022b, 2023). The urbanization of agricultural lands, changes in agricultural methodology, land usage, loss of habitat, lack of documentation, harvesting, and genetic diversity of medicinal plants result into a threat of extinction of ethnobotanical knowledge (Arya et al., 2021). For conservation and utilization of biological resources, there is a compelling need for documentation of indigenous knowledge by conducting ethnobotanical surveys (Howes et al., 2020). Conservation of antimalarial plant species requires authentic data on its distribution and level of usage. In sub-Saharan Africa, malaria is the leading cause of death, ill health, and poverty. Members of the *Plasmodium* genus of parasitic protozoa, which are transferred to the human host by the female *Anopheles* mosquito, are responsible for the sickness. *Plasmodium falciparum*, which causes malignant malaria, is one of the deadliest species of *Plasmodium* that infect people with severe symptoms and may lead to death (Asafo-Agyei et al., 2019). The effects of malaria on public health are still catastrophic. According to the World Health Organization (WHO), there were around 212 million cases worldwide in 2015, which resulted in 429,000 fatalities, the majority of which were in children under the age of 5 in Africa (WHO, 2016). Despite significant advancements in the nation's health coverage, there are still significant disparities in service coverage and the distribution of health facilities between urban and rural areas. Therefore, the public has the choice of using conventional herbal remedies. Additionally, a key problem that demands action is the global expansion of drug resistance to the majority of the effective and reasonably priced antimalarial medications.

The issue with the conservation of medicinal plant species is not just related to overexploitation but also to how the species are harvested. Root digging, root cutting, and bark removal are the basic methods of harvesting (Rabgyal and Pelden, 2020). Additionally, the survival and durability of medicinal plants are possibly threatened by changing climate conditions and human population. Sustainable harvesting should be thoroughly researched because choosing permanent plant parts like roots, bark, tubers, and reproductive parts for use as medicine sources can jeopardise plant populations or species sustainability, resulting in death, wilting, stagnant growth, and species extinction. Alternative tree planting around the home and farm, agroforestry, agricultural practises, spreading knowledge, and education are all conservation strategies that can ease the burden of harvesting to maintain plant stability. It is important to gather medicinal plants ethically. A method that satisfies the requirements of the present without jeopardising the ability of future generations to satisfy their own needs has been dubbed 'sustainable' and has been utilized for centuries in the management of natural resources. One of the main causes of the loss of plant resources in the wild are destructive collection practices (Rabgyal and Pelden, 2020).

DOI: 10.1201/9781003378396-4

Destructive collection occurs for a variety of reasons which include lack of knowledge about proper collection techniques, increasing industrial demand for wild resources, tax regulations, monitoring systems for management, competition among local collectors, a lack of better prices for primary collectors, and an inadequate policy environment.

4.2 CONSERVATION OF MEDICINAL PLANTS

Harvesting of medicinal plant from its wild habitat has increased in large quantity. In recent years, Asia, Europe, and North America have seen an increase in the demand (8–15% annually) for medicinal plant (Barboza et al., 2009). It is necessary to understand the biological traits and geographic distribution of medicinal plants in order to guide conservation efforts (Sahoo et al., 2022). The treatment of malaria was carried out in Ghana using a variety of plant species, such as *Azadiracta indica* A. Juss (*Meliaceae*), *Nauclea latifolia* Sm. (*Rubiaceae*), *Ocimum viride* Willd. (*Lamiaceae*), *Alstonei boonei* De Willd (*Apocynaceae*), *Cryptolepis sanguinolenta* (Lindl.) Schttr. (*Asclepidaceae*) (Mshana, 2000). *Cyperus scariosus* (Nagarmotha) and *Leucas aspera* (Goma or Drona pushpin) are on the verge of extinction, primarily as a result of the alteration in the structure of water reservoirs (ponds) and changes in the pattern of land use. *Adhatoda vasica, Cassia fistula,* and *Swertia chirata* are the most popular plant species used by 82% of people for curing malaria in three districts, that is, Kushinagar, Maharajganj, and Gorakhpur, of eastern Uttar Pradesh (Qayum et al., 2016). A total of 34,000 plants species have been given the 'endangered' designation by the International Union for Conservation of Nature and Natural Resources (IUCN) Red Data Book. Thus, maintaining biodiversity has emerged as one of the most urgent environmental concerns. Figure 4.1 enlisting the methods of conservation of nature and natural resources.

4.2.1 *In Situ* Conservation of Plants

The 'on-site conservation' of the wild genetic variability in its natural environment is the objective of the *in situ* conservation technique. National parks, biosphere reserves, and wildlife sanctuaries are being conserved through protected areas. Rules, regulations, and the probable compliance of

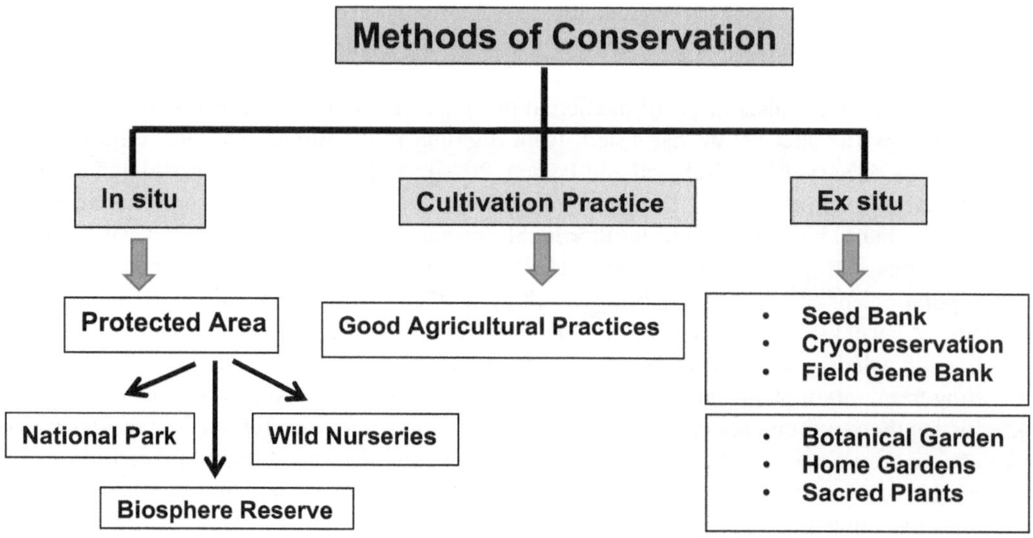

FIGURE 4.1 Methodological system for conserving medicinal plants.

medicinal plants within their growth environments are necessary for successful *in situ* conservation (Volis and Blecher, 2010). *In situ* conservation involves the conservation of whole communities and is used to preserve the complicated web of natural communities that exist among communities. The proportion of diversity can be increased under *in situ* conservation, which also deepens the connection between resource protection and sustainable use. *In situ* conservation involves the development of protected areas and adopting an ecosystem-focused strategy. Human populations that live in or around protected forest areas can actively support and participate in the *in situ* conservation of medicinal plants. The local population will feel more responsible for ecosystem resilience and sustainable use for their personal well-being. The various methods of *in situ* conservation of antimalarial plant species are as follows.

4.2.1.1 National Parks

In order to maintain and restore biodiversity, national parks were established as protected regions of significant wild resources. Around 13.2 million km^2, or 8.81% of the Earth's land surface, are covered by more than 12,700 protected areas that have been established globally (Huang et al., 2002). The contribution and ecological functions of each habitat must be analyzed in order to conserve medicinal plants through the protection of important natural habitats.

4.2.1.2 Biosphere Reserves

By limiting reckless human interference in the ecosystem, biosphere reserves seek to protect it while also preserving endemic and threatened species. In India, there are 18 biosphere reserves. Three of the world's 34 biodiversity hotspots – the Western Ghats, the Eastern Himalayas, and the Indo-Burma region – are located in India (Kadam and Pawar, 2020).

4.2.1.3 Wild Nurseries

In a protected region, a natural habitat, or a location that is only a short distance from where the plants naturally grow, a wild nursery is constructed for the species-focused cultivation and domestication of endangered medicinal plants. Overexploitation, habitat destruction, and invasive species have severely harmed the populations of many wild species. Wild nurseries can offer a practical solution for the *in situ* conservation of medicinal plants that are endangered and endemic (Chen et al., 2016).

4.2.2 *Ex Situ* Conservation of Plants

Ex situ conservation refers to the 'off-site conservation' of wild genetic diversity in their native habitat. *Ex situ* conservation, which is usually undertaken to conserve medicinal plant resources, strives to nurture and naturalise threatened species to ensure their continuous survival and to produce significant quantities of medicinal plants required for the manufacturing of drugs (Kadam and Pawar, 2020). *Ex situ* conservation of species should be prioritized to destroyed or unprotected habitats. *Ex situ* conservation is a complementary strategy that lessens pressure on wild areas and increases the availability of raw materials by preserving genetic variety. Another benefit of *ex situ* conservation is that it is typically simpler to obtain plant material for the purposes of propagation, reintroduction, agronomic improvement, research, and education.

4.2.2.1 Botanical Gardens

Botanical gardens are best adapted to carry out the *ex situ* conservation of medicinal plants. By creating guidelines for propagation and culture as well as implementing domestication and variety breeding programmes, botanical gardens can contribute further to the conservation of medicinal plants. Botanical gardens have an essential role in research and monitoring hubs for populations of wild plants (Maunder et al., 2001). They can keep the ecosystems in good condition to help rare and endangered plant species survive. There are currently 1,775 botanical gardens and arboreta around the world, according to the Botanical Gardens Conservation International (BGCI) database.

4.2.2.2 Seed Banks

Seed banks are a better alternative to botanical gardens for *ex situ* storage of the genetic diversity of many therapeutic plants. Species and individuals can be stored in very high numbers in a seed bank, but seed banks are prone to power outages, need careful monitoring, and do require time-consuming regeneration (Schoen and Brown, 2001). Plant samples can be quickly accessed from seed banks to assess their qualities (Botanical Survey of India, BSI *ex situ* conservation). Restoration of the plant species into the wild and active support of the restoration of wild populations are the tough issues of seed banking. Other institutions, like national and regional crop gene banks, have seed banks with medicinal plant varieties in various nations.

4.2.2.3 Field Gene Banks

Field gene banks use biotechnological tools for conservation of plants. It includes *in vitro* propagation and molecular masker technology. To preserve vegetatively propagated seed-producing species, a variety of *in vitro* procedures have been devised. Land and labour are needed in order to conserve the genetic material of vegetative propagated plants and forest species in field gene banks, whereas *in vitro* storage requires minimal space in growth rooms for maintaining thousands of genotypes together (Kadam and Pawar, 2020).

Cryopreservation involves storing tissue and cell cultures at extremely low temperatures, sometimes as low as -196°C (Kadam and Pawar, 2020). The steps listed in what follows are part of the new cryopreservation procedure.

1. Dehydration and encapsulation
2. Vitrification
3. Desiccation
4. Pre-growth
5. Desiccation
6. Freezing of droplets

Cryopreservation of *Plasmodium* sporozoites has been successfully carried out for use in malarial research and vaccine development (Bowers et al., 2022). In another study, *Cryptolepis sanguinolenta* is used as an antimalarial treatment in Ghana. Freshly harvested seeds of *Cryptolepis sanguinolenta* were found to be most suitable for cryopreserving them in order to improve the germinability of seeds (Opoku-Agyemang et al., 2022).

4.2.3 Cultivation Practices

Cultivation practices are used to employ new ways to address issues that arise during the cultivation of medicinal plants. Cultivation practices face dominant issues, such as the presence of poisonous components, low concentrations of active substances, pesticide contamination, and incorrect botanical provenance (Chen at al., 2016). Cultivation practices depend on many factors for increase in product yields, such as optimal amount of nutrients, water, and additives, and environmental elements like temperature, light, and humidity are provided throughout the cultivation processes in order to increase product yields. The good agriculture and collection practices (GACP) programme assures that the best collection techniques are used for both small- and large-scale gathering of fresh medicinal plants while preserving the long-term survival of wild populations and their environment. For each species of medicinal plant and plant part used, GACP specifies appropriate collection techniques. The good agricultural practice (GAP) rules guarantee that the plant materials are completely protected from the elements that degrade the quality of medicinal plants (Raina et al., 2011). The recommended guidelines are provided by WHO for achievement of quality assurance of herbal medicine. WHO aims to improve the standards of medicinal plant, starting with collection, cultivation, and post-harvest procedures. GAP implementation will encourage

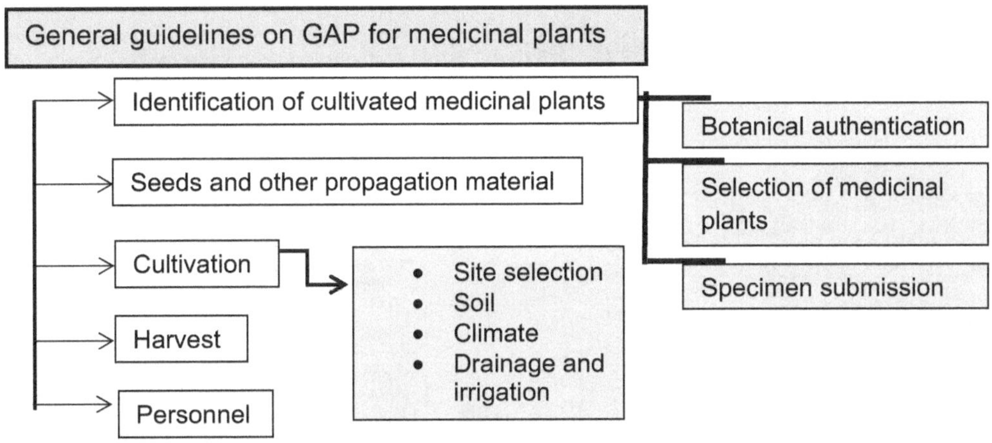

FIGURE 4.2 Guidelines on GAP for medicinal plants.

the efficient use of water resources, herbicides, and fertilisers. Also, GAP implementation would produce environment-friendly agriculture practices (Preet and Ashish, 2018). Protecting the health of agricultural workers from the improper use of herbicides and pesticides is another aspect of its social responsibility. Figure 4.2 represents general guidelines by WHO 2003 for medicinal plants on GAP. As per the guidelines, botanical species selected for cultivation should match those that are listed in the national pharmacopoeia of the country. Each therapeutic plant being grown should have its botanical identity, including its scientific name (genus, species, subspecies, author, and family), which is validated and recorded. A voucher botanical specimen should be submitted to a regional or national herbarium for authentication and identification. All essential information regarding the identity, quality, and performance of seeds and other propagation materials should be specified by the suppliers. Cultivation of medicinal plants requires intense management and care. Selection of site, ecological factors like soil, climate, and irrigation practices are the major factors for cultivating medicinal plant. After cultivation, the harvest time varies, depending on the part used. To ensure the best quality of medicinal plant components and processed herbal products, it is important to harvest medicinal plants at the right time of year or season. Properly trained personnel (producers and growers) are required for harvesting and post-harvest maintenance of medicinal plants.

4.3 CONSERVATION THROUGH TISSUE CULTURE

Cells, tissues, organs, or whole plants can be cultured *in vitro* under regulated nutrient and environmental conditions (Figure 4.3). Plant species have different therapeutic effects, depending on the active ingredient. Tissue cultures are used to conserve plants that produce bioactive compounds responsible for disease treatment. Tissue culture is used to propagate plant species whose regeneration via standard techniques is challenging. Tissue culture is a method that can be used to preserve plant species whose numbers have declined owing to overexploitation (Thorpe, 2007). An efficient and cost-effective *in vitro* propagation method has been developed for *Artemisia annua* L., an important antimalarial drug plant in providing enormous potential for manufacturing artemisinin (Sharma et al., 2011). *In vitro* propagation is still regarded as a crucial method in conservation since it offers an option for plant species that have poor reproductive capabilities. For the final recovery of transgenic plants, a tissue culture step is a requirement for the majority of current transformation techniques. Several plants species effective against malaria have been used for *in vitro* propagation (Table 4.1) and also elicitated for their respective bioactive compounds.

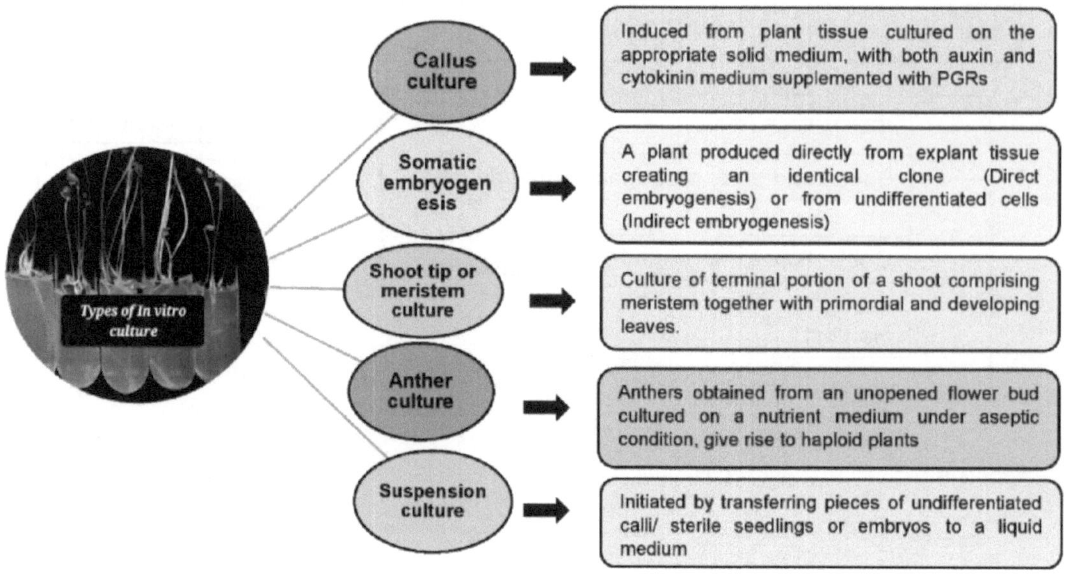

FIGURE 4.3 Types of *in vitro* culture for conserving various antimalarial plant species.

TABLE 4.1
Some Important Examples of *in Vitro* Propagation of Antimalarial Plant Species

S. No.	Plant Species	Explants	Media	Key Reference
1.	*Aristolochia indica* L.	Nodal and internodal explant	Murashige and Skoog (1962) (MS medium)	Prabha et al., 2008
2.	*Aristolochia saccata* Wall., *A. cathcartii* Hook. f.	Nodal explant	MS	Sarma et al., 2017
3.	*Artemisia annua* L.	Nodal explant	MS	Sharma et al., 2011
4.	*Artemisia annua* L.	Leaf explants	MS	Alejos-Gonzalez et al., 2013; Paul et al., 2012; Zayova et al., 2020
5.	*Artemisia annua* L.	Nodal stem explant	MS	Hailu et al., 2013; Janarthanam et al., 2012; Gaurav et al., 2011
6.	*Artemisia vulgaris* L.	Nodal explant	MS	Jogam et al., 2020
7.	*Aspidosperma polyneuron* Müll. Arg.	Epicotyl and hypocotyl nodal segments	Woody plant medium (WPM)	Ribas et al., 2017
8.	*Cinchona ledgeriana* Moens	Nodal segment	MS	Mayerni et al., 2015
9.	*Cinchona officinalis* L.	Nodal segment	Gamborg B5	Armijos et al., 2021
10.	*Croton lechleri* (Muell. Arg.)	Nodal segment and apical bud	Quoirin andLeproivre medium	Oliveira et al., 2022
11.	*Cryptolepis sanguinolenta* Lindl. Schltr.	Nodal explant	MS	Monney et al., 2016
12.	*Momordica charantia* L.	Shoot tips	MS	Minnu et al., 2022
13.	*Momordica cymbalaria* Fenzl.	Young leaves, nodes, and shoot tips	MS	Chaitanya et al., 2020

(*Continued*)

TABLE 4.1 (*Continued*)
Some Important Examples of *in Vitro* Propagation of Antimalarial Plant Species

S. No.	Plant Species	Explants	Media	Key Reference
14.	*Ophiorrhiza* sp.	Shoot tips or axillary buds	MS	Pandey et al., 2022
15.	*Spilanthes acmella* L.	Leaf explant	MS	Pandey and Agrawal, 2009
16.	*Stachytarpheta jamaicensis* (L.) vahl.	Shoot tips	MS	Rajender et al., 2017
17.	*Tabebuia donnell-smithii* Rose	Axillary shoots	MS	Gonzalez et al., 2010

4.4 SUSTAINABLE HARVESTATION

Sustainable harvesting is defined ecologically as harvesting that enables population numbers to be sustained or grown over time. Sustainable harvesting emphasizes the preservation of species populations in the wild despite strong demand around the globe (Deepa et al., 2018). Destructive harvesting typically leads to resource exhaustion. Destructive harvesting leads to the extinction for medicinal plants with low abundance and slow growth. Hence, it is very important to think about the sustainable use of medicinal plants and to develop good harvesting procedures (Larsen and Olsen, 2007). Compared to harvesting of leaves, flowers, or buds, medicinal plants (such as herbs, shrubs, and trees) suffer more damage from root and whole-plant harvesting (Table 4.2).

4.4.1 Principles of Sustainable Harvesting

4.4.1.1 Resource Mapping, Estimation, and Value Analyses

Through the use of this principle, the local community can comprehend the availability of resources and their conditions in the wild in a better way. Also, this principle is quite useful to reveal about the potential quantity that can be collected and the effects of destructive harvesting. Furthermore, it offers the opportunity to comprehend the worth of products obtained through the value chain, the advantages enjoyed by the various stakeholders, and their duties in resource conservation. This principle allows the local community to keep an eye on the effects of destructive harvesting and already-established value chains. Through such initiatives, they are able to create alternative value chain models, harvest selected species sustainably, and create a participative management plan for its execution based on an adaptive management strategy.

4.4.1.2 Participatory Approach Involving Local Community in Conservation

The local community can use this principle to identify and choose species and locations for sustainable collection. It is based on the criteria that develop the approach for sustainable harvesting and take traditional knowledge into account. The following factors are taken into consideration when choosing a species: (a) species that offer greater livelihood opportunities, encouraging the local population to participate in the sustainable management of the resources; (b) endemic species; (c) species with multiple harvestable parts; (d) species collected only from the wild; and (e) highly traded species.

4.4.1.3 Documentation, Assimilation, and Application of Traditional Harvesting Knowledge

This principle aids the local population in the documentation of traditional knowledge concerning harvesting practices, ensures the rights of forest dwellers through their exclusive participation

TABLE 4.2
General Recommendations for Time and Method of Collection of Different Plant Parts

S. No.	Plant Parts	Time of Collection	Method of Collection
1.	Bark	Spring or autumn	Remove the bark in long vertical strips using a thin flexible blade/knife. Avoid ring barking, which involves cutting off entire tree rings.
2.	Leaves	In dry weather during flowering period	Individual leaves are taken out instead of leaf stripping. Harvesting of leaves should occur prior to or at the beginning of flowering.
3.	Roots/rhizomes	Before flowering for annuals During autumn/winter for biennials after growth of 1 year and after 2 or more years for perennials	Dig the root at least 30 cm away from the main stem or taproot. Collect lateral roots only.
4.	Flowers	During dry weather and early morning	Pick flowers gently to avoid damaging the main stem of the plant. To capture the aroma of the flower, it must be picked right after it blooms or soon after.
5.	Bulbs	After flowering and fruiting/late autumn	Gather the large mature bulbs, leaving the small ones for regrowth. Bulbs should be collected only after seed shedding.
6.	Fruits and seeds	Collect fully ripened fruits	It is advisable to collect a little earlier in order to prevent seed dispersal. To make it easier to collect the fruits and seeds, branches of the tree or shrub should not be clipped.

Source: WHO (2003).

without the use of outside labour forces. It combines scientific knowledge with traditional knowledge to create good harvesting practises and acknowledges the relationship between community and resources.

4.4.1.4 Development of Sustainable Harvesting Protocols, Tools, and Field Implementation

This principle allows creating species-specific sustainable harvesting practises (SHPs), which are thorough and all-encompassing and include ecological knowledge of the species, their collection, value addition, and marketing. Species and location profiles, distribution, phenology, morphology, threat status, medicinal uses, propagation methods, resource assessments, best collecting procedures, and post-harvesting methods are among the elements of SHP. The five basic core elements (Figure 4.4) for sustainable harvesting are (Deepa et al., 2018):

1. Part to be harvested
2. Stage of collection
3. Time of collection (season as well as frequency)
4. Method of harvesting
5. Quantity of collection

4.4.1.5 Harvest of Different Plant Parts

For the production of herbal medicines, various plant parts, including roots, rhizomes, bulbs, bark, leaves, fruit, and seeds, are collected and processed for further usages. Harvesting methods and timing depend on the species and plant sections. Table 4.2 provides general recommendations on technique and timing for harvesting various plant sections.

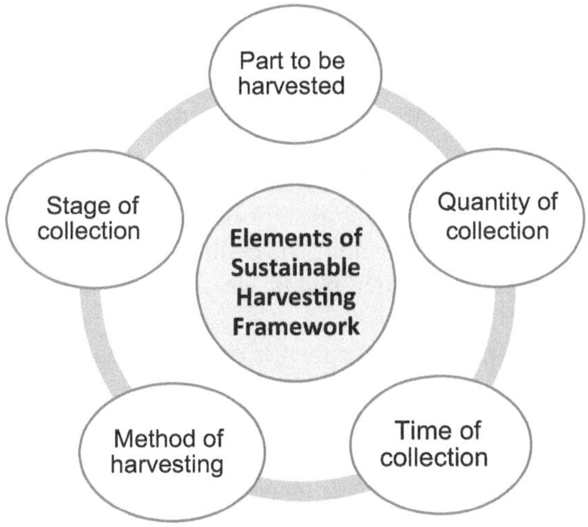

FIGURE 4.4 Five elements of sustainable harvesting.

4.4.1.6 Training, Capacity Building, and Dissemination

This principle facilitates the dissemination of methodology-related information. Its success lies in being able to solicit feedback from others and revise the adaptive management plan. At the local level, interpersonal communication is helpful, whereas group/mass communication can effectively reach other stakeholders.

4.4.1.7 Linking Trade, Price, and Conservation Principles and Practices

The economic value chain is enhanced by increasing the skills of various stakeholders regarding the significance of sustainable harvesting, post-harvesting procedures, marketing, and certification. The financial incentives directly benefit the primary stakeholders and the dependent community. One of the crucial factors in obtaining higher pricing at domestic and international markets is certification of the sustainably harvested products. An additional issue in the trading of natural resources is price fixing and benefit sharing. Giving the dependent local community more control over these matters will benefit them economically and contribute to resource conservation.

4.4.1.8 Interventions for Post-Harvesting Techniques and Market Linkages

Techniques used after harvesting, such as drying, storing, adding value, labelling, and branding, bring more money and create jobs for the local population. To guarantee the supply of high-quality raw materials and to provide health security, value-added goods are directly related to the herbal industries. As the only financial incentive available to the stakeholders to reward their participation in the sustainable collection operations, marketing is a crucial intervention for sustainably gathered items.

4.5 GUIDELINES FOR SUSTAINABLE HARVESTING OF MEDICINAL PLANTS (WHO, 2003)

One of the most serious environmental issues today is biodiversity loss, which is also being held increasingly responsible for the Anthropocene. The biggest problem that the world is facing currently is the increased biodiversity loss rate. There are more threatened species than the resources needed to save them. Conservationists have become keen to make efforts about the need to prioritise conservation of the species (Tali et al., 2019). In growing nations, almost 80% of the population relies on resources from plants for their indigenous medical practices. Most traded medicinal plants in India are taken from the wild, more commonly in a damaging and unsustainable way. The genetic stock that is currently available and the variety of medicinal plants across the Indian Himalayan region have both been put under serious threat. Moreover, studies from the area have revealed that medicinal plants are rapidly disappearing as a result of being over-harvested for commerce (Pandey et al., 2019).

Keeping in mind the threat to medicinal plants due to overexploitation and unscientific harvesting, there is a need to focus more on developing and shifting toward the sustainable scientific methods to harvest medicinal plants without harming the plant (Raina and Gautam, 2020). WHO in 2003 gave the guidelines for harvesting medicinal plants, which are as follows:

- Sustainable practices should be followed while collecting/harvesting the plants.
- Harvesting should be done under favourable conditions. Harvesting in wet/moist environment should be avoided. If harvesting is done in wet situations, the harvested material should be dried immediately in indoor conditions to prevent the harmful impacts of moisture, like growth of pathogenic microbes favoured in moist conditions.
- After harvesting, the plant material should be prevented from direct soil contact. In case any underground part like bulb/rhizome or root is harvested, the soil particles attached to the material should be removed immediately after harvesting to control the microbial load on the harvested material.
- The material used for storage of harvested medicinal plants should be free from any contamination, including the remaining parts of previously collected plants. After being collected, the plant materials may go through the necessary preliminary processing, such as removing pollutants and washing (to remove extra soil) unwanted materials, cutting, and sorting.
- Transportation should be done fast under contamination- and moisture-free conditions. They can be transported to a central location for transit to the processing facility in moisture-free bags, clean baskets, trailers, or other well-aerated containers. The raw medicinal plant materials may need to be air- or sun-dried before shipping if the collection site is some distance from the processing facilities.
- The various plant parts should be collected separately and transported in separate containers if more than one component of the medicinal plant is to be collected. Contamination from other parts must always be prevented. Avoid any compaction or mechanical damage of the raw medicinal plant components that could lead to degradation of the quality. During harvest, it is important to identify and remove any decomposed medicinal plant elements to prevent microbial contamination and product quality loss.
- Each container used for collection/storage of medicinal plant should be contamination-free. Moisture should be checked if plastic containers are used, as moisture can favour the growth of moulds in the container. If the containers are not in use, then they must be stored under dry conditions, and the storing area must be inaccessible to insects, rodents, or any other pests.
- Birds, domestic animals, rodents, insects, and other pests should be kept away from collected plant material as they can damage the plant material.
- The tools like shears, mechanical tools, and machetes used in harvesting of medicinal plant material should be clean, free from rust and contamination. These tools should also be stored under dry conditions, which must be away from the reach of insects or pests. The part of the tools which comes in contact with the plant should not be excessively greased.

4.6 CONCLUSION AND FUTURE REMARKS

Conservation and domestication methods are advised for antimalarial plant species that sustain large-scale cultivation in order to maintain a sustainable supply of plant materials and avoid its extinction. In-depth ethnobotanical studies are required to document the regional therapeutic use of newly discovered antimalaria and other medicinal plants. Therefore, it is essential that endangered antimalarial plants are managed properly through conservation in order to ensure their long-term availability. In order to prevent the extinction of antimalarial plants and the exploitation of indigenous populations, individuals, societies, socio-groups, and governmental as well as nongovernmental organizations should develop plans that could aid in their conservation. Conservation assessment and management plans (CAMP) provide strategic and comprehensive guidance for conservation and management of threatened plant species. It is necessary to raise awareness among key stakeholders of the need for resource development, restoration, and conservation, owing to the effects of climate change. Using a variety of precautions and techniques, sustainable harvesting is achievable. Broadly speaking, protecting plants from fire and grazing, fostering young regeneration, controlling extraction, and popularizing various medicinal plant uses are some methods to sustainably harvest plants for their therapeutic properties. Furthermore, large-scale biosynthesis of natural products is now feasible due to genetic engineering, and developments in medicinal plant fermentation and tissue culture have created new opportunities for the highly effective and large-scale manufacture of desirable bioactive chemicals against malaria.

REFERENCES

Alejos-Gonzalez, F., Perkins, K., Winston, M. I., & Xie, D. Y. 2013. Efficient somatic embryogenesis and organogenesis of self-pollination *Artemisia annua* progeny and artemisinin formation in regenerated plants. *American Journal of Plant Science* 4:2206–2217.

Armijos-González, R., Espinosa-Delgado, L., & Cueva-Agila, A. 2021. Indirect shoot regeneration using 2, 4-D induces somaclonal variations in *Cinchona officinalis*. *Floresta e Ambiente* 28:1–8.

Arya, A., Kumar, S., Paul, R., Suryavanshi, A., Kain, D., & Sahoo, R. N. 2021. Ethnopharmacological survey of indigenous medicinal plants of Palampur, Himachal Pradesh in north-western Himalaya, India. *Advances in Traditional Medicine* 9:1–44.

Asafo-Agyei, T., Blagogee, H. R., Mintah, S. O., Archer, M. A., Ayertey, F., Sapaty, A. C., & Appiah, A. A. 2019. Ethnobotanical studies of medicinal plants used in traditional treatment of malaria by some herbalists in Ghana. *Journal of Medicinal Plants Research* 13(16):370–383.

Barboza, G. E., Cantero, J. J., Núñez, C., Pacciaroni, A., & Ariza Espinar, L. 2009. Medicinal plants: A general review and a phytochemical and ethnopharmacological screening of the native Argentine Flora. *Kurtziana* 34:7–365.

Bowers, C., Hancox, L., Peissig, K., Shiau, J. C., Vantaux, A., Witkowski, B., & Kurup, S. P. 2022. Cryopreservation of plasmodium sporozoites. *Pathogens* 11(12):1487.

Chaitanya, G., Suvarchala, V., Ramakrishna, D., Sunitha, D., & Shasthree, T. 2020. In vitro plant regeneration of *Momordica cymbalaria* Fenzl. and assessment of genetic fidelity using ISSR primers. *Research Journal of Biotechnology* 15:103–110.

Chen, S. L., Yu, H., Luo, H. M., Wu, Q., Li, C. F., & Steinmetz, A. 2016. Conservation and sustainable use of medicinal plants: Problems, progress, and prospects. *Chinese Medicine* 11:1–10.

Deepa, G. B., Jagannatha Rao, R., Suresh, H. M., Mark, A. S., & Pisupati, B. 2018. *Principles and Practices of Sustainable Use and Sustainable Harvesting*. FLEDGE, Bengaluru, India.

"Ex-Situ Conservation." *Botanical Survey of India*. Government of India. bsi.gov.in/content/256_1_Exsitu Conservation.aspx.

Gaurav, S., Vinay, S., & Veena, A. 2011. An efficient micropropagation protocol of an elite clone EC-353508 of *Artemisia annua* L., an important antimalarial plant. *International Journal of Pharma and Bio Sciences* 2:205–214.

Gonzalez, J. A., Ramírez-Garduza, F., Robert, M., O'Connor, A., & Peña-Ramírez, Y. 2010. Adventitious shoot induction from adult tissues of the tropical timber tree yellow Ipé primavera (*Tabebuia donnell-smithii* rose [Bignoniaceae]). *In Vitro Cellular & Developmental Biology – Plant* 6:411–421.

Hailu, T., Abera, B., & Mariam, G. 2013. In vitro mass propagation of Artemisia (*Artemisia annua* L.) cv: Anamed. *Plant Tissue Culture Biotechnology* 23:165–176.

Howes, M. J. R., Quave, C. L., Collemare, J., Tatsis, E. C., Twilley, D., Lulekal, E., . . . & Nic Lughadha, E. 2020. Molecules from nature: Reconciling biodiversity conservation and global healthcare imperatives for sustainable use of medicinal plants and fungi. *Plants, People, Planet* 2(5):463–481.

Huang, H., Han, X., Kang, L., Raven, P., Jackson, P.W., & Chen, Y. 2002. Conserving native plants in China. *Science* 297:935–936.

Husen, A. 2021. *Traditional Herbal Therapy for the Human Immune System*. CRC Press, Boca Raton, FL. https://doi.org/10.1201/9781003137955

Husen, A. 2022a. *Herbs, Shrubs and Trees of Potential Medicinal Benefits*. CRC Press, Boca Raton, FL. https://doi.org/10.1201/9781003205067

Husen, A. 2022b. *Exploring Poisonous Plants: Medicinal Values, Toxicity Response and Other Potential Uses*. CRC Press, Boca Raton, FL. https://doi.org/10.1201/b23017

Husen, A. 2023. *Antidiabetic Medicinal Plants and Herbal Treatments*. CRC Press, Boca Raton, FL.

Janarthanam, B., Rashmi, P., & Sumathi, E. 2012. Rapid and efficient plant regeneration from nodal explants of *Artemisia annua* L. *Plant Tissue Culture Biotechnology* 22:33–39.

Jogam, P., Sandhya, D., Shekhawat, M. S., Alok, A., Manokari, M., Abbagani, S., & Allini, V. R. 2020. Genetic stability analysis using DNA barcoding and molecular markers and foliar micro-morphological analysis of in vitro regenerated and in vivo grown plants of *Artemisia vulgaris* L. *Industrial Crops and Products* 151:112476.

Kadam, S. T., & Pawar, A. D. 2020. Conservation of medicinal plants: A review. *International Ayurvedic Medical Journal* 8:3890–3895.

Larsen, H. O., & Olsen, C. S. 2007. Unsustainable collection and unfair trade? Uncovering and assessing assumptions regarding Central Himalayan medicinal plant conservation. *Biodiversity and Conservation* 16:1679–1697.

Maunder, M., Higgens, S., & Culham, A. 2001. The effectiveness of botanic garden collections in supporting plant conservation: A European case study. *Biodiversity and Conservation* 10:383–401.

Mayerni, R., Eka Pratiwi, E., & Warnita, W. 2015. Shoot multiplication of quinine plant (*Cinchona ledgeriana* Moens) with several concentrations of kinetin on in vitro. *International Journal on Advanced Science, Engineering and Information Technology* 5(2):57–61.

Minnu, A. J., Kumar, T. P., Reshmika, P. K., Mathew, D., Veni, K., & Varun, R. C. 2022. Characterization and maintenance of promising gynoecious bitter gourd line through hormonal regulation and micropropagation. *Indian Journal of Horticulture* 79(3):287–295.

Monney, M. A. D., Amissah, N., & Blay, E. 2016. Influence of BA and IBA or NAA combinations on micropropagation of *Cryptolepis sanguinolenta*. *American Journal of Plant Sciences* 7(3):572–580.

Mshana, N. R. 2000. *Traditional Medicine and Pharmacopoeia: Contribution to the Revision of Ethnobotanical and Floristic Studies in Ghana*. Organization of African Unity/Scientific, Technical & Research Commission, Accra.

Murashige, T., & Skoog, F. 1962. A revised medium for rapid growth and bioassays with tobacco tissue cultures. *Physiologia Plantarum* 15:473–497.

Oliveira Junior, J. B. D., Pessoa, C. M. P., Scherwinski-Pereira, J. E., da Silva Lopes, H., & da Silva Costa, F. H. 2022. A simple, alternative and efficient sealing system to improve natural ventilation in culture vessels and the morphophysiological and anatomical quality of *Croton lechleri* (Muell. Arg.) grown in vitro. *Biologia* 77(10):2945–2954.

Opoku-Agyemang, F., Dodoo, J. N. O., Hlomador, T. E., Gilday, K., & Amissah, J. N. 2022. Conservation and sustainable use of Crytolepis sanguinolenta. In *Herbs and Spices-New Advances*. IntechOpen, London, United Kingdom. doi: 10.5772/intechopen.108249

Pandey, A., Chandra Sekar, K., Joshi, B., & Rawal, R. S. 2019. Threat assessment of high-value medicinal plants of cold desert areas in Johar valley, Kailash Sacred Landscape, India. *Plant Biosystems-An International Journal Dealing with All Aspects of Plant Biology* 153(1):39–47.

Pandey, D. K., Konjengbam, M., Ghorai, M., Dwivedi, P., Roy, D., Kant, N., & Dey, A. 2022. Biotechnology for micropropagation and camptothecin production in *Ophiorrhiza* sp. *Applied Microbiology and Biotechnology* 106(11): 3851–3877.

Pandey, V., & Agrawal, V. 2009. Efficient micropropagation protocol of *Spilanthes acmella* L. possessing strong antimalarial activity. *In Vitro Cellular & Developmental Biology-Plant* 45:491–499.

Paul, S., Gupta, M. M., & Khanuja, S. P. S. 2012. Maintaining the artemisinin content through direct and indirect in vitro regeneration and their assessment of variations with the field grown mother plants of *Artemisia annua* L. *Natural Products Bioprospecting* 2:194–199.

Prabha, A., Subbu, R., & Kulloli, S. 2008. Micropropagation of *Aristolochia indica L.* (Aristolochiaceae). *Ecology, Environment and Conservation* 14:409–411.

Preet, A. S., & Ashish, B. 2018. Good agricultural practices: A prequisite approach for enhancing the quality of Indian Herbal Medicines. *Biomedical Journal of Scientific & Technical Research* 5: 1–4.

Qayum, A., Arya, R., & Lynn, A. M. 2016. Ethnobotanical perspective of antimalarial plants: Traditional knowledge based study. *BMC Research Notes* 9:1–20.

Rabgyal, J., & Pelden, K. 2020. *Sustainable Harvesting Practices for Endangered Medicinal Plants of Bhutan*. Agriculture Research and Development Centre, Yusipang, Thimphu, Bhutan.

Raina, R., Chand, R., & Sharma, Y. P. 2011. Conservation strategies of some important medicinal plants. *International Journal of Medicinal and Aromatic Plant* 1: 342–347.

Raina, R., & Gautam, K. (2020). Conservation and Utilization of High-Altitude Threatened Medicinal Plants. In: Rajasekharan, P., Wani, S. (eds) *Conservation and Utilization of Threatened Medicinal Plants*. Springer, Cham. https://doi.org/10.1007/978-3-030-39793-7_13

Rajender, K., Prasad, B., Rohela, G., Pendli, S., & Reuben, T. C. 2017. In vitro micropropagation of *Stachytarpheta jamaicensis* l.vahl., an ethnomedicinal plant and confirmation of genetic fidelity of the plantlets using RAPD markers. *International Journal of Advanced Research* 5:1494–1502.

Ribas, L. L. F., Guerra, M. P., Kulchetscki, L., & Zanette, F. 2017. Micropropagation of *Aspidosperma polyneuron* Müll. Arg. From in vitro germinated seedlings. *Ciência Florestal* 27:391–402.

Sahoo, R. N., Kumar, S., Suryavanshi, A., Kain, D., Arya, A., & Chaudhry, B. 2022. Venerated and medicinal aspects of plants used in India: An ethnobotanical review. *Journal of Drug Research in Ayurvedic Sciences* 6(3):128–140.

Sarma, B., & Tanti, B. 2017. In vitro regeneration of plantlets from nodal explants of *Aristolochia saccata* and *Aristolochia cathcartii*. *European Journal of Biological Research* 7:191–201.

Schoen, D. J., & Brown, A. H. 2001. The conservation of wild plant species in seed banks: Attention to both taxonomic coverage and population biology will improve the role of seed banks as conservation tools. *BioScience* 51(11):960–966.

Sharma, G., Shankar, V., & Agrawal, V. 2011. An efficient micropropagation protocol of an elite clone EC-353508 of *Artemisia annua* L., an important antimalarial plant. *International Journal of Pharma and Bio Sciences* 2:205–214.

Tali, B. A., Khuroo, A. A., Nawchoo, I. A., & Ganie, A. H. 2019. Prioritizing conservation of medicinal flora in the Himalayan biodiversity hotspot: An integrated ecological and socioeconomic approach. *Environmental Conservation* 46(2):147–154.

Thorpe, T. 2007. History of plant tissue culture. *Journal of Molecular Biotechnology* 37:169–180.

Volis, S., & Blecher, M. 2010. Quasi In situ: A bridge between ex situ and in situ conservation of plants. *Biodiversity and Conservation* 19:2441–2454.

World Health Organization. 2003. *Guidelines on Good Agricultural and Collection Practices (GACP) for Medicinal Plants*. World Health Organization, Geneva.

World Health Organization. 2016. *World Malaria Report*. World Health Organization, Geneva, Switzerland.

Zayova, E., Nedev, T., Petrova, D., Zhiponova, M., Kapchina, V., & Chaneva, G. 2020. Tissue culture applications of *Artemisia annua* L. callus for indirect organogenesis and production phytochemical. *Plant Tissue Culture Biotechnology* 30:97–106.

5 Toxic Compounds Associated with Some Antimalarial Plant Species

Shivansh Negi, Noor Sayed, and Manu Pant

5.1 INTRODUCTION

Plasmodium is the protozoan parasite responsible for the fatal disease malaria. *Plasmodium vivax* and *Plasmodium falciparum* are two of the different causative species that are known to be the deadliest for human life, causing the deaths of more than 6 lakh people annually. Over the years, chemical-based treatment in allopathy has established itself as the most effective line of treatment to cure malaria (Kar and Kar, 2010). The drugs used for the purpose are either single agents or combination treatments. However, continual and extensive use of allopathic medications has resulted in rising drug resistance to these treatments (Ijarotimi et al., 2010). Consequently, an herbal line of treatment has emerged as a possible solution to the problem. Many plants have been mentioned as malaria treatments in traditional systems and folk remedies. The utilization of plant-based antimalarial drug dates back to as long as 1632, when cinchona bark was used to treat malaria patients, leading to the first plant-based drug, quinine (Adebayo and Krettli, 2011). Another important plant is *Artemisia annua*, from which the drug artemisinin is prepared and which has proven to be a very effective drug to treat malaria. Subsequently, the exploration of antimalarial properties of plants gained momentum, and over 2,000 phytochemicals are now reported to be effective against malarial infection.

Despite the fact that herbal drugs are more readily available and well-accepted by users, they cannot be characterized as completely non-toxic and non-debilitating (Voravuth et al., 2016; Husen, 2023). It is important to understand that the phytochemical composition of a medicinally active plant largely depends on its genotype, physiological parameters of growth, plant part used to prepare the extract, and the harvesting and post-harvesting methods (Husen, 2021, 2022a, 2022b). Variations in these parameters have been observed to alter the pharmacokinetics of plant-based drugs (Eseyin et al., 2014). The cytotoxicity of these plant extracts has been the subject of another significant study (Husen, 2023). Individual studies have reported the lethal effect of phytoconstituents on mammalian cell lines, indicating their toxicity to living cells (Table 5.1).

The present chapter is a compilation of information on the cytotoxicity of medicinally active plants that are known to be useful in curing malaria. Prominent antimalarial plant species are being discussed whose ingredients have been shown to be toxic to some type of live cells. The present knowledge will be useful in the utilization of these plants for the preparation of safe antimalarial drugs, either alone or in combination with other potential antimalarial plant species.

5.2 *ALLIUM* SPP.

The genus *Allium* consists of pungent bulbous herbs which belong to the family Amaryllidaceae. Plants from this genus are widely distributed throughout the world except in Australia, New Zealand, and other tropical regions of the world. They are mainly used as food crops and ornamental plants around the world.

DOI: 10.1201/9781003378396-5

TABLE 5.1
Cytotoxicity of Some Medicinal Plants Known to Have Antimalarial Property

Genus	Species	Plant Part	Cytotoxic Compound	Cytotoxicity	Cell Line
Allium	*sativum*	Garlic pearls	Allicin	Moderately cytotoxic	Human embryonic kidney cell line HEK 293
Allium	*paradoxum*	Aerial parts and bulb			
Aspidosperma	*pyrifolium* and *megalocarpon*	–	Aspidolimidine	Cytotoxic	Fibroblast cell line NIH 3T3
Aspidosperma	*olivaceum*	Stem bark	Uleine and apparacine	Moderately cytotoxic	Human hepatoma cells (HepG2) and normal kidney glomerular cells (BGM)
Aspidosperma	*vargasii* and *olivaceum*	–	Ellipticine and olivacine	Very low cytotoxicity	Mouse macrophages
Bixa	*orellana*	Roots	Shwarane, ellagic acid, δ-tocotrienol, bixin, stigmasterol, β-sitosterol, inositol, ursolic acid, maslinic acid, and arjunolic acid, out of which ishwarane, δ-tocotrienol, stigmasterol	Slightly cytotoxic	Fibroclasts (BJ cells), human embryonic kidney (HEK-293) cells
Canella	*winterana*	Leaves	Muzigadial	Highly cytotoxic	Chinese hamster ovary (CHO) cells
			9-epideoxymuzigadial, 9-deoxymuzigadial, 3-β-acetoxypolygodial, muzigodiol	Moderately cytotoxic	
Citrus	spp.	Peels	Limonene	Slightly cytotoxic	Kidney cell line (LLC-MK2)
Holarrhena	*antidysenterica*	Stem bark	Conessine	Slightly cytotoxic	Mononucleate myoblast (L-6 cell line)
Hypercium	*lanceolatum*	Stem bark	Betulinic acid	Mild cytotoxicity	Monkey kidney epithelial cell line (LLC-MK2)
Kigelia	*africana*	Stem bark	Atranorin	Moderately cytotoxic	
			2β, 3β, 19αtrihydroxy-urs12-en-28-oic acid	Slightly cytotoxic	
Ocimum	*tenuiflorum*	Roots, leaves	Eugenol	Significantly cytotoxic	Human gingival fibroblasts, human oral cells
Ocimum	*gratissimum*	Leaves			
Piper	*tricuspe*	Whole plant	Dycthiochromenol, 3-farnesyl p-hydroxy benzoic acid, 2′E,6′E 2-farnesyl hydroquinone	Significantly cytotoxic	Mononucleate myoblast (L-6 cells)
Piper	*coruscans*	Aerial parts	Strobopinin 7-methyl ether, 5-hydroxy-7-methoxy6, 8-dimethyl flavanone, desmethoxymatteucinol, alpinetin, pinocembrin, dimethyl cryptostrobin	Slightly cytotoxic	Primary human dermal fibroblasts

(Continued)

TABLE 5.1 (*Continued*)
Cytotoxicity of Some Medicinal Plants Known to Have Antimalarial Property

Genus	Species	Plant Part	Cytotoxic Compound	Cytotoxicity	Cell Line
Rumex	*crispus*	Leaves, root, fruit	Nepodin	Moderately cytotoxic	Vero cells
Stephania	*dielsiana*	Leaves	Oxostephanine and thailandine	Cytotoxic	Human mammary epithelial cell (MCF10A)
Stephania	*venosa*	Tubers	Stephanine	Highly cytotoxic	Non-cancer breast cell line (184B5)
Stephania	*zippeliana*	Leaves	Xylopine	Low cytotoxicity	Human embryonic kidney cells (HEK 293)
Tebebuia	spp.	Leaves	β-lapachone	Highly cytotoxic	Pig kidney epithelial cell lines
			α-lapachone	Highly cytotoxic	
			α-lapachone	Slightly cytotoxic	Mouse neural crest–derived cell lines
			Atovaquone	Highly cytotoxic	
Teclea	*trichocarpa*	Leaves	Arborinine, skimmianine	Moderately cytotoxic	Rat skeletal muscle myoblast (L-6) cell line
Zingiber	*officinale*	Rhizome	Gingerol	Slightly cytotoxic	Normal mammary epithelial cells (MCF-10A)
			Zerumbone	Slightly cytotoxic	Vero cell line
	zerumbet	Rhizome	Zerumbone	Slightly cytotoxic	Vero cell line

Allium sativum, commonly called garlic, is a plant famous for its taste in culinary dishes and also for its pharmacological properties. It is commonly used in treating several disorders. The ability of garlic to fight infection has been known and utilized by China and India for a long time (Coppi et al., 2006). Its use in treatment of malaria is also documented. Garlic extract in combination with artemisinin and artemether have been shown to be effective against *Plasmodium berghei* in mice models (Vathsala and Murthy, 2020).

Another plant species of this genus, *Allium paradoxum*, has been used as an herbal medicine in many areas of the world. It is also called few-flowered garlic. This plant is originally native to the Mazandaran province of Iran. It is mainly used for the treatment of digestive ailments. It has shown anticancer, antioxidant, and antibacterial properties. Hence, its antimalarial properties were studied in a separate study. The extract of its aerial part and bulb showed significant activity against *Plasmodium falciparum* and *Plasmodium bergei* in mice model (Elmi et al., 2021). This plant also contains allicin (Khanabadi et al., 2022), which is abundantly present in *Allium sativum*.

Garlic is rich in thiosulfinate allicin, which is produced when the enzyme allinase reacts with the compound alliin only when the clove is crushed. The component accounts for the major healing properties of the plant (Satyal et al., 2017). The antimalarial activity of allicin has been reported in vivo in the mice model of *Plasmodium berghei* (Coppi et al., 2006) and *Plasmodium yoelii* (Feng et al., 2012). In a separate research work, the microtubule-disrupting action of allicin was reported to be cytotoxic towards NIH-3T3 fibroblasts, which is a type of murine embryonic fibroblast cell line. This compound significantly restricted migration and polarization and induced microtubule disassembly of these cells (Khoutorsky et al., 2007). Moreover, the studies confirmed that high levels of allicin can be toxic to the organism. In the mice model, a high dose of allicin shows toxicity with 50% lethal dose of 60 mg/kg after intravenous injection (Coppi et al., 2006). Besides, allicin has also been reported to be cytotoxic to human embryonic kidney (HEK-293) cells, exhibiting phenomena like DNA damage, DNA fragmentation, and increased nitrate and nitrite concentration in the cells on allicin treatment (Narain et al., 2022).

5.3 *ASPIDOSPERMA* SPP.

The *Aspidosperma* genus belongs to the family Apocynaceae, and the plants are extensively used as an astringent, antipyretic, antidiarrhoeal, and antimalarial agent. Many plants of this genus have been reported as an antimalarial medicine for a long time in various parts of Brazil, South America, and other tropical regions (Chierrito et al., 2014; Paula et al., 2014).

The cytotoxicity of different alkaloids of on *Aspidosperma pyrifolium* (Jetichuriqui, Bolivia) and *Aspidosperma megalocarpon* has been studied against *Plasmodium falciparum*. Twelve compounds were studied for their antiplasmodial activity and cytotoxicity on the human cell line NIH 3T3. The compounds studied were apisdospermine, 10-methoxy-aspidospermidine, N-formyl-aspidospermidine, aspidospermine, vallesine, dimethyl-aspidospermine, demethoxy-aspidospermine, palosine, haplocine, fendlerine, aspidoalbine, and aspidolimidine. Among these 12 compounds studied, aspidolimidine was found to show high cytotoxicity (Mitaine-Offer et al., 2002). In another study conducted *in vitro* against *Plasmodium falciparum* K1 and 3D7 strains, it was found that the compound ellipticine in *Aspidoperma* spp. reversibly inhibited mammalian and bacterial cells at low concentrations, and at very high concentrations, the cytotoxic effect was irreversible (Frausin et al., 2015). In another study conducted on the stem bark of *Aspidosperma olivaceum in vitro* against *Plasmodium falciparum* (clone W2), six compounds – carpine, demethoxy-aspidospermine, uleine, apparicne, N-methyl-tetrahydrolivacine, and olivacine – were isolated, and among these bioactive compounds, uleine and apparacine were reported to be moderately toxic when studied on two cell lines: normal kidney glomerular cells (BGM) and human hepatoma cells (HepG2) (Chierrito et al., 2014). The ethanolic extract of *Aspidosperma vargasii* alkaloid compound ellipticine and *Aspidosperma olivaceum* compound olivacine has also been tested against

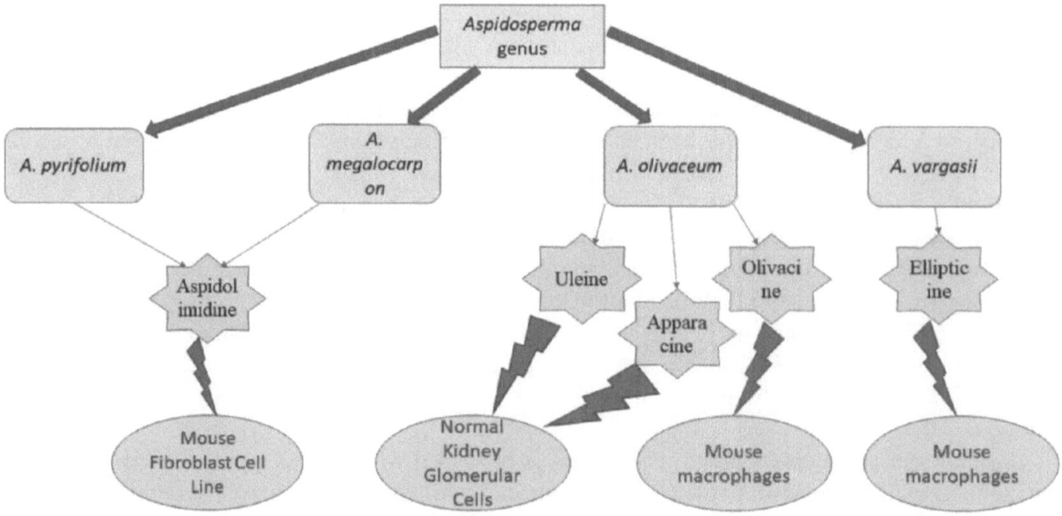

FIGURE 5.1 Cytotoxicity of *Aspideoperma* spp.

Plasmodium falciparum 3D7b and has been shown to have very low cytotoxicity on mouse macrophages (Silva et al., 2012) (Figure 5.1).

5.4 *CITRUS*

Citrus plants belong to the family Rutaceae. They are believed to be native to tropical and subtropical South-East Asia. Apart from the culinary use of *Citrus*, its fruits are used as a supplement of vitamin C, vitamin A, folic acid, and others. Because of their antioxidant potential, their fruits are widely used for the treatment of a plethora of diseases, including scurvy, cancer, etc.

One of the plants, *Citrus maxima*, which is commonly called pomelo, is used as an antioxidant, antidiabetic, hepatoprotective, anti-aging, and antidepressant agent. The leaves and fruit of this plant are used in the preparation of herbal medicine for the treatment of malaria, and its extract has shown significant antiplasmodial activity against *Plasmodium falciparum* 3D7 and RKL-9 (Gogoi et al., 2021). Several other plants of this genus have been used in the herbal preparation for the treatment of malaria, for example, *Citrus aurantifolia* (lime), commonly called Osan wewe in South-West Nigeria, whose root, bark, stem, twigs, leaves, and fruit are used in the preparation of antimalarial medicine. Studies have shown significant antimalarial activity of the flower extract of *C. aurantifolia* against *Plasmodium falciparum* (Bapna et al., 2017). Though in another study on mice model, it was reported that intake of the essential oil of lime induces cellular injuries, particularly necrosis, and its continuous exposure could result in malfunctioning of organs; for example, in liver, sinusoidal dilatation, leucocytosis, congestion of central vein, inflammation, and pyknotic nuclei in medium to high doses were observed, while necrotic lesions were observed in the glomeruli in the kidney. Accumulation of hemosiderin was observed in the spleen and vascular congestion, and influx of inflammatory cells was observed in the lungs (Adokoh et al., 2019). Other plants from this genus, such as sour lime, or *Citrus aurantium* (common name: osanganinganin), and grapefruit, or *Citrus paradisi* (common name: osangerepu), are also used as antimalarial herbal medicine in South-West Nigeria, using plant parts like fruits, root, leaves, stem, twigs, etc. (Odugbemi et al., 2007). The essential oils of *Citrus* spp. are known to be very rich in limonene, which ranges from about 30 to 40% in *C. maxima* and *C. aurantifolia* (Pino and Rosado, 2001; Singh et al., 2010) to about 90% in the essential oil of peels of *Citrus aurantium* (Boussaada and Chemli, 2006). Also, about 91% of limonene was found in the peels of *Citrus paradisi* (Njoroge et al., 2005). In a study conducted on humans, it was observed that when limonene is inhaled into the lungs at the highest concentration (450 mg/m^3), it significantly decreased the lung vital capacity of the subjects tested (Filipsson et al., 1993). Interestingly, this compound has been reported as mildly toxic against Rhesus monkey epithelial cells (LLC MK-2) (Arruda et al., 2009), indicating its potential cytotoxicity.

5.5 *OCIMUM* SPP.

These plants belong to the family Lamiaceae and are well-known for their medicinal and aromatic properties. *Ocimum tenuiflorum* is an herb native to the Indian subcontinent known for its various medicinal uses. In Ayurveda, the plant extracts have been used in preparation of antimalarial medicines (Raina et al., 2013). *In vivo* studies (mouse model of *Plasmodium bergei*) have confirmed the inhibitory activity of the plant towards adult mosquitoes and their larvae (Banyal et al., 2015). The antimalarial activity of the plant is linked to its main phytochemical constituent, eugenol (Banyal et al., 2015). Similarly, antimalarial effects of *Ocimum gratissimum* have been studied (Florence et al., 2022). The essential oil of this plant also contains high amounts of eugenol (about 68%) (Matasyoh et al., 2007). However, in a separate study, pure eugenol was reported to be significantly cytotoxic towards human gingival fibroblasts, and it was also claimed that its cytotoxic effect depends on its concentration and time of contact with cells (Gerosa et al., 1996). This compound was also claimed to be cytotoxic against human oral cells (Sakagami et al., 2018), as it has shown inhibition of the citric acid cycle and induction apoptotic cell death.

5.6 *PIPER* SPP.

For a very long time, several species of the family Piperaceae have been used as herbal medicine in treatment of several ailments, like cardiovascular, gastrointestinal, bacterial infections, malaria, etc. There is an abundance of biologically active phytochemicals in the family of Piperaceae, out of which secondary metabolites like alkaloids and amides from the genus *Piper* have shown medicinal effects, apart from their use for culinary purposes. One of the most common species used as herbal medicine against malaria is *Piper nigrum*, commonly called black pepper, popular for its culinary applications in Asia and for its use in many traditional medicines. Among the bioactive compounds present in *P. nigrum*, compounds like piperine, pellitorine, guineensine, pipnoohine, trichostachine, and piperonal are known to have potential medicinal properties. Out of these compounds, piperine is the major component responsible for the spicy, pungent flavour and other pharmacological properties of the genus (Rao et al., 2011). Piperine has been reported as a potent lead for the development of new antimalarial drugs because of its significant antimalarial activity (Thiengsusuk et al., 2018). *P. nigrum* has been widely used as prophylactic treatment of malaria in many parts of the world. The seed extract of the plant has been confirmed to have shown significant activity against *Plasmodium falciparum* (Kamaraj et el., 2012). Fruits of the plant are used as antiperiodics in malaria fever and taken with *Allium cepa* orally in some parts of India. *P. nigrum* is also recommended to be taken with juice of the leaves of *Lucas linifolia* and *Ocimum sanctum* for curative effects (Marques et al., 2015). Despite the therapeutic potential of piperine, its toxicity is also an important factor to be considered. Studies have shown the reproductive toxicity of piperine in Swiss Albino mice, where an increase in the diestrus phase leads to reduced fertility and mating performance, when administered orally at concentrations of 10 and 20 mg/kg body weight. In the same study, anti-implantation activity and anti-prostaglandin activity of piperine were also reported (Daware et al., 2000). Another study confirmed toxicity of piperine on male reproductive system of albino mice at a dose of 10 mg/kg body weight of piperine for 60 days (Chinta et al., 2017). Continual intake of piperine resulted in decreased size of reproductive organs, increased gonadosomatic index (which is inversely proportional to reproductive efficiency), increased in testicular coefficient (which is directly proportional to reproductive toxicity), decreased serum testosterone levels, and increased levels of follicle-stimulating hormone and luteinizing hormone in male rats.

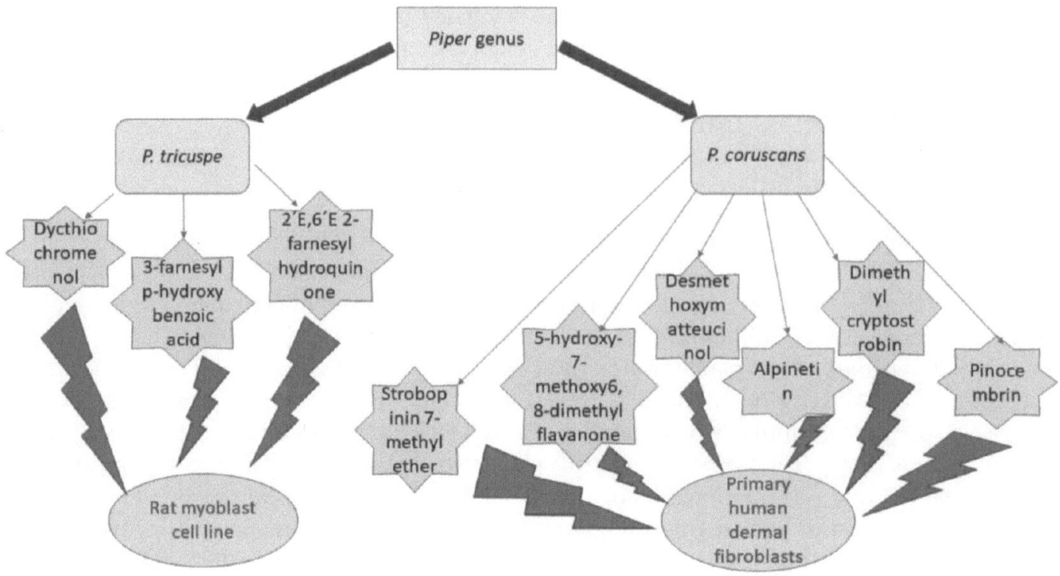

FIGURE 5.2 Cytotoxicity of *Piper* spp.

Piper tricuspe, commonly called Hojamano or Costeno, is found in the Colombian pacific zone and has been used traditionally in the treatment of snakebite and malaria. Studies have confirmed the antimalarial potential of the plant extract against *Plasmodium falciparum*. However, plant metabolites have also shown to exhibit cytotoxicity. Three active metabolites of the plant, namely, dycthiochromenol, 3-farnesyl p-hydroxy benzoic acid, and 2´E,6´E 2-farnesyl hydroquinone, have been reported to have antimalarial activity, and all of them were also found to be significantly cytotoxic against mammalian cell line, the rat skeletal muscle myoblast L-6 cells (Vega et al., 2008).

Piper coruscans is a plant traditionally used in the treatment of malaria in the Amazon region due to its antiplasmodial activity. However, the plant has also been traditionally used as poison. A recent study confirmed the cytotoxicity of plant extracts AB943 primary human dermal fibroblasts. All the cytotoxic components of the plant extract were flavanones, which included strobopinin 7-methyl ether, 5-hydroxy-7-methoxy6, 8-dimethyl flavanone, desmethoxymatteucinol, alpinetin, pinocembrin, and dimethyl cryptostrobin (Ocmín et al., 2022) (Figure 5.2).

5.7 *STEPHANIA* SPP.

Stephania genus belongs to the family Menispermaceae and is native to Africa, South-East Asia, and the northern and eastern parts of Australia (Blanchfield et al., 2003). *Stephania* is a popular traditional and folk medicine to treat malaria in various parts of South-East Asia (Desgrouas et al., 2014). In a study conducted on tubers of *Stephania rotunda*, the antiplasmodium activity of eight compounds – cepharanthine, palmatine, pseudopalmatine, roemerine, stephanine, tetrahydropalmatine, vireakine, and xylopinine – was confirmed on *Plasmodium falciparum* type W2 strain, with cepharanthine and pseudopalmatine showing cytotoxic behaviour against lymphoblast cells. Among these two cytotoxic compounds, cepharanthine showed more cytotoxicity on K652S cells, and pseudopalmatine showed low cytotoxicity to K562S cells (Baghdikian et al., 2013). Another study conducted on crude methanol extract from *S. dielsiana* leaves confirmed that antimalarial potential of the plant could be attributed to seven compounds – crebanine, o-methylbulbocapnine, oxostephanine, palmatine, stephanine, tetrahydropalmatine, and thailandine – which were isolated

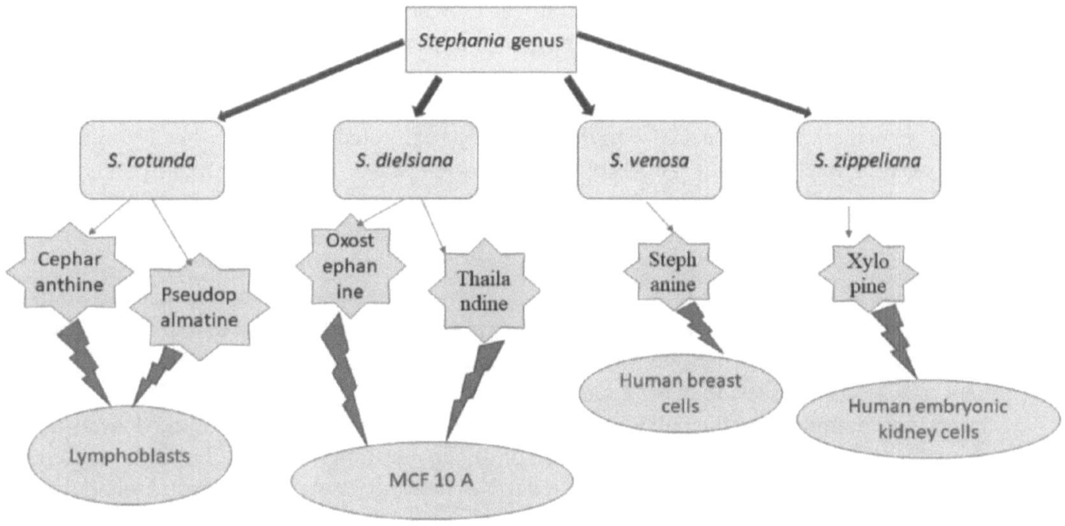

FIGURE 5.3 Cytotoxicity of *Stephania* spp.

in pure form. Among these compounds, oxostephanine was found to be the most cytotoxic towards non-cancer models MCF10A and 184B5 human breast epithelial cell lines, while crebanine, o-methylbulbocapine, and palmatine chloride showed low cytotoxicity (Knockleby et al., 2020).

In *Stephania venosa*, crebanine, dehydrocrebanine, o-methybulbocapnine stephanine, and tetrahydropalmatine were isolated from the dichloromethanol extract of tubers, which exhibited antiplasmodial activity against two strains, 3D7 and W2, of *P. falciparum*. Out of these compounds, stephanine showed high cytotoxicity on non-cancer breast cell line 184B5 (Mai Le et al., 2017).

In *Stephania zippeliana*, lirodenine and xylopine were isolated using methanol/water extract and showed antiplasmodial activity. Among these two compounds, xylopine was found to have low cytotoxicity when studied against the mammalian (human embryonic kidney) cell line HEK 293 (Fernandez et al., 2010) (Figure 5.3).

5.8 *ZINGIBER* SPP.

The members of the ginger family (Zingiberaceae) are famously used as spices and in traditional and herbal medications. *Z. officinale***,** commonly known as true ginger, canton, canton ginger, is a widely famous spice across the world because of its pungent flavour and characteristic aroma. It is believed to have originated from tropical regions of India. It has been used as the herbal remedy of several ailments, like fever, common cold, cough, stomach infections, etc. It is widely distributed in South-Eastern Asia, China, Japan, Austria, Latin America, Jamaica, and Africa (Hasan et al., 2012). The rhizome of this plant is reported to be used as an herbal medicine for malaria in South-West Nigeria, where it is commonly called Ajo and Ata-ila (Odugbemi et al., 2007). Antimalarial activity of the essential oils of the rhizome of *Z. officinale* against *Plasmodium falciparum* has also been experimentally confirmed (Héritier et al., 2018). Ginger rhizomes are believed to be rich in 10-gingerol (a phenolic lipid) (Hasan et al., 2012), which imparts therapeutic potential to the plant. However, it has been reported to inhibit the growth of MCF-10A normal mammary epithelial cells (Ediriweera et al., 2020), confirming its cytotoxicity. Another terpenoid, zerumbone, found in the plant (Oh et al., 2018) has also been shown to impart antimalarial property to the plant. However, the cytotoxicity of zerumbone to normal cell lines (Vero cells) has also been confirmed (Sriphana et al., 2013)

Another species, *Z. zerumbet*, commonly known as wild ginger or shampoo ginger, is widely distributed in the Indo-Malayan region and also grown in many tropical regions (Baby et al., 2009). The rhizome of this plant has been used in malaria treatment by locals since ancient times in some parts of southern India (Kamaraj et al., 2012). However, the toxicity of the plant has also been studied. In a report, ethanol extract of *Zingiber zerumbet* exhibited significant toxicity in the brine shrimp assay (Hossain et al., 2012). This is owing to the presence of zerumbone in the plant rhizome.

Some other antimalarial plant species with tested cytotoxicity on different cell lines are discussed next.

Bixa orellana.The plant belongs to family Bixaceae and is a very common species in South America. It is commonly called achiote in Latin America and annatto in North America. Apart from its use as a condiment, it is also used for its medicinal properties in the treatment of various diseases, including malaria (Ruiz et al., 2011, Zhai et al., 2014). In a study, several compounds from the hairy root culture of *B. orellana* were isolated, like ishwarane, ellagic acid, δ-tocotrienol, bixin, stigmasterol, β-sitosterol, inositol, ursolic acid, maslinic acid, and arjunolic acid, out of which ishwarane, δ-tocotrienol, and stigmasterol showed promising antimalarial activity. However, the study confirmed the cytotoxicity of extract against normal cell lines like BJ cells and HEK-293 cells (Zhai et al., 2014).

Canella winterana.The plant is found in the tropical region of the Caribbean and some areas of Florida Keys. It is commonly called wild cinnamon because its bark is used as a spice, as it is similar to true cinnamon. This plant consists of drimane sesquiterpenoids dialdehydes, which is a group of characteristic compounds of Canellaceae (Grace et al., 2010). The leaf extracts have been shown to contain five drimane-type sesquiterpenoids, like 9-epideoxymuzigadial, muzigadial, 9-deoxymuzigadial, 3-β-acetoxypolygodial, and muzigodiol. Muzigadial has been shown to have high antiplasmodial activity as well as high cytotoxicity towards CHO (Chinese hamster ovary) cells, while other isolated compounds have moderate antimalarial activity as well as cytotoxicity towards CHO cells (Grace et al., 2010), indicating their cytotoxic nature.

Holarrhena antidysenterica. Commonly called Kurchi, *Holarrhena antidysentrica* belongs to the family Apocynaceae (Gopinath et al., 2020). These plants are traditionally used in North-West Himalayan regions to treat malaria (Verma et al., 2011). In an *in vitro* antiplasmodial study conducted on compound conessine isolated from the stem bark of *Holarrhena antidysentrica*, it was observed that conessine was cytotoxic against the *Plasmodium falciparum* K1 strain and L-6, rat skeleton muscle myoblasts (Dua et al., 2013).

Hypercium lanceolatum. *Hypercium lanceolatum* (family Guttiferae) is known to have antimalarial properties (Tchakam et al., 2012). A study conducted against *Plasmodium falciparum* strain W2mef using the stem bark of *H. lanceolatum* showed that the plant extract was rich in betulinic acid (HLT1), 2,2',5,6'-tetrahydroxybenzophenone (HLT2), 5-hydroxy-3-methoxyxanthone (HLT3), 3-hydroxy-5-methoxyxanthone (HLT4), and HLT0 (unidentified). Among these compounds, betulinic acid exhibited cytotoxicity against *Plasmodium* strains and monkey kidney epithelial cells (Zofou et al., 2011b)

Kigelia africana. Many antimalarial compounds have been isolated from the family Bignoniaceae. One of them is *Kigelia africana*, which is commonly used in the treatment of malaria in African countries. This plant is commonly found in Central, West, and South Africa. Since this plant bears huge fruits, it is commonly called sausage tree or cucumber tree. In Cameroon, West Africa, the stem bark of this plant has been used for the preparation of antimalarial herbal medicine. The bark is boiled in water, or its powdered extract mixed with palm oil is taken orally in malarial fever (Zofou et al., 2011a). In South-Western Nigeria, this plant is locally called Iyan, Pandoro Somunroro, and there its stem bark is infused with the stem bark of other plants, which are *Anarcardium occidentalis*, *Mangifera indica*, and *Terminalia catapa*; this infusion is taken orally for malaria treatment (Oladele and Adewunmi, 2008). Active compounds isolated from the crude bark extracts of *K. africana* have been tested against *Plasmodium falciparum*, which showed significant antiplasmodial activity. In the same study, the cytotoxicity of these isolated compounds was also evaluated, in which out of the five isolated compounds (Atranorin, 2β, 3β, 19αtrihydroxy-urs12-en-28-oic acid, 2βhydroxyoleanolic acid, Specicoside, and p-hydroxycinnamoic acid), two compounds (Atranorin, 2β, 3β, 19αtrihydroxy-urs12-en-28-oic acid) were found to be significantly active against *Plasmodium falciparum* while also exhibiting cytotoxicity against monkey kidney epithelial cells. Atranorin was moderately cytotoxic, while 2β, 3β, 19αtrihydroxy-urs12-en-28-oic acid was mildly cytotoxic against this particular mammalian cell line (Zofou et al., 2011a).

Rumex crispus. *Rumex crispus* is an unpleasant weed that is commonly found in Asian countries, especially in China and Korea. It is taken orally as herbal medicine to treat a variety of ailments, like fungal and peptide disorders, including malaria (Lee and Rhee, 2013). This plant is commonly called curled dock. It is a well-known traditional medicine in the Eastern Cape, South Africa (Idris et al., 2019). Nepodin is the primary active ingredient of the plant, which accounts for its strong antimalarial property. However, moderate cytotoxicity against normal cell lines is also reported (Lee and Rhee, 2013).

***Tebebuia* spp.** *Tebebuia* belongs to the largest genus of Bignoniaceae. The plants are native to the Amazon forest and Latin America (Zhang et al., 2020). Since ancient times, these plants have been used in the treatment of fever, malaria, fungal infections, bacterial infections, and skin diseases. Commonly known as trumpet trees, the plants are rich in lapachol, α- and β-lapachone, and atovaquone, which account for their antimalarial properties. A recent study was conducted to test the antiplasmodial and cytotoxicity of the natural naphthoquinones lapochol, α- and β-lapachone, which were obtained from a plant of the *Tabebuia* genus. The cytotoxicity of these compounds was tested on the A549 (human lung adenocarcinoma cells), HepG2 (human hepatoblastoma cells), LLC-PK 1 (porcine kidney tubular epithelial cells), MRC-5 cells (the human foetal lung fibroblast cell line), neuro-2A (mouse neuroblastoma cells) against *Plasmodium falciparum* W2 strain. It was observed that β-lapachone was the most cytotoxic in all the cells tested. α-lapachone was found to be more cytotoxic to kidney cells LLC-PK1 and less cytotoxic to neuro-2A and MRC-5 cells, and atovaquone was observed to be the most cytotoxic to neuro-2A and MRC-5 cells (Nascimento et al., 2020).

Teclea trichocarpa. Some *Teclea* species are used for the preparation of herbal remedies for malaria. One of these species is *Teclea trichocarpa*. It is used by the Akambia tribe of East Africa for the treatment of malaria. The active compounds of the plant extracts, like beta-sitosterol, alpha-amyrin, melicopicine, normelicopicine, arborinine, and skimmianine, have been shown to exhibit mild to moderate cytotoxicity towards the L-6 muscle cell line (Mwangi et al., 2010).

5.9 CONCLUSION

The increasing resistance to antimalarial chemical drugs like sulfadoxine and chloroquine has resulted in pathogenic resistance and loss of drug efficacy, leading to the wide usage of herbal drugs against malaria. The plant-based drugs have been largely popular due to their low cost, high accessibility, and being more acceptable among users. However, these drugs are not without any side effects. One of the most crucial factors is the proven cytotoxic effect of phytoingredients of some of the commonly used antimalarial plants. Another important factor is the increased cytotoxicity of these drugs when many plants are used to prepare combinational drugs without effective testing, as some phytochemicals might prove to be more toxic when used in concoctions. The plant-based medicine preparation and dosage requirements for safe use are also debatable as phytoconstituents vary with several physical parameters.

It, therefore, becomes very important to realize all these factors before herbal drug designing and preparation, making it a must to perform extensive clinical trials on the safety and efficacy of plant-based antimalarial drugs.

REFERENCES

Adebayo JO, Krettli AU (2011). Potential Antimalarials from Nigerian Plants: A Review. *Journal of Ethnopharmacology* 133: 289–302

Adokoh CK, Asante DB, Acheampong DO, Kotsuchibashi Y, Armah FA, Sirikyi IH, Kimura K, Gmakame E, Rauf SA (2019) Chemical Profile and *In Vivo* Toxicity Evaluation of Unripe *Citrus Aurantifolia* Essential Oil. Toxicology Reports 6: 692–702

Arruda DC, Miguel DC, Yokoyama YJKU, Katzin AM, Uliana SRB (2009) Inhibitory Activity of Limonene Against Leishmania Parasites in Vitro and in Vivo. *Biomedicine & Pharmacotherapy* 63: 643–649

Baby S, Dan M, Thaha ARM, Johnson AJ, Kurup R, Balakrishnapillai P, Lim CK (2009) High Content of Zerumbone in Volatile Oils of Zingiber Zerumbet from Southern India and Malaysia. *Fragrance and Flavour Journal* 24: 301–308

Baghdikian B, Mahiou LV, Bory S, Bun SS, Dumetre A, Mabrouki F, Hutter S, Azas N, Ollivier E (2013) New Antiplasmodial Alkaloids from *Stephania Rotunda*. *Journal of Ethnopharmacolpgy* 145: 381–385

Banyal HS, Sharma R, Devi N (2015) Antimalarial Effect of *Ocimum Sanctum* Linn. and Bauhinia Variegata Linn. on *Plasmodium Berghei*. *IOSR Journal of Pharmacy and Biological Sciences* 10: 70–72

Bapna S, Satvekar T, Jadhav P, Sawant MG (2017) Evaluation of *In Vitro* Antimalarial Activity of *Calotropis Gigantia* and *Citrus Aurantifolia* flower Extract Against *Plasmodium Falciparum* 3D7. *World Journal of Pharmaceutical Research* 6: 1619–1626

Blanchfield JT, Sands DPA, Kennard CHL, Byriel KA, Kitching W (2003) Characterisation of Alkaloids from Some Australian *Stephania* (Menispermaceae) Species. *Phytochemistry* 63: 711–720

Boussaada O, Chemli R (2006) Chemical Composition of Essential Oils from Flowers, Leaves and Peel of *Citrus Aurantium* L. var. Amara from Tunisia. *Journal of Essential Oil-Bearing Plants* 9: 133–139

Chierrito TPC, Aguiar ACC, De Andrade IM, Ceravolo IP, Goncalves RAC, De Oliveira AJB, Krettli AU (2014) Anti-Malarial Activity of Indole Alkaloids Isolated from *Aspidosperma Olivaceum*. *Malarial Journal* 13: 142

Chinta G, Coumar MS, Periyasamy L (2017) Reversible Testicular Toxicity of Piperine on Male Albino Rats. *Pharmacognosy Magazine* 13: S525–S532

Coppi A, Cabinian M, Mirelman D, Sinnis P (2006) Antimalarial Activity of Allicin, a Biologically Active Compound from Garlic Cloves. *Antimicrobial Agents and Chemotherapy* 50: 1731–1737

Daware MB, Mujumdar AM, Ghaskabdi S (2000) Reproductive Toxicity of Piperine in Swiss Albino Mice. *Planta Medica* 66: 231–236

De Paula RC, Dolabela MF, De Oliveira AB (2014) *Aspidosperma* Species as Sources of Antimalarials. Part III. A Review of Traditional Use and Antimalarial Activity. *Plant Medica: Journal of Medicinal Plant and Natural Product Research* 80(5): 378–386

Desgrouas C, Taudaon N, Bun SS, Baghdikian B, Bory S, Parzy D, Ollivier E (2014) Ethnobotany, Phytochemistry and Pharmacology of *Stephania Rotunda* Lour. *Journal of Ethnopharmacology* 154: 537–563

Dua VK, Verma G, Singh B, Rajan A, Bagai U, Agarwal DD, Gupta NC, Kumar S, Rastogi A (2013) Anti-Malarial Property of Steroidal Alkaloid Conessine Isolated from the Bark of *Holarrhena Antidysenterica*. *Malarial Journal* 12: 194

Ediriweera MK, Moon JY, Nguyen YTK, Cho SK (2020) 10-Gingerol Targets Lipid Rafts Associated PI3K/Akt Signaling in Radio-Resistant Triple Negative Breast Cancer Cells. *Molecules* 25: 3164–3177

Elmi T, Hajialiani F, Asadi MR, Sadeghi S, Namazi MJ, Tabatabaie F, Zamani Z (2021) Antimalarial Effects of the Hydroalcoholic Extract of *Allium Paradoxum in Vitro* and *in Vivo*. *Journal of Parasitic Diseases* 45: 1055–1064

Eseyin OA, Sattar MA, Rathore HA (2014) A Review of the Pharmacological and Biological Activities of the Aerial Parts of *Telfairia occidentalis* Hook. F. (Cucurbitaceae). *Tropical Journal of Pharmaceutical Research* 13(10): 1761–1769

Feng Y, Zhu X, Wang Q, Jiang Y, Shang H, Cui L, Cao Y (2012) Allicin Enhances Host Pro-inflammatory Immune Responses and Protects Against Acute Murine Malaria Infection. *Malaria Journal* 11: 268–277

Fernandez LS, Sykes ML, Andrews KT, Avery VM (2010) Antiparasitic Activity of Alkaloids from Plant Species of Papua New Guinea and Australia. *International Journal of Antimicrobial Agents* 36: 275–279

Filipsson FA, Lof A, Hagberg M, Hjelm EW, Wang Z (1993) D- Limonene Exposure to Humans by Inhalation: Uptake, Distribution, Elimination, and Effects on the Pulmonary Function. *Journal of Toxicology and Environmental Health: Current Issues* 38: 77–88

Florence CC, Nonye NE, Josephine EOO (2022) *In Vitro* Antiplasmodial and *in Vivo* Toxicity Potential of *Mentha Piperita* and *Ocimum Gratissimum* Essential Oils and Their Synergistic Effect with Conventional Antimalarial Drugs Against Plasmodium Falciparum. *International Journal of Mosquito Research* 9: 114–122

Frausin G, Hidalgo AF, Lima RBS, Kinupp VF, Ming LC, Pohlit AM, Milliken W (2015) An Ethnobotanical Study of Anti-Malarial Plants Among Indigenous People on the Upper Negro River in the Brazilian Amazon. *Journal of Ethnopharmacology* 174: 238–252

Gerosa R, Borin M, Menegazzi G, Puttini M, Cavalleri G (1996) *In Vitro* Evaluation of the Cytotoxicity of Pure Eugenol. *Journal of Endodontics* 22: 532–534

Gogoi N, Gogoi B, Chetia D (2021) In Vitro Antimalarial Activity Evaluation of Two Ethnomedicinal Plants Against Chloroquine Sensitive and Resistant Strains of *Plasmodium Falciparum. Clinical Phytoscience* 7: 42–52

Gopinath G, Thirumal M, Kumar PR (2020) *Holarrhena Antidysentrica* Linn.-A Review. *Research Journal of Pharmacy and Technology* 13

Grace MH, Lategan C, Mbeunkui F, Graziose R, Smith PJ, Raskin I, Lila MA (2010) Antiplasmodial and Cytotoxic Activities of Drimane Sesquiterpenes from *Canella Winterana. Natural Product Communications* 5: 1869–1872

Hasan HA, Raauf AMR, Razik BMA, Hassan BAR (2012) Chemical Composition and Antimicrobial Activity of the Crude Extracts Isolated from *Zingiber Officinale* by Different Solvents. *Pharmaceutical Analytica Acta* 3: 184–189

Héritier VNV, Arthur DN, Augustin MM, Nadège NK, Roger KV, Eric et SZLP, Thaddée MNJ (2018) Antibacterial and Antiplasmodial Potentials of Essential Oils from Two Plants of Tangawisi Products: *Zingiber Officinalis* Roscoe and Monodora Myristica (Gaertn) Dunal. *Journal of Pharmacognosy and Phytochemistry* 7: 643–648

Hossain S, Kader G, Nikkon F, Yeasmin T (2012) Cytotoxicity of the Rhizome of Medicinal Plants. *Asian Pacific Journal of Tropical Biomedicine* 2: 125–127

Husen A (2021) *Traditional Herbal Therapy for the Human Immune System*. CRC Press, Boca Raton, FL. https://doi.org/10.1201/9781003137955

Husen A (2022a) *Herbs, Shrubs and Trees of Potential Medicinal Benefits*. CRC Press, Boca Raton, FL. https://doi.org/10.1201/9781003205067

Husen A (2022b) *Exploring Poisonous Plants: Medicinal Values, Toxicity Response and Other Potential Uses*. CRC Press, Boca Raton, FL. https://doi.org/10.1201/b23017

Husen A (2023) *Antidiabetic Medicinal Plants and Herbal Treatments*. CRC Press, Boca Raton, FL

Idris OA, Wintola OA, Afolayan AJ (2019) Evaluation of the Bioactivities of *Rumex Crispus* L. Leaves and Root Extracts Using Toxicity, Antimicrobial, and Antiparasitic Assays. *Evidence-Based Complementary and Alternative Medicine* 2019: 1–13

Ijarotimi SO, Agbedahunsi JM, Onyeji CO, Adewumi CO (2010) Chemotherapeutic Interaction Between Khaya Grandifolia (WELW) CDC Stem Bark Extract and Two Antimalarial Drugs in Mice. *African Journal of Traditional, Complementary and Alternative Medicines* 7(4): 370–376

Kamaraj C, Kaushik NK, Rahuman AA, Mohanakrishnan D, Bagavan A, Elango G, Zahir AA, Santhoshkumar T, Marimuthu S, Jayaseelan C, Kirthi AV, Rajakumar G, Velayutham K, Sahal D (2012) Antimalarial Activities of Medicinal Plants Traditionally Used in the Villages of Dharmapuri Regions of South India. *Journal of Ethnopharmacology* 141: 796–802

Kar S, Kar S. (2010) Control of Malaria. *Nature Reviews Drug Discovery* 9: 511–512

Khanabadi F, Badirzadeh A, Hesari AK, Akbariqomi M, Torkashvand H, Didehdar M, Elmi T (2022) *In Vivo* Anti-Malarial Activity and Toxicity of *Allium Ursinum* (Wild Garlic) Hydroalcoholic Extract. *Journal of Applied Biotechnology Reports* 9: 781–789

Khoutorsky PM, Goncharov I, Rabinkov A, Mirelman D, Geiger B, Bershadsky AD (2007) Allicin Inhibits Cell Polarization, Migration and Division via Its Direct Effect on Microtubules. *Cell Motility and the Cytoskeleton* 64: 321–337

Knockleby J, Pradines B, Gendrot M, Mosnier J, Nguyen TT, Trinh TT, Lee H, Mai LP (2020) Cytotoxic and Anti-Plasmodial Activities of *Stephania Dielsiana* Y. C. Wu Extracts and the Isolated Compounds. *Molecules* 25: 3755

Lee KH, Rhee KH (2013) Antimalarial Activity of Nepodin Isolated from *Rumex Crispus. Archives of Pharmacal Research* 36: 430–435

Mai LP, Srivastava V, Nguyen TT, Pradines B, Madamet M, Mosnier J, Trinh TT, Lee H (2017) Stephanine from *Stephania Venosa* (Blume) Spreng Showed Effective Antiplasmodial and Anticancer Activities, the Latter by Inducing Apoptosis Through the Reverse of Mitotic Exit. *Phytotherapy Research* 31: 1357–1368

Marques AM, Peixoto ACC, de Paula RC, Nascimento MFA, Soares LF, Velozo LSM, Guimaraes EF, Kaplan MAC (2015) Phytochemical Investigation of Anti-plasmodial Metabolites from Brazilian Native Piper Species. *Journal of Essential Oil Bearing Plants* 18: 74–81

Matasyoh LG, Matasyoh JC, Wachira FN, Kinyua MG, Muigai AWT, Mukiama TK (2007) Chemical Composition and Antimicrobial Activity of the Essential Oil of *Ocimum Gratissimum* L. Growing in Eastern Kenya. *African Journal of Biotechnology* 6: 760–765

Mitaine-Offer AC, Sauvain M, Valentin A, Callapa J, Mallie M, Zeches-Hanrot M (2002) Antiplasmodial Activity of *Aspidosperma* Indole Alkaloids. *Journal of Phytomedicine* 9: 142–145

Mwangi ESK, Keriko JM, Machocho AK, Wanyonyi AW, Malebo HM, Chhabra SC, Tarus PK (2010) Antiprotozoal Activity and Cytotoxicity of Metabolites from Leaves of *Teclea Trichocarpa*. *Journal of Medicinal Plants Research* 9: 726–731

Narain NM, Amoako DG, Somboro AM, Arhin I, Mhlongo NN, Kumalo HM, Khan RB (2022) Allicin Induces Post-Translational Modifications of p53, DNA Damage, and Oxidative Stress in Human Embryonic Kidney Cells. *Journal of Biotech Research* 13: 90–102

Nascimento MFA, Borgati TF, De Souza LCR, Tagliati CA and De Oliveira AB (2020) *In Silico, in Vitro* and *in Vivo* Evaluation of Natural Bignoniaceous Naphtoquinones in Comparison with Atovaquone Targeting the Selection of Potential Antimalarial Candidates. *Toxicology and Applied Pharmacology* 401

Njoroge SM, Koaze H, Karanja PN, Sawamura M (2005) Volatile Constituents of Redblush Grapefruit (Citrus paradisi) and Pummelo (Citrus grandis) Peel Essential Oils from Kenya. *Journal of Agriculture and Food Chemistry* 53: 9790–9794

Ocmín VPG, Gallard JF, Baelen ACV, Leblanc K, Cojean S, Mouray E, Grellier P, Philippe G, Guerra CAA, Beniddir MA, Evanno L, Figadère B, Maciuk A (2022) Bio Dereplication of Antiplasmodial Extracts: Application of the Amazonian Medicinal Plant *Piper coruscans* Kunth. *Molecules* 27: 7638–7656

Odugbemi TO, Akinsulire OR, Odunayo R, Aibinu IE, Fabeku PO (2007) Medicinal Plants Useful for Malaria Therapy in Okeigbo, Ondo State, Southwest Nigeria. *African Journal of Traditional, Complementary and Alternative Medicines* 4: 191–198

Oh TI, Jung HJ, Lee YM, Lee S, Kim GH, Kan SY, Kang H, Oh T, Ko HM, Kwak KC, Lim JH (2018) Zerumbone, a Tropical Ginger Sesquiterpene of Zingiber Officinale Roscoe, Attenuates α-MSH-Induced Melanogenesis in B16F10 Cells. *International Journal of Molecular Sciences* 19: 3149–3166

Oladele AT, Adewunmi CO (2008) Medicinal Plants Used in the Management of Malaria Among the Traditional Medicine Practitioners (TMP'S) in South Western Nigeria. *African Journal of Infectious Diseases* 2: 51–59

Pino JA, Rosado A (2001) Comparative Investigation of the Distilled Lime Oils (*Citrus aurantifolia* Swingle and *Citrus lafifolia* Tanaka) from Cuba. *Journal of Essential Oil Research* 13: 179–180

Raina AP, Kumar A, Dutta M (2013) Chemical Characterization of Aroma Compounds in Essential Oil Isolated from "Holy Basil" (*Ocimum tenuiflorum* L.) Grown in India. *Genetic Resources and Crop Revolution* 60: 1727–1735

Rao VRS, Raju SS, Sarma VU, Sabine F, Babu KH, Babu KS, Rao JM (2011) Simultaneous Determination of Bioactive Compounds in *Piper Nigrum* L. and a Species Comparison Study Using HPLC-PDA. *Natural Product Research* 25: 1288–1294

Ruiz L, Ruiz L, Maco M, Cobos M, Choquevilca ALG, Roumy V (2011) Plants Used by Native Amazonian Groups from the Nanay River (Peru) for the Treatment of Malaria. *Journal of Ethnopharmacology* 133: 917–921

Sakagami H, Sugimoto M, Kanda Y, Murakami Y, Amano O, Saitoh J, Kochi A (2018) Changes in Metabolic Profiles of Human Oral Cells by Benzylidene Ascorbates and Eugenol. *Medicines* 5: 116–135

Satyal P, Craft JD, Dosoky NS, Setzer WN (2017) The Chemical Compositions of the Volatile Oils of Garlic (*Allium sativum*) and Wild Garlic (*Allium vineale*). *Foods* 6: 63–73

Silva LFR, Montoia A., Amorim RCN., Melo MR, Henrique MC, Nunomura SM, Costa MRF, Neto VFA, Costa DC, Dantas G, Lavrado J, Moreira R, Paulo A, Pintp AC, Tadei WP, Zacardi RS, Eberlin MN, Pohlit AM (2012) Comparative in Vitro and in Vivo Antimalarial Activity of the Indole Alkaloids Ellipticine, Olivacine, Cryptolepine and a Synthetic Cryptolepine Analog. *Journal of Phytomedicine* 20: 71–76

Singh P, Shukla R, Prakash B, Kumar A, Singh S, Mishra PK, Dubey NK (2010) Chemical Profile, Antifungal, Antiaflatoxigenic and Antioxidant Activity of *Citrus Maxima* Burm. and *Citrus Sinensis* (L.) Osbeck Essential Oils and Their Cyclic Monoterpene, DL-Limonene. *Food and Chemical Toxicology* 48: 1734–1740

Sriphana U, Pitchuanchom S, Kongsaeree P, Yenjai C (2013) Antimalarial Activity and Cytotoxicity of Zerumbone Derivatives. *Science Asia* 39: 95–99

Tchakam PD, Lunga PL, Kowa TK, Lonfouo AHN, Wabo HK, Tapondjou LA, Tane P, Kuiate JR (2012) Antimicrobial and Antioxidant Activities of the Extracts and Compounds from the Leaves of *Psorospermum Aurantiacum* Engl. and *Hypercium Lanceolatum* Lam. *BMC Complementary and Alternative Medicine* 12: 136

Thiengsusuk A, Muhamad P, Chaijaroenkul W, Bangchang KN (2018) Antimalarial Activity of Piperine. *Journal of Tropical Medicine* 7: 1–7

Vathsala PG, Murthy PK (2020) Immunomodulatory and Antiparasitic Effects of Garlic-Arteether Combination via Nitric Oxide Pathway in *Plasmodium Berghei*-Infected Mice. *Journal of Parasitic Diseases* 44: 49–61

Vega AS, Rojano B, Blair S, Segura C, Figadere B, Seon B, Grellier P, Sáez J (2008) Antimalarials and Antioxidants Compounds from *Piper Tricuspe* (Piperaceae). *Pharmacologyonline* 1: 1–8

Verma G, Dua VK, Agarwal DD, Atul PK (2011) Anti-Malarial Activity of *Holarrhena Antidysenterica* and *Viola Canecens*, Plants Traditionally Used Against Malaria in the Garhwal Region of North-West Himalaya. *Malarial Journal* 10: 20

Voravuth S, Preeyanuch B, Chokdee K, Kittiyaporn D, Panatda B, Butsarat S (2016) Antimalarial Properties of Aqueous Crude Extracts of *Gynostemma Pentaphyllum* and *Moringa Oleifera* Leaves in Combination with Artesunate in *Plasmodium Berghei*-Infected Mice. *Hindawi Publishing Corporation Journal of Tropical Medicine*, Article ID 8031392, 6 pages. http://doi.org/10.1155/2016/8031392

Zhai B, Clark J, Ling T, Connelly M, Bolivar FM, Rivas F (2014) Antimalarial Evaluation of the Chemical Constituents of Hairy Root Culture of *Bixa orellana* L. *Molecules* 19: 756–766

Zhang J, Hunto ST, Yang Y, Lee J, Cho JY (2020) *Tabebuia Impetiginosa*: A Comprehensive Review on Traditional Uses, Phytochemistry, and Immunopharmacological Properties. *Molecules* 25(18)

Zofou D, Kengne ABO, Tene M, Ngemenya MN, Tane P, Titanji VPK (2011a) *In Vitro* Antiplasmodial Activity and Cytotoxicity of Crude Extracts and Compounds from the Stem Bark of *Kigelia Africana* (Lam.) Benth (Bignoniaceae). *Parasitology Research* 108: 1383–1390

Zofou D, Kowa TK, Wabo HK, Ngemenya MN, Tane P, Titanji VPK (2011b) *Hypercium Lanceolatum* (Hypericaceae) as a Potential Source of New Anti-Malarial Agents: A Bioassay-Guided Fractionation of the Stem Bark. *Malarial Journal* 10: 167

6 Plant-Based Secondary Compounds and Nutrients as Therapeutic Agents in the Management of Malaria Infection

Ahmed Olatunde, Olalekan Bukunmi Ogunro, Habibu Tijjani, and Shakirdeen Mayowa Obidola

6.1 INTRODUCTION

Malaria has been the leading cause of death in about 100 countries of approximately 2.4 billion residents, mostly dominant in sub-Saharan Africa. Even though a sharp decline in mortalities related to malaria has been recorded in Africa, children below the age of 5 and pregnant women in sub-Saharan region of the continent have the bigger share of malaria-related deaths. Globally, approximately 3.3 billion individuals are endangered with malaria, out of which 1.2 billion are posed to greater threat (Kendie et al., 2021). Based on this, malaria continue to be a serious global threat that requires immediate and potent interventions.

One of the contributing factors to the global burden caused by malaria is the onerous or difficult concern of antimalarial use in parasites-dominated regions with tremendous increase in the rates of human movement and all the year-round malaria transmission (Ezenyi and Salawu, 2016). The deadly malaria parasite *Plasmodium falciparum* is responsible for most of these related deaths, which majorly occur through the development of resistance to several categories of synthetic antimalarial agents, causing severe complications of the disease. Apart from the *P. falciparum*, there are other less deadly strains of the malaria parasites, such as *P. vivax* and *P. ovale*, and they have the ability to deteriorate health for weeks to months following the initial infection by inactive hypnozoites in the liver. The malaria parasites have also been prognosticated to have a wider coverage (attributed to global warming increase) due to increase in the species of *Anopheles* and the malaria vectors that are resistant to most of the readily available antimalaria drugs (Laporta et al., 2015). It is therefore of utmost importance to have antimalarial agents with diverse actions and effective enough as alternative strategies against all strains of malaria parasites and malaria-induced infections.

Over the last years, significant successes have been made in identifying alternative therapies in managing malaria (Hemingway et al., 2016). Even as that, the mortality rate associated with malaria is still on the high side, which makes exploration of antimalarial agents among the top priority lists for researchers. Fortunately, plant-based chemicals and nutrients have been described as a great reservoir for antimalarial drugs, many of which are still of clinical relevance for preventing and treating severe cases of malaria. From the initial isolation of quinine (first successful antimalarial drug from *Cinchona* tree of South American origin) to the synthetic antimalarial chloroquine (which later became resistant to the parasite) and to the isolation of artemisinin (from the leaves of *Artemisia annua* of Chinese origin), robust scientific interest has been put on the continued exploration of natural products as antiplasmodial agents (Tajuddeen and Van Heerden, 2019).

DOI: 10.1201/9781003378396-6

Like in the case of secondary metabolites, the role of nutrition cannot be overlooked, and the lack or absence of effective nutrition makes the body system susceptible to malaria infection (Sakwe et al., 2019). In addition to this, the absence of proper nutrition can cause about 50% of malaria mortality in African regions (UNICEF, 2022). Malnutrition decreases the body system immunity, characterized by defective phagocytic functions, complement system, production of cytokine, and others (Munteanu and Schwartz, 2022; Mrimi et al., 2023). Lack of micronutrients has significant effects on the immune systems and exposes the body to infectious diseases, such as malaria (Sakwe et al., 2019). Therefore, the aspect of nutrients is also vital in curtailing the detrimental effects of malarial infection. The uninterrupted search for novel antimalarial principles persists to be a top priority, and because malaria chemotherapy is historically linked to natural products, this chapter therefore provides a comprehensive information on the role and efficacy of chemicals and nutrients of plant origin as therapeutics in the management of malaria.

6.2 PLANT-BASED SECONDARY COMPOUNDS AS ANTIMALARIA AGENTS

The medicinal properties of plants are normally dependent on certain secondary metabolites or active principles. These secondary metabolites do not directly influence the organisms in their growth, development, and reproduction but mediate their actions indirectly. Their absence does not necessarily cause instant death, but long-term deficiency may cause impairment of the well-being.

Plants have a long historical link in herbal medicine for treating various diseases and maladies because they produce a diverse range of chemical agents that interact with targets in the biological system (Husen, 2021, 2022a, 2022b, 2023; Rahman and Husen, 2023a, 2023b). Approximately 15% of plant species have been subjected to scientific investigation of their phytochemical constituents, while only 6% of this have been considered for their pharmacological efficacy. There are therefore several active agents in plants yet to be discovered for the potency in combating diseases, including malaria. As regards malaria parasites, several drugs derived from chemical agents from plants have been devised in modern therapy as antimalarials to treat or control the disease.

For synthesizing new and potent antimalarial agents, medicinal plants have played a critical role. The potency of plant extract and isolated agents ascertained through several pharmacological studies has been able to substantiate the effectiveness of most plants in combating malaria parasites. In most tested cases, resistance to the parasites has not been recorded, which can be attributed to the synergistic action of the constituents in plant. This alone presents a robust efficacy of extracts from plants against malaria resistance. The eradication of malaria in most parts of the world, especially in Africa, may also rely on the uniqueness of plant agents in being able to lessen the burden of the disease without getting rid of the continued vulnerability to infection important to sustain immunity. Extraction, isolation, and characterization of a lot of active phytochemicals (secondary metabolites, mineral elements, and amino acids) from plant sources have given way for antimalarial drugs. For instance, gedunin, febrifugine, ellipticine, olivacine, uvaretin, diuvaretin, bartericin A, 4-Hydroxylonchocarpin, lanaroflavonem, ineupatorolide A, 6α,7β-Diacetoxyvouacapane, neosergeolide, (+)-Catechin-3-gallate, and (+)-Catechin-5-gallate are isolated antiplasmodial compounds from different plant sources (Ezenyi and Salawu, 2016). These bioactive phytochemical agents are the main characteristics of plants which exhibit physiological actions in humans. Natural products with antiplasmodial principles generally include alkaloids, cucurbitacin glycosides, terpenoids, benzophenones, essential oils, triketones, sesquiterpene lactones, xanthanoids, polyphenol, iridoid glycosides, coumarins, anthraquinones, tetrahydropyrone derivatives, limonoids, tetranortriterpenoids, flavonoids, triketones, cembranolides, triterpene hydrocarbon derivatives, poliothrysoside, saponins, tannins, chalcones, cembranolides, xanthanoids, and phloroglucinols but have been classified majorly into alkaloids, quinones and polyketides, terpenes, and polyphenols in this chapter based on their chemical structure, composition, solubility in solvents, or the pathway of their synthesis. These antimalaria compounds present in plant extracts are summarized in Table 6.1, and the major classes described thus.

TABLE 6.1
Plant-Based Chemical Compounds with Various Mechanisms of Antimalarial Actions

S. No.	Natural Product Class	Natural Product Subclass	Active Compound	Source	Action/Inference	Reference
1	Alkaloids	Monoindole	Strychnobrasiline	*Strychnos myrtoides*	Promising agent for resistance reversal when combined with quinoline-containing drugs	Rafatro et al., 2000; Ramanitrahasimbola et al., 2006
			Malagashanine		Aided the accumulation of chloroquine in drug-resistant parasites	Rafatro et al., 2000
		Naphthylisoquinolines	Ancistectorines A1, Ancistectorines *N*-methyl A_1, Ancistectorines A_2, Ancistectorines 5-epi-A2, Ancistectorines A_3, Ancistectorines C_1	Twigs of Chinese *Ancistrocladus tectorius*	Suppressed *P. falciparum* (K1 strain) without cytotoxicity to the cells	Bringmann et al., 2013
		Benzylisoquinolines and other isoquinolines	Cassiarin J	Extract of *Cassia siamea* (Fabaceae) leaf	Inhibited growth of 3D7 *P. falciparum in vitro*	Deguchi et al., 2012
			Cassiarin K			Deguchi et al., 2012
			Cassiarin A		Promising antiplasmodial activity agent	Morita et al., 2007
		Phenanthridine, phenanthroindolizidine, and phenanthrene alkaloids, derivatives of henanthrene	2-hydroxyatherosperminine	Extract of *Cryptocarya nigra* (Lauraceae) bark	Potency against the K1 strain of *P. falciparum*	Nasrullah et al., 2013
			Dehydrotylophorine	Methanol twigs extract of *Ficus septica* (Moraceae)	*In vitro* inhibition of 3D7 strain of *P. falciparum*	Kubo et al., 2016
			Dehydroantofine			Kubo et al., 2016
			Tylophoridicine			Kubo et al., 2016
			Nitidine	*Zanthoxylum rhoifolium* and *Zanthoxylum chalybeum* (Rutaceae)	Strong activity against *P. falciparum* sensitive to chloroquine	Gakunju et al., 1995; Bouquet et al., 2012
			Dihydronitidine	*Zanthoxylum heitzii* bark	Demonstrated drug resistance against 3D7 strain of parasites	Goodman et al., 2016

Terpenylindoles	Uleine	Brazilian tree *Aspidosperma parvifolium* (Apocynaceae) trunk bark	Better potency against strain of W2 > 3D7	Dolabela et al., 2015
	N-acetylpolyveoline	*Greenwayodendron suaveolens* (Annonaceae) root bark	Selectively (SI = 10.6) cytotoxic against MRC-5 cells	Muganza et al., 2016
	Polyalthenol		Cytotoxic against MRC-5 cells	Muganza et al., 2016
Bisindoles and related indoles	Flinderole A	Australian tree *Flindersia acuminata* (Rutaceae) bark	Antiplasmodial activity opposed to *P. falciparum* selective	Fernandez et al., 2008, 2010
	Isoborreverine			Fernandez et al., 2008, 2010
	Flinderoles B	*Flindersia amboinensis* from Papua New Guinea	Potent against *P. falciparum* trophozoites, while parasites showed differences in the morphology of digestive vacuole and a decreased hemozoin constitution	Fernandez et al., 2008, 2010
	Flinderoles C			Fernandez et al., 2008, 2010
	Dimethylisoborreverine			Fernandez et al., 2008, 2010
	Pimentelamine C	*Flindersia pimenteliana*	Dd2 and 3D7 strain of the parasite suppressed and exhibited minimal toxicity to HEK-293 cells	Robertson et al., 2017
	Borreverine			Robertson et al., 2017
	4-methylborreverine			Robertson et al., 2017
	Tabernaelegantine B	*Muntafara sessilifolia* (Apocynaceae) stem bark	FcB1 plasmodial strain inhibited, while cytotoxic noted against cells of MRC-5 and L6	Girardot et al., 2012
	Tabernaelegantine D			Girardot et al., 2012
	C-3′ Tabernaelegantine B		Selectivity against P. falciparum demonstrated	Girardot et al., 2012
	C-3′ Tabernaelegantine D			Girardot et al., 2012
β-Carbolines	Hyrtiosulawesine	Root of *Aristolochia cordigera* (Aristolochiaceae)	Inhibition of P. falciparum (3D7) *in vitro*	Liew et al., 2014
	Marinacarboline A	*Marinactinospora thermotolerans* SCSIO 00652	Activity against chloroquine-sensitive 3D7 strain 18 times better Multidrug-resistant Dd2	Huang et al., 2011
	Opacalin A	New Zealand ascidian *Pseudodistoma opacum*	Low cytotoxicity to cells but selectively repressed P. falciparum (K1 strain)	Chan et al., 2011

(Continued)

TABLE 6.1 (*Continued*)
Plant-Based Chemical Compounds with Various Mechanisms of Antimalarial Actions

S. No.	Natural Product Class	Natural Product Subclass	Active Compound	Source	Action/Inference	Reference
		Pyridone, pyrimidine, and piperidine alkaloids	Carpine	Decoction of *Carica papaya* (Caricaceae) leaves	51% inhibition demonstrated antiplasmodial activity *in vitro* (at 4.8 μg/mL) but was inactive *in vivo* from 11.9% inhibition at 5 mg/kg	Julianti et al., 2014
			Carpamic acid		*In vitro* suppressive (51%) action (at 4.8 μg/mL)	Julianti et al., 2014
			Methyl carpamate			Julianti et al., 2014
			Cassine	*Senna spectabilis* leaf	Inhibitory efficacy on *P. falciparum* (3D7) *in vitro*	Pivatto et al., 2014
			Spectaline			Pivatto et al., 2014
			Ingamine A	Marine sponge *Petrosid Ng5, Sp5*	Inhibitory action on *P. falciparum* (W2 and D6), but with no cytotoxicity in the tested doses	Ilias et al., 2012
			(22S)-hydroxyingamine A			Ilias et al., 2012
			Dihydroingenamine D			Ilias et al., 2012
			Haliclonacyclamine A	*Heliclona* sp. From Solomon Islands (ethanol extract) marine sponges	10 mg/kg/day suppressed *Plasmodium vinckei* petter by 45%	Mani et al., 2011
		Pyrroles	Dibromopalau'amine	*Agelas* and *Axinella* genera sponges	Antiplasmodial efficacy on *P. falciparum* (K1)	Scala et al., 2010
			Spongiadicin B			Davis et al., 2012
			Dispacamide B			Davis et al., 2012
			Bispyrroloiminoquinone tsitsikammamine C	Marine sponge *Zyzzya* sp.	Antiplasmodial (nanomolar) potency against *P. falciparum strains* of 3D7 and Dd2	Davis et al., 2012
			Makaluvamines G		Parasitaemia suppressed by 48% at 8 mg/kg/day in *berghei*-infected mice without any sign of toxicity	Davis et al., 2012
			Makaluvamines J		Also potent against both trophozoites and schizonts parasites	Davis et al., 2012
			Makaluvamines K			Davis et al., 2012
			Makaluvamines L			Davis et al., 2012
			Damirone A		Potent antiplasmodial activity	Davis et al., 2012

			Specicoside	Extract of *Kigelia africana* (Bignoniaceae)	Suppresses W2mef strain; synergistic action with artemether, but no antagonistic action with *p*-hydroxycinnamic acid noticed	Zofou et al., 2011a, 2012
2	Terpenes	Sesquiterpenes	1β-(*p*-coumaroyloxy) polygodial	Stem bark extract of *Drimys brasiliensis* (Winteraceae)	Suppresses FcR3 strain of *P. falciparum*	Claudino et al., 2013
			Sesquiterpenoid sporogen-AO1	Soil fungus *Penicillium copticola* PSURSPG138	Repressive efficacy on K1 strain parasites but exhibited cytotoxicity against Vero cells and human oral epidermoid carcinoma (KB)	Daengrot et al., 2015
			Phomoarcherin B	*Phomopsis archeri* isolated from cortex stem of *Vanilla albidia* as endophytic fungus	Antiplasmodial activity demonstrated	Hemtasin et al., 2011
			2-isocyanoclovene 2-isocyanoclovane Spiro heterocycle 3,4-dehydrotheaspirone	Australian nudibranch *Phyllidia ocellata*	Selective suppression of 3D7 and Dd2 strains	White et al., 2015
		Sesquiterpene lactones	Artemisinin	*Artemisia annua* L (Asteraceae)	Antiplasmodial activity	Liu et al., 2014 Misra et al., 2014
			Fortunilide A Sarglabolide J Chlorajaponilide C	*Sarcandra glabra* (Chloranthaceae)	Inhibition of *P. falciparum* compared well with artemisinin, with no cytotoxic effect	Zhou et al., 2016
			Urospermal A 15-*O*-acetate	*Dicoma tomentosa* (Asteraceae)	Inhibited the 3D7 and W2 strains with no haemolytic side effect	Jansen et al., 2012
			Acanthospermolide-type Sesquiterpene lactones	*Acanthospermum hispidum* (Asteraceae), aerial parts extract	Suppressed the 3D7 and W2 strain of *P. falciparum* parasites	Bero et al., 2009
			Vernopicrin	*Vernonia guineensis* (Asteraceae) leaves	No action demonstrated against Dd2 and Hb3 strains of *P. falciparum*	Toyang et al., 2013

(*Continued*)

TABLE 6.1 (*Continued*)
Plant-Based Chemical Compounds with Various Mechanisms of Antimalarial Actions

S. No.	Natural Product Class	Natural Product Subclass	Active Compound	Source	Action/Inference	Reference
		Diterpene	Caesalsappanins G, H, I	*Caesalpinia sappan* (Fabaceae) seeds extracted with chloroform extract	Active against K1 IC50 = 0.38 μg/mL	Ma et al., 2015
			Caesalsappanins H			Ma et al., 2015
			Caesalsappanins I			Ma et al., 2015
			Norcaesalpin D	Root extract (dichloromethane) of *Caesalpinia bonducella*	Suppressed the strains of *P. falciparum* (3D7 and Dd2) as well as IPC 5202 and Battambang and IPC 4912 Mondolkiri-Cambodia resistant to artemisinin	Nondo et al., 2016, 2017
			Amphadilactones A-F, H-I	*Aphanamixis grandifolia* (Meliaceae) leaves	Demonstrated antiplasmodial activity by inhibiting the Dd2 strain	Yin et al., 2014
			Cupacinoside	Extracts (dichloromethane and hexane) of *Cupania cinerea* (Sapindaceae) bark of Ecuador origin	Suppressed the *P. falciparum* K1 strain but demonstrated cytotoxicity against L6 cells	Gachet et al., 2010, 2011
			6′-de-*O*-acetylcupacinoside		Cytotoxic to L6 cells and suppressed the *P. falciparum* K1 strain	Gachet et al., 2010, 2011
			Serrulatane Diterpenoid	*Eremophila microtheca* (Scrophulariaceae) aerial parts	*P. falciparum* not inhibited at 10 μM	Kumar et al., 2017
			8-isocyano-15-formamidoamphilect-11(20)-ene	Marine sponge of *Stylissa cf. massa* of Thailand origin	Suppressed K1 strain of *P. falciparum* with no cytotoxicity in the cells	Chanthathamrongsiri et al., 2012
			Monamphilectines B	Caribbean marine sponge *Svenzea flava* from Puerto Rico	The parasitic 3D7 strain was suppressed	Avilés et al., 2015
			Monamphilectines C			Avilés et al., 2015
			8,15-diisocyano-11(20)-amphilectene			Avilés et al., 2015
			Monamphilectine A		The parasitic W2 strain suppressed	Avilés and Rodríguez, 2010
			Bifurcatriol	Irish brown alga *Bifurcaria bifurcata*	K1 strain of the parasites inhibited with no significant cytotoxicity to the cells	Smyrniotopoulos et al., 2017

	Diterpene-derived phenanthrenone drypetenone D	*Drypetes gerrardii* var *gerrardii* (Putranjivaceae) stem extract	Inhibitory effect on the parasites, but with no *in vivo* efficacy in mice infected with *P. berghei*	Hata et al., 2014
Triterpenes	Betulin coumaroyl esters	Leaf, fruits, and twigs extract (chloroform) of *Buxus cochinchinensis* (Buxaceaea)	Inhibitory activity of Dd2, HB3, and NHP1337 *P. falciparum* strains with no cytotoxicity to the cells	Cai et al., 2016
	Lupane esters; (*Z-* or *E-;* C-3 or C-23) coumaroyl; (*Z-* or *E-;* C-3 or C-23) feruloyl esters of betulin	Supercritical fluid extract (methanol) of *B. sempervirens*	Suppressed the HB3, Dd2, and NHP1337 strains *P. falciparum* with no cytotoxic effect to the cells	Cai et al., 2016
	Betulone	Bark of *Cupania cinerea* (Sapindaceae)	Inhibitory effect on the K1 strain of the parasite with moderate cytotoxic action on cells	Gachet et al., 2011
	Arototiamins A–G	Cameroonian *Entandrophragma congoense* (Meliaceae) bark	Inhibited the NF54 *P. falciparum* strain, but demonstrated different cellular toxicity cells	Happi et al., 2015 Bickiia et al., 2007; Happi et al., 2015
	Isoflindissone lactone	*Boswellia serrata* (Burseraceae) oleo-gum resin – DCM extract	*P. falciparum* strain inhibited	Greve et al., 2017
	Triterpenoid	*Kigelia africana* (Bignoniaceae) stem extract (ethyl acetate)	Inhibited the W2 strain of *P. falciparum* as well as CAM10 and SHF4 isolates; combination with artemether on the W2mef resulted to synergistic effect without antagonistic effect with quinine	Zofou et al., 2011a, 2012
	Balsaminol F	*Momordica balsamina* (Cucurbitaceae) aerial parts extract (methanol)	Both 3D7 and Dd2 strains of *P. falciparum* inhibited	Ramalhete et al., 2011a
	Balsaminol F derivative	Triacetyl semi-synthetic	Demonstrated better activity with to cytotoxic effect to cells compared to the parent compound	Ramalhete et al., 2011a
	Karavilagenin C	*Momordica balsamina*	Selective antiplasmodial activity	Ramalhete et al., 2011b

(*Continued*)

TABLE 6.1 (*Continued*)
Plant-Based Chemical Compounds with Various Mechanisms of Antimalarial Actions

S. No.	Natural Product Class	Natural Product Subclass	Active Compound	Source	Action/Inference	Reference
			Neomacrolactone	Ugandain *Neoboutonia macrocalyx* (Euphorbiaceae)	FcB1 *P. falciparum* inhibited, but cytotoxicity against KB and MRC-5 cells generally demonstrated	Namukobe et al., 2014
			22α-acetoxyneomacrolactone			Namukobe et al., 2014
			6-hydroxyneomacolactone			Namukobe et al., 2014
			22α-acetoxy-6-hydroxyneomacrolactone			Namukobe et al., 2014
			4-methyleneneomacrolactone			Namukobe et al., 2014
			22-de-*O*-acetyl-26-deoxyneoboutomellerone			Namukobe et al., 2014
			Neomacroin			Namukobe et al., 2014
			Aszonapyrone A	*Neosartoryatatenoi* KKU-2NK23 extracted by ethyl acetate as soil fungus	Suppressed *P. falciparum*	Yim et al., 2014
		Steroids	Physalins B	*Physalis angulata* (Solanaceae)	*In vitro* suppressive action on the parasitic W2 strain, but with cytotoxic effect	Sá et al., 2011
			Physalins D		65% inhibition in mice infected with *P. berghei* at 100 mg/kg	Sá et al., 2011
			Physalins F		Repressive action on W2 parasite *in vitro*	Sá et al., 2011
			3-*O*-[β-glucopyranosy-(1→2)-*O*-β-xylopyranosyl] stigmasterol	*n*-butanol fraction of extract (stem bark) from *Caesalpinia volkensii* (Fabaceae)	D6 and W2 strains of *P. falciparum* both inhibited	Ochieng et al., 2013
			Halymeniaol	Marine red alga *Halymenia floresii*	3D7 of *P. falciparum* inhibited, but demonstrated cytotoxic effect	Meesala et al., 2018
			Pandaroside G	Caribbean sponge *Pandaros acanthifolium*	Repressive efficacy on K1 strain of parasite but cytotoxic to the cells	Regalado et al., 2010
			Pandaroside G methyl ester			Regalado et al., 2010
		Limonoids	Grandifolione	Seed petroleum ether extract of *Khaya anthotheca* (Meliaceae)	Inhibited the K1 strain	Obbo et al., 2013
			7-deacetylkhivorin			Obbo et al., 2013

		Quassinoids	Simalikalactone D	Leaf extract of *Quassia amara* (Simaroubaceae)	*In vitro* and *in vivo* suppression of FcB1 *P. falciparum* and not toxic overall	Bertani et al., 2005, 2006
			Quassin	Stem bark extract of *Quassia amara* (Simaroubaceae)	Repressive action on the MRC-pf-303 and MRC-pf-20 strains	Mishra et al., 2010
			Neo-quassin			Mishra et al., 2010
3	Polyphenols	Biflavonoids	Biflavanones GB1a	Seed extracts of *Garcinia kola* (Clusiaceae)	Repressive action on the FCR3 strain and exhibited low cytotoxicity	Konziase, 2015
			Biflavanones GB1		52% inhibitory efficacy on *P. berghei* at 100 mg/kg *in vivo*	Konziase, 2015
			Biflavanones GB2			Konziase, 2015
			Volkensiflavone	*Allanblackia floribunda* (Clusiaceae)	*P. falciparum* strains (F32 and FcM29) inhibited	Konziase, 2015
		Prenylated flavonoids	Styracifolin B	*Artocarpus styracifolius* (Moraceae) stem bark extract (ethyl acetate)	FcB1 *P. falciparum* strain inhibited at 10 μg/mL by 87%	Atilaw et al., 2017
			Artonin B		Varying degree of suppressed FcB1 parasites with cytotoxicity to KB cells	Atilaw et al., 2017
			Artonin F			Atilaw et al., 2017
			Heterophyllin			Atilaw et al., 2017
			Artoheterophyllin B			Atilaw et al., 2017
		Lignans	(7′*R*,8*S*,8′*R*)-4,5-Dimethoxy-3′,4′-methylenodioxy-2,7′-cyclolignan-7-one	n-hexane extract of *Holostylis reniformis*		de Andrade-Neto et al., 2007
			7′*R*,8*S*,8′*R*)-3′,4,4′,5-tetramethoxy-2,7′-cyclolignan-7-one			de Andrade-Neto et al., 2007
			(7′*R*,8*R*,8′*S*)-3′,4,4′,5-tetramethoxy-2,7′-cyclolignan-7-one			de Andrade-Neto et al., 2007
			(7′*R*,8*S*,8′*S*)-3′,4,4′,5-tetramethoxy-2,7′-cyclolignan-7-one			de Andrade-Neto et al., 2007
			(7′*R*,8*S*,8′*S*)-3′,4′-dimethoxy-4,5-methylenodioxy-2,7′-cyclolignan-7-one			de Andrade-Neto et al., 2007
			(7′*R*,8*S*,8′*S*)-4,5-dimethoxy-3′,4′-methylenodioxy-2,7′-cyclolignan-7-one			de Andrade-Neto et al., 2007

(*Continued*)

TABLE 6.1 (*Continued*)
Plant-Based Chemical Compounds with Various Mechanisms of Antimalarial Actions

S. No.	Natural Product Class	Natural Product Subclass	Active Compound	Source	Action/Inference	Reference
		Coumarins and lactones	3′-decanoyl-*cis*-khellactone 4′-decanoyl-*cis*khellactone	*Angelica purpuraefolia*	Selective inhibition of D10 strain of the parasite and no cytotoxic effect on the cells	Chung et al., 2010a
			6-acetyl-2-isopropenyl-8-methoxy-1,3-benzodioxin-4-one	Extract of *Carpesium divaricatum* (Asteraceae)	Antiplasmodial activity	Chung et al., 2010b
			Malleastrumolide A	*Malleastrum* sp. (Meliaceae), dichloromethane extract	Parasites of Dd2 strain inhibited but had cytotoxic effect to the cells	Du et al., 2017a
		Phenolic acids, phenylethanoids, phenylpropanoids, and other shikimic acid–derived metabolites	Methyl gallate	Leaf/stem bark extract of *Dacryodes edulis* (Burseraceae)	Suppressed the 3D7 and Dd2 parasites and had synergistic action with quinine	Zofou et al., 2011b
			2,3,4-trihydroxy-2-methylbutyl gallate	*Limonium leptophyllum* (Plumbaginaceae) aerial parts	D6 and W2 suppressef in double fold as those resistant to chloroquine	Gadetskaya et al., 2017
			4-*O*-(3′-methylgalloyl) norbergenin	Stem bark extract of *Diospyros sanza-minika* (Ebenaceae)	Demonstrated antiplasmodial activity	Tantangmo et al., 2010
			Ellagic acid	Methanol extract of *Anogeissus leiocarpus* (Combretaceae) bark	Inhibited 3D7 *P. falciparum* and demonstrated synergistic action when combined with artesunate and chloroquine	Soh et al., 2009; Ndjonka et al., 2012
			Ococymosin	Hexane stem extract of *Ocotea cymosa* (Lauraceae)	Active metabolite against the Dd2 strain	Rakotondraibe et al., 2015
			Robustasides D	Leaf and twigs extract of *Grevillea* (Poorinda Queen, Proteaceae)	Suppressed the *P. falciparum* resistant to chloroquine than those strains sensitive to chloroquine and as well as suppressed both TM93-C1088 and TM90-C2B strains	Ovenden et al., 2011
			Robustasides G		Suppressed the *P. falciparum* resistant to chloroquine than those strains sensitive to chloroquine, while it also suppressed both Dd2 and K1 strains	Ovenden et al., 2011

			9-methoxystrobilurins G			
			Oudemansin B			
		Xanthones	Pentadexanthone	*Pentadesma butyracea* (Clusiaceae) extract of the fruit pericarp	Active against *P. falciparum* W2 strain	Lenta et al., 2011
			Cratoxylone		Active against *P. falciparum* W2 strain	Lenta et al., 2011
			Garcinone E		Repressive action on W2 and FcB1 strains, but with cytotoxic effect	Zelefack et al., 2009; Lenta et al., 2011
			α-mangostin		Active against *P. falciparum* W2 strain	Lenta et al., 2011
		Phloroglucinol derivatives	Mallotojaponins B Mallotojaponins C	Extract (inflorescence and leaf) *Mallotus oppositifolius* (Euphorbiaceae)	Dd2 strain of the parasite suppressed, with cytotoxic effect to the cell	Harinantenaina et al., 2013
		β-Triketones	Rhodomyrtosone F	*Syncarpia glomulifera* (Myrtaceae) stem bark extract	Inhibition of Dd2 *P. falciparum*	Su et al., 2016
			Rhodomyrtone	Flower extract of *Angophora woodsiana* (Myrtaceae)	Both 3D7 and Dd2 strain of the parasites inhibited, though cytotoxic effect moderate	Senadeera et al., 2017
			Tomentosone A	Leaf extract (DCM) of *Rhodomyrtus tomentosa* (Myrtaceae)	The Dd2 and 3D7 of the parasites' strain inhibited and without any cytotoxic effect to the cell	Hiranrat et al., 2012
		Other polyphenols	Atranorin	Bark extract *Kigelia africana* (Bignoniaceae)	The W2, SHF4, and CAM10 strains, together with the W2mef, inhibited, while it demonstrated synergistic reaction with artemether when combined	Zofou et al., 2011a, 2012
4	Quinones and polyketides	Anthraquinones	Chryslandicin	*Kniphofia ensifolia* (Asphodelaceae) dichloromethane extract	Active against the Dd2 strain of the parasites	Dai et al., 2014
			10-(chrysophanol-7-yl)-10-hydroxy-chrysophanol anthrone		Suppressed Dd2 *P. falciparum*, but with selective cytotoxicity to the cells	Wube et al., 2005; Dai et al., 2014
			Knipholone		Inhibited the 3D7 strain, but without cytotoxic effect to the cells	Wube et al., 2005; Dai et al., 2014
			3,4-di*O*-methylcaffeoyl ester		Inhibition of the *P. falciparum* without any cytotoxic action	Dai et al., 2014
			Urdamycinone G			
			Urdamycin E			
		Naphthoquinones	Pentalongin	*Pentas longiflora* (Rubiaceae) root extract	Inhibitory efficacy on D6 and W2 strains	Endale et al., 2012
			Psychorubrin			Endale et al., 2012
			Furanonaphthoquinones	Stem bark of *Markhamia tomentosa* (Bignoniaceae) extract (ethyl acetate)	Inhibited the K1 and W2 parasites without being cytotoxic to the cells	Tantangmo et al., 2010

6.2.1 Alkaloids

Alkaloids represent the largest class of plant secondary metabolites having the highest number of compounds that exhibit antiplasmodial activity. They are also templates from which a lot of the quinolone-based antimalarial drugs are synthesized. Alkaloids that exhibit potent antiplasmodial efficacy occur as naphthylisoquinolines; benzylisoquinolines and other isoquinolines; phenanthridine, phenanthroindolizidine, and phenanthrene alkaloids, derivatives of phenanthrene; terpenylindoles; bisindoles and related indoles; β carbolines pyridone, pyrimidine, and piperidine alkaloids; and pyrroles, among others. Naphthylisoquinolines alkaloids isolated from the root bark of Congolese *Ancistrocladus ileboensis*, including dioncophyllines C2, dioncophyllines F, ancistrocladisine A, and 5′-O-methyldioncophylline D, as well as dioncophylline D2 from the leaves of same plant, are all highly effective against malaria (Li et al., 2017). Quinine is also an important drug in severe cases of chloroquine-resistant malaria, even though it is limited due to its side effects.

Following the successful discovery of quinine, natural antiplasmodial alkaloids have been isolated. There are alkaloids that inhibit fatty acid biosynthesis in the parasite, while some act as resistance reversers (Table 6.1). Strychnobrasiline and malagashanine isolated from *Strychnos myrtoides* enhanced the actions of quinolones, chloroquine, halofantrine, and aminoacridines (Rafatro et al., 2000; Ramanitrahasimbola et al., 2006). While malagashanine did not exhibit any cytotoxic action, it helps in the accumulation of chloroquine in drug-resistant parasites by improving influx of chloroquine and preventing its efflux from the parasites (Ramanitrahasimbola et al., 2006).

6.2.2 Terpenes

Since the discovery of artemisinin, an endoperoxide sesquiterpene lactone, more efforts have been put on terpene compounds, especially sesquiterpenoids, because of their intrinsic antiplasmodial activity. Terpenes also provide the opportunity to have desired physicochemical attributes for enhanced potency by modification of the chemical constituents. β-(p-coumaroyloxy) polygodial is a potent antiplasmodial sesquiterpene isolated from the stem bark extract of *Drimys brasiliensis* (Winteraceae). Ineupatorolide A is another active antiplasmodial sesquiterpene that has been isolated. Other potent terpenes with antiplasmodial principles belong to the subclass of monoterpenes, sesquiterpene lactones, diterpene; triterpenes, steroids, limonoids, and quassinoids (Table 6.1). For instance, the steroid 3-*O*-[β-glucopyranosy-(1→2)-*O*-β-xylopyranosyl] stigmasterol from *n*-butanol fraction of extract (stem bark) from *Caesalpinia volkensii* (Fabaceae) exhibited high antiplasmodial activity against the D6 and W2 strains of *P. falciparum*. Zhou et al. (2016) reported the isolation of fortunilide A, sarglabolide J, and chlorajaponilide C from *Sarcandra glabra* (Chloranthaceae), which inhibited *P. falciparum* K1 strain (Table 6.1).

6.2.3 Quinones

Quinones, not limited to anthraquinones, naphthoquinones, perylenequinones, and other polyketides, also exhibit several pharmacological properties, including antimalarial activity. Quinones isolated from plants like *Kniphofia ensifolia*, *Kniphofia foliosa* root, *Pentas longiflora* (Rubiaceae) root extract, and ethylacetate extract of *Markhamia tomentosa* (Bignoniaceae) stem bark inhibited the growth of malaria parasites. The antiplasmodial activities of these compounds have helped validate their medicinal use in the control of malaria.

Wube et al. (2005) investigated the dichloromethane *Kniphofia foliosa* Hochst (Asphodelaceae) from Ethiopia against chloroquine-sensitive 3D7 strain of *P. falciparum* (ED50 3.8 μg/mL) with low cytotoxic effect against KB cells with an ED50 value of 35.2 μg/mL. The isolated compounds 10-(chrysophanol-7′-yl)-10-(ξ)-hydroxychrysopanol-9-anthrone and chryslandicin inhibited the growth of *P. falciparum* and were confirmed as promising antimalarial compounds with very little host cell toxicity. The other antimalarial quinones identified were 10-(chrysophanol-7-yl)-10-methoxychrysophanol anthrone (Abdissa et al., 2013) as well as pentalongin and psychorubrin from *Pentas longiflora* (Rubiaceae) root extract (Endale et al., 2012). These compounds demonstrated inhibitory efficacy against D6 and W2 strains (Table 6.1). Bringmann et al. (1999), in another study, described knipholone and three other derivatives from the same plant, as well as seven compounds related structurally. The phenylanthraquinones compounds exhibited antiplasmodial activity against two chloroquine-resistant strains of *P. falciparum in vitro* (K1 and NF 54) with mild-level cytotoxicity.

6.2.4 Polyphenols

Phenolic compounds are frequent in plants and are distributed within numerous phytochemical classes. They are compounds majorly derived from the shikimate/phenylpropanoid and/or the polyketide pathway. They are deprived of nitrogen-based function and feature more than one phenolic unit. When this functional structure is repeated, the compound is named a polyphenol. Polyphenols, therefore, include, but are not limited to, phenolic acids, coumarins, flavonoids, stilbenes, lignans, tannins, and stilbenes.

Phenolic compounds demonstrate their antimalarial efficacy by being able to easily undergo oxidation reactions, making them strong radical scavengers, which is one of the pathways of antimalarials. In addition, polyphenols and their derivatives conserve simple phenolic moieties and are, therefore, related to several mechanisms of action. They are capable of inhibiting haemoglobin digestion and hemozoin formation, disturbing the redox homeostasis and, possibly, iron chelating (Rakotondraibe et al., 2015).

Flavonoids are a ubiquitous chemical agent in higher plants where they function to protect plants from pathogenic invasion and as growth modulators. They also function to inhibit the fatty acid biosynthesis (FAS II) pathway, which is only unique to the apicoplast of parasite and not in human hosts. Luteolin-7-O-β-d-glucopyranoside was the first natural product–isolated flavonoids known to target plasmodial FAB I enzyme that modulates the FAS II pathway (Ezenyi and Salawu, 2016). There are, however, reports on the inhibition of myoinositol and l-glutamine influx into infected erythrocytes by some flavonoids. Flavonoids may also interfere with hemin degradation, while chalcones inhibit plasmepsin II and cyclin-dependent protein kinases (Ezenyi and Salawu, 2016). Biflavanones GB1a, biflavanones GB1, and biflavanones GB2 are flavonoids isolated from seed extracts of *Garcinia kola* (Clusiaceae), and the compounds have been reported to have repressive action on the FCR3 strain and exhibited low cytotoxicity, 52% inhibitory efficacy on *P. berghei* at 100 mg/kg *in vivo* (Konziase, 2015).

Xanthones is an interesting class of antiplasmodials. They are dibenzo-c-pyrone compounds formed by cyclization of a benzophenone. Allanxanthone C, tovophyllin A, α-Mangostin, macluraxanthone (from *Allanblackia monticola* Staner), demethylcalaaxanthone, 6-deoxy-γ-mangostin (from *Calophyllum caledonicum*), garcinone E (from *Pentadesma butyracea* Sabine), and symphonin (from *Symphonia globulifera*) are reported xanthones with antiplasmodial activities. These compounds demonstrated antiplasmodial activity in chloroquine or multidrug-resistant strains with an advantage against the resistant parasite, and a unique attribute of Clusiaceae, found in tropical regions. Prenylation at key carbons with 2-methylpent-2-ene groups may be responsible for the characteristics of the compounds against the parasites. Lenta et al. (2011) presented the inhibition of FcB1 *P. falciparum* by Pentadexanthone, Cratoxylone, Garcinone E, and α-mangostin all isolated from the fruit pericarp extract of *Pentadesma butyracea* (Clusiaceae) (Table 6.1).

Lignans are compounds with high antiplasmodial potential and varied selective index. For instance, (7′*R*,8*S*,8′*R*)-3′,4,4′,5-Tetramethoxy-2,7′-cyclolignan-7-one, (7′*R*,8*R*,8′*S*)-3′,4,4′,5-Tetramethoxy-2,7′-cyclolignan-7-one, (7′*R*,8*S*,8′*S*)-3′,4,4′,5-Tetramethoxy-2,7′-cyclolignan-7-one, (7′*R*,8*S*,8′*S*)-3′,4′-Dimethoxy-4,5-methylenodioxy-2,7′-cyclolignan-7-one, (7′*R*,8*S*,8′*S*)-4,5-Dimethoxy-3′,4′-methylenodioxy-2,7′-cyclolignan-7-one, and (7′*R*,8*S*,8′*R*)-4,5 Dimethoxy-3′,4′-methylenodioxy-2,7′-cyclolignan-7-one isolated from the n-hexane extract of *Holostylis reniformis* root were tested *in vivo* against *P. berghei* and *in vitro* against a chloroquine-resistant strain of *P. falciparum*

(Table 6.1). These compounds demonstrated potency for the development of antimalarial drugs (de Andrade-Neto et al., 2007). It is, however, not certain about the mechanism of action of lignans in the general sense.

Coumarins are derivatives of benzopyrone (1,2-benzopyrones or 2H-1-benzopyran-2-ones) which are largely distributed in nature. Other derivatives, such as pyranocoumarins and furanocoumarins, can be derived from modification, such as cyclization, O- or C-prenylation. Coumarins are known to exhibit antiplasmodial efficacy. For instance, (+)-4′-decanoyl-*cis*-khellactone, 1-O-galloyl-6-O-luteoyl-R-D-glucose, *trans*-avicennol, and clausarin are all isolated coumarins with reported inhibitory activity against malaria parasites. Chung et al. (2010a) demonstrated the selective inhibition of D10 strain of the parasite with no cytotoxic effect on the cells of 3′-decanoyl-cis-khellactone and 4′-decanoyl-ciskhellactone from *Angelica purpuraefolia* (Table 6.1).

6.3 NUTRIENTS AS POTENTIAL ANTIMALARIAL AGENTS

6.3.1 Vitamin A

Vitamin A promotes the normal function of immunity, and some studies have proposed its role in the management of malaria (Shankar, 2000). It was reported that lack of vitamin A increases the occurrence of malaria infection (Roos et al., 1946). In rats and mice deficient in vitamin A, susceptibility to malaria was elevated, and this was ameliorated with supplements of the vitamin (Krishnan et al., 1976; Stoltzfus et al., 1989). In another study, a genetic locus, which involves cellular retinol-binding protein 1, was indicated to control malaria parasitaemia and mortality in mice (Foote et al., 1997). In an *in vitro* model, free retinol addition to *P. falciparum* cultures decreased replication of the parasite (Davis et al., 1998); however, contrary results were obtained in another study (Samba et al., 1992). In a recent study, vitamin A supplementation was shown to confer immunoregulatory action in C57BL/6J experimental mice infected with *P. berghei* and *P. yoelii* through the reduction of parasitaemia, upregulation of MHCII expression in dendritic cells, and suppression in the production of IFN-γ and TNF-α. They concluded that the supplementation of vitamin A decreased cerebral malaria inflammatory response; however, the supplementation did not confer an immunoregulatory action against P. yoelii (Chen et al., 2019). Similarly, George and coworkers stated that vitamin A (60 mg/kg) might exhibit antimalarial activity, as observed from their study (George et al., 2019).

In a systematic review and meta-analysis involving pregnant women and children not older than 5 years, it was shown that supplementation with vitamin A did not have any significant beneficial effect on placental malarial infection in pregnancy and did not decrease malaria mortality or peripheral parasitaemia incidence in children under the age of 5 (Yakoob et al., 2018).

6.3.2 Folate

Folate is an important nutrient involved in the growth of cells and production of red blood cells. Many antimalarial agents target folate metabolism of the malarial parasite, and the relationship between host organism folic acid supplementation and status in malaria-endemic regions is of interest (Shankar, 2000). A study of prophylactic folic acid supplementation in Gambian preschool pupils indicated no detrimental effects in malaria treatment. In a similar study on folic acid supplementation in pregnant women, it was found that there are no detrimental effects on parasitaemia, although there was increase in the counts for reticulocyte (Fuller et al., 1988; Gail and Herms, 1969). In total, some studies proposed that deficiency of folic acid might elevate the morbidity caused by malaria, even though the routine folic acid supplements used in regions endemic with malaria has been recommended in the prevention and treatment of anaemia (Fleming and Werblinska, 1982).

6.3.3 Vitamin C

Ascorbic acid (vitamin C) can serve as an antioxidant, particularly in the protection of the immune system from reactive radicals (Wintergerst et al., 2006). In support of this, the plasma level of vitamin C was shown to be strongly reduced in acute and chronic malaria disease (Mohammad, 2012; D'Souza and D'Souza, 2006). A study on the synergistic antimalarial action of vitamin C and certain ketones was reported to be equivocal. In this work, the vitamin augments the antimalarial activities of exofone, a ketone, against *P. falciparum* via its antioxidant action in an *in vitro* model (Winter et al., 1997). On the contrary, a poor synergistic action was documented against *P. berghei* in mice between vitamin C and some ketones (Mahajan et al., 2005), although the work stated that a lower dose of the ketones used in the combined therapy may be the reason for the poor synergistic effect. In another combined therapy, chloroquine (25 mg/kg) was combined with vitamin C (25 mg/kg), and it was observed to attenuate oxidative stress in female mice infected with *P. berghei* (Iyawe et al., 2006). In a recent work, vitamin C (15 and 7.5 mg/kg) was reported to display antimalarial activity via reduction of parasitaemia and amelioration of impaired liver markers associated with the infection (Mgbemena and Ifeyinwa, 2020). It was shown that vitamin C abrogates blood-stage *P. berghei* and *P. falciparum* through oxidative stress (Shi et al., 2021). In another recent work, the administration of vitamin C together with artesunate-amodiaquine increases the mRNA levels of antioxidant enzymes in the cells of bone marrow and alleviates impaired biochemical markers in mice infected with *P. berghei* (Ebohon et al., 2021).

6.3.4 Iron

The demerits of both malaria and iron deficiency fall mainly on pregnant women and preschool children, and supplementation with iron to this group of individuals is the major way of preventing and managing anaemia and anaemia-related diseases (Gibson, 1990; Wyler, 1993; Shankar, 2000). Also, several works have documented the role of iron in the management of malaria, although some studies have shown that iron supplementation elevates the risk of developing malarial infection, whereas others showed no significant detrimental influence (Oppenheimer et al., 1986; Smith et al., 1989). Briefly, it was reported that prophylactic supplementation of iron was linked to elevations in some malariometric parameters. But this seems to be relatively negligible. In the other way, enhancement of hematologic status subsequent to supplementation of iron was significant. Also, even though some studies proposed more unwanted effects for individuals staying in regions with moderate malaria transmission levels, the main effects of supplementation of iron in some settings could not be estimated. In total, a meta-analysis proposed that programmes involving supplementation of iron in relation to international guidelines should be proposed to iron-deficient individuals staying in regions endemic to malaria (Shankar, 2000). Contrary to these studies, it was stated that a high content of iron in the hepatocytes can promotes the growth of *Plasmodium* and induce oxidative damage in the cells. This study then shows that an iron-chelating compound called deferiprone-resveratrol hybrid chelates high levels of iron and leads to protection against malaria infection–induced liver damage (Chuljerm et al., 2022).

6.3.5 Zinc

Zinc, a micronutrient, is needed for normal functioning of the immune system and is required for different functions of lymphocytes implicated in malaria resistance, such as formation of tumour necrosis factor-α, immunoglobin G, interferon-γ, macrophages microbicidal action (Shankar and Prasad, 1998; Good et al., 1998). A placebo-controlled study of zinc supplementation in Gambia preschool children indicated a 30% decrease in health centre attendance as a result of *P. falciparum* (Bates et al., 1993), even though this was not significant statistically. In mice, supplementation with zinc reduced oxidative stress markers in *P. berghei* infection (Arif et al., 1987). In addition, it was shown that moderate deficiency of zinc leads to 40% death from the *P. yoelii*–treated rodent (Shankar et al., 1995). In another recent study, zinc and vitamin C diet supplements administered with artemether drug enhanced the survival rate and haematological profile of *P. berghei*–infected mice (Ekeh et al., 2019). The authors hypothesized that zinc–vitamin C supplements play a vital role in managing malaria infection and complementing the antimalarial action of artemether.

6.4 CONCLUSION AND GENERAL REMARKS

Secondary metabolites and nutrients derived from plants are important in the prevention and management of malarial infection. The lack of micronutrients increases the prevalence of malaria infection. However, plant secondary products and nutrients aid in the control of malarial infection. Some of the mechanisms of action of these plant-derived compounds include modulation of immune system, antioxidant action, and action against the parasites of malarial infection. Although a disparity in the action of vitamin C against *Plasmodium* parasite was stated, the ameliorative role seems to be greater. Thus, plant secondary compounds and nutrients could serve as a combined therapy in the prevention and management of malarial infection.

REFERENCES

Abdissa, N., Induli, M., Akala, H. M., Heydenreich, M., Midiwo, J. O., Ndakala, A., & Yenesew, A. (2013). Knipholone cyclooxanthrone and an anthraquinone dimer with antiplasmodial activities from the roots of Kniphofia foliosa. *Phytochemistry Letters*, *6*(2), 241–245. https://doi.org/10.1016/j.phytol.2013.02.005

Arif, A. J., Mathur, P. D., Chandra, S., Singh, C., & Sen, A. B. (1987). Effect of zinc diet on xanthine oxidase activity of liver of mice infected with Plasmodium berghei. *Indian Journal of Malariology*, *24*, 59–63.

Atilaw, Y., Duffy, S., Heydenreich, M., Muiva-Mutisya, L., Avery, V. M., Erdélyi, M., & Yenesew, A. (2017). Three chalconoids and a pterocarpene from the roots of tephrosia aequilata. *Molecules (Basel, Switzerland)*, *22*(2), 318. https://doi.org/10.3390/molecules22020318

Atilaw, Y., Muiva-Mutisya, L., Ndakala, A., Akala, H. M., Yeda, R., Wu, Y. J., Coghi, P., Wong, V. K. W., Erdélyi, M., & Yenesew, A. (2017). Four prenylflavone derivatives with antiplasmodial activities from the stem of tephrosia purpurea subsp. leptostachya. *Molecules (Basel, Switzerland)*, *22*(9), 1514. https://doi.org/10.3390/molecules22091514

Avilés, E., Prudhomme, J., Le Roch, K. G., & Rodríguez, A. D. (2015). Structures, semisyntheses, and absolute configurations of the antiplasmodial α-substituted β-lactam monamphilectines B and C from the sponge *Svenzea flava*. *Tetrahedron*, *71*(3), 487–494. https://doi.org/10.1016/j.tet.2014.11.060

Avilés, E., & Rodríguez, A. D. (2010). Monamphilectine A, a potent antimalarial β-lactam from marine sponge Hymeniacidon sp: Isolation, structure, semisynthesis, and bioactivity. *Organic Letters*, *12*(22), 5290–5293. https://doi.org/10.1021/ol102351z

Bates, C. J., Evans, P. H., Dardenne, M., et al. (1993). A trial of zinc supplementation in young rural Gambian children. *British Journal of Nutrition*, *69*, 243–255.

Bero, J., Ganfon, H., Jonville, M. C., Frédérich, M., Gbaguidi, F., DeMol, P., Moudachirou, M., & Quetin-Leclercq, J. (2009). *In vitro* antiplasmodial activity of plants used in Benin in traditional medicine to treat malaria. *Journal of Ethnopharmacology*, *122*(3), 439–444. https://doi.org/10.1016/j.jep.2009.02.004

Bertani, S., Bourdy, G., Landau, I., Robinson, J. C., Esterre, P., & Deharo, E. (2005). Evaluation of French Guiana traditional antimalarial remedies. *Journal of Ethnopharmacology*, *98*(1–2), 45–54. https://doi.org/10.1016/j.jep.2004.12.020

Bertani, S., Houël, E., Stien, D., Chevolot, L., Jullian, V., Garavito, G., Bourdy, G., & Deharo, E. (2006). Simalikalactone D is responsible for the antimalarial properties of an Amazonian traditional remedy made with Quassia amara L. (Simaroubaceae). *Journal of Ethnopharmacology*, *108*(1), 155–157. https://doi.org/10.1016/j.jep.2006.04.017

Bouquet, J., Rivaud, M., Chevalley, S., Deharo, E., Jullian, V., & Valentin, A. (2012). Biological activities of nitidine, a potential anti-malarial lead compound. *Malaria Journal*, *11*(1), 67. https://doi.org/10.1186/1475-2875-11-67

Bringmann, G., Menche, D., Bezabih, M., Abegaz, B. M., & Kaminsky, R. (1999). Antiplasmodial activity of knipholone and related natural phenylanthraquinones. *Planta Medica*, *65*(8), 757–758. https://doi.org/10.1055/s-2006-960859

Bringmann, G., Zhang, G., Olschläger, T., Stich, A., Wu, J., Chatterjee, M., & Brun, R. (2013). Highly selective antiplasmodial naphthylisoquinoline alkaloids from Ancistrocladus tectorius. *Phytochemistry*, *91*, 220–228. https://doi.org/10.1016/j.phytochem.2012.02.017

Cai, S., Risinger, A. L., Nair, S., Peng, J., Anderson, T. J., Du, L., Powell, D. R., Mooberry, S. L., & Cichewicz, R. H. (2016). Identification of compounds with efficacy against malaria parasites from common North American Plants. *Journal of Natural Products*, *79*(3), 490–498. https://doi.org/10.1021/acs.jnatprod.5b00874

Chan, S. T., Pearce, A. N., Page, M. J., Kaiser, M., & Copp, B. R. (2011). Antimalarial β-carbolines from the New Zealand ascidian Pseudodistoma opacum. *Journal of Natural Products*, *74*(9), 1972–1979. https://doi.org/10.1021/np200509g

Chanthathamrongsiri, N., Yuenyongsawad, S., Wattanapiromsakul, C., & Plubrukarn, A. (2012). Bifunctionalized amphilectane diterpenes from the sponge Stylissa cf. massa. *Journal of Natural Products*, *75*(4), 789–792. https://doi.org/10.1021/np200959j

Chen, G., Du, Y.-T., Liu, J.-H., Li, Y., Zheng, L., Qin, X.-S., & Cao, Y.-M. (2019). Modulation of anti-malaria immunity by vitamin A in C57BL/6J mice infected with heterogenic plasmodium. *International Immunopharmacology*, *76*, 105882.

Chuljerm, H., Maneekesorn, S., Punsawad, C., Somsak, V., Ma, Y., Ruangsuriya, J., Srichairatanakool, S., & Koonyosying, P. (2022). Deferiprone-resveratrol hybrid, an iron-chelating compound, acts as an antimalarial and hepatoprotective agent in Plasmodium berghei-infected mice. *Bioinorganic Chemistry and Applications*, *2022*, 3869337.

Chung, I. M., Ghimire, B. K., Kang, E. Y., & Moon, H. I. (2010a). Antiplasmodial and cytotoxic activity of khellactone derivatives from Angelica purpuraefolia Chung. *Phytotherapy Research: PTR*, *24*(3), 469–471. https://doi.org/10.1002/ptr.3056

Chung, I. M., Seo, S. H., Kang, E. Y., Park, W. H., Park, S. D., & Moon, H. I. (2010b). Antiplasmodial activity of isolated compounds from Carpesium divaricatum. *Phytotherapy Research: PTR*, *24*(3), 451–453. https://doi.org/10.1002/ptr.2881

Claudino, V. D., da Silva, K. C., Cechinel Filho, V., Yunes, R. A., Delle Monache, F., Giménez, A., Salamanca, E., Gutierrez-Yapu, D., & Malheiros, A. (2013). Drimanes from Drimys brasiliensis with leishmanicidal and antimalarial activity. *Memorias do Instituto Oswaldo Cruz*, *108*(2), 140–144. https://doi.org/10.1590/0074-0276108022013002

Daengrot, C., Rukachaisirikul, V., Tansakul, C., Thongpanchang, T., Phongpaichit, S., Bowornwiriyapan, K., & Sakayaroj, J. (2015). Eremophilane sesquiterpenes and diphenyl thioethers from the soil fungus penicillium copticola PSU-RSPG138. *Journal of Natural Products*, *78*(4), 615–622. https://doi.org/10.1021/np5005328

Dai, Y., Harinantenaina, L., Bowman, J. D., Da Fonseca, I. O., Brodie, P. J., Goetz, M., Cassera, M. B., & Kingston, D. G. (2014). Isolation of antiplasmodial anthraquinones from Kniphofia ensifolia, and synthesis and structure-activity relationships of related compounds. *Bioorganic & Medicinal Chemistry*, *22*(1), 269–276. https://doi.org/10.1016/j.bmc.2013.11.032

Davis, R. A., Buchanan, M. S., Duffy, S., Avery, V. M., Charman, S. A., Charman, W. N., White, K. L., Shackleford, D. M., Edstein, M. D., Andrews, K. T., Camp, D., & Quinn, R. J. (2012). Antimalarial activity of pyrroloiminoquinones from the Australian marine sponge Zyzzya sp. *Journal of Medicinal Chemistry*, *55*(12), 5851–5858. https://doi.org/10.1021/jm3002795

Davis, T. M. E., Skinner-Adam, T. S., & Beilby, J. (1998). In vitro growth inhibition of Plasmodium falciparum by retinol at concentrations present in normal human serum. *Acta Tropica*, *69*, 111–119.

de Andrade-Neto, V. F., da Silva, T., Lopes, L. M., do Rosário, V. E., de Pilla Varotti, F., & Krettli, A. U. (2007). Antiplasmodial activity of aryltetralone lignans from Holostylis reniformis. *Antimicrobial Agents and Chemotherapy*, *51*(7), 2346–2350. https://doi.org/10.1128/AAC.01344-06

Deguchi, J., Hirahara, T., Hirasawa, Y., Ekasari, W., Widyawaruyanti, A., Shirota, O., Shiro, M., & Morita, H. (2012). New tricyclic alkaloids, cassiarins G, H, J, and K from leaves of Cassia siamea. *Chemical & Pharmaceutical Bulletin*, *60*(2), 219–222. https://doi.org/10.1248/cpb.60.219

Dolabela, M. F., Póvoa, M. M., Brandão, G. C., Rocha, F. D., Soares, L. F., de Paula, R. C., & de Oliveira, A. B. (2015). Aspidosperma species as sources of anti-malarials: Uleine is the major anti-malarial indole alkaloid from Aspidosperma parvifolium (Apocynaceae). *Malaria Journal*, *13*(Suppl 1), 498. https://doi.org/10.1186/s12936-015-0997-4

D'Souza, V., & D'Souza, B. (2006). Comparative study on lipid peroxidation and antioxidant vitamins E and C in falciparum and vivax malaria. *Indian Journal of Clinical Biochemistry*, *21*(2), 103–106.

Du, Y., Abedi, A. K., Valenciano, A. L., Fernández-Murga, M. L., Cassera, M. B., Rasamison, V. E., Applequist, W. L., Miller, J. S., & Kingston, D. G. I. (2017a). Isolation of the new antiplasmodial butanolide, malleastrumolide A, from Malleastrum sp. (Meliaceae) from Madagascar. *Chemistry & Biodiversity*, *14*(12). https://doi.org/10.1002/cbdv.201700331

Du, Y., Pearce, K. C., Dai, Y., Krai, P., Dalal, S., Cassera, M. B., Goetz, M., Crawford, D., & Kingston, D. G. I. (2017b). Antiplasmodial sesquiterpenoid lactones from trichospira verticillata: Structure elucidation by spectroscopic methods and comparison of experimental and calculated ECD data. *Journal of Natural Products*, *80*(5), 1639–1647. https://doi.org/10.1021/acs.jnatprod.7b00247

Ebohon, O., Irabor, F., Omoregie, E. S. (2021). Ascorbic acid coadministration with artesunate – amodiaquine, up-regulated antioxidant enzymes gene expression in bone marrow cells and elicited biochemical changes in Plasmodium berghei-infected mice. *SN Applied Sciences*, *3*, 6.

Ekeh, F. N., Ekechukwu, N. E., Chukwuma, C. F., Aguzie, I. O. N., Ohanu, C. M., Ebido, C., & Oluah, S. N. (2019). Mixed vitamin C and zinc diet supplements co-administered with artemether drug improved haematological profile and survival of mice infected with Plasmodium berghei. *Food Science and Human Wellness*, *8*, 275–282.

Endale, M., Alao, J. P., Akala, H. M., Rono, N. K., Eyase, F. L., Derese, S., Ndakala, A., Mbugua, M., Walsh, D. S., Sunnerhagen, P., Erdelyi, M., & Yenesew, A. (2012). Antiplasmodial quinones from Pentas longiflora and Pentas lanceolata. *Planta Medica*, *78*(1), 31–35. https://doi.org/10.1055/s-0031-1280179

Ezenyi, I. C., & Salawu, O. A. (2016). Approaches, challenges and prospects of antimalarial drug discovery from plant sources. *Current Topics in Malaria*. https://doi.org/10.5772/65658

Fernandez, L. S., Buchanan, M. S., Carroll, A. R., Feng, Y. J., Quinn, R. J., & Avery, V. M. (2008). Flinderoles A-C: Antimalarial bis-indole alkaloids from Flindersia species. *Organic Letters*, *11*(2), 329–332. https://doi.org/10.1021/ol802506n

Fernandez, L. S., Sykes, M. L., Andrews, K. T., & Avery, V. M. (2010). Antiparasitic activity of alkaloids from plant species of Papua New Guinea and Australia. *International Journal of Antimicrobial Agents*, *36*(3), 275–279. https://doi.org/10.1016/j.ijantimicag.2010.05.008

Fleming, A. F., & Werblinska, B. (1982). Anaemia in childhood in the quinea savanna of Nigeria. *Annals of Tropical Paediatrics*, *2*, 161–173.

Foote, S. J., Burt, R. A., Baldwin, R. M., et al. (1997). Mouse loci for malaria-induced mortality and the control of parasitaemia. *Nature Genetics*, *17*, 380–381.

Fuller, N. J., Bates, C. J., Hayes, R. J., et al. (1988). The effects of antimalarials and folate supplements on haematological indices and red cell folate levels in Gambian children. *Annals of Tropical Paediatrics*, *8*, 61–67.

Gachet, M. S., Kunert, O., Kaiser, M., Brun, R., Muñoz, R. A., Bauer, R., & Schühly, W. (2010). Jacaranone-derived glucosidic esters from Jacaranda glabra and their activity against Plasmodium falciparum. *Journal of Natural Products*, *73*(4), 553–556. https://doi.org/10.1021/np900528m

Gachet, M. S., Kunert, O., Kaiser, M., Brun, R., Zehl, M., Keller, W., Muñoz, R. A., Bauer, R., & Schuehly, W. (2011). Antiparasitic compounds from Cupania cinerea with activities against Plasmodium falciparum and Trypanosoma brucei rhodesiense. *Journal of Natural Products*, *74*(4), 559–566. https://doi.org/10.1021/np100415m

Gadetskaya, A. V., Mohamed, S. M., Tarawneh, A., Mohamed, N. M., Ma, G., Ponomarev, B. N., Zhusupova, G. E., Cantrell, C. L., Cutler, S. J., & Ross, S. A. (2017). Phytochemical characterization and biological activity of secondary metabolites from three Limonium species. *Medicinal Chemistry Research*, *26*, 2743–2750. https://doi.org/10.1007/s00044-017-1973-z

Gail, K., & Herms, V. (1969). Influence of pteroylglutamic acid (folic acid) on parasite density (Plasmodium falciparum) in pregnant women in West Africa. *Z Tropenmedizin Parasitol*, *20*, 440–450.

Gakunju, D. M., Mberu, E. K., Dossaji, S. F., Gray, A. I., Waigh, R. D., Waterman, P. G., & Watkins, W. M. (1995). Potent antimalarial activity of the alkaloid nitidine, isolated from a Kenyan herbal remedy. *Antimicrobial Agents and Chemotherapy*, *39*(12), 2606–2609. https://doi.org/10.1128/AAC.39.12.2606

George, B. O., Atalor, B., & Okpoghono, J. (2019). Role of vitamin E, vitamin A and aqueous extract of Aframomum sceptrum in mice infected with malaria parasites. *Nigerian Journal of Nutritional Sciences*, *40*(1), 113–117.

Gibson, R. S. (1990). *Principles of Nutritional Assessment*. Oxford University Press, New York.

Girardot, M., Deregnaucourt, C., Deville, A., Dubost, L., Joyeau, R., Allorge, L., Rasoanaivo, P., & Mambu, L. (2012). Indole alkaloids from Muntafara sessilifolia with antiplasmodial and cytotoxic activities. *Phytochemistry*, *73*(1), 65–73. https://doi.org/10.1016/j.phytochem.2011.09.012

Good, M. F., Kaslow, D. C., & Miller, L. H. (1998). Pathways and strategies for developing a malaria blood-stage vaccine. *Annual Review of Immunology*, *16*, 57–87.

Goodman, C. D., Austarheim, I., Mollard, V., Mikolo, B., Malterud, K. E., McFadden, G. I., & Wangensteen, H. (2016). Natural products from Zanthoxylum heitzii with potent activity against the malaria parasite. *Malaria Journal*, *15*, 481. https://doi.org/10.1186/s12936-016-1533-x

Greve, H. L., Kaiser, M., Brun, R., & Schmidt, T. J. (2017). Terpenoids from the oleo-gum-resin of boswellia serrata and their antiplasmodial effects in vitro. *Planta Medica*, *83*(14–15), 1214–1226. https://doi.org/10.1055/s-0043-116943

Happi, G. M., Kouam, S. F., Talontsi, F. M., Zühlke, S., Lamshöft, M., & Spiteller, M. (2015). Minor secondary metabolites from the bark of Entandrophragma congoënse (Meliaceae). *Fitoterapia*, *102*, 35–40. https://doi.org/10.1016/j.fitote.2015.01.018

Harinantenaina, L., Bowman, J. D., Brodie, P. J., Slebodnick, C., Callmander, M. W., Rakotobe, E., Randrianaivo, R., Rasamison, V. E., Gorka, A., Roepe, P. D., Cassera, M. B., & Kingston, D. G. (2013). Antiproliferative and antiplasmodial dimeric phloroglucinols from Mallotus oppositifolius from the Madagascar Dry Forest (1). *Journal of Natural Products*, *76*(3), 388–393. https://doi.org/10.1021/np300750q

Hata, Y., De Mieri, M., Ebrahimi, S. N., Mokoka, T., Fouche, G., Kaiser, M., Brun, R., Potterat, O., & Hamburger, M. (2014). Identification of two new phenathrenones and a saponin as antiprotozoal constituents of Drypetes gerrardii. *Phytochemistry Letters*, *10*, cxxxiii–cxl. https://doi.org/10.1016/j.phytol.2014.05.005

Hemingway, J., Shretta, R., Wells, T. N., Bell, D., Djimdé, A. A., Achee, N., & Qi, G. (2016). Tools and strategies for malaria control and elimination: What do we need to achieve a grand convergence in Malaria?. *PLoS Biology*, *14*(3), e1002380. https://doi.org/10.1371/journal.pbio.1002380

Hemtasin, C., Kanokmedhakul, S., Kanokmedhakul, K., Hahnvajanawong, C., Soytong, K., Prabpai, S., & Kongsaeree, P. (2011). Cytotoxic pentacyclic and tetracyclic aromatic sesquiterpenes from Phomopsis archeri. *Journal of Natural Products*, *74*(4), 609–613. https://doi.org/10.1021/np100632g

Hiranrat, A., Mahabusarakam, W., Carroll, A. R., Duffy, S., & Avery, V. M. (2012). Tomentosones A and B, hexacyclic phloroglucinol derivatives from the Thai shrub Rhodomyrtus tomentosa. *The Journal of Organic Chemistry*, *77*(1), 680–683. https://doi.org/10.1021/jo201602y

Huang, H., Yao, Y., He, Z., Yang, T., Ma, J., Tian, X., Li, Y., Huang, C., Chen, X., Li, W., Zhang, S., Zhang, C., & Ju, J. (2011). Antimalarial β-carboline and indolactam alkaloids from Marinactinospora thermotolerans, a deep sea isolate. *Journal of Natural Products*, *74*(10), 2122–2127. https://doi.org/10.1021/np200399t

Husen, A. (2021). *Traditional Herbal Therapy for the Human Immune System*. CRC Press, Boca Raton, FL. https://doi.org/10.1201/9781003137955

Husen, A. (2022a). *Herbs, Shrubs and Trees of Potential Medicinal Benefits*. CRC Press, Boca Raton, FL. https://doi.org/10.1201/9781003205067

Husen, A. (2022b). *Exploring Poisonous Plants: Medicinal Values, Toxicity Response and Other Potential Uses*. CRC Press, Boca Raton, FL. https://doi.org/10.1201/b23017

Husen, A. (2023). *Antidiabetic Medicinal Plants and Herbal Treatments*. CRC Press, Boca Raton, FL.

Ilias, M., Ibrahim, M. A., Khan, S. I., Jacob, M. R., Tekwani, B. L., Walker, L. A., & Samoylenko, V. (2012). Pentacyclic ingamine alkaloids, a new antiplasmodial pharmacophore from the marine sponge Petrosid Ng5 Sp5. *Planta Medica*, *78*(15), 1690–1697. https://doi.org/10.1055/s-0032-1315213

Iyawe, H. O. T., Onigbinde, A. O., & Aina, O. O. (2006). Effect of chloroquine and ascorbic acid interaction on the oxidative stress status of Plasmodium berghei infected mice. *International Journal of Pharmacology*, *2*(1), 1–4.

Jansen, O., Tits, M., Angenot, L., Nicolas, J. P., de Mol, P., Nikiéma, J. B., & Frédérich, M. (2012). Antiplasmodial activity of Dicoma tomentosa (Asteraceae) and identification of urospermal A-15-O-acetate as the main active compound. *Malaria Journal*, *11*, 289–289. https://doi.org/10.1186/1475-2875-11-289

Julianti, T., De Mieri, M., Zimmermann, S., Ebrahimi, S. N., Kaiser, M., Neuburger, M., Raith, M., Brun, R., & Hamburger, M. (2014). HPLC-based activity profiling for antiplasmodial compounds in the traditional Indonesian medicinal plant Carica papaya L. *Journal of Ethnopharmacology*, *155*(1), 426–434. https://doi.org/10.1016/j.jep.2014.05.050

Kendie, F. A., Hailegebriel W/Kiros, T., Nibret Semegn, E., & Ferede, M. W. (2021). Prevalence of malaria among adults in ethiopia: A systematic review and meta-analysis. *Journal of Tropical Medicine*, *2021*, 8863002. https://doi.org/10.1155/2021/8863002

Konziase, B. (2015). Protective activity of biflavanones from Garcinia kola against Plasmodium infection. *Journal of Ethnopharmacology*, *172*, 214–218. https://doi.org/10.1016/j.jep.2015.06.038

Krishnan, S., Krishnan, A. D., Mustafa, A. S., Talwar, G. P., & Ramalingaswami, V. (1976). Effect of vitamin A and undernutrition on the susceptibility of rodents to a malarial parasite Plasmodium berghei. *Journal of Nutrition*, *106*, 784–791.

Kubo, M., Yatsuzuka, W., Matsushima, S., Harada, K., Inoue, Y., Miyamoto, H., Matsumoto, M., & Fukuyama, Y. (2016). Antimalarial phenanthroindolizine alkaloids from ficus septica. *Chemical & Pharmaceutical Bulletin*, *64*(7), 957–960. https://doi.o10.1248/cpb.c16-00181

Kumar, R., Duffy, S., Avery, V. M., & Davis, R. A. (2017). Synthesis of antimalarial amide analogues based on the plant serrulatane diterpenoid 3,7,8-trihydroxyserrulat-14-en-19-oic acid. *Bioorganic & Medicinal Chemistry Letters*, *27*(17), 4091–4095. https://doi.org/10.1016/j.bmcl.2017.07.039

Laporta, G. Z., Linton, Y. M., Wilkerson, R. C., Bergo, E. S., Nagaki, S. S., Sant'Ana, D. C., & Sallum, M. A. (2015). Malaria vectors in South America: Current and future scenarios. *Parasites & Vectors*, *8*, 426. https://doi.org/10.1186/s13071-015-1038-4

Lenta, B. N., Kamdem, L. M., Ngouela, S., Tantangmo, F., Devkota, K. P., Boyom, F. F., Rosenthal, P. J., & Tsamo, E. (2011). Antiplasmodial constituents from the fruit pericarp of Pentadesma butyracea. *Planta Medica*, *77*(4), 377–379. https://doi.org/10.1055/s-0030-1250384

Li, J., Seupel, R., Feineis, D., Mudogo, V., Kaiser, M., Brun, R., Brünnert, D., Chatterjee, M., Seo, E. J., Efferth, T., & Bringmann, G. (2017). Dioncophyllines C2, D2, and F and related naphthylisoquinoline alkaloids from the Congolese Liana *Ancistrocladus ileboensis*, with potent activities against *Plasmodium falciparum* and against multiple myeloma and leukemia cell lines. *Journal of Natural Products*, *80*(2), 443–458. https://doi.org/10.1021/acs.jnatprod.6b00967

Liew, L. P., Fleming, J. M., Longeon, A., Mouray, E., Florent, I., Bourguet-Kondracki, M., & Copp, B. R. (2014). Synthesis of 1-indolyl substituted β-carboline natural products and discovery of antimalarial and cytotoxic activities. *Tetrahedron*, *70*, 4910–4920. https://doi.org/10.1016/j.tet.2014.05.068

Liu, S., Tian, N., & Liu, Z. (2014). Extraction, purification, and quantification of artemisinin and its analogs from *Artemisia annua* L. In: Aftab, T., Ferreira, J., Khan, M., & Naeem, M. (eds) *Artemisia Annua – Pharmacology and Biotechnology*. Springer, Berlin and Heidelberg. https://doi.org/10.1007/978-3-642-41027-7_9

Ma, G., Wu, H., Chen, D., Zhu, N., Zhu, Y., Sun, Z., Li, P., Yang, J., Yuan, J., & Xu, X. (2015). Antimalarial and antiproliferative cassane diterpenes of caesalpinia sappan. *Journal of Natural Products*, *78*(10), 2364–2371. https://doi.org/10.1021/acs.jnatprod.5b00317

Mahajan, S. S., Kamath, V. R., & Ghatpande, S. S. (2005). Synergistic antimalarial activity of ketones with rufigallol and vitamin C. *Parasitol*, *131*, 459–466.

Mani, L., Jullian, V., Mourkazel, B., Valentin, A., Dubois, J., Cresteil, T., Folcher, E., Hooper, J. N., Erpenbeck, D., Aalbersberg, W., & Debitus, C. (2012). New antiplasmodial bromotyrosine derivatives from Suberea ianthelliformis Lendenfeld, 1888. *Chemistry & Biodiversity*, *9*(8), 1436–1451. https://doi.org/10.1002/cbdv.201100309

Meesala, S., Gurung, P., Karmodiya, K., Subrayan, P., & Watve, M. G. (2018). Isolation and structure elucidation of halymeniaol, a new antimalarial sterol derivative from the red alga Halymenia floresii. *Journal of Asian Natural Products Research*, *20*(4), 391–398. https://doi.org/10.1080/10286020.2017.1342636

Mgbemena, I., & Ifeyinwa, C. (2020). The efficacy of antioxidant (Vitamin C) in the treatment and management of malaria. *Advances in Biotechnology & Microbiology*, *15*(4), 555919.

Mishra, K., Chakraborty, D., Pal, A., & Dey, N. (2010). Plasmodium falciparum: In vitro interaction of quassin and neo-quassin with artesunate, a hemisuccinate derivative of artemisinin. *Experimental Parasitology*, *124*(4), 421–427. https://doi.org/10.1016/j.exppara.2009.12.007

Misra, H., Mehta, D., Mehta, B. K., & Jain, D. C. (2014). Extraction of artemisinin, an active antimalarial phytopharmaceutical from dried leaves of Artemisia annua L., using microwaves and a validated HPTLC-visible method for its quantitative determination. *Chromatography Research International*, *2014*(5), 1–11. https://doi.org/10.1155/2014%2F361405

Mohammad, A. (2012). Effect of serum antioxidant ascorbic acid concentration by malarial infection. Man. *The Experiment*, *3*(4), 214–215.

Morita, H., Oshimi, S., Hirasawa, Y., Koyama, K., Honda, T., Ekasari, W., Indrayanto, G., & Zaini, N. C. (2007). Cassiarins A and B, novel antiplasmodial alkaloids from Cassia siamea. *Organic Letters*, *9*(18), 3691–3693. https://doi.org/10.1021/ol701623n

Mrimi, E. C., Palmeirim, M. S., Minja, E. G., Long, K. Z., & Keiser, J. (2023). Correlation of cytokines with parasitic infections, undernutrition and micronutrient deficiency among school children in rural Tanzania: A cross-sectional study. *Nutrients*, *15*, 1916. https://doi.org/10.3390/nu15081916

Muganza, D. M., Fruth, B., Nzunzu, J. L., Tuenter, E., Foubert, K., Cos, P., Maes, L., Kanyanga, R. C., Exarchou, V., Apers, S., & Pieters, L. (2016). In vitro antiprotozoal activity and cytotoxicity of extracts and isolated constituents from Greenwayodendron suaveolens. *Journal of Ethnopharmacology*, *193*, 510–516. https://doi.org/10.1016/j.jep.2016.09.051

Munteanu, C., & Schwartz, B. (2022). The relationship between nutrition and the immune system. *Frontiers in Nutrition*, *9,* 1082500. doi: 10.3389/fnut.2022.1082500

Namukobe, J., Kiremire, B. T., Byamukama, R., Kasenene, J. M., Dumontet, V., Guéritte, F., Krief, S., Florent, I., & Kabasa, J. D. (2014). Cycloartane triterpenes from the leaves of Neoboutonia macrocalyx L. *Phytochemistry*, *102*, 189–196. https://doi.org/10.1016/j.phytochem.2014.02.005

Nasrullah, A. A., Zahari, A., Mohamad, J., & Awang, K. (2013). Antiplasmodial alkaloids from the bark of Cryptocarya nigra (Lauraceae). *Molecules (Basel, Switzerland)*, *18*(7), 8009–8017. https://doi.org/10.3390/molecules18078009

Ndjonka, D., Bergmann, B., Agyare, C., Zimbres, F. M., Lüersen, K., Hensel, A., Wrenger, C., & Liebau, E. (2012). In vitro activity of extracts and isolated polyphenols from West African medicinal plants against Plasmodium falciparum. *Parasitology Research*, *111*(2), 827–834. https://doi.org/10.1007/s00436-012-2905-y

Nondo, R. S., Erasto, P., Moshi, M. J., Zacharia, A., Masimba, P. J., & Kidukuli, A. W. (2016). *In vivo* antimalarial activity of extracts of Tanzanian medicinal plants used for the treatment of malaria. *Journal of Advanced Pharmaceutical Technology & Research*, *7*(2), 59–63. https://doi.org/10.4103/2231-4040.179748

Nondo, R. S., Moshi, M. J., Erasto, P., Masimba, P. J., Machumi, F., Kidukuli, A. W., Heydenreich, M., & Zofou, D. (2017). Anti-plasmodial activity of Norcaesalpin D and extracts of four medicinal plants used traditionally for treatment of malaria. *BMC Complementary and Alternative Medicine*, *17*(1), 167. https://doi.org/10.1186/s12906-017-1673-8

Obbo, C. J., Makanga, B., Mulholland, D. A., Coombes, P. H., & Brun, R. (2013). Antiprotozoal activity of Khaya anthotheca, (Welv.) C.D.C. a plant used by chimpanzees for self-medication. *Journal of Ethnopharmacology*, *147*(1), 220–223. https://doi.org/10.1016/j.jep.2013.03.007

Ochieng, C. O., Manguro, L. A., Owuor, P. O., & Akala, H. (2013). Voulkensin C-E, new 11-oxocassane-type diterpenoids and a steroid glycoside from Caesalpinia volkensii stem bark and their antiplasmodial activities. *Bioorganic & Medicinal Chemistry Letters*, *23*(10), 3088–3095. https://doi.org/10.1016/j.bmcl.2013.03.007

Oppenheimer, S. J., Gibson, F. D., Macfarlane, S. B., et al. (1986). Iron supplementation increases prevalence and effects of malaria: Report on clinical studies in Papua New Guinea. *Transactions of the Royal Society of Tropical Medicine and Hygiene*, *80*, 603–612.

Ovenden, S. P., Cobbe, M., Kissell, R., Birrell, G. W., Chavchich, M., & Edstein, M. D. (2011). Phenolic glycosides with antimalarial activity from Grevillea "Poorinda Queen". *Journal of Natural Products*, *74*(1), 74–78. https://doi.org/10.1021/np100737q

Pivatto, M., Baccini, L. R., Sharma, A., Nakabashi, M., Danuello, A., Viegas Júnior, C., Garcia, C. R. S., & Bolzani, V. S. (2014). Antimalarial activity of piperidine alkaloids from *Senna spectabilis* and semisynthetic derivatives. *Journal of the Brazilian Chemical Society*, *25*, 1900–1906. https://doi.org/10.5935/0103-5053.20140195

Rafatro, H., Ramanitrahasimbola, D., Rasoanaivo, P., Ratsimamanga-Urverg, S., Rakoto-Ratsimamanga, A., & Frappier, F. (2000). Reversal activity of the naturally occurring chemosensitizer malagashanine in Plasmodium malaria. *Biochemical Pharmacology*, *59*(9), 1053–1061. https://doi.org/10.1016/s0006-2952(99)00400-1

Rahman, S., & Husen, A. (2023a). *Trigonella foenum-graecum* L.: A useful medicinal plant for diabetes and other diseases. In: Husen, A. (ed) *Antidiabetic Medicinal Plants and Herbal Treatments*. CRC Press, Boca Raton, FL.

Rahman, S., & Husen, A. (2023b). Role of Green Chiretta [*Andaphis paniculata* (Burm. F.)] nees in diabetes and other ailments. In: Husen, A. (ed) *Antidiabetic Medicinal Plants and Herbal Treatments*. CRC Press, Boca Raton, FL.

Rakotondraibe, L. H., Graupner, P. R., Xiong, Q., Olson, M., Wiley, J. D., Krai, P., Brodie, P. J., Callmander, M. W., Rakotobe, E., Ratovoson, F., Rasamison, V. E., Cassera, M. B., Hahn, D. R., Kingston, D. G., & Fotso, S. (2015). Neolignans and other metabolites from Ocotea cymosa from the Madagascar rain forest and their biological activities. *Journal of Natural Products*, *78*(3), 431–440. https://doi.org/10.1021/np5008153

Ramalhete, C., da Cruz, F. P., Lopes, D., Mulhovo, S., Rosário, V. E., Prudêncio, M., & Ferreira, M. J. (2011a). Triterpenoids as inhibitors of erythrocytic and liver stages of Plasmodium infections. *Bioorganic & Medicinal Chemistry*, *19*(24), 7474–7481. https://doi.org/10.1016/j.bmc.2011.10.044

Ramalhete, C., Lopes, D., Molnár, J., Mulhovo, S., Rosário, V. E., & Ferreira, M. J. (2011b). Karavilagenin C derivatives as antimalarials. *Bioorganic & Medicinal Chemistry*, *19*(1), 330–338. https://doi.org/10.1016/j.bmc.2010.11.015

Ramanitrahasimbola, D., Rasoanaivo, P., Ratsimamanga, S., & Vial, H. (2006). Malagashanine potentiates chloroquine antimalarial activity in drug resistant Plasmodium malaria by modifying both its efflux and influx. *Molecular and Biochemical Parasitology*, *146*(1), 58–67. https://doi.org/10.1016/j.molbiopara.2005.10.018

Regalado, E. L., Tasdemir, D., Kaiser, M., Cachet, N., Amade, P., & Thomas, O. P. (2010). Antiprotozoal steroidal saponins from the marine sponge Pandaros acanthifolium. *Journal of Natural Products*, *73*(8), 1404–1410. https://doi.org/10.1021/np100348x

Robertson, L. P., Duffy, S., Wang, Y., Wang, D., Avery, V. M., & Carroll, A. R. (2017). Pimentelamines A-C, indole alkaloids isolated from the leaves of the Australian tree flindersia pimenteliana. *Journal of Natural Products*, *80*(12), 3211–3217. https://doi.org/10.1021/acs.jnatprod.7b00587

Roos, A., Hegsted, D. M., & Stare, F. J. (1946). Nutritional studies with the duck. IV. The effect of vitamin deficiencies on the course of P. lophurae infection in the duck and the chick. *Journal of Nutrition*, *32*, 473–484.

Sá, M. S., de Menezes, M. N., Krettli, A. U., Ribeiro, I. M., Tomassini, T. C., Ribeiro dos Santos, R., de Azevedo, W. F., Jr, & Soares, M. B. (2011). Antimalarial activity of physalins B, D, F, and G. *Journal of Natural Products*, *74*(10), 2269–2272. https://doi.org/10.1021/np200260f

Sakwe, N., Bigoga, J., Ngondi, J., Njeambosay, B., Esemu, L., Kouambeng, C., et al. (2019). Relationship between malaria, anaemia, nutritional and socio-economic status amongst under-ten children, in the North Region of Cameroon: A cross-sectional assessment. *PLoS ONE*, *14*(6), e0218442. https://doi.org/10.1371/journal.pone.0218442

Samba, D. C., Basco, L. K., Bleiberg-Daniel, F., Lemmonier, D., & Le Bras, J. (1992). Absence of effect of retinol on the in vitro development of Plasmodium falciparum. *International Journal for Vitamin and Nutrition Research*, *62*, 99–100.

Scala, F., Fattorusso, E., Menna, M., Taglialatela-Scafati, O., Tierney, M., Kaiser, M., & Tasdemir, D. (2010). Bromopyrrole alkaloids as lead compounds against protozoan parasites. *Marine Drugs*, *8*(7), 2162–2174. https://doi.org/10.3390/md8072162

Senadeera, S. P. D., Duffy, S., Avery, V. M., & Carroll, A. R. (2017). Antiplasmodial β-triketones from the flowers of the Australian tree Angophora woodsiana. *Bioorganic & Medicinal Chemistry Letters*, *27*(11), 2602–2607. https://doi.org/10.1016/j.bmcl.2017.03.065

Shankar, A. H. (2000). Nutritional modulation of malaria morbidity and mortality. *The Journal of Infectious Diseases*, *182*(Suppl 1), S37–S53.

Shankar, A. H., Kumar, N., & Scott, A. L. (1995). Zinc-deficiency exacerbates experimental malaria infection in mice. *FASEB Journal*, *9*, A4269.

Shankar, A. H., & Prasad, A. S. (1998). Zinc and immune function: The biological basis of altered resistance to infection. *The American Journal of Clinical Nutrition*, *68*, 447S–463S.

Shi, X., Wei, M., Xu, Z., Liu, Y., Zhang, M., Lv, L., & Wang, Q. (2021). Vitamin C inhibits blood-stage plasmodium parasites via oxidative stress. *Frontiers in Cell and Developmental Biology*, *9*, 639944.

Smith, A. W., Hendrickse, R. G., Harrison, C., Hayes, R. J., & Greenwood, B. M. (1989). The effects on malaria of treatment of iron-deficiency anaemia with oral iron in Gambian children. *Annals of Tropical Paediatrics*, *9*, 17–23.

Smyrniotopoulos, V., Merten, C., Kaiser, M., & Tasdemir, D. (2017). Bifurcatriol, a new antiprotozoal acyclic diterpene from the brown alga bifurcaria bifurcata. *Marine Drugs*, *15*(8), 245. https://doi.org/10.3390/md15080245

Soh, P. N., Witkowski, B., Olagnier, D., Nicolau, M. L., Garcia-Alvarez, M. C., Berry, A., & Benoit-Vical, F. (2009). In vitro and in vivo properties of ellagic acid in malaria treatment. *Antimicrobial Agents and Chemotherapy*, *53*(3), 1100–1106. https://doi.org/10.1128/AAC.01175-08

Stoltzfus, R. J., Jalal, F., Harvey, P. W. J., & Nesheim, M. C. (1989). Interactions between vitamin A deficiency and Plasmodium berghei infection in the rat. *Journal of Nutrition*, *119*, 2030–2037.

Su, Q., Dalal, S., Goetz, M., Cassera, M. B., & Kingston, D. G. (2016). Antiplasmodial phloroglucinol derivatives from Syncarpia glomulifera. *Bioorganic & Medicinal Chemistry*, *24*(11), 2544–2548. https://doi.org/10.1016/j.bmc.2016.04.020

Tajuddeen, N., & Van Heerden, F. R. (2019). Antiplasmodial natural products: An update. *Malaria Journal*, *18*, 404. https://doi.org/10.1186/s12936-019-3026-1

Tantangmo, F., Lenta, B. N., Boyom, F. F., Ngouela, S., Kaiser, M., Tsamo, E., Weniger, B., Rosenthal, P. J., & Vonthron-Sénécheau, C. (2010). Antiprotozoal activities of some constituents of Markhamia tomentosa (Bignoniaceae). *Annals of Tropical Medicine and Parasitology*, *104*(5), 391–398. https://doi.org/10.1179/136485910X12743554760180

Toyang, N. J., Krause, M. A., Fairhurst, R. M., Tane, P., Bryant, J., & Verpoorte, R. (2013). Antiplasmodial activity of sesquiterpene lactones and a sucrose ester from Vernonia guineensis Benth. (Asteraceae). *Journal of Ethnopharmacology*, *147*(3), 618–621. https://doi.org/10.1016/j.jep.2013.03.051

UNICEF. (2022). Malnutrition. Available at: https://data.unicef.org/topic/nutrition/malnutrition/. Accessed June 10, 2022.

White, A. M., Pierens, G. K., Skinner-Adams, T., Andrews, K. T., Bernhardt, P. V., Krenske, E. H., Mollo, E. & Garson, M. J. (2015). Antimalarial isocyano and isothiocyanato sesquiterpenes with tri- and bicyclic skeletons from the nudibranch phyllidia ocellata. *Journal of Natural Products*, *78*(6), 1422–1427. https://doi.org/10.1021/acs.jnatprod.5b00354

Winter, R. W., Ignatushchenko, M., Ogundahunsi, O. A., Cornell, K. A., Oduola, A. M., Hinrichs, D. J., & Riscoe, M. K. (1997). Potentiation of an antimalarial oxidant drug. *Antimicrob Agents Chemother*, *41*(7), 1449–1454.

Wintergerst, E. S., Maggini, S., & Hornig, D. H. (2006). Immune enhancing role of vitamin C and zinc and effect on clinical conditions. *Annals of Nutrition and Metabolism*, *50*, 85–94.

Wube, A. A., Bucar, F., Asres, K., Gibbons, S., Rattray, L., & Croft, S. L. (2005). Antimalarial compounds from Kniphofia foliosa roots. *Phytotherapy Research: PTR*, *19*(6), 472–476. https://doi.org/10.1002/ptr.1635

Wyler, D. J. (1993). Malaria: Overview and update. *Clinical Infectious Diseases*, *16*, 449–458.

Yakoob, M. Y., Murad, Q., & Hany, E.-O. (2018). Vitamin A supplementation for prevention and treatment of malaria during pregnancy and childhood: A systematic review and meta-analysis. *Journal of Epidemiology and Global Health*, *8*(1–2), 20–28.

Yim, T., Kanokmedhakul, K., Kanokmedhakul, S., Sanmanoch, W., & Boonlue, S. (2014). A new meroterpenoid tatenoic acid from the fungus Neosartorya tatenoi KKU-2NK23. *Natural Product Research*, *28*(21), 1847–1852. https://doi.org/10.1080/14786419.2014.951353

Yin, J. P., Gu, M., Li, Y., & Nan, F. J. (2014). Total synthesis of aphadilactones A-D. *The Journal of Organic Chemistry*, *79*(13), 6294–6301. https://doi.org/10.1021/jo501117k

Zelefack, F., Guilet, D., Fabre, N., Bayet, C., Chevalley, S., Ngouela, S., Lenta, B. N., Valentin, A., Tsamo, E., & Dijoux-Franca, M. G. (2009). Cytotoxic and antiplasmodial xanthones from Pentadesma butyracea. *Journal of Natural Products*, *72*(5), 954–957. https://doi.org/10.1021/np8005953

Zhou, B., Wu, Y., Dalal, S., Cassera, M. B., & Yue, J. M. (2016). Euphorbesulins A-P, structurally diverse diterpenoids from euphorbia esula. *Journal of Natural Products*, *79*(8), 1952–1961. https://doi.org/10.1021/acs.jnatprod.6b00205

Zofou, D., Kengne, A. B., Tene, M., Ngemenya, M. N., Tane, P., & Titanji, V. P. (2011a). *In vitro* antiplasmodial activity and cytotoxicity of crude extracts and compounds from the stem bark of Kigelia africana (Lam.) Benth (Bignoniaceae). *Parasitology Research*, *108*(6), 1383–1390. https://doi.org/10.1007/s00436-011-2363-y

Zofou, D., Tene, M., Ngemenya, M. N., Tane, P., & Titanji, V. P. (2011b). *In vitro* antiplasmodial activity and cytotoxicity of extracts of selected medicinal plants used by traditional healers of Western cameroon. *Malaria Research and Treatment*, *2011*, 561342. https://doi.org/10.4061/2011/561342

Zofou, D., Tene, M., Tane, P., & Titanji, V. P. (2012). Antimalarial drug interactions of compounds isolated from Kigelia africana (Bignoniaceae) and their synergism with artemether, against the multidrug-resistant W2mef Plasmodium falciparum strain. *Parasitology Research*, *110*(2), 539–544. https://doi.org/10.1007/s00436-011-2519-9

7 Antioxidants from Antimalarial Plants and Their Applications

Ishaan Lakhani, Gunjan Sharma, Hina Alim, Ahmad Ali, and Nimisha Patel

7.1 INTRODUCTION

Malaria can be spread by the bite of a female *Anopheles* mosquito that has been infected with *Plasmodium* parasites. Although unfavourable outcomes may be prevented with early diagnosis and effective treatment, malaria is still the leading cause of mortality globally. Contrary to the industrialized world, where it is imported from endemic areas, malaria is the most common disease in Africa and a few Asian countries (Talapko et al., 2019). Malaria has a mortality rate of 0.3% to 2.2% worldwide, and in tropical regions, it can have a severe case incidence of 11–30% (White et al., 2014). The cause of malaria is a microscopic protozoan from the *Plasmodium* species which have multiple subspecies. Human disease is brought on by some *Plasmodium* species (White et al., 2014; Walker et al., 2017).

There are 172 different *Plasmodium* species in total, but only 5 of them have a negative impact on human health. They involve *P. knowlesi*, *P.vivax*, *P. ovale*, *P. malariae, and P. falciparum. P. knowlesi* is known to cause zoonotic malarial disease in South-East Asia. According to Walker et al. (2017), Antinori et al. (2012), Ashley et al. (2018), other species hardly ever infect humans. *P. falciparum*, which is responsible for more than 90% of all malaria infections, is considered the most dangerous strain of the illness in Africa. This makes it the species that is most characterized (Roca-Feltrer et al., 2010). However, due to evidence of extensive malaria *Plasmodium vivax* infection in recent years, *P. vivax* infections are becoming more significant (Baird, 2007). With reports mostly from Oceania, South and South-East Asia, the Horn of Africa, Latin America, and the Middle East, *Plasmodium vivax* has also recently drawn a lot of attention because it is the species with the widest global range (Battle et al., 2014).

In the early 1940s, chloroquine, a red blood cell schizophrenic agent against malaria parasites, was developed in the West. Primaquine and pyrimethamine, synthesized in 1946 and 1951, respectively, were used for the curative and causative prophylaxis of malaria (Chen, 2014). Malaria epidemics are thought to have existed since the dawn of mankind (Lehrer, 1981). The first descriptions of malaria date to the ancient texts dated about 1550 century and ancient Chinese medical history, both of which date around 2700 BC (White et al., 2014).

Several of the chemicals historically essential in the battle against malaria are sourced from plants or patterned after structures obtained from plant-isolated lead compounds. Quinolone antimalarials like quinine and chloroquine are examples, as are artemisinin and its derivatives (Schalkwyk, 2015). Plant-based treatments are commonly used to treat numerous ailments in many countries of Africa, Asia, and the Americas, including malaria and malaria-like disorders. According to the World Health Organization (WHO), herbal supplements are used by more than 70% of the African population for health care (WHO, 2005). Several medicinal plants are utilized in Ghana to treat malaria and its symptoms (Asase et al., 2005). In Ghana, four medicinal plants – *Grosseria vignei*,

 DOI: 10.1201/9781003378396-7

Celtis Africana, *Physalis micrantha*, and *Stachytarpheta angustifolia* – are frequently used to treat malaria. Although artesunate is thought to be the most effective antimalarial medication for severe infections, artemisinin-based combination therapy (ACT) is one of the most frequently suggested treatments for *Plasmodium falciparum* (Roussel et al., 2017; Wicht et al., 2020). Essential oils (EOs), in particular those made from plants native to north-eastern Brazil, are thought to belong to a large family of naturally occurring substances with antimalarial effects. These substances comprise low-molecular-weight components that are abundant in monoterpenes and sesquiterpenes (Mota et al., 2012).

7.2 DISCOVERY OF MALARIA

The first mentions of malaria date from 2700 BC and are found in the Ebers Papyrus and ancient Chinese medical documents (White et al., 2014). Alexander the Great, a military monarch, died of malaria (Lehrer, 1981). The fact that George Washington, Cesare Borgia, Albrecht Dürer, and Christopher Columbus all suffered from this ailment demonstrates how pervasive it was in society (Moss et al., 2008).

7.2.1 Malaria: Life Cycle

The malaria parasite's sexual and asexual life cycles, which include mosquitoes as vectors and vertebrates as hosts, are both extremely intricate. The sexual phase of its life cycle takes place in the vector carrying the parasite, that is, mosquito. The life cycle of the malarial intermediate male host includes an asexual stage (Vuk et al., 2008; Soulard et al., 2015). The *Plasmodium* parasite life cycle has several phases and includes both mosquitoes and humans as hosts. Following are the stages of *Plasmodium* parasite life cycle:

1. Sporozoites. *Plasmodium* parasite's infectious stage.
2. Liver Stage. Sporozoites multiply asexually within the liver cells.
3. Blood Stage. Red blood cells become infected by merozoites, which reproduce asexually and cause the cells to burst, releasing additional merozoites.
4. Mosquito Infection. As a mosquito feeds on an infected person, it consumes gametocytes, the parasite's sexual stage that develops in red blood cells. Male and female gametes develop from gametocytes in the mosquito's intestines.
5. Fertilization and Development. In order to form an oocyst, a male gamete must fertilize a female gamete.
6. Release. The mosquito is now parasite-infected, and when it bites someone else, it can spread the disease.

(Centers for Disease Control and Prevention (CDC), 2020, Global Health, Division of Parasitic Diseases and Malaria)

7.3 BIOACTIVE COMPOUNDS OF PLANTS, THEIR PROPERTIES AND MECHANISM OF ACTION

A substance that lowers radicals *in vitro* does not always act as an antioxidant in a living organism. This is because of how quickly free radical (FR) diffuses and spreads. Given that certain antioxidants have extremely short lives in the order of nanoseconds, it might be difficult to have antioxidants present when and where oxidative damage occurs. A second-order reaction also occurs when antioxidants and FR interact. Therefore, in addition to parameters related to antioxidant and free radical abundance, it also depends on factors related to the medium, reaction conditions, and chemical structure of both reagents (Francenia Santos-Sánchez et al., 2019).

7.3.1 Characteristics of Bioactive Compounds/Antioxidants

By stabilizing the generated radical, a chemical or antioxidant system helps avoid or detect a chain of oxidative propagation, preventing oxidative harm to the body (Butnariu & Grozea, 2012). The two main categories of antioxidants are primary (which scavenge free radicals and interrupt chain reactions) and secondary, or preventive. In addition to limiting the creation of undesired volatiles, deactivating metals, replenishing primary antioxidants, and eliminating singlet oxygen to inhibit lipid hydroperoxides are all examples of secondary antioxidant processes (Vitale et al., 2000).

According to the US Food and Drug Administration, antioxidants are compounds used to preserve food by postponing oxidation-related deterioration, rancidity, or discolouration (FDA). The food industry needs antioxidants because they stop rancidity. Antioxidants are also valued by researchers and medical professionals because they may shield the body from diseases brought on by the risks of ROS by inhibiting ROS-related enzymes (Bagchi & Puri, 1998). Antioxidants can protect the body's cells and organs from the harmful effects of oxidative stress through a variety of protective mechanisms, including enzymatic as well as nonenzymatic processes that work together and interact synergistically. To prevent lipid oxidation, nonenzymatic antioxidants are routinely added to food (Pandey & Rizvi, 2010; Kurutas, 2016).

7.3.2 Phenolic Compounds

The signalling and defence processes of plants depend heavily on phenolic chemicals. These substances work to reduce the stress that pathogenic organisms and predators cause. These substances have a variety of purposes within plants, including acting as building blocks form ore complex chemicals, playing roles in mechanisms that regulate and control plant development, and acting as defences increase. By preventing the oxidation of low-density lipoproteins, phenolic compounds can act as hydrogen donors and bind metal ions like iron and copper (LDL) (Paran et al., 2009). These phenolic component qualities are associated with a decreased incidence of neurological illnesses, as well as cardiovascular diseases, gastrointestinal malignancies, colon cancer, breast and ovarian cancers, as well as helps cure leukaemia (Ren et al., 2001; Teofili et al., 1992; Yoshida et al., 1990; Yoshida et al., 1992). Moreover, phenolic compounds have promise as vasorelaxants and anti-allergens (Sakakibara et al., 2003). Phenolic compounds stop the oxidation of LDL *in vitro* (Riemersma et al., 2001). Phenolic compounds prevent or lessen the production of free radicals by transferring a hydrogen atom from a hydroxyl group. A hydrogen cation is coordinatedly transported from a phenolic molecule to a peroxyl radical (ROO•), resulting in the formation of an H-O bond with one electron. If a solvent that might form hydrogen bonds with the phenolic molecule is present in the reaction medium, the phenolic component's capacity to function as an antioxidant is significantly reduced. For instance, alcohol doubles the rate of phenol and peroxyl radical interaction. As hydrogen bond acceptors, alcohols operate. Anionic phenoxides, on the other hand, may react quickly with peroxyl radicals by electron transfer; therefore, they favour ionization of phenols over those compounds. The final compound's level of ionization significantly affects the solvent's overall impact on the antioxidant activity of phenolic compounds (Foti, 2007).

7.3.3 Carotenoids

Three different ways explain how carotenoids act as antioxidants. The first is SET (Single electron transfer), these condis adduct formation, and the third is HAT (hydrogen atom transfer). Carotenoids' strong ability to donate electrons is often linked to their antioxidant capabilities. Via SET, carotenoid and $NO2^{\bullet}$ interact. Researchers have also looked on carotenoid reactivity in the presence of unreactive benzyl peroxyl radicals, indicating that the reaction mechanism in this situation involves, among others, adduct formation (Everett et al., 1995). However, HAT-mediated responses are of little importance (Mortensen, 2002).

7.3.4 Vitamin C

Being water-soluble, vitamin C has the chemical ability to act as an antioxidant and interact with the most physiologically important ROS. Vitamin C's antioxidant response mechanism is based on HAT (hydrogen atom transfer) to peroxy radicals, singlet oxygen deactivation, and molecular oxygen scavenging (Lee et al., 2004; Abraham, 2014).

7.3.5 Flavonoids

The bulk of phenolic substances found in plants are flavonoids, which are small molecules. The 15 carbon atoms that make them up are arranged in a C6-C3-C6 configuration. Due to their high redox potential and capacity to function as reducing agents, hydrogen donors, and singlet oxygen quenchers, flavonoids are particularly significant antioxidants. It can chelate metals as well (Walter & Marchesan, 2011).

7.4 PARASITIC INFECTION AND OXIDATIVE STRESS

Reactive oxygen or nitrogen atoms with unpaired electrons in their outer shells cause oxidative stress. Cells frequently produce two forms of reactive species: reactive oxygen species (ROS) and reactive nitrogen species (RNS). These radicals are oxidants that can harm cells even though they are involved in crucial physiological processes, such intracellular signalling and oxidative burst in cells of innate immunity (Liguori et al., 2018). By increasing the production of reactive oxygen species and reactive nitrogen species (ROS/RNS), which weaken the host's natural antioxidant defences, parasite infections cause oxidative stress, which is characterized by a host redox imbalance. The host immunological response that results in ROS/RNS in phagocytic cells (the respiratory burst process) and the involvement of the systemic inflammatory response (SIR) allow the parasite to utilize damaged haemoglobin as a source of amino acids for his diet (Percário et al., 2012). Nevertheless, oxidative stress has a more complex role in infectious disorders since it harms the host's molecules while simultaneously assisting in the eradication of invasive infections. It is generally recognized that viral diseases cause organ damage and metabolic disruption, which results in higher levels of ROS and RNS that support the immune response (Pohanka, 2013). Malaria causes an increase in oxidative stress, which damages cells and tissues throughout the body, including the brain and lungs. Based on the parenchymal lesions in the brain, this redox imbalance is linked to the most severe types of malaria, including cerebral malaria, and can cause physical and mental issues (Reis et al., 2010). In order to combat the free radicals created by infection and halt oxidative alterations, host cells have an innate defence mechanism that utilizes enzymes and tiny antioxidant molecules (Percário et al., 2012). Nonetheless, these defences frequently fall short of what the host needs (Ezzi, 2017). Studies to support the host's antioxidant defence system are looking at supplemental antioxidant therapy (Gomes et al., 2015; Dwivedi et al., 2016; Ighodaro et al., 2018; Ekeh et al., 2019).

7.4.1 Causes of Oxidative Stress

Oxidative stress in malaria can be harmful or beneficial during plasmodial infection, depending on the stage of the illness (Atiku et al., 2019; Götz et al., 2019; Velagapudi et al., 2019). At various stages of infection, oxidative stress is involved. In fact, higher amounts of malondialdehyde (MDA), a product of lipid peroxidation and a sign of significant oxidative stress, have been seen in instances of malaria (Narsaria et al., 2012).

7.4.1.1 Heme

In malaria, heme contributes to oxidative stress, which affects both the host and the parasite. Haemoglobin-infected red blood cells provide food in the form of amino acids to the parasite

Plasmodium (Becker et al., 2004). When haemoglobin breaks down, heme is liberated, and the iron atom in heme enables it to generate reactive oxygen species. By the Fenton reaction, heme iron can encourage the creation of hydroxyl radicals, which is harmful, including both parasites and hosts (Percário et al., 2012). To counteract the deleterious effects of heme, the malaria parasite converts heme into hemozoin, an intracellular accumulation where heme molecules unite to form crystalline structures. Hemozoin doesn't harm the host or the parasite through oxidative stress (Becker et al., 2004).

7.4.1.2 Phagocytic Oxidative Burst

One of the first defence cells that appear during the early stages of infection are neutrophils. When activated, they can kill malaria parasites in a variety of ways, but the main one is oxidative stress–induced generation of ROS and RNS (Aitken et al., 2008, Figure 7.1). According to Cunnington et al. (2012), neutrophils in patients with malaria behave differently than in uninfected patients, and because of their high plasma heme levels, patients with *Plasmodium falciparum* have diminished oxidative burst, which increases their risk of cerebrovascular illness. (Feintuch et al., 2016).

7.4.1.3 Induced by Antimalarial Drugs

It is not surprising that numerous antimalarial drugs take advantage of *Plasmodium* parasites' susceptibility to ROS-mediated damage in order to limit their development within human hosts. By preventing infected erythrocytes from converting free heme into hemozoin, quinolones such as chloroquine and amodiaquine increase oxidative stress against the malaria parasite (Kavishe et al., 2017). Dihydroartemisinin and artesunate are two important members of the artemisinins class of antimalarial drugs (Kavishe et al., 2017). Moreover, the mentioned medications cause the parasite to experience oxidative stress, which renders them antiparasitic. One of artemisinin's hypothesized modes of action is to generate free radicals when interacting with iron (Cui & Su, 2009). The host may experience negative effects even if boosting oxidative stress is an effective strategy to get rid

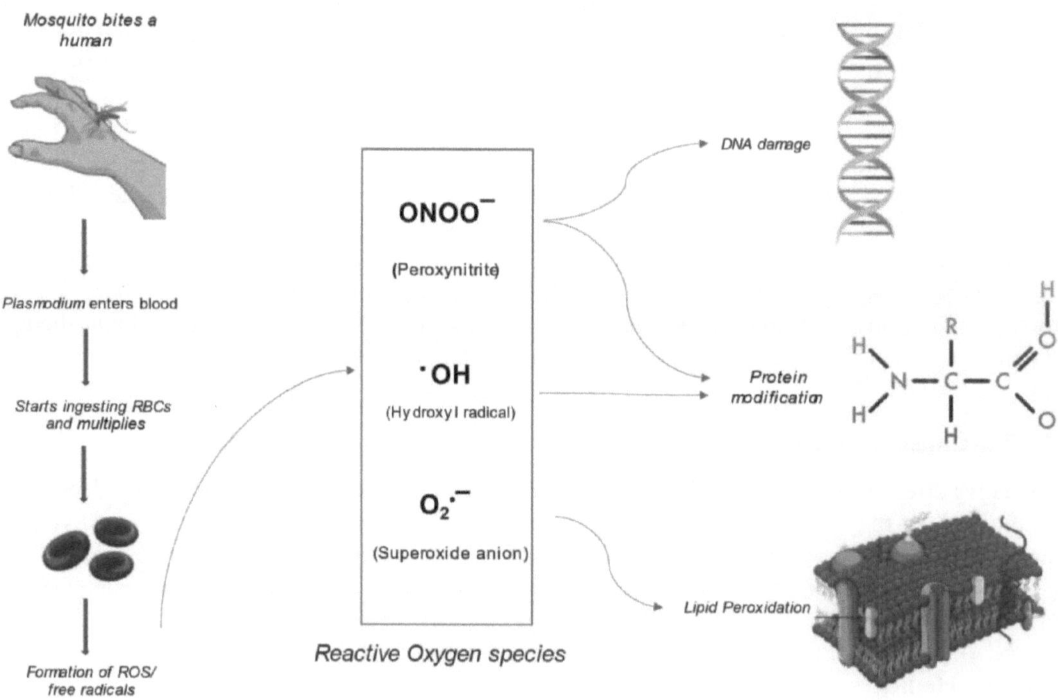

FIGURE 7.1 Oxidative damage caused by the ROS/RNS during malarial infection.

of parasites. Invitro oxidative damage caused by artesunate was seen in human cell DNA (Berdelle et al., 2011).

7.5 ANTIMALARIAL PLANTS

Malaria is brought on by the potentially fatal *Plasmodium* parasite, which is spread to people via the bite of an *Anopheles* mosquito. Although there are a number of effective antimalarial drugs, renewed interest in traditional herbs that have been used for centuries to treat and prevent malaria has been driven by the increasing emergence of drug-resistant *Plasmodium* strains. By restricting the production of free radicals, halting their development, oxidizing already-existing free radicals, and changing poisonous chemicals into less-dangerous ones, antioxidants defend cells on several levels (Kumar et al., 2014). The most likely culprits for the free radical scavenging action are flavonoids and tannins. The primary antioxidants are also substances called phenols that are present in plants.

Also, a sizable amount of India's natural resources is comprised of healing and therapeutic plants. According to many studies, the antioxidant properties of phenolic and flavonoid effects of substances include anti-inflammation, antibacterial, antiviral, antiproliferative, and immunomodulatory properties (Shahidi & Ambigaipalan, 2015; Comunian et al., 2017). Following are a few examples of plants whose potential antimalarial characteristics have been researched:

Artemisia annua (sweet wormwood). For thousands of years, fever has been treated using *Artemisia annua* in traditional Chinese medicine. Its major component, artemisinin, is a strong antimalarial medication that has been used in combination with other medications to treat *Plasmodium* strains that are drug-resistant. Artemisinin is a sesquiterpenoid produced in the glandular trichomes of the Chinese medicinal herb sweet wormwood (*Artemisia annua L.*), a major source for artemisinin combination therapies (ACT) (Klayman, 1985). However, the low concentration of artemisinin in *A. annua* (between 0.01% and 0.8% per dry weight of tissue) severely restricts the drug's ability to be commercialized (Abdin et al., 2003). Hence, the existing artemisinin supply chain would immediately benefit from enhanced *A. annua* varieties for farmers in underdeveloped nations by lowering production costs, stabilizing supplies, and boosting producers' faith in the crop (Arrow et al., 2004). *Artemisia annua*, which is mainly recognized for its ability to treat malaria, has been investigated and found to have other biological activities, such as antitumour, antimicrobial, and immunomodulatory abilities (Malaterre et al., 2020).

Cinchona officinalis (nilgiris). The bark of *Cinchona officinalis*, which contains quinine, was first discovered in the 19th century and is still utilized in some antimalarial medications today. It is a medicinal plant and one of numerous *Cinchona* species used for the synthesis of quinine, among other compounds. Quinine is a substance that reduces fever and is particularly helpful in the treatment and prevention of malaria (Gachelin et al., 2017).

Azadirachta indica (neem). The leaves, bark, and seeds of this neem tree have all been used in conventional therapy to treat a variety of diseases, including malaria. Neem extracts have been proven to have antimalarial qualities and may be useful against drug-resistant forms of *Plasmodium*, according to studies (Emmanuel et al., 2013).

Zingiber officinale* Roscoe *(African ginger). It is generally used in Ethiopia as an antimalarial traditional natural drug against *Plasmodium* parasite like *Plasmodium berghei* (Biruksew et al., 2018).

Allium sativum (garlic). Garlic is not a common antimalarial treatment, like ginger, but research has shown that its active components have antimalarial effects and may be useful against drug-resistant strains of *Plasmodium*.

Malaria can be averted by allicin, a naturally occurring compound created by crushed garlic bulbs. Allicin appears to be an inhibitor of the sporozoite cysteine protease(s) required for infectivity in the mammalian host since it inhibited cell invasion and proteolytic cleavage of cysteine protease(s) (CSP) but did not impact gliding motility. Allicin's effects on sporozoites are similar to those of the highly selective cysteine protease inhibitor E-64 at low dosages (Coppi et al., 2006).

Lantana camara (kongini). The Verbenaceae family member, also referred to as wild or red sage, has recently been the subject of extensive phytochemical investigation (Thorat et al., 2021). It can also be used as a tonic, antispasmodic, carminative, antiemetic, and diaphoretic. Leaf infusions are used to treat eruptions, eczema, and bilious fever (Chaitali et al., 2022).

Although being a significant source of novel antimalarial compounds, not all plants that are traditionally used to treat malaria have been scientifically proven to be safe and effective. Much scientific research must be done in order to identify the active components in these plants and evaluate their efficacy and safety. A list of a few plants with antimalarial chemicals that have been used to treat malaria is provided in Table 7.1.

The potential advantages of supplementary antioxidants coming from meals, vitamins, and medications have been emphasized in several research. These substances would work by altering the host's antioxidant and immune response to malaria. This response may involve subtly eliminating the parasite or enhancing the body's natural antioxidant defence against oxidative stress. *Terminalia albida* stem bark has been used to reduce ROS/RNS and improve survival in mice with

TABLE 7.1
List of Plants with Antimalarial Activity

Plant Species	Compound	Activity	References
Artemisia annua	Artemisinin	Prevents haemoglobin digestion at blood stage	Krishna et al., 2008
Azadirachta indica, Moringa oleifera	Flavonoids	Reducing agents and oxygen quenchers	Walter & Marchesan, 2011
Bixa orellana, Cymbopogon citratus	Flavonoids, terpenoids, and phenolics	Counteracts oxidative damage caused by malaria parasites at significant toxicity levels	Saleh et al., 2015
Carica papaya, Azadirachta indica, Moringa oleifera	Vitamin C	Prevents lipid peroxidation by eliminating peroxyl radicals	Arredondo, 2016
Chasmanthera dependens	Flavanoids, saponins, and carotenoids	Able to scavenge free radicals	Enenebeaku et al., 2022
Cleome gynandra	Lipoic acids	Oxidation decarboxylation ofα-keto acids, free radicals' quenchers	Packer et al., 1995; El Barky et al., 2017
Calotropis procera, Lantana camara	Flavonoids and phenolics	Controls malaria parasite invasion into host erythrocytes and reduces subsequent oxidative stress	Joshi et al., 2022
Eremophila microtheca	Quinoline	Polymerization of free heme to non-toxic hemozoin	Kumar et al., 2018
Ocimum tenuiflorum	Carotenoids	Free radical scavenger, prevents lipid peroxidation	Rahman, 2007
Parkia biglobosa	Phenols	Prevents the oxidation of LPL proteins; anti-allergenic properties	Paran et al., 2009; Sakakibara et al., 2003
Sophora exigua	Exiguaflavanone A and B; sophoraflavanone G	Inhibits blood-stage propagation of parasite	Kaewdana et al., 2021

cerebral malaria (Camara et al., 2019). Gallicacid–rich plants, such as *Terminalia avicennioides* (Ouattaraetal.,2014) and *Acacianilotica* leaf, fruit, and skin extracts showed antimitotic activity against *P. falciparum in vitro* (Cai et al., 2017; Li et al., 2018).

7.5.1 Antioxidants: Mode of Action

7.5.1.1 Inhibition of Oxidation of Lipids

The oxidative degradation of lipids with C-C double bonds, such as polyunsaturated fatty acids (PUFAs), glycolipids, cholesterol, cholesterol esters, and phospholipids, is known as lipid peroxidation. Reactive peroxidation cascade reactions are started when ROS attacks methylene CH2 groups and unsaturated fatty acids with plenty of double bonds (Porter et al., 1995). Peroxide radicals can be immediately quenched, and the chain reaction can be stopped by antioxidants. Many chronic diseases, including cancer and atherosclerosis, are connected to DNA damage and lipid peroxidation. Because antioxidants may scavenge ROS and peroxide radicals, they can be used to prevent or treat several pathogenic diseases (Krishnamurthy & Wadhwani, 2012).

7.5.2.2 Inhibition of Protein Alteration

Proteins can be altered by amino acid nitration or chlorination, in addition to lipid peroxidation and DNA damage, which is another effect of ROS. Peroxynitrite, also known as ONOO, is a potent nitrating and oxidizing agent that is created in cells when oxygen and nitric oxide free radicals interact in a diffusion-controlled manner. DNA and proteins are only two examples of the materials that ONOO – a considerably more potent oxidant than O2 –can harm in living things. Direct oxidation modifications can occur during one- or two-electron oxidation cycles when ONOO and its protonated product peroxynitrite (ONOOH) are present (Pacher et al., 2007). Proteins are protected from ONOO alterations by the use of antioxidants and antioxidant enzymes. H_2O_2 is removed by antioxidants or enzymes like catalase (CAT), which also stops HOCl (hypochlorous acid) from forming. Antioxidants like SOD (superoxide dismutase) and polyphenols may also scavenge O_2 and stop the production of ONOO (Krishnamurthy & Wadhwani, 2012).

7.5.3.3 Inhibition of DNA Degeneration

Nitric oxide and oxygen radicals ($OH^•$ and $ONOO^-$) were produced in vivo by directly interacting with plasmid DNA macromolecules, and oxygen can cleave DNA strands and cause oxidative DNA damage. Cancer, aging, neurological disease, and heart disease are all associated with cell death and mutations due to DNA damage. As a result, antioxidant research and development has focused on DNA or plasmid damage and used them as models (Zhao et al., 2007).

7.6 APPLICATIONS

Antimalarial plants are those that have substances that can treat or prevent malaria. These plants have been used for generations in conventional medicine to cure a variety of ailments, including malaria. Compounds from antimalarial plants have strong antimalarial effects. Malaria treatments can be made from these chemicals. For instance, the plant *Artemisia annua* contains the powerful antimalarial agent artemisinin. Malaria can also be avoided by using antimalarial herbs. For instance, the neem tree (*Azadirachta indica*) has substances that ward off mosquitoes, the main carrier of malaria. These plants are a plentiful supply of organic substances that can be exploited to create brand-new antimalarial medications (Olotu & Muhammed, 2019; Cowman & Healer, 2017). Current interest has been focused on the development of semi-synthetic artemisinin in plants thanks to low-cost herbal artemisinin combination treatment (ACT), because the antimalarial activity of this therapy is stronger and artemisinin resistance was overcome by the interaction between artemisinin and the plant matrix (Muangphrom et al., 2016)

As a first- and second-line therapy for *Plasmodium falciparum* and chloroquine-resistant *vivax*, ACT, is advised by the WHO. In the first three days of therapy, the artemisinin compound reduces the number of parasites (reduction in parasite biomass), and the partner medication's job is to get rid of any residual parasites, which promotes faster recovery (WHO, 2018b). Only the suppressive prevention of malaria is presently treated with piperaquine, caused by *Plasmodium vivax* in certain populations such as migrant workers, border officials, travellers, etc.

However, this medicine should not be taken continuously for more than four months as it may accumulate in the liver (Department of Disease Control, China, 2007). Naphthoquinone was first produced in China in 1981. When tested on mice infected with *P. berghei*, the drug demonstrated cross-resistance to chloroquine (Li et al., 1982).

In order to increase the effectiveness and lessen negative effects of traditional antimalarial medications, antimalarial herbs can be utilized as a supplemental therapy. A significant obstacle to treat malaria is drug resistance. Drug-resistant malaria can be controlled by using antimalarial botanicals, which offer different therapy choices. Local communities can benefit economically from the growth and processing of antimalarial plants, which also supports community-based healthcare.

REFERENCES

Abdin, M. Z., Israr, M., Rehman, R. U. & Jain, S. K. 2003. Artemisinin, a novel antimalarial drug: Biochemical and molecular approaches for enhanced production (review). *Planta Medica* 69(4):289–299.

Abraham, S. E. 2014. Biochemistry of free radicals and antioxidants. *Scholars Academic Journal of Biosciences* 2(2):110–118.

Aitken, E. H., Alemu, A. & Rogerson, S. J. 2008. Neutrophils and malaria. *Frontier Immunology* 9:3005.

Antinori, S., Galimberti, L., Milazzo, L. & Corbellino, M. 2012. Biology of human malaria plasmodia including *Plasmodium knowlesi*. *Mediterranean Journal of Hematology and Infectious Diseases* 4:e2012013.

Arredondo, M. L. 2016. Relationship between vitamin intake and total antioxidant capacity in elderly adults. *Universitas Scientiarum* 21(2):167–177.

Arrow, K. J., et al. 2004. Antimalaria drug and drug resistance. In *Saving Lives, Buying Time: Economics of Malaria Drugs in an Age of Resistance*, pp. 252–300. National Academies Press.

Asase, A., Oteng-Yeboah, A. A., Odamtten, G. T. & Simmonds, M. S. 2005. Ethnobotanical study of some Ghanaian anti-malarial plants. *Journal of Ethnopharmacology* 99(2):273–279.

Ashley, E. A., Pyae Phyo, A. & Woodrow, C. J. 2018. Malaria. *The Lancet* 391(10130):1608–1621.

Atiku, S. M., Louise, N. & Kasozi, D. M. 2019. Severe oxidative stress in sickle cell disease patients with uncomplicated *Plasmodium falciparum* malaria in Kampala, Uganda. *BMC Infectious Diseases* 19:600.

Bagchi, K. & Puri, S. 1998. Free radicals and antioxidants in health and disease: A review. *Eastern Mediterranean Health Journal* 4(2):350–360.

Baird, J. K. 2007. Neglect of *Plasmodium vivax* malaria. *Trends in Parasitology* 23(11):533–539.

Battle, K. E., Karhunen, M. S., Bhatt, S., Gething, P. W., Howes, R. E., Golding, N., Van Boeckel, T. P., Messina, J. P., Shanks, G. D. & Smith, D. L. 2014. Geographical variation in Plasmodium vivax relapse. *Malaria Journal* 13:144.

Becker, K., Tilley, L., Vennerstrom, J. L., Roberts, D., Rogerson, S. & Ginsburg, H. 2004. Oxidative stress in malaria parasite infected erythrocytes: Host – parasite interactions. *International Journal for Parasitology* 34(2):163–189.

Berdelle, N., Nikolova, T., Quiros, S., Efferth, T. & Kaina, B. 2011. Artesunate induces oxidative DNA damage, sustained DNA double-strand breaks, and the ATM/ATR damage response in cancer cells. *Molecular Cancer Therapeutics* 10(12):2224–2233.

Biruksew, A., Zeynudin, A., Alemu, Y., Golassa, L., Yohannes, M., Debella, A., Urge, G., De Spiegeleer, B. & Suleman, S. 2018. *Zingiber officinale* Roscoe and *Echinops Kebericho Mesfin* showed antiplasmodial activities against *Plasmodium berghei* in a dose-dependent manner in Ethiopia. *Ethiopia Journal Health Science*:655–664.

Butnariu, M. & Grozea, I. 2012. Antioxidant (antiradical) compounds. *Journal of Bioequivalence and Bioavailability* 4(6):4–6.

Cai, T. Y., Zhang, Y. R., Ji, J. B. & Xing, J. 2017. Investigation of the component in Artemisia annua L. leading to enhanced antiplasmodial potency of artemisinin via regulation of its metabolism. *Journal of Ethnopharmacology* 207:86–91.

Camara, A., Haddad, M., Reybier, K., Traoré, M. S., Baldé, M. A., Royo, J. & Aubouy, A. 2019. Terminalia albida treatment improves survival in experimental cerebral malaria through reactive oxygen species scavenging and anti-inflammatory properties. *Malaria Journal* 18:431.

Centers for Diseases Control and Prevention (CDC). 2020. *Global Health, Division of Parasitic Diseases and Malaria*. www.cdc.gov/malaria/about/biology/index.html

Chaitali, A., Rao, P. & Vikhe, S. R. 2022. Phytochemical and pharmacological activities of lantana camara-review. *Research Journal of Pharmacognosy and Phytochemistry* 14(2):128–131.

Chen, C. 2014. Development of antimalarial drugs and their application in China: A historical review. *Infectious Diseases of Poverty* 3(1):9.

Comunian, T. A., Ravanfar, R., de Castro, I. A., Dando, R., Favaro-Trindade, C. S. & Abbaspourrad, A. 2017. Improving oxidative stability of echium oil emulsions fabricated by Microfluidics: Effect of ionic gelation and phenolic compounds. *Food Chemistry* 15(233):125–134.

Coppi, A., Cabinian, M., Mirelman, D. & Sinnis, P. 2006. Antimalarial activity of allicin, a biologically active compound from garlic cloves. *Antimicrobial Agents and Chemotherapy*:1731–1737.

Cowman, A. F. & Healer, J. (2017). Marrying artemisinins with plant natural products to tackle malaria. *Chemistry & biology* 24(11):1096–1098.

Cui, L. & Su, X. Z. 2009. Discovery, mechanisms of action and combination therapy of artemisinin. *Expert Review Anti-Infectious Therapy* 7(8):999–1013.

Cunnington, A. J., Njie, M., Correa, S., Takem, E. N., Riley, E. M. & Walther, M. 2012. Prolonged neutrophil dysfunction after *Plasmodium Falciparum* malaria is related to hemolysis and heme oxygenase-1 induction. *Journal of Immunology* 189(11):5336–5346.

Department of Disease Control. 2007. Ministry of health of the People's Republic of China. In *The Manual of Malaria Control*, pp. 274–277. Department of Disease Control.

Dwivedi, H., Singh, S. K., Chauhan, B. S., Gunjan, S. & Tripathi, R. 2016. Potential cerebral malaria therapy: Intramuscular arteether and vitamin D co-administration. *Parasitology* 143:1557.

Ekeh, F. N., Ekechukwu, N. E., Chukwuma, C. F., Aguzie, I. O. N., Ohanu, C. M., Ebido, C. & Oluah, S. N. 2019. Mixed vitamin C and zinc diet supplements co-administered with artemether drug improved haematological profile and survival of mice infected with *Plasmodium berghei. Food Science and Human Wellness* 8:275–282.

El Barky, A. R., Hussein, S. A. & Mohamed, T. M. 2017. The potent antioxidant alpha lipoic acid. *Journal of Plant Chemistry and Ecophysiology* 2(1):1–5.

Emmanuel, H., et al. 2013. Antimalarial effect of Neem leaf and Neem stem bark extracts on *plasmodium berghei* infected in the pathology and treatment of malaria. *International Journal of Research in Biochemistry and Biophysics* 3:7–14.

Enenebeaku, C. K., Ogukwe, C. E., Nweke, C. O., et al. 2022. Antiplasmodial and in vitro antioxidant potentials of crude aqueous and methanol extracts of Chasmanthera dependens (Hochst). *Bulletin of the National Research Centre* 46:33.

Everett, S. A., Kundu, S. C., Maddix, S. & Willson, R. L. 1995. Mechanism of free-radical scavenging by the nutritional antioxidant β-carotene. *Biochemical Society Transactions* 23:230S.

Ezzi, A. A., Salahy, M. B., Shnawa, B. H., Abed, G. H. & Mandour, A. M. 2017. Changes in levels of antioxidant markers and status of some enzyme activities among *falciparum* malaria patients in Yemen. *Journal of Microbiology & Experimentation* 4(6):00131.

Feintuch, C. M., Saidi, A., Seydel, K., Chen, G., Goldman-Yassen, A., Mita-Mendoza, N. K., Daily, J. P. 2016. Activated neutrophils are associated with pediatric cerebral malaria vasculopathy in Malawian children. *MBio* 7:e01300–e01315.

Foti, M. C. 2007. Antioxidant properties of phenols. *The Journal of Pharmacy and Pharmacology* 59(12):1673–1685.

Francenia Santos-Sánchez, N., Salas-Coronado, R., Villanueva-Cañongo, C., & Hernández-Carlos, B. 2019. *Antioxidant Compounds and Their Antioxidant Mechanism*. IntechOpen. London, United Kingdom, doi: 10.5772/intechopen.85270

Gachelin, G., Garner, P., Ferroni, E., Tröhler, U. & Chalmers, I. 2017. Evaluating Cinchona bark and quinine for treating and preventing malaria. *Journal of the Royal Society of Medicine* 110(1):31–40. https://doi.org/10.1177/0141076816681421

Gomes, A. R. Q., Moreira, D. R., Dolabela, M. F., Santos, R. S. & Percário, S. 2015. N-acetyl cysteine and mushroom *Agaricus sylvaticus* supplementation decreased parasitaemia and pulmonary oxidative stress in a mice model of malaria. *Malaria Journal* 14:202.

Götz, A., Ty, M. C. & Rodriguez, A. 2019. Oxidative stress enhances dendritic cell responses to *Plasmodium falciparum. Immuno Horizons* 3:511–518.

Ighodaro, O. M. & Akinloye, O. A. 2018. First line defence antioxidants-superoxide dismutase (SOD), catalase (CAT) and glutathione peroxidase (GPX): Their fundamental role in the entire antioxidant defence grid. *Alexandria Journal of Medicine* 54:287–293.

Joshi, U., Jani, D., George, L. B. & Highland, H. 2022. Antioxidant efficacy of selected plant extracts debilitates the plasmodium invasion through erythrocytic membrane stabilisation – An in vitro study. *Infectious Diseases* 2:6.

Kaewdana, K., Chaniad, P., Jariyapong, P., et al. 2021. Antioxidant and antimalarial properties of *Sophora exigua* Craib. root extract in *Plasmodium berghei*-infected mice. *Tropical Medicine and Health* 49:24.

Kavishe, R. A., Koenderink, J. B. & Alifrangis, M. 2017. Oxidative stress in malaria and artemisinin combination therapy: Pros and cons. *Federation of European Biochemical Societies Journal* 284(16):2579–2591.

Klayman, D. L. 1985. Qinghaosu (artemisinin): An anti-malarial drug from China. *Science* 228:1049–1055.

Krishna, S., Bustamante, L., Haynes, R. K. & Staines, H. M. 2008. Artemisinins: their growing importance in medicine. *Trends in Pharmacological Sciences*, 29(10):520–527. https://doi.org/10.1016/j.tips.2008.07.004

Krishnamurthy, P. & Wadhwani, A. 2012. Antioxidant enzymes and human health. In *Antioxidant Enzyme*, pp. 3–18. In Tech.

Kumar, R., Duffy, S., Avery, V. M., Carroll, A. R. & Davis, R. A. 2018. Microthecaline A, a quinoline serrulatane alkaloid from the roots of the Australian desert plant *Eremophila microtheca*. *Journal of Natural Products* 81(4):1079–1083.

Kumar, S., Sandhir, R. & Ojha, S. 2014. Evaluation of antioxidant activity and total phenol in different varieties of Lantana camara leaves. *BMC Research Notes* 7(1):1–9.

Kurutas, E. B. 2016. The importance of antioxidants which play the role in cellular response against oxidative/nitrosative stress: Current state. *Kurutas Nutrition Journal* 15:71–93.

Lee, J., Koo, N. & Min, D. B. 2004. Reactive oxygen species, aging, and antioxidative nutraceuticals. *Comprehensive Reviews in Food Science and Food Safety* 3(1):21–33.

Lehrer S. 1981. *Vitezi Medicine*. Prosveta; Beograd, Serbia.

Li, F. L., Wang, L. H., Ding, D. B, Yang, J. D. & Gao, X. S. 1982. Studies on antimalarials: Synthesis of 4-arylamino-tert-butylaminomethyl phenols. *Acta Pharmaceutica Sinica B* 17:77–79.

Li, J., Zhang, C., Gong, M. & Wang, M. 2018. Combination of artemisinin-based natural compounds from *Artemisia annua* L. for the treatment of malaria: Pharmacodynamic and pharmacokinetic studies. *Phytotherapy Research* 32:1415–1420.

Liguori, I., Russo, G., Curcio, F., Bulli, G., Aran, L. & Della-Morte, D., et al. 2018. Oxidative stress, aging, and diseases. *Clinical Interventions in Aging* 13:757–772.

Malaterre, A. S., Rakoto, M. L., Marodon, C., Bedoui, Y., Nakab, J., Simon, E., Hoarau, L., Savriama, S., Strasberg, D., Guiraud, P., Selambarom, J. & Gasque, P. 2020. *Artemisia annua*, a traditional plant brought to light. *International Journal of Molecular Sciences* 21:4986.

Mortensen, A. 2002. Scavenging of benzylperoxyl radicals by carotenoids. *Free Radical Research* 36:211–216.

Moss, W. J., Shah, S. N., & Morrow, R. H. 2008. The history of malaria and its control. In *International Encyclopedia of Public Health* (pp. 389–398). Elsevier Inc., https://doi.org/10.1016/B978-012373960-5.00374-9

Mota, M. L., Lobo, L. T., Costa, J. M., Costa, L. S., Rocha, H. A., Rocha e Silva, L. F., et al. 2012. In vitro and in vivo antimalarial activity of essential oils and chemical components from three medicinal plants found in north-eastern Brazil. *Planta Medica* 78:658–664.

Muangphrom, P., Seki, H., Fukushima, E. O., et al. 2016. Artemisinin-based antimalarial research: Application of biotechnology to the production of artemisinin, its mode of action, and the mechanism of resistance of Plasmodium parasites. *Journal of Natural Medicines* 70:318–334.

Narsaria, N., Mohanty, C., Das, B. K., Mishra, S. P. & Prasad, R. 2012. Oxidative stress in children with severe malaria. *Journal of Tropical Pediatrics* 58:147–150.

Olotu, A. & Muhammed, A. M. 2019. Antimalarial plants: Efficacy and mechanism of action. *International Journal of Biological and Chemical Sciences* 13(4):1624–1634.

Ouattara, L. P., Sanon, S., Mahiou-Leddet, V., Gansané, A., Baghdikian, B., Traoré, A. & Sirima, S. B. 2014. In vitro antiplasmodial activity of some medicinal plants of *Burkina Faso*. *Parasitology Research* 113: 405–416.

Pacher, P., Beckman, J. B. & Liaudet, L. 2007. Nitric oxide and peroxynitrite in health and disease. *Physiological Review* 87:315–424.

Packer, L., Witt, E. H. & Tritschler, H. J. 1995. Alpha lipoic acid as a biological antioxidant. *Free Radical Biology and Medicine* 19(2):227–250.

Pandey, K. B. & Rizvi, S. I. 2010. Markers of oxidative stress in erythrocytes and plasma during aging in humans. *Oxidative Medicine and Cellular Longevity* 3(1):2–12.

Paran, E., Novack, V., Engelhard, Y. N. & Hazan-Halevy, I. 2009. The effects of natural antioxidants from tomato extract in treated but uncontrolled hypertensive patients. *Cardiovascular Drugs and Therapy* 23(2):145–151.

Percário, S., Moreira, D., Gomes, B., et al. 2012. Oxidative stress in malaria. *International Journal of Molecular Sciences* 13(12):16346–16372.

Pohanka, M. 2013. Role of oxidative stress in infectious diseases: A review. *Folia Microbiologica* 58(6):503–513.

Porter, N. A., Caldwell, S. E. & Mills, K. A. 1995. Mechanisms of free radical oxidation of unsaturated lipids. *Lipids* 30:277–290.

Rahman, K. 2007. Studies on free radicals, antioxidants, and co-factors. *Clinical Interventions in Aging* 2(2):219–236.

Reis, P. A., Comim, C. M., Hermani, F., Silva, B., Barichello, T., Portella, A. C., & Castro-Faria-Neto, H. C. 2010. Cognitive dysfunction is sustained after rescue therapy in experimental cerebral malaria and is reduced by additive antioxidant therapy. *PLoS Pathogens* 6:e1000963.

Ren, W., Qiao, Z. & Wang, H. 2001. Tartary buckwheat flavonoid activates caspase 3 and induces HL-60 cell apoptosis. *Methods and Findings in Experimental and Clinical Pharmacology* 23:427–432.

Riemersma, R. A., Rice-Evans, C., Tyrrell, R. M., Clifford, M. N. & Lean, M. E. 2001. Tea flavonoids and cardiovascular health. *Quarterly Journal of Medicine* 94:277–282.

Roca-Feltrer, A., Carneiro, I., Smith, L., Schellenberg, J. R. A., Greenwood, B. & Schellenberg, D. 2010. The age patterns of severe malaria syndromes in sub-Saharan Africa across a range of transmission intensities and seasonality settings. *Malaria Journal* 13(9):282.

Roussel, C., Caumes, E., Thellier, M., Ndour, P. A., Buffet, P. A. & Jauréguiberry, S. 2017. Artesunate to treat severe malaria in travellers: Review of efficacy and safety and practical implications. *Journal of Travel Medicine* 24.

Sakakibara, H., Honda, Y., Nakagawa, S., Ashida, H. & Kanazawa, K. 2003. Simultaneous determination of all polyphenols in vegetables, fruits, and teas. *Journal of Agricultural and Food Chemistry* 51: 571–581.

Saleh, Z., Zuhud, E. A. M. & Sari, R. K. 2015. Phytochemical screening and antioxidant activity of ethanolic extract of *Rhizanthes deceptor* (Rafflesiaceae) and its host *Tetrastigma papillosum*. *Research Journal of Medicinal Plant* 9:293–299.

Schalkwyk, D. A. Van. 2015. *History of Antimalarial Agents*. Wiley, pp. 1–5.

Shahidi, F. & Ambigaipalan, P. 2015. Phenolics and polyphenolics in foods, beverages and spices: Antioxidant activity and health effects – A review. *Journal of Functional Foods* 1(18):820–897.

Soulard, Vale´rie, Bosson-Vanga, Henriette, Lorthioi, Audrey, Roucher, Clémentine, Franetich, Jean-François, Zanghi, Gigliola, Bordessoulles, Mallaury, Tefit, Maurel, Thellier, Marc, Morosan, Serban Le Naour, Gilles, Capron, Frédérique, Suemizu, Hiroshi, Snounou, Georges, Moreno-Sabater, Alicia & Mazier, Dominique. 2015. Plasmodium falciparum full life cycle and Plasmodium ovale liver stages in humanized mice. *Nature Communications* 6(1):7690. DOI: 10.1038/ncomms8690

Talapko, J., Škrlec, I., Alebić, T., Jukić, M. & Včev, A. 2019. Malaria: The past and the present. *Microorganisms* 7(6):179.

Teofili, L., Pierlli, L. & Lovino, M. S. 1992. The combination of quercetin and cytosine arabinoside synergistically inhibits leukemic cell growth. *Leukemia Research* 16(5):497–503.

Thorat, V. H., Tamboli, F. A., Jadhav, A. S. & Chavan, R. 2021. Phytochemical analysis and antimicrobial activity of *Lantana camara*. *International Journal of Pharmaceutical Chemistry and Analysis* 8(4): 171–173.

Velagapudi, R., Kosoko, A. M. & Olajide, O. A. 2019. Induction of neuroinflammation and neurotoxicity by synthetic hemozoin. *Cell Molecular Neurobiology* 39:1187–1200.

Vitale, M., Di Matola, T. & Ďascoli, F. 2000. Iodide excess induces apoptosis in thyroid cells trough a p53-independent mechanism involving oxidative stress. *Endrocrinology Society* 141:598–605.

Vuk, I., Rajic, Z. & Zorc, B. 2008. Malaria, and antimalarial drugs. *Farm Glas* 64:51–60.

Walker, N., Nadjm, B. & Whitty, C. 2017. Malaria. *Medicine* 42:52–58.

Walter, M. & Marchesan, E. 2011. Phenolic compounds and antioxidant activity of rice. *Brazilian Archives of Biology and Technology* 54(1):371–377.

White, N. J. N., Pukrittayakamee, S., Hien, T. T. T., Faiz, M. A., Mokuolu, O. A. O. & Dondorp, A. A. M. 2014. Malaria. *The Lancet* 383:723–735.

WHO. 2018b. *Artemisinin Resistance and Artemisinin-Based Combination Therapy Efficacy: Status Report*. World Health Organization.

Wicht, K. J., Mok, S. & Fidock, D. A. 2020. Molecular mechanisms of drug resistance in *Plasmodium Falciparum* malaria. *Annual Review of Microbiology* 74:431–454.

World Health Organization. 2005. *National Policy on Traditional Medicine and Regulation of Herbal Medicines: Report of a WHO Global Survey*. World Health Organization.

Yoshida, M., Sakai, T., Hosokawa, N., Marui, N., Matsumoto, K., Fujioka, A., et al. 1990. The effect of quercitin on cell cycle progression and growth of human gastric cancer cells. *Federation of European Biochemical Societies* 260(1):10–13.

Yoshida, M., Yamamoto, M. & Nikaido, T. 1992. Quercitin arrests human leukemic T-cells in late G1 phase of the cell cycle. *Cancer Research* 52:6676–6681.

Zhao, F., Tang, Y. Z. & Liu, Z. Q. 2007. Protective effect of icariin on DNA against induced oxidative damage. *Journal of Pharmacy and Pharmacology* 59:1729–1732.

8 Traditional Claims and Scientific Validations of the Antimalarial Plant Species

Jamal Akhtar, Shah Alam, and Fouzia Bashir

8.1 INTRODUCTION

Malaria continues to cause morbidity and mortality on a large scale in tropical countries. It is an endemic disease that affected 229 million people and caused 409,000 deaths in 2019. There are four different species of *Plasmodium* capable of causing malaria in humans: *Plasmodium falciparum*, *Plasmodium vivax*, *Plasmodium ovale*, and *Plasmodium malariae*. *P. falciparum* is responsible for most of the mortality and morbidity associated with malaria disease. Disease control is based on early diagnosis and specific treatment with antimalarial drugs since no vaccines are commercially available to prevent the disease. Malarial vaccine has not yet been developed because the parasite keeps changing the interaction of the metabolic pathways during its life cycle. Drug chemotherapy has a strong historical link to the use of traditional plant infusions and other natural products in various cultures. The research based on such knowledge has yielded quinine from *Cinchona* species and artemisinin from *Artemisia annua*. Resistance to the mostly commonly used drug, artemisinin, the first-line drug class as antimalarial, is a threat to mankind. The alarming rate at which the parasite, particularly *Plasmodium falciparum*, has developed resistance to currently used antimalarial drugs makes it imperative to search for newer, more effective therapeutic agents. It must be noted that despite recent successes in rational drug design and synthetic chemistry techniques by pharmaceutical companies, natural products and, particularly, medicinal plants remain an important source of new drugs. Further traditional knowledge and usage of medicinal plant extracts to treat febrile illnesses offer the advantage of increased rate of success in identifying new antimalarial compounds. Also, it is both convenient and economical to have readily available cures in the vicinity of patients afflicted by a disease.

Plants are the principal source for about one-*fourth* of the prescribed drugs globally. For health care, compared to drugs, herbs are more preferably used as an ideal mode of treatment. Still others, along with conventional pharmaceuticals, use herbs as adjunct therapy. Several developing societies utilize traditional medicine, which is the sole system of health care that is accessible and inexpensive. Those making use of herbal medicines should be certain that the products they procure are harmless and contain specific contents, may it be a particular herb or amount of a specific herbal element (Wachtel-Galor and Benzie, 2011; Husen, 2021, 2022a, 2022b, 2023; Rahman and Husen, 2021, 2022, 2023a, 2023b). The existing traditional medicines and cultural practices for treating malaria and its associated symptoms are listed in Table 8.1.

DOI: 10.1201/9781003378396-8

TABLE 8.1
Existing Traditional Medicines and Cultural Practices for Treating Malaria

Plant	Traditional Use	Reference
Aegle marmelos	The decoction of leaves is to be taken orally.	Taek et al. (2018)
Allium sativum	Bulb of this plant can be used by preparation of concoction, tincture, and infusion	Odugbemi et al. (2007)
Alstonia boonei	Stem, root, leaves, and bark can be used for treating malaria. Infusion, tincture, and decoction can be prepared.	Odugbemi et al. (2007), Ogbuehi and Ebong (2015)
Anacadum occidentallis	Leaves are washed and boiled in water for 20–50 min.; 300 ml of the same is taken once a day for 1 week or until the symptoms disappear.	Omosun et al. (2013)
Andrographis paniculata	Decoction of whole plant can be used, and the same can also be used to have bath with.	Taek et al. (2018)
Andropogon schœnanthus/nardis	The leaf decoction is consumed with plenty of sugar, and the fumes are inhaled as a febrifuge.	Randrianarivelojosia et al. (2003)
Aspilla Africana, *Bambusa vulgaris*	Leaves are washed and pounded or squeezed, and the juice is then filtered. Until the fresh symptom disappears, the juice is rubbed on the forehead.	Omosun et al. (2013)
Azadirachta indica	Decoction of stem bark and leaves is used.	Odugbemi et al. (2007), Ogbuehi and Ebong (2015)
Azadirachta indica, *Carica papaya*, *Psidium guajava*, *Eucalyptus citsodora*, *Mangifera indica*	Leaves are sun-dried and then pounded and mixed with water, then filtered. For 7 days, 300 ml of filtrate is taken twice.	Omosun et al. (2013)
Azadirachta indica, *Ocimum gratissium*	Leaves are washed and pounded or squeezed, and the juice is filtered. The filtrate is consumed twice daily.	Omosun et al. (2013)
Azadirachta indica, *Psidium guajava*, *Carica papaya*	Leaves are washed and boiled in water for 30–50 min. For 7 days, 300 ml of the concoction is taken twice.	Omosun et al. (2013)
Bambusa vulgaris, *Uwaria chamae*, *Cymbopogon citratus*	Aqueous decoction of mixed fresh leaves is prepared, and 300 ml of the same is taken 4 times daily until malarial symptoms disappear.	Omosun et al. (2013)
Cactus opunita	Leaves are sun-dried and pounded; the mixture is poured into cold water and left to stand for 5 min and then filtered, and 300 ml of the decoction is taken daily.	Omosun et al. (2013)
Carica papaya	Fruit, seeds, and leaves of this plant are used as infusion.	Odugbemi et al. (2007), Ogbuehi and Ebong (2015)
Carica papaya, *Mangifera indica*	Leaves are washed and boiled in water for 30 min, and 300 ml of the concoction is taken twice daily for 10 days.	Omosun et al. (2013)
Carica papaya, *Psiduim guajava*, *Eucalyptus*, *Mangifera indica*	Leaves are washed and boiled in water for 20–50 min, and 300 ml of the concoction is taken two times for 10 days and the water is used for bathing through the duration of treatment.	Omosun et al. (2013)
Cinnamosma fragrans	A decoction of the leaf and bark is drunk (1 bowl, 3–4 times daily) to reduce malarial symptoms, tiredness, and muscular aches.	Randrianarivelojosia et al. (2003)
Citrus aurantifolia	Leaves are washed and boiled in water. The juice from the fruit is squeezed separately, and 300 ml of tisane is taken once a day and 10 ml spoonful of the squeezed juice is taken 3 times daily.	Omosun et al. (2013)

Citrus aurantifolia, *Cymbopogon citrates*, *Eucapyptus citrissodora*, *Psidium guajava*	Leaves are washed and boiled for 50 min, and 300 ml of the concoction is taken once a day for 7 days and the water used for bathing during the duration of treatment.	Omosun et al. (2013), Ogbuehi and Ebong (2015)
Curcuma longa	Rhizome of this plant is used in the form of tincture and decoction to reduce the malarial symptoms.	Odugbemi et al. (2007)
Cymbopogen citrates, *Eucalyptus citriodora*, *Phyllanthus amarus*	Fresh leaves are to be macerated in water, and 300 ml of extract is taken 3 or 4 times daily for 10 days.	Omosun et al. (2013)
Cymbopogon citrates	Leaves are washed, and boiled water is added to it and allowed to stand for 10–50 min. Then, 300 ml of the infusion is taken twice daily until symptoms disappear.	Omosun et al. (2013)
Enantia chlorantha	Decoction, tincture, and infusion of bark are used.	Odugbemi et al. (2007)
Ficus megapoda	A decoction of the leaf and bark is drunk (1 bowl, 3–4 times daily) to alleviate malarial symptoms.	Randrianarivelojosia et al. (2003)
Funtumia africana	Root of this plant can be used as decoction.	Odugbemi et al. (2007)
Garuga floribunda	Bark of this plant can be used as decoction.	Taek et al. (2018)
Gongronema latifolium	Leaves are washed and pounded or squeezed, and the juice is filtered. Then, 300 ml of the filtrate is taken until symptoms disappear.	Omosun et al. (2013)
Gongrononama latifolium	Leaves are washed and boiled in water for 20–50 min. The same is to be consumed 3 times daily until symptoms disappear.	Omosun et al. (2013)
Khaya grandifoliola	Bark can be used to make decoction and infusion.	Odugbemi et al. (2007)
Leconiodiscus cupanioides	Decoction of roots is used.	Odugbemi et al. (2007)
Magnifera indica, *Citrus aurantifolia*	Roots and leaves are boiled in water for 50 min, and 300 ml of extract is taken 3 or 4 times daily for 5 days.	Omosun et al. (2013), Ogbuehi and Ebong (2015)
Manihot esculentum	Roots are washed and eaten raw. The same is to be taken once daily for 5 days.	Omosun et al. (2013)
Morinda lucida	Stem, roots, and leaves are used as tincture and infusion.	Odugbemi et al. (2007), Ogbuehi and Ebong (2015)
Nauclea latifolia	Tincture and decoction of bark and roots are to be consumed.	Odugbemi et al. (2007)
Ocimum gratssimum	Leaves are washed and pounded or squeezed, and the juice is filtered. Filtrate is to be consumed until malaria symptoms disappear.	Omosun et al. (2013), Ogbuehi and Ebong (2015)
Pterocarpus spp.	Leaves are washed and boiled in water for 20–30 min, and 300 ml of the filtered tisane is to be taken 3 times daily for 3 days.	Omosun et al. (2013)
Pterocarpus santalinoides, *Caesalpinia*	Leaves are washed and boiled in water for 20–30 min, and the filtrate is taken twice daily.	Omosun et al. (2013)
Rauvolfia vomitoria	Decoction of roots is used to reduce malarial symptoms.	Odugbemi et al. (2007)
Sida acuta	Leaves and/or tender stem are boiled for 30 min. The tisane is strained, and half a glass of the same is to be consumed twice a day.	Ogbuehi and Ebong (2015)
Solanum spp.	Leaves are washed and pounded or squeezed, and the juice is filtered or eaten raw. The filtrate is taken twice daily or consumed raw for 1 week.	Omosun et al. (2013)
Solanum spp., *Pterocarpus santalimoides*	Leaves are washed and boiled in water for 5 min, and 300 ml of the concoction is taken for 2 weeks.	Omosun et al. (2013)

(Continued)

TABLE 8.1 (*Continued*)
Existing Traditional Medicines and Cultural Practices for Treating Malaria

Plant	Traditional Use	Reference
Tithonia diversifolia	Infusion of leaves is used.	Odugbemi et al. (2007)
Tristellateia madagascariensis	The leaf decoction is drunk (2–3 glasses a day) to treat malaria. Inhalation of steam from the decoction is also used as a febrifuge.	Randrianarivelojosia et al. (2003)
Uvaria chamae, *Caesalpinia pulcherima*	Roots and bark are boiled in water for 30–40 min, and 300 ml of the tisane is taken 2 times daily for 7 days.	Omosun et al. (2013)
Vernonia amydalina	Decoction of leaves is used to reduce the symptoms of malaria.	Odugbemi et al. (2007), Ogbuehi and Ebong (2015)
Vernonia amygdalina, *Gongronema latifolium*	Leaves are washed and pounded or squeezed, and the juice is filtered. Then, 300 ml of the concoction is taken for 10 days.	Omosun et al. (2013)
Zanthoxylum tsihanimposa	A decoction of the leaf and bark is used singly or mixed with a selection of plants from the following: *Vepris ampody, Dracaena reflexa, Gambeya boiviniana, Peddia involucrata*. The decoction is drunk (1 bowl, 3–4 times daily) to alleviate malarial symptoms, tiredness, muscular aches and pains, and poisoning.	Randrianarivelojosia et al. (2003)

8.2 EVIDENCE-BASED RESEARCH ON ANTIMALARIAL DRUGS

8.2.1 Antimalarial Activity *in Vitro* and *in Vivo*

Khalid et al. (1986) reported that the leaf and immature bark extracts of *Aegle marmelos* exhibited *in vitro* antimalarial activity (IC50 = 48.2 g/mL) against *Plasmodium falciparum*, and a new anthraquinone, 1-methyl-2-(3′-methyl-but-2′-enyloxy)-anthraquinone, isolated from seeds of *Aegle marmelos* was found to have significant antifungal activity (Mishra et al., 2010). Dichloromethane and methanol extracts of *Lantana camara* have been reported to exhibit *in vitro* antimalarial activity against the 3D7 (IC50 = 5.7 g/mL) and W2 (IC50 = 14.1 g/mL) strains of *Plasmodium falciparum* (Jonville et al., 2008). The leaves' ethanolic extracts of *Lantana* sp. have been reported to display good activity (IC50 < 17.5 g/mL) against *Plasmodium falciparum* (Valadeau et al., 2009). Flavonoids (linaroside and lantanoside) isolated from *Lantana camara* were found to be active against *Mycobacterium tuberculosis* (Begum et al., 2008). In previous reports, leaf ethyl acetate and methanol extract of *Leucas aspera* showed good antiplasmodial activity (IC50 = 7.81 and 22.76 g/mL) against *chloroquine* (CQ)-sensitive (3D7) strain of *Plasmodium falciparum* which was cultured following the candle jar method (Bagavan et al., 2011a), and chloroform extract of *Leuca scephalotes* (IC50 =3.96 g/mL) that showed good antiplasmodial activity against erythrocytic stages of *Plasmodium falciparum* was determined using a modified [3H]-hypoxanthine incorporation assay with the chloroquine- and pyrimethamine-resistant K1 strain (Dua et al., 2011). Sadhu et al. (2003) isolated eight lignans and four flavonoids from whole plant of *Leuca saspera* having prostaglandin inhibitory and antioxidant activities. Earlier, authors Mesia et al. (2008) reported that the methanol extract of *Momordica charantia* (whole plant) showed good antiprotozoal activity against *Trypanosoma brucei* (IC50 = 7g/mL) and *Trypanosoma cruzi* (IC50 = 37.0g/mL), whereas the same was not potent against *Plasmodium falciparum* (IC50 > 64 g/mL). In contrast, *Momordica charantia* aerial part of aqueous extract displayed a moderate *in vivo* activity (LD70 = 904 mg/kg of body weight) on the rodent malaria against *Plasmodium vinckeipetteri* (279BY) (Munoz et al., 2000). However, Amorim et al. (1991) reported that the oral administration of crude dried leaves ethanolic extracts of *Momordica charantia* was ineffective in lowering the parasitaemia level against *Plasmodium berghei*–infected mice at 500 mg/kg. Ling et al. (2009) isolated three compounds, (19S, 23E)-5beta,19-epoxy-19-methoxy-cucurbita-6,23-dien-3beta and 25-diol; (19R, 23E)-5beta,19-epoxy-19-methoxy-cucurbita-6,23-dien-3beta and 25-diol; and 3beta, 7beta,25-trihydroxycucurbita-5,23-dien-19-al-3-O-beta-d-glucopyranoside, from *Momordica charantia* having inhibitory effects against pests (*Liriomyza sativae*).

Boiling water extract of *Phyllanthus amarus* has been previously reported to be marginally active *in vitro* and *in vivo* against *Plasmodium falciparum* (IC50 80.11 g/mL) and *Plasmodium berghei*–infected mice (ED50 112.78 ± 32.32 mg/kg) (Traore et al., 2008). However, subsequent results indicate good antimalarial activity (IC50 = 15 g/mL) for the methanolic extract of leaves of *Phyllanthus amarus*. This difference in activity may be related to the solvents used in the two studies. The methanol extracts of *Phyllanthus niruri* have earlier been found to have significant *in vitro* antiplasmodial activity against *chloroquine*-resistant (FCR-3) and *chloroquine*-sensitive (D-10) strains of *Plasmodium falciparum* (IC50 = 2.3–3.9 and 3.2–2.5g/mL), respectively, and also significant *in vivo* activity against *Plasmodium berghei*–infected mice with an ED50 of 9.1 mg/kg/day (Mustofa Sholikhah and Wahyuono, 2007). Earlier, antiplasmodial activity was reported in the leaf ethyl acetate and methanol extract of *Phyllanthus emblica* (IC50 = 7.25 and 3.125g/mL, respectively) (Bagavan et al., 2011b). Finding of the presence of antimalarial activity in *Phyllanthus amarus* is supported by previous reports which indicated ellagic acid (formed by hydrolysis of tannins) as the active constituent of an ethanolic extract of the same plant (Dhooghe et al., 2011). Xiang et al. (2011) isolated a polyphenolic compound 1,2,4,6-tetra-O-galloyl-β-D-glucose (1246TGG) from *Phyllanthus emblica* having *in vitro* antiviral activities against herpes simplex virus type 1 (HSV-1) and type 2 (HSV-2). *Piper nigrum* has been used by South Indian traditional healers to treat fevers in general, malaria asthma, cold, intermittent fever, cholera, colic pain, and diarrhoea (Simpson and Ogorzaly, 1995). In our observation, the ethyl acetate seed extract of *Piper nigrum* showed promising *in vitro* antiplasmodial activity against *Plasmodium falciparum* 3D7 and INDO strains with IC50 values of 12.5 and 12.0 g/mL, respectively, with a low cytotoxicity (TC50 = 87.0 g/mL) and a significant therapeutic index of 7.0. Alkaloids piperine, guineensine, pipericide, N-feruloyltyramine and N-isobutyl-2E, and 4E-dodecadienamide have been isolated from *Piper nigrum*, and piperine has been reported as a stimulator of *in vitro* melanocyte proliferation (Lin et al., 2007). *Cassia auriculata* (flower), *Datura metel* (leaf), *Diospyros melanoxylon* (leaf), *Eclipta prostrata* (leaf), *Nelumbo nucifera* (leaf), and *Trachyspermum ammi* (seed) displayed moderate antimalarial activities. Previous reports demonstrate the antiplasmodial (Kamaraj et al., 2011), antidiabetic (Gupta et al., 2010), larvicidal (Kamaraj et al., 2009), and anticancer (Prasanna et al., 2009) effects of leaf extracts of *Cassia auriculata*. Juan-Badaturuge et al. (2011) isolated the antioxidant kaempferol-3-O-rutinoside, kaempferol, quercetin, and luteolin from *Cassia auriculata*. *Datura metel* has been well-known for its use in traditional Indian and Chinese systems of medicine for centuries as a narcotic antispasmodic useful in the treatment of cough, asthma, diarrhoea, and it has also been used as an analgesic, to control fever and kill parasites (Bonde, 2001; Parrotta, 2001; Rajesh Sharma, 2002). The aqueous extracts of *Datura metel* are known to exhibit moderate larvicidal activity against *filariasis* and avian malaria vector of *Culex quinque fasciatus* (Rahuman et al., 2008; Tandon and Sirohi, 2010). Three new withanolide glycosides, daturametelins, daturataturinA, and 7,27-dihydroxy-1-oxowitha-2,5,24-trienolide, were isolated from the methanol extract of the aerial parts of *Datura metel*. Withanolide 7,27-dihydroxy-1-oxowitha-2,5,24-trienolide has been reported to have anti–human colorectal carcinoma (HCT116) activity (IC50 = 3.2 M) (Ma et al., 2006). Kantamreddi and Wright (2008) reported four Indian *Diospyros* species, viz, *Diospyros melanoxylon*, *Diospyros peregrine*, *Diospyros sylvatica*, and *Diospyros tomentosa*, in which methanolic extracts were found to have significant activity (IC50 = 16.5–92.9 g/mL) against chloroquine-sensitive 3D7 strain of *Plasmodium falciparum* and similar activities against chloroquine-resistant K1 strains (IC50 = 20.5–121.6 g/mL). The compound 4-O-(3´-methylgalloyl) norbergenin isolated from the stem bark of *Diospyrossanza minika* was highly active against *Plasmodium falciparum* (IC50 = 0.6g/mL), whereas 4-O-galloylnorbergenin (IC50 = 3.9 g/mL), 11-O-phydroxy-benzoyl-norbergenin (IC50 = 4.9 g/mL), and norbergenin and 4-O-syringoylnorbergenin (IC50 ≥ 32g/mL) were less active (Tangmouo et al., 2010). *Eclipta prostrata* showed moderate antiplasmodial activity (IC50 = 30 g/mL) in the study. In previous reports, *Eclipta prostrata* has been studied for antileishmanial and larvicidal activities. Dasyscyphin C compound isolated from *Eclipta prostrata* showed good leishmanicidal activity (Khanna et al., 2009), and the leaf extracts showed good mosquito larvicidal effect (Khanna and Kannabiran, 2007; Elango et al., 2009). Bapna et al. (2007) reported moderate antimalarial activity at 750 mg/kg/day of methanolic leaf extract of *Eclipta alba* against *Plasmodium berghei* ANKA strain in mice. Orobol, isolated from *Eclipta prostrata* (Tewtrakul et al., 2011), exhibited anti-inflammatory effect. In the present investigation, leaf ethyl acetate extract *Nelumbo nucifera* showed low activity whereas methanol extract showed moderate activity against chloroquine-sensitive (3D7) strain of *Plasmodium falciparum* (IC50 = 69 and 35g/mL) when compared with earlier report. *Nelumbo nucifera* has been used for various medicinal purposes in oriental medicine, and two compounds, (R)-roemerine and N-methylasimilobine, isolated from its leaf exhibited good *in vitro* antimalarial activity (IC50 = 0.2 g/mL) (Agnihotri et al., 2008). *Armepavine*, a compound from *Nelumbo nucifera*, was found to exert immunosuppressive effects on T-lymphocytes in lupus nephritic mice (Weng et al., 2009). Liensinine, negferine, and isoliensinine isolated from the leaves and embryo of *Nelumbo nucifera* showed potent anti-HIV activity (IC50: 0.8g/mL) (Kashiwada et al., 2005). Seed ethyl acetate extract of *Trachyspermum ammi* has shown moderate *in vitro* antiplasmodial activity against CQ-sensitive strain of *Plasmodium falciparum* (3D7), with IC50 values of 25g/mL. Previous authors also reported the seed cyclohexane, methylene chloride, and methanol extract of *Trachyspermum ammi* to have good *in vitro* antimalarial activity (IC50: 8.75–9.23 g/mL) (Chenniappan and Kadarkarai, 2010). A novel compound, (4aS, 5R, 8aS) 5,8a-di-1-propyl-octahydronaphthalen-1-(2H)-one, isolated from the seed of *Trachyspermum ammi* showed good antibiofilm and antiadherence activities (Khan et al., 2010), whereas a phenolic monoterpene isolated from the same source was found to have antifilarial activity (Mathew et al., 2008). Leaf methanolic extracts of *Andrographis lineata*, leaf ethyl acetate extracts of *Argemone Mexicana, Nelumbo nucifera*, stem of *Phyllanthus amarus*, and rhizome of *Zingiberzer umbet*, seed ethyl acetate extract of *Acacia concinna*, leaf methanol extract of *Adhatoda vasica*, leaf ethyl acetate extract of *Cassia nictitans*, leaf methanol extract of *Cocculus hirsutus*, seed methanol extract of *Coriandrum sativum*, leaf ethyl acetate extract of *Cuminum cyminum*, methanol extract of leaf and fruit of *Solanum trilobatum* displayed comparatively low antimalarial activity.

A. squamosa L. (Annonaceae), commonly known as custard apple, is a native of West Indies and is cultivated throughout India mainly for its edible fruit. The major constituents identified in the members of the Annonaceae are, typically, acetogenins (Bermejo et al., 2005; Liu et al., 1999; Zeng et al., 1996) that are known to exhibit *in vitro* antimalarial activities (Rakotomanga et al., 2004). Earlier moderate antiplasmodial activity (IC50~30 mg/ml) was reported in the ethyl acetate and methanol extracts of the leaves of *A. squamosa* (Asq) (Bagavan et al., 2011c). Similarly, the methanolic extracts of the leaves of have been reported to show high to moderate antiplasmodial activity against the chloroquine-sensitive strain 3D7 (IC50: 2 mg/ml) and chloroquine-resistant strain Dd2 (IC50: 30 mg/ml) of *P. falciparum* (El Tahir et al., 1999). Later on, high antiplasmodial activity has been observed in ethanolic extract of *A. squamosa* against both CQ-sensitive 3D7 (IC50: 2.1 mg/ml) and CQ-resistant INDO (IC50: 3.3 mg/ml) strains of *P. falciparum*, with resistance index of 1.57.

A. indica is widely used as a mosquito repellent; moderate antiplasmodial activity has been reported in leaf extract (IC50 32 mg/ml) and seed oil (IC50: 62 mg/ml) of *A. indica*. *B. pinnatum* (Lam.) Oken (Crassulaceae) is a perennial herb native to Madagascar and used in folkloric medicine in tropical Africa, tropical America, India, China, and Australia. In traditional medicine, the leaves of *B. pinnatum* (Bp) have been used for their antimicrobial and antifungal properties. Singh et al. (2015) observed good antiplasmodial activity in leaf ethanol extracts of *B. pinnatum* {IC50 14 mg/ml (Pf3D7) and 29 mg/ml (Pf INDO)}. *C. procera* (Aiton) Dryand. belongs to Apocynaceae, the milkweed family. *C. procera* (Cp) continues to be in traditional use as purgative and medicine for intermittent fevers. Chemical analysis of the plant shows the presence of cardenolides like Calotropin. The *in vitro P. falciparum* schizonticidal activity of the ethanolic extracts of different parts of *C. procera* has been found to be moderate to poor (IC50 0.11–1.2 mg/ml) (Sharma and Sharma, 2000). However, in the present study, it is observed that the ethanolic leaf extract of *C. procera* has high antiplasmodial activity (IC50: 2.5 mg/ml) with low toxicity (TC50: 10 mg/ml). It is interesting to note that *Calotropis gigantea*, another species of the same genus, has also been reported to be effective against *P. falciparum* 3D7 strain (Wong et al., 2011). *C. carvi* L., also known as caraway, meridian fennel, Shahi-Jeera, etc., is a biennial plant of family Apiaceae and is native to western Asia, Europe, and Northern Africa. Fruits of *C. carvi* (Cc) are used in Indian families as part of spices in various cuisines. Traditional healers recommend it for treatment of inflammation, anorexia, febrile diseases, etc. The oil of caraway derived from Cc fruits is known to possess insecticidal (Pitasawat et al., 2007) and antiplasmodial (Fujisaki et al., 2012) properties. In the present study, Singh et al. (2015) found that the ethanolic extract of caraway seed possesses good antiplasmodial activity (IC50: 27 mg/ml (Pf3D7), 30 mg/ml (PfINDO)), with resistance index of 1.1. *D. sissoo* DC (Leguminosae) possesses immense traditional applications. It is widely used as analgesic, anti-inflammatory, antipyretic, antimicrobial, antidiarrhoeal, antiulcerogenic, spermicidal, larvicidal, and mosquito repellent in the traditional system of medicines. However, heartwood flavonoids of *Dalbergia louvelii* species of genus *Dalbergia* are found to have antiplasmodial activity (IC50 5.8–8.7 mM) (Beldjoudi et al., 2003). Here it is interesting that for the first time it is reported by Singh et al. (2015) that ethanolic root extract of *D. sissoo* (Ds) possesses good antiplasmodial activity {IC50 20 mg/ml (Pf3D7), 14.5 mg/ml (PfINDO)}, with resistance index of 0.73. In contrast, the leaf ethanolic extract of Ds was found to be inactive (IC50: 4,100 mg/ml). *C. citriodora* (synonym *Eucalyptus citriodora*), also called lemon eucalyptus, belongs to the family Myrtaceae. Its lemon-scented oil has insecticidal, antifungal, antiseptic, and anti-inflammatory properties.

Chemical analysis of the plant shows the anti-inflammatory eucalyptol/cineole to be the major constituent (Gbenou et al., 2013). In the present study by Singh et al. (2015), for the first time, promising antiplasmodial activity {IC50 5 mg/ml (Pf 3D7)} has been observed in ethanolic extract of the leaves of *C. citriodora*. *E. prostrata* (L.) L. (synonym *Eclipta alba*), popularly known as 'Bhringaraj' in India, plays a significant role in the Ayurvedic and Unani systems of medicine. It has been reported that 750 mg/kg/day of methanol extract of *E. alba* increased the survival time of *P. berghei* ANKA–infected mice (Bapna et al., 2007). In the present study, Singh et al. (2015) observed that ethanol extract of Ep inhibits the growth of *P. falciparum* {IC50 23 mg/ml(Pf 3D7)}, indicating direct action on the parasite. *H. pubescens* Wall. ex G. Don (synonym *Holarrhena antidysenterica*) (Apocynaceae) is well-known traditionally for curing dysentery and diarrhoea. Chemical analysis of *H. pubescens* (Hp) shows mostly a rich abundance of alkaloids. The principal alkaloid is conessine, having structural similarity to quinine and quinidine. The whole plant chloroform extract of *H. antidysenterica* has been found to show *in vitro* antiplasmodial activity (IC50: 5.5 µg/ml) (Verma et al., 2011). Singh et al. (2015) found that the ethanolic extract of Hp bark is slightly more potent {IC50 4.5 µg/ml (Pf 3D7)} than the ethanolic extract of Hp leaves {IC50 7 µg/ml(Pf 3D7)}. However, the potency order was reversed in the case of Cq-resistant Pf INDO strain, where the leaves ethanol extract (IC50 8.5 µg/ml) was more potent than the bark ethanol extract (IC50 11 µg/ml). *M. longifolia* var. latifolia (Roxb.) A. Chev. (synonym *Madhuca indica*) (Mahua) (Sapotaceae) is a forest-based tree that is virtually the lifeline of the tribal belt in Central India. *M. longifolia* (Ml) is culturally the most identified with Indian life in the plains. The cake of Ml is insecticidal and is also used as bait for fishing. This plant is also endowed with antifungal (Sidhu et al., 2009) and antioxidant (Chaudhary et al., 2012) activity. It was interesting to observe antiplasmodial activity in the bark, leaf, and seed of Ml. The bark was found to be most potent {IC50 24.1 µg/ml (Pf 3D7)}, followed by the leaves {IC50 55 µg/ml (Pf 3D7)} and seeds {(IC50 92 µg/ml (Pf 3D7)}. Seed oil was found not active up to 100 µg/ml. *M. pudica* L. (Leguminosae), commonly called touch-me-not, or Chuimui, is used traditionally as antidote for scorpion sting and as an antidiabetic. According to the Unani system of medicine, *M. pudica* (Mp) root is useful in diseases arising from blood impurities and bile, bilious fevers, piles, jaundice, leprosy, etc. (Azmi et al., 2011). Laboratory experiments have demonstrated that Mp has antimicrobial, antimyotoxic, wound healing, antiasthmatic, anticancer, and anti-mumps activities. Singh et al. (2015), in the present study, found Mp to have moderate antiplasmodial activity (IC50: 51 mg/ml). *P. pinnata* (L.) Pierre (Leguminosae) is a handsome flowering tree with foliage like the beech. Sanskrit writers call it Karanja and Naktamāla or Naktamālaka, 'garland of the night'. Flavonoids isolated from Pp have exhibited moderate antileishmanial activity (IC50: 32 mM) (Kapingu et al., 2006). Ethanolic extract of Pp has also been reported to have moderate antiplasmodial activity against CQ-sensitive Pf3D7 (IC50: 24–40 mg/ml) (Simonsen et al., 2001). Singh et al. (2015), in the present study, found that the leaf ethanolic extract of Pp is nearly equipotent against both CQ-sensitive 3D7 and CQ-resistant INDO strain (IC50s: 22.8 and 20.4 mg/ml, respectively). In contrast, the ethanolic bark extract is more potent against the sensitive than the resistant strain (IC50s: 9.5 and 22.5 mg/ml, respectively). *P. hysterophorus* L. is a wild-growing weed belonging to the family Compositae and is a common invasive species in India. Hydroethanolic (80% ethanol) extract of *P. hysterophorus* (Ph) has been reported to have weak antiplasmodial activity (IC50: 45.2 mg/ml) (Valdes et al., 2010). However, the present study, by Singh et al. (2015) found that the whole plant ethanol extract of Ph has promising antiplasmodial activity against both CQ-sensitive {IC50 6 µg/ml (Pf 3D7)} and CQ-resistant strain {IC50 2.1 µg/ml (Pf INDO)}, with resistance index of 0.35. In view of its impressive antiplasmodial potency, the ethanolic extract of Ph was subjected to activity-guided purification using RPHPLC. A well-resolved chromatogram together with assessment of antiplasmodial activity of fractions helped identify the fractions with promising antiplasmodial activity and low toxicity to two mammalian cell lines. Evaluation for stage specificity of action by two of the most potent fractions, viz, 14 and 19, revealed that they were inhibitory to the parasite growth in a dose-dependent, stage-specific manner. Thus, while 3.125 mg/ml was sufficient to inhibit the growth of trophozoites, the growth inhibition of rings was seen only at 6.25 mg/ml. Furthermore, schizonts treated with 6.25 mg/ml of fractions 14/19 showed delayed egress, resulting in merozoites that were invasion-incompetent. Finally, it was interesting to observe in drug exposure followed by drug withdrawal experiments that 14/19 killed all stages of the malaria parasite. The molecule in the dominant peak representing fraction 14 is likely to be amphipathic since its RPHPLC retention time is approximately midway between the retention times expected of very polar versus very nonpolar molecules. Further, its 20-fold greater absorbance at 214 nm over the absorbance at 254 nm suggests that the molecule may be aliphatic in nature. Indeed, chemical analysis of Ph shows sesquiterpene lactones to be the major chemical constituent in all parts, including pollens and trichomes (Picman et al., 1979).

Some of the other chemical constituents found in Ph and reported to have antiprotozoan, antibacterial, and antitumour properties are parthenin/hymenin, coronopilin, ambrosin, tetraneurin A, and charminarone. Since a large number of fractions of Ph have been found to exhibit promising antiplasmodial activity, it is quite likely that the fractions may have some of the compounds listed (Pandey, 2009). *S. indica* L (Ashoka) is one of the most legendary and sacred trees of India. It is an evergreen tree used extensively as a tonic by local people, as a uterine sedative and for women-related problems like menorrhagia (scant menses) and dysmenorrhoea (painful menses, menstrual cramps), and depression, bleeding haemorrhoids, and uterine fibroids. A study by Singh et al. (2015) on the antiplasmodial activity of the ethanolic extracts of Si against Pf 3D7 found the leaf extract to be more potent (IC50 21 µg/ml) than the bark extract (IC50 38 µg/ml). *C. roseus* (L.) G. Don (synonym *Vinca rosea*) (Apocynaceae) is also known as Nayantara and is widely used by traditional healers as cure of many ailments. While the hot-water extract of dried entire plant is taken orally by human adults for cancer and menorrhagia, the hot-water extract of dried leaves is taken orally for Hodgkin's disease. Leaves methanol extract of Cr has been reported to be not active against *Plasmodium* up to a concentration of 100 mg/ml (Kamaraj et al., 2012). However, the present study found that the ethanolic extract of the whole plant possesses good antiplasmodial activity (IC50 16 mg/ml) against chloroquine-sensitive strain Pf3D7. *Z. jujuba* Mill. (synonym Zizyphus mauritiana) (also known as Ber) belongs to Rhamnaceae. Phytochemical investigations have established the genus *Ziziphus* to be a rich source of cyclopeptide alkaloids (El-Seedi et al., 2007). Cyclopeptide macrocycles of *Ziziphus* species showed interesting biological properties, including sedative, analgesic, antibacterial, antifungal, and antiplasmodial (Suksamrarn et al., 2006) activities. Cyclopeptide alkaloids isolated from the root of *Z. mauritiana* have been found to exhibit potent antiplasmodial activity (IC50s: 3.7–10.3 µM). Singh et al. (2015) tested the antiplasmodial activity of ethanolic extracts of *Z. jujuba* leaves and bark and found that while the leaves extract is inactive (IC50 Pf 3D7: 4,100 µg/ml), the bark extract is active {IC50: 25 µg/ml (Pf 3D7), 16 µg/ml(Pf INDO)}.

Antimalarial potential of *Alstonia scholaris* has been proven scientifically by testing it against *Plasmodium falciparum* K1 strain, with a very high activity category. The bisalole alkaloids, namely, villalstonine and macrocarpamine, are the bioactive components of these plants responsible for these activities (Keawpradub et al., 1999). Rezeki et al. (2012) showed the use of *A. scholaris* in effectively reducing the levels of malaria parasites in some patients clinically. This treatment was also proven to reduce the clinical symptoms of malaria, such as fever, headache, muscle aches, and diarrhoea.

Teng et al. (2019) carried out a scientific research on the potential use of *Carica papaya* as an antimalarial agent, testing it against *P. falciparum* 3D7 parasites and Dd2 strains. The results showed an antimalarial activity in the medium category. The compound responsible for this activity is carpaine, which belongs to the alkaloids group. Several studies have also proven the plant to have some anti-inflammatory, antitumour, and antihypertensive properties as well.

Andrographis paniculata is one of the herbs within the family of Acanthaceae, widely cultivated in South and South-East Asia. The andrographolide and its derivatives are the bioactive chemical contents of *A. paniculata*. Andrographolide is a diterpenic lactone compound known with its antiviral, anti-inflammatory, and anticancer properties (Bharati et al., 2011). According to Mishra et al.

(2011), andrographolide compounds have antimalarial activity, both *in vitro* (*P. falciparum*) and *in vivo* (*P. berghei*).

8.3 FUTURE REMARKS

This chapter examined the ethnobotanical uses with brief phytochemistry and pharmacological activities and clinical trials of two traditional medicinal plants, *Glycyrrhiza glabra* L. (liquorice) and *Gymnema sylvestre* R. Br. (Gurmar). The plant *G. glabra* has a strong ethnobotanical history.

In Europe and in Eastern countries, both root and rhizome are used as folklore medicine, whereas in Unani system of medicine, only rhizome is used as medicine. It possesses a wide range of pharmacological actions, like anticancer, antimutagenic, antiulcer, antifungal, and antioxidant, due to the presence of active chemical constituents like glycyrrhizin, glabridin, glycyrrhetic acid, etc.

This plant has been in use since the ancient times and has been mentioned in traditional pharmacopoeias. The other important medicinal plant is *Gymnema sylvestre* R. Br., which has antidiabetic, anti-inflammatory, antioxidant, hepatoprotective, antibacterial, and lipid-lowering activities, and the chief chemical constituents are gymnemanol, gurmarin, saponins, flavonol, and glycosides. This plant is a cheap substitute of antidiabetic drug, so people are using it without any prior knowledge of which part of this plant is to be used, which leads to the redundant ruining of the whole plant. Due to overexploitation and overusage to meet the requirement, this plant has been disappearing very fast. Hence, legal production and thorough monitoring of medicinal preparation should be done. Further comprehensive scientific studies regarding the mechanism of action of these medicinal plants, their phytochemical compounds, drug interaction, effective dose determination, and side effects, are required.

This chapter also explored the antimalarial potential of various medicinal plants with their traditional usage in malaria. Plants like *Allium sativum*, *Andrographis paniculata*, *Azadirachta indica*, *Ocimum gratissium*, etc. have been used for medicinal purposes from the beginning of time, and it is referenced in conventional pharmacopoeias for various diseases, including malaria also. It is necessary to do further in-depth scientific research on the mechanism of action of antimalarial plants, their phytochemical components, medication interactions, effective dose estimation, and adverse effects.

REFERENCES

Agnihotri, V.K., ElSohly, H.N., Khan, S.I., Jacob, M.R., Joshi, V.C., Smillie, T., Khan, I.A., Walker, L.A., 2008. Constituents of Nelumbo nucifera leaves and their antimalarial and antifungal activity. *Phytochemistry Letters* 1, 89–93.

Amorim, C.Z., Marques, A.D., Cordeiro, R.S., 1991. Screening of the antimalarial activity of plants of the Cucurbitaceae family. *Memórias do Instituto Oswaldo Cruz* 2, 177–180.

Azmi, L., Singh, M.K., Akhtar, A.K., 2011. Pharmacological and biological overview on Mimosa pudica Linn. *International Journal of Pharmacy & Life Sciences* 2, 1226–1234.

Bagavan, A., Rahuman, A.A., Kamaraj, C., Kaushik, N.K., Mohanakrishnan, D., Sahal, D., 2011a. Antiplasmodial activity of botanical extracts against Plasmodium falciparum. *Parasitology Research* 108, 1099–1109.

Bagavan, A., Rahuman, A.A., Kaushik, N.K., Sahal, D., 2011b. In vitro antimalarial activity of medicinal plant extracts against Plasmodium falciparum. *Parasitology Research* 108, 15–22.

Bagavan, A., Rahuman, A.A., Kaushik, N.K., Sahal, D., 2011c. In vitro antimalarial activity of medicinal plant extracts against Plasmodium falciparum. *Parasitology Research* 108, 15–22.

Bapna, S., Adsule, S., Shirshat Mahendra, S., Jadhav, S., Patil, L.S., Deshmukh, R.A., 2007. Anti-malarial activity of Eclipta alba against Plasmodium berghei infection in mice. *The Journal of Communicable Diseases* 39, 91–94.

Begum, S., Wahab, A., Siddiqui, B.S., 2008. Antimycobacterial activity of flavonoids from Lantana camara Linn. *Natural Product Research* 22, 467–470.

Beldjoudi, N., Mambu, L., Labaïed, M., Grellier, P., Ramanitrahasimbola, D., Rasoanaivo, P., Martin, M.T., Frappier, F., 2003. Flavonoids from Dalbergia louvelii and their antiplasmodial activity. *Journal of Natural Products* 66, 1447–1450.

Bermejo, A., Figadere, B., Zafra-Polo, M.C., Barrachina, I., Estornell, E., Cortes, D., 2005. Acetogenins from annonaceae: Recent progress in isolation, synthesis and mechanisms of action. *Natural Product Reports* 22, 269–303.

Bharati, B.D., Sharma, P.K., Kumar, N., Dudhe, R., Bansal, V., 2011. Pharmacological activity of Andrographis paniculata: A brief review. *Pharmacologyonline* 2, 1–10.

Bonde, K., 2001. The genus *Datura*: From research subject to powerful hallucinogen. *Journal of Ethanobotanical Leaflet* 29, 335–336.

Chaudhary, A., Bhandari, A., Pandurangan, A., 2012. Antioxidant potential and total phenolic content of methanolic bark extract of Madhuca indica (koenig) Gmelin. *Ancient Science of Life* 31, 132–136.

Chenniappan, K., Kadarkarai, M., 2010. In vitro antimalarial activity of traditionally used Western Ghats plants from India and their interactions with chloroquine against chloroquine-resistant *Plasmodium falciparum*. *Parasitology Research* 107, 1351–1364.

Dhooghe, L., Meert, H., Cimanga, R.K., Vlietinck, A.J., Pieters, L., Apers, S., 2011. The quantification of ellagic acid in the crude extract of Phyllanthus amarus Schum. & Thonn. (Euphorbiaceae). *Phytochemical Analysis* 22, 361–366.

Dua, V.K., Verma, G., Agarwal, D.D., Kaiser, M., Brun, R., 2011. Antiprotozoal activities of traditional medicinal plants from the Garhwal region of North West Himalaya, India. *Journal of Ethnopharmacology* 136, 123–128.

El-Seedi, H.R., Zahra, M.H., Goransson, U., Verpoorte, R., 2007. Cyclopeptide alkaloids. *Phytochemistry Reviews* 6, 143–165.

El Tahir, A., Satti, G.M., Khalid, S.A., 1999. Antiplasmodial activity of selected Sudanese medicinal plants with emphasis on Maytenus senegalensis (Lam.) Exell. *Journal of Ethnopharmacology* 64, 227–233.

Elango, G., Rahuman, A.A., Bagavan, A., Kamaraj, C., Zahir, A.A., Venkatesan, C., 2009. Laboratory study on larvicidal activity of indigenous plant extracts against *Anopheles subpictus* and *Culex tritaeniorhynchus*. *Parasitology Research* 104, 1381–1388.

Fujisaki, R., Kamei, K., Yamamura, M., Nishiya, H., Inouye, S., Takahashi, M., Abe, S., 2012. In vitro and in vivo anti-plasmodial activity of essential oils, including hinokitiol. *The Southeast Asian Journal of Tropical Medicine and Public Health* 43, 270–279.

Gbenou, J.D., Ahounou, J.F., Akakpo, H.B., Laleye, A., Yayi, E., Gbaguidi, F., BabaMoussa, L., Darboux, R., Dansou, P., Moudachirou, M., 2013. Phytochemical composition of Cymbopogon citratus and Eucalyptus citriodora essential oils and their anti-inflammatory and analgesic properties on Wistar rats. *Molecular Biology Reports* 40, 1127–1134.

Gupta, S., Sharma, S.B., Singh, U.R., Bansal, S.K., Prabhu, K.M., 2010. Elucidation of mechanism of action of Cassia auriculata leaf extract for its antidiabetic activity in streptozotocin-induced diabetic rats. *Journal of Medicinal Food* 13, 528–534.

Husen, A., 2021. *Traditional Herbal Therapy for the Human Immune System*. CRC Press, Boca Raton, FL. https://doi.org/10.1201/9781003137955

Husen, A., 2022a. *Herbs, Shrubs and Trees of Potential Medicinal Benefits*. CRC Press, Boca Raton, FL. https://doi.org/10.1201/9781003205067

Husen, A., 2022b. *Exploring Poisonous Plants: Medicinal Values, Toxicity Response and Other Potential Uses*. CRC Press, Boca Raton, FL. https://doi.org/10.1201/b23017

Husen, A., 2023. *Antidiabetic Medicinal Plants and Herbal Treatments*. CRC Press, Boca Raton, FL.

Jonville, M.C., Kodja, H., Humeau, L., Fournel, J., De Mol, P., Cao, M., Angenot, L., Frédérich, M., 2008. Screening of medicinal plants from Reunion Island for antimalarial and cytotoxic activity. *Journal of Ethnopharmacology* 120, 382–386.

Juan-Badaturuge, M., Habtemariam, S., Thomas, M.J.K., 2011. Antioxidant compounds from a South Asian beverage and medicinal plant, *Cassia auriculata*. *Food Chemistry* 125, 221–225.

Kamaraj, C., Bagavan, A., Rahuman, A.A., Zahir, A.A., Elango, G., Pandiyan, G., 2009. Larvicidal potential of medicinal plant extracts against *Anopheles subpictus* Grassi and Culex tritaeniorhynchus Giles (Diptera: Culicidae). *Parasitology Research* 104, 1163–1171.

Kamaraj, C., Kaushik, N.K., Mohanakrishnan, D., Elango, G., Bagavan, A., Zahir, A.A., Rahuman, A.A., Sahal, D., 2011. Antiplasmodial potential of medicinal plant extracts from Malaiyur and Javadhu hills of South India. *Parasitology Research*. https://doi.org/10.1007/s00436-011-2457-6.

Kamaraj, C., Kaushik, N.K., Mohanakrishnan, D., Elango, G., Bagavan, A., Zahir, A.A., Rahuman, A.A., Sahal, D., 2012. Antiplasmodial potential of medicinal plant extracts from Malaiyur and Javadhu hills of South India. *Parasitology Research* 111, 703–715.

Kantamreddi, V.S., Wright, C.W., 2008. Investigation of Indian *Diospyros* species for antiplasmodial properties. *Evidence-based Complementary and Alternative Medicine* 5, 187–190.

Kapingu, M.C., Mbwambo, Z.H., Moshi, M.J., Magadula, J.J., Cos, P., Berghe, D.V., Maes, L., Theunis, M., Apers, S., Pieters, L., Vlietinck, A., 2006. A novel isoflavonoid from Millettia puguensis. *Planta Medica* 72, 1341–1343.

Kashiwada, Y., Aoshima, A., Ikeshiro, Y., Chen, Y.P., Furukawa, H., Itoigawa, M., Fujioka, T., Mihashi, K., Cosentino, L.M., Morris-Natschke, S.L., Lee, K.H., 2005. Anti-HIV benzyl-isoquinoline alkaloids and flavonoids from the leaves of *Nelumbo nucifera*, and structure–activity correlations with related alkaloids. *Bioorganic and Medicinal Chemistry* 13, 443–448.

Keawpradub, N., Kirby, G.C., Steele, J.C.P., Houghton, P.J., 1999. Antiplasmodial activity of extracts and alkaloids of three Alstonia species from Thailand. *Planta Medica* 65(8), 690–694.

Khalid, S.A., Farouk, A., Geary, T.G., Jensen, J.B., 1986. Potential antimalarial candidates from African plants: An in vitro approach using Plasmodium falciparum. *Journal of Ethnopharmocology* 15, 201–209.

Khan, R., Zakir, M., Khanam, Z., Shakil, S., Khan, A.U., 2010. Novel compound from *Trachyspermum ammi* (Ajowan caraway) seeds with anti-biofilm and anti-adherence activities against *Streptococcus mutans*: A potential chemotherapeutic agent against dental caries. *Journal of Applied Microbiology* 109, 2151–2159.

Khanna, V.G., Kannabiran, K., 2007. Larvicidal effect of *Hemidesmus indicus, Gymnema sylvestre,* and *Eclipta prostrata* against *Culex qinquifaciatus* mosquito larvae. *African Journal of Biotechnology* 6, 307–311.

Khanna, V.G., Kannabiran, K., Getti, G., 2009. Leishmanicidal activity of saponins isolated from the leaves of *Eclipta prostrata* and *Gymnema sylvestre*. *Indian Journal of Pharmacology* 41, 32–35.

Lin, Z., Liao, Y., Venkatasamy, R., Hider, R.C., Soumyanath, A., 2007. Amides from Piper nigrum L. with dissimilar effects on melanocyte proliferation in vitro. *Journal of Pharmacy and Pharmacology* 59, 529–536.

Ling, B., Xiang, Y.L., Wang, G.C., Chen, S.H., Zhang, M.X., 2009. Antifeedant and antioviposition activities of Momordica charantia leaf ethanol extract against Liriomyza sativae. *The Journal of Applied Ecology* 20, 836–842.

Liu, X.X., Alali, F.Q., Pilarinou, E., McLaughlin, J.L., 1999. Two bioactive monotetrahydrofuran acetogenins, annoglacins A and B, from Annona glabra. *Phytochemistry* 50, 815–821.

Ma, L., Xie, C.M., Li, J., Lou, F.C., Hu, L.H., 2006. Daturametelins H, I, and J: Three new withanolide glycosides from Datura metel L. *Chemistry and Biodiversity* 3, 180–186.

Mathew, N., Misra-Bhattacharya, S., Perumal, V., Muthuswamy, K., 2008. Antifilarial lead molecules isolated from *Trachyspermum ammi*. *Molecules* 13, 2156–2168.

Mesia, G.K., Tona, G.L., Nanga, T.H., Cimanga, R.K., Apers, S., Cos, P., Maes, L., Pieters, L., Vlietinck, A.J., 2008. Antiprotozoal and cytotoxic screening of 45 plant extracts from Democratic Republic of Congo. *Journal of Ethnopharmacology* 115, 409–415.

Mishra, B.B., Kishore, N., Tiwari, V.K., Singh, D.D., Tripathi, V., 2010. A novel antifungal anthraquinone from seeds of Aegle marmelos Correa (family Rutaceae). *Fitoterapia* 81, 104–107.

Mishra, K., Dash, A.P., Dey, N., 2011. Andrographolide: A novel antimalarial diterpene lactone compound from Andrographis paniculata and its interaction with curcumin and artesunate. *Journal of Tropical Medicine* [Internet] 2011, 579518. http://www.ncbi.nlm.nih.gov/pubmed/21760808

Munoz, V., Sauvain, M., Bourdy, G., Callapa, J., Rojas, I., Vargas, L., Tae, A., Deharo, E., 2000. The search for natural bioactive compounds through a multidisciplinary approach in Bolivia. Part II. Antimalarial activity of some plants used by Mosetene Indians. *Journal of Ethnopharmacology* 69, 139–155.

Mustofa Sholikhah, E.N., Wahyuono, S., 2007. In vitro and in vivo antiplasmodial activity and cytotoxicity of extracts of Phyllanthus niruri L. herbs traditionally used to treat malaria in Indonesia. *Southeast Asian Journal of Tropical Medicine and Public Health* 38, 609–615.

Odugbemi, T.O., Akinsulire, O.R., Aibinu, I.E., Fabeku, P.O., 2007. Medicinal plants useful for malaria therapy in Okeigbo, Ondo State, Southwest Nigeria. *African Journal of Traditional, Complementary and Alternative Medicines* 4, 191–198. https://doi.org/10.4314/ajtcam.v4i2.31207

Ogbuehi, I.H., Ebong, O.O., 2015. Traditional medicine treatment of malaria in Onitsha, South East Nigeria. *Greener Journal of Medical Sciences* 5, 011–018. https://doi.org/10.15580/gjms.2015.1.051114384

Omosun, G., Okoro, I.A., Ekundayo, E., Ojimelukwe, P., Ibe, O., 2013. Ethnobotanical study of medicinal plants useful for malaria therapy in eight local government areas of Abia State, Southeast Nigeria. *Advancement in Medicinal Plant Research* 1, 39–44.

Pandey, D., 2009. Allelochemicals in Parthenium in response to biological activity and the environment. *Indian Journal of Weed Science* 41, 111–123.

Parrotta, J.A., 2001. *Healing Plants of India*. CABI Publishing, Wallingford, pp. 917–919.

Picman, A.K., Rodriguez, E., Towers, G.H., 1979. Formation of adducts of parthenin and related sesquiterpene lactones with cysteine and glutathione. *Chemico Biological Interactions* 28, 83–89.

Pitasawat, B., Champakaew, D., Choochote, W., Jitpakdi, A., Chaithong, U., Kanjanapothi, D., Rattanachanpichai, E., Tippawangkosol, P., Riyong, D., Tuetun, B., Chaiyasit, D., 2007. Aromatic plant-derived essential oil: An alternative larvicide for mosquito control. *Fitoterapia* 78, 205–210.

Prasanna, R., Harish, C.C., Pichai, R., Sakthisekaran, D., Gunasekaran, P., 2009. Anticancer effect of Cassia auriculata leaf extract in vitro through cell cycle arrest and induction of apoptosis in human breast and larynx cancer cell lines. *Cell Biology International* 33, 127–134.

Rahman, S., Husen, A., 2021. Potential role of medicinal plants in the cure of liver and kidney diseases. In: *Non-Timber Forest Products* (Eds. Husen, A., Bachheti, R.K., Bachheti, A.). Springer International Publishing AG, Cham, pp. 229–254. https://doi.org/10.1007/978-3-030-73077-2_10

Rahman, S., Husen, A., 2022. *Asparagus racemosus* (Shatavari) and *Dioscorea villosa* (Wild yams). In: *Herbs, Shrubs, and Trees of Potential Medicinal Benefits* (Ed. Husen, A.). CRC Press, Boca Raton, FL.

Rahman, S., Husen, A., 2023a. *Trigonella foenum-graecum* L.: A useful medicinal plant for diabetes and other diseases. In: *Antidiabetic Medicinal Plants and Herbal Treatments* (Ed. Husen, A.). CRC Press, Boca Raton, FL.

Rahman, S., Husen, A., 2023b. Role of Green Chiretta [*Andaphis paniculata* (Burm. F.)] nees in diabetes and other ailments. In: *Antidiabetic Medicinal Plants and Herbal Treatments* (Ed. Husen, A.). CRC Press, Boca Raton, FL.

Rahuman, A.A., Venkatesan, P., Gopalakrishnan, G., 2008. Mosquito larvicidal activity of oleic and linoleic acids isolated from Citrullus colocynthis (Linn.) Schrad. *Parasitology Research* 103, 1383–1390.

Rajesh Sharma, G.L., 2002. Studies on antimycotic properties of Datura metel. *Journal of Ethnopharmacology* 80, 193–197.

Rakotomanga, M., Razakantoanina, V., Raynaud, S., Loiseau, P.M., Hocquemiller, R., Jaureguiberry, G., 2004. Antiplasmodial activity of acetogenins and inhibitory effect on Plasmodium falciparum adenylate translocase. *Journal of Chemotherapy* 16, 350–356.

Randrianarivelojosia, M., Rasidimanana, V.T., Rabarison, H., Cheplogoi, P.K., Ratsimbason, M., Mulholland, D.A., Mauclere, P., 2003. Plants traditionally prescribed to treat tazo (malaria) in the eastern region of Madagascar. *Malaria Journal* 2, 1–9. https://doi.org/10.1186/1475-2875-2-1.

Rezeki, R.S., Saragih, A., Bahri, S., 2012. Observasi klinis seduhan serbuk kulit batang kayu susu (Alstonia scholaris (L.) R. Br.) sebagai antimalaria di Manokwari. *Journal of Pharmacy and Pharmacology* [Internet] 1(2), 95–103. http://download.port algaruda.org/article.php?article58845&val4145.

Sadhu, S.K., Okuyama, E., Fujimoto, H., Ishibashi, M., 2003. Separation of Leucas aspera, a medicinal plant of Bangladesh, guided by prostaglandin inhibitory and antioxidant activities. *Chemical and Pharmaceutical Bulletin* 51, 595–598.

Sharma, P., Sharma, J.D., 2000. In-vitro schizonticidal screening of Calotropis procera. *Fitoterapia* 71, 77–79.

Sidhu, O.P., Chandra, H., Behl, H.M., 2009. Occurrence of aflatoxins in mahua (Madhuca indica Gmel.) seeds: Synergistic effect of plant extracts on inhibition of Aspergillus flavus growth and aflatoxin production. *Food and Chemical Toxicology: An International Journal Published for the British Industrial Biological Research Association* 47, 774–777.

Simonsen, H.T., Nordskjold, J.B., Smitt, U.W., Nyman, U., Palpu, P., Joshi, P., Varughese, G., 2001. In vitro screening of Indian medicinal plants for antiplasmodial activity. *Journal of Ethnopharmacology* 74, 195–204.

Simpson, B. B., & Ogorzaly, M. C., 1995. Economic Botany: Plants in Our World. New York, USA: McGraw-Hill, Inc.

Singh, N., Kaushik, N.K., Mohanakrishnan, D., Tiwari, S.K., Sahal, D., 2015. Antiplasmodial activity of medicinal plants from Chhotanagpur plateau, Jharkhand, India. *Journal of Ethnopharmacology* 165, 152–162.

Suksamrarn, S., Panseeta, P., Kunchanawatta, S., Distaporn, T., Ruktasing, S., Suksamrarn, A., 2006. Ceanothane- and lupane-type triterpenes with antiplasmodial and antimycobacterial activities from Ziziphus cambodiana. *Chemical and Pharmaceutical Bulletin* 54, 535–537.

Taek, M.M., Ew, B.P., Agil, M., 2018. Plants used in traditional medicine for treatment of malaria by Tetun ethnic people in West Timor Indonesia. *Asian Pacific Journal of Tropical Medicine* 11, 630–637. https://doi.org/10.4103/1995-7645.246339.

Tandon, P., Sirohi, A., 2010. Assessment of larvicidal properties of aqueous extracts of four plants against *Culex quinquefasciatus* Larvae. *Jordan Journal of Biological Sciences* 3, 1–6.

Tangmouo, J.G., Ho, R., Matheeussen, A., Lannang, A.M., Komguem, J., Messi, B.B., Maes, L., Hostettmann, K., 2010. Antimalarial activity of extract and norbergenin derivatives from the stem bark of *Diospyros sanza-minika* A. *Chevalier* (Ebenaceae). *Phytotherapy Research* 24, 1676–1679.

Teng, W.C., Chan, W., Suwanarusk, R., Ong, A., Ho, H.-K., Russell, B., et al., 2019. In vitro antimalarial evaluation and cytotoxicity investigations of Carica papaya leaves and carpaine. *Natural Product Communications* 14(1), 33–36.

Tewtrakul, S., Subhadhirasakul, S., Tansakul, P., Cheenpracha, S., Karalai, C., 2011. Antiinflammatory constituents from Eclipta prostrata using RAW264.7 macrophage cells. *Phytotherapy Research*, doi:10.1002/ptr.3383.

Traore, M., Diallo, A., Nikièma, J.B., Tinto, H., Dakuyo, Z.P., Ouédraogo, J.B., Guissou, I.P., Guiguemdé, T.R., 2008. In vitro and in vivo antiplasmodial activity of 'saye', an herbal remedy used in Burkina Faso traditional medicine. *Phytotherapy Research* 22, 550–551.

Valadeau, C., Pabon, A., Deharo, E., Albán-Castillo, J., Estevez, Y., Lores, F.A., Rojas, R., Gamboa, D., Sauvain, M., Castillo, D., Bourdy, G., 2009. Medicinal plants from the Yanesha (Peru): Evaluation of the leishmanicidal and antimalarial activity of selected extracts. *Journal of Ethnopharmacology* 123, 413–422.

Valdes, A.F., Martinez, J.M., Lizama, R.S., Gaiten, Y.G., Rodriguez, D.A., Payrol, J.A., 2010. In-vitro antimalarial activity and cytotoxicity of some selected Cuban medicinal plants. *Revista do Instituto de Medicina Tropical de Sao Paulo* 52, 197–201.

Verma, G., Dua, V.K., Agarwal, D.D., Atul, P.K., 2011. Anti-malarial activity of Holarrhena antidysenterica and Viola canescens, plants traditionally used against malaria in the Garhwal region of north-west Himalaya. *Malaria Journal* 10, 20.

Wachtel-Galor, S., Benzie, I.F.F., 2011. Herbal medicine: An introduction to its history, usage, regulation, current trends, and research needs. In: *Herbal Medicine: Biomolecular and Clinical Aspects* (Eds. Benzie, I.F.F., Wachtel-Galor, S.). CRC Press Llc, Boca Raton, FL.

Weng, T.C., Shen, C.C., Chiu, Y.T., Lin, Y.L., Kuo, C.D., Huang, Y.T., 2009. Inhibitory effects of armepavine against hepatic fibrosis in rats. *Journal of Biomedical Science* 2, 16–78.

Wong, S.K., Lim, Y.Y., Abdullah, N.R., Nordin, F.J., 2011. Assessment of antiproliferative and antiplasmodial activities of five selected Apocynaceae species. *BMC Complementary and Alternative Medicine* 11, 3.

Xiang, Y., Pei, Y., Qu, C., Lai, Z., Ren, Z., Yang, K., Xiong, S., Zhang, Y., Yang, C., Wang, D., Liu, Q., Kitazato, K., Wang, Y., 2011. In vitro anti-herpes simplex virus activity of 1,2,4,6-tetra-O-galloyl-d-glucose from Phyllanthus emblica L. (Euphorbiaceae). *Phytotherapy Research*. https://doi.org/10.1002/ptr.3368

Zeng, L., Ye, Q., Oberlies, N.H., Shi, G., Gu, Z.M., He, K., McLaughlin, J.L., 1996. Recent advances in Annonaceous acetogenins. *Natural Product Reports* 13, 275–306.

9 Antimalarial Response, Traditional and Other Potential Uses of *Aristolochia* Genera

Uttra Chauhan, Harshika Joshi, Gaurav Pant, and Manu Pant

9.1 INTRODUCTION

Malaria is a life-threatening and one of the deadliest infectious diseases in developing countries. It is a disease caused by the bite of female *Anopheles* mosquitoes infected with *Plasmodium* parasites. Five types of parasites cause malaria in human beings, and two of these species, *P. vivax* and *P. falciparum*, pose the greatest threat while *P. knowles*, *P. ovale*, and *P. malariae* pose a lesser threat. *P. falciparum* is the most common and lethal malaria parasite on the continent of Africa. *P. vivax* is the most common malaria parasite outside of sub-Saharan Africa. In 2021, there were an estimated 247 million cases of malaria worldwide (WHO, 2023, www.who.int/news-room/fact-sheets/detail/malaria). Many researches have been done for the treatment of malaria, and the use of medicinal plants is also an alternative in treating and preventing malaria. In places where malaria is endemic, medicinal plants are utilized as an alternative to cure a variety of illnesses, including malaria (Mathew et al., 2020).

Medicinal plants have been in use in the prevention and treatment of a wide range of diseases since the ancient times throughout the world because of their therapeutic effects (Husen, 2021, 2022a, 2022b, 2023; Rahman and Husen, 2021, 2022, 2023a, 2023b). They have a lot of beneficial effects, and the practice of using medicinal plants for treatment is increasing day by day. The bioactive chemicals found in medicinal plants have disease-preventing properties, like antimicrobial, anticancer, anti-inflammatory, antioxidant activities, etc. They are in demand in developed and developing countries and are used as alternative medicines as they have long-lasting effects, have less side effects, are easily available, and are less costly. The phytoconstituents derived from medicinal plants are alkaloids, saponins, tannins, glycosides, and phenolics and are a major source for drug production (Borah et al., 2021).

Aristolochia is one of the largest genera of the Aristolochiaceae family having medicinal values and used for treating and preventing a variety of diseases, like malaria, stomach ache, hypertension, dysentery, body ache, urinary tract infections, headache, and sexual dysfunction (Borah et al., 2021). The word *Aristolochia* is from Greek language, meaning 'aristos' – best or most appropriated – and 'lochia' – delivery (Babu et al., 2021). The genus has about 550 species, from which 20 species are found in India. In the world's subtropical, Mediterranean, tropic, and temperate regions, including Africa, America, and Asia, these species are found (Nath et al., 2022). *Aristolochia maxima* is found in Western Ghats (Tilari Ghats) of Maharashtra. This is an indigenous species of South and Central America. In tropical, subtropical, and Mediterranean countries, including Nepal, India, and Bangladesh, *Aristolochia indica* is found. *Aristolochia cathcartii*, *Aristolochia indica* Linn., and *Aristolochia saccata* Wall. are found in eastern India (Nath et al., 2022). Several Asian nations, including China, Japan, Bangladesh, Thailand, India, Sri Lanka, and Korea, have used *Aristolochia* species as medicine in their traditional medical systems (Mariyammal et al., 2023). The major and active compound of this genus is

DOI: 10.1201/9781003378396-9

aristolochic acid, which is extracted from the roots. It has pharmacological activities, like antifertility, anticancerous, antidiabetic, anti-inflammatory, anti-venom, antidiarrheal, and anti-pruritic (Borah et al., 2021). The entire *Aristolochia* plant is thought to provide a variety of therapeutic benefits in Indian traditional medicine. The roots are known to be anti-inflammatory, diuretic, and cardiotonic due to the presence of aristolochin, which is an alkaloid. The leaves of *Aristolochia* are useful in treating cholera and intermittent fever in children, and the seeds are traditionally used to treat biliousness, inflammation, and dry cough (Nath et al., 2022).

9.1.1 Botanical Description

Aristolochia is an evergreen or deciduous herbaceous and woody vine perennial. Alternate, sharp, deeply cordate, and with a narrow sinus, the leaves measure 8 in long, 4.5 in wide. The flowers are axillary in position, irregular, stalked distinctly and can be found in solitary form or in clusters. Flowers are bisexual and zygomorphic. Perianth is purple in colour, having tubular of three to seven lobes, and it has a fly trap mechanism as its special feature. Chironomid fly (Diptera) insects pollinate it. Stamens are six in number, and anther is sessile. Fruit is long-stalked, enclosed in capsule. Seeds are flattened and broad-winged (Rajani et al., 2020; Nath et al., 2022).

9.1.2 Traditional Use

In India, approximately 3,000 plants species are reported to have medicinal properties (Vineeta et al., 2022). *Aristolochia* species are one of these medicinal plants which have therapeutic values and used by different communities throughout the world in traditional medicine (Borah et al., 2021). The use of accessible natural resources for everyday sustenance, such as fodder, food, shelter, and healthcare, has allowed ethnic communities around the world to develop and maintain traditional knowledge. In many developing countries, indigenous rural people have a strong tradition of using herbs or plants as a kind of traditional or folk medicine. Throughout the country, Ayurveda, Siddha, and Unani are still important primary rural healthcare systems. The Indian traditional medical system is also a combination of several folk and regional customs based on a wide collection of knowledge about various plants (Vineeta et al., 2022). Different parts of this plant, such as leaves, roots, stem, bark, fruit, and whole plant, are used for the treatment and prevention of certain diseases, like malaria, dysentery, stomach ache, body pain, etc. Some of the traditional uses of *Aristolochia* species are listed in Table 9.1. For poisonous bites from scorpions and snakes, the roots of the plant species are used, and the rheumatic area is massaged with the paste of the roots by many tribal communities. It is also used to increase the production of breastmilk and for the treatment of fever, skin diseases, high blood pressure, beriberi and swollen feet, stomach pain, chest pain, snakebite, female infertility, and malaria. Leaf juice is orally consumed to treat snakebite. Paste of leaves is consumed for stomach

TABLE 9.1
Traditional Use of *Aristolochia* Species

Species	Traditional Use	Key Reference
A. indica	Heal wounds, enhance fertility, skin disease, diarrhoea, mosquito bite, analgesic, anti-inflammatory, antibiotic, antidote against poison, anthelmintic	Kumar et al., 2019; Taid et al., 2014
A. assamica	Malaria, dysentery, high blood pressure, body pain	Borah et al., 2020
A. saccate	Fever, body pain, malaria fever, stomach ache, constipation, piles	Teron, 2019; Sarmah and Saikia, 2014
A. acuminata	Diarrhoea, dysentery, snakebite, antipyretic, anti-inflammatory	Rao, 2019
A. cathcartii	Stomach ache, urogenital disorders, food poisoning, insect repellent	Paul et al., 2011

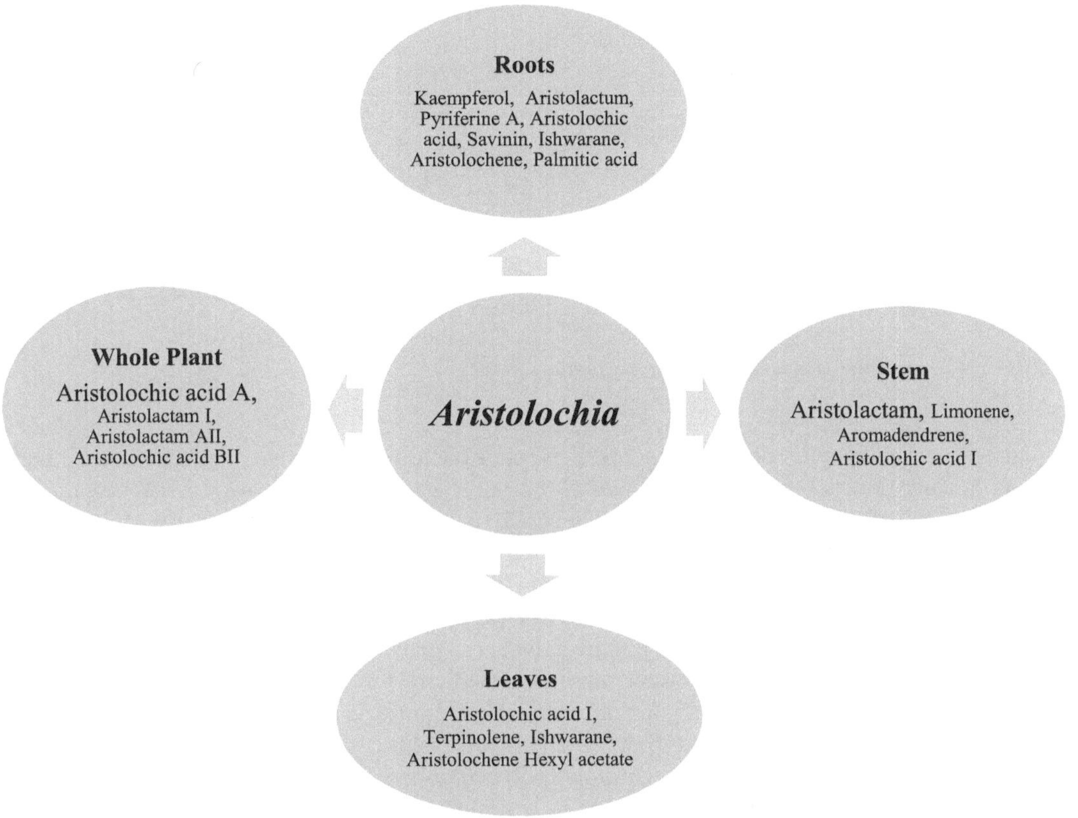

FIGURE 9.1 Phytoconstituents present in different parts of *Aristolochia* genus plant.

ache, vomiting, and diarrhoea. The paste of the plant is used to cure abdominal pain, headache, cold, fever, gynaecological disorders, and malaria (Padhy, 2021).

9.1.3 Phytoconstituents

The term 'phytochemical' derives from the Greek word 'phyto', which means 'plant', and they are physiologically active substances that are beneficial to human health (Saxena et al., 2013). Many studies have proved that medicinal plants are a rich source of phytoconstituents (Figure 9.1).

The secondary metabolites of this genus include aristolochic acids and its derivatives, protoberberines, aporphines, isoquinolines, aristolactams, benzylisoquinolines, flavonoids, diphenyl ethers, coumarins, amides, lignans, alkaloids, saponins, tannins, and phenols, steroids, tetralones, terpenoids (Borah et al., 2021). Aristolochic acid is known to be cytotoxic, apoptotic, antitumour, anticancer, and nephrotoxic (Hadem and Sen, 2018). Allantonin, alkaloid aristolodin, ishwarane aristolochine, sesquiterpene hydrocarbon, isoaristolochic acid, alcohol ishwararol, and essential oil containing carboxyl compounds are present in the root and rhizomes. It contains the presence of alkaloids, saponins, flavonoid, phenol, and tannin (Rajani et al., 2020). The chief active compound in the stem and root contains alkaloid and aristolochic acids (Mathew et al., 2020).

9.1.4 Pharmacological Uses

Traditional communities majorly rely on the naturally obtained herbs as a medicine for humans because of their cultural adaptability, similarity, and suitability of the plant with fewer side effects

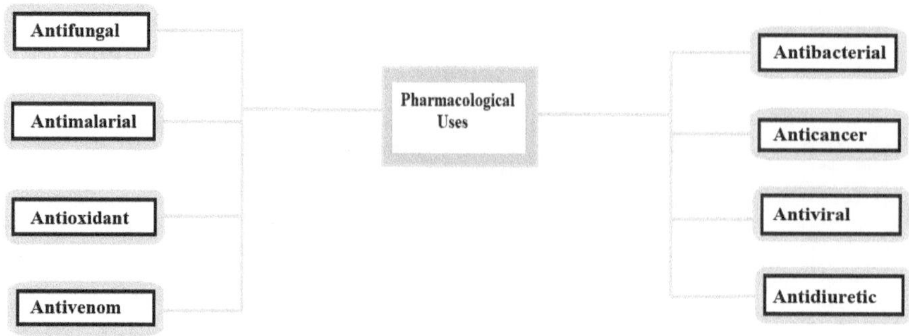

FIGURE 9.2 Pharmacological uses of *Aristolochia* genus.

(Figure 9.2). The benefits of these medicinal plants over allopathic drugs are lesser side effects, lasting healing effects, and encouragement of a healthy system, and they are developed through well-skilled knowledge (Oladeji, 2016). The genera *Aristolochia*, a member of the Aristolochiaceae family, is a well-known medicinal plant that has been used for therapeutic purposes in India for a very long time. Its often-used Tamil name is 'Perumarundhu'. In recent years, the Indian subcontinent and the rest of the world have recognized the exceptional medicinal benefits of the genus *Aristolochia* (Babu et al., 2021). *Aristolochia indica* is a medicinal herb that has been used since prehistoric times, and its species have been practiced as an alternative medicine for anti-inflammation conditions like chronic inflammation, rheumatism, gout, etc. and are still in use as a natural medicine. A variety of *Aristolochia indica* types is utilized in the southern part of India to treat skin disorders and snake-bites. Besides treating inflammation, the roots of *Aristolochia indica* are used to treat dry coughs, biliousness, intermittent fever (as it claims to be a stimulant and febrifuge), diarrhoea, dropsy, and lack of appetite. Along with its anti-inflammatory effects, it also serves anticancer, antiviral, antibacterial, antifungal, antianalgesic, and antidiuretic activities (Mathew et al., 2020).

9.1.5 Characteristic Properties

9.1.5.1 Antibacterial Activity

The extracts of acetone, ethyl acetate, and methanol from the roots showed the most effective and promising antimicrobial activity on gram-positive and gram-negative bacteria, such as *Staphylococcus lentus* and *Bacillus Cereus*, and are therefore proved to be helpful in treating various bacterial infections as they show a broad spectrum of activity (Lerma-Herrera et al., 2022). Throughout Europe, a number of *Aristolochia* species were utilized as medicinal herb (Bartha et al., 2019). In phytochemical screening of hexane, the extract of *A. bracteolata* leaves exhibits the presence of flavonoids, terpenoids, and carbohydrates and the absence of glycosides, tannins, alkaloids, and proteins. Similarly, ethyl extract shows the presence of steroids, glycosides, and carbohydrates, and the absence of phytoconstituents was confirmed. The result of leaf extract analysis showed the relative distribution of secondary metabolites, which can be responsible for antibacterial activity. Polyphenols and flavonoids that are responsible for its antibacterial activity are present in the organic extract of *A. longa*. In addition to inhibiting proteins and biofilms, the release of cell material into the environment results in a loss of cell integrity and cell membrane (Lerma-Herrera et al., 2022).

9.1.5.2 Anticancer Activity

The root extract from *A. acuminata* reduced the levels of TNF (tumour necrosis factor-alpha) and nuclear factor kappa beta activation, which in turn reduced the elevated activities of these marker enzymes. When tested against six human cancer cell lines, extracts from the plant's roots and leaves demonstrated potential for chemoprevention. The *A. foetida* leaves and stem extract used in the cell line induces late apoptotic cell death through the intrinsic route. The primary compounds that are responsible for cytotoxic activity are methyl octadeca-9,12,15-trienoate, ethyl hexadecanoate, hexadecenoic acid, and 2-butoxyethyl dodecanoate. The presence of phenolic compounds in the extract leads to the inhibition of enzymes that disrupt the cell cycle, which is responsible for cell death against cancerous cells (Lerma-Herrera et al., 2022).

9.1.5.3 Antidiuretic Activity

The experimental findings demonstrated that aerial parts of *A. indica* have considerable antidiabetic properties. The group of enzymes that break down the dietary carbohydrates into simple monosaccharides is called glucosidase, and its inhibitor like acarbose reduces the rate of carbohydrates digestion and delays the carbohydrate absorption from the digestive tract. Therefore, many studies have found that preventing the excessive rise of blood glucose level by natural sources is effective in real life as well (Das et al., 2016).

9.1.5.4 Antimalarial Activity

One of the major health problems that has been going around for a while is malaria, which is the most parasitic and prevalent disease. In order to fight against the *Plasmodium* species, attention was derived in the development of drug that shows high efficacy, is low cost, and is safe (Mathew et al., 2020). The root extract of *Aristolochia albida* produces significant dose-dependent decrease in the parasitaemia levels in the extract-treated group of *Plasmodium*. The usage of *A. albida* is safe and contains antimalaria activity, which justifies their usage as antimalaria remedies. The wide range of antimalaria chemotherapeutic medicines already in use is dependent on natural materials, and this fact indicates that, if properly utilized, new antimalaria treatments might continuously develop from tropical plant sources (Usman et al., 2018).

9.1.5.5 Antioxidant Activity

According to various studies, free radicals significantly contribute to the development of cancer, heart disease, ageing, cataracts, and immune system damage. These unstable free radicals can be eliminated by antioxidants, which slow down oxidation and protect cells from damage. For the prevention and treatment of oxidative stress-related illnesses such as diabetes, Alzheimer's disease, atherosclerosis, stroke, and cancer, antioxidant-containing medications are utilized. Nowadays, herbal medicines are being consumed, as they have less side effects and price than allopathic medicines (El Omari et al., 2019). The methanolic roots extract from *A. acuminata* was tested and concluded to have greater significant value than standard compound ascorbic acid. Therefore, it showed higher scavenging activity in comparison to chloroform leaf extract. The presence of phenols and flavonoids in the plant concludes that it exhibits various biological properties, like anti-inflammatory, antitumour, cytotoxicity; therefore, they can act as a powerful antioxidant, protecting the human body from reactive oxygen species and free radicals (Das et al., 2016).

9.1.5.6 Anti-Venom Activity

Aristolochia has been utilized in anti-venom treatments since the ancient times. Its chemical components, ishwarone, aristolochene, ishwarol, aristolinidquinone, and aristolochic acid, have anti-inflammatory, anti-phospholipase A2 (which causes haemorrhaging), anti-haemorrhagic activity, enzyme inhibition, and anti-venomous activity (Dhivya et al., 2018) by modifying the functions of proteins and enzymes, chemical components such aristolochic acid, sesquiterpenes, and aristololide, which are responsible for anti-venom properties (Borah et al., 2021). Hexanoic extracts from *Aristolochia elegans* mast roots were tested for their ability to suppress smooth muscle contractions induced by scorpion venom. Because of their carcinogenic properties, aristolochic acids were not considered safe by the World Health Organization (WHO, 2023), yet they were used as a potential venom cure in spite of these problems.

9.1.5.7 Anti-Inflammatory Activity

A. acuminata root extracts in ethanol and ethyl acetate also have anti-inflammatory properties. In THP-1 cells treated with LPS, aristolactam I and (-) hinokinin from *A. indica* also have anti-inflammatory actions and reduced the generation of IL-6 and TNF-α. There are many phytoconstituents isolated from the plant from which 'kaempferol' is responsible for anti-inflammatory activity (Borah et al., 2021). In a metabolic arthritis rat model, *Aristolochia bracteolata* exhibited a promising hyperuricaemia and strong anti-inflammatory and proliferative effects on human dermal and fibroblasts, resulting in an *in vitro* wound healing action (Mathew et al., 2020).

9.1.6 Ethnomedical Uses

Aristolochia, the most prevalent genus in the Aristolochiaceae family, is used as a source of traditional medicine for analgesics, sedatives, muscle relaxants, anticancers, antihistamines, and antiparasitics, used in the treatment of various issues, like malaria, bites and stings from insects and animals, skin infections, abdominal pain, rheumatism, etc. (Borah et al., 2021). This practice of using plants extract is beneficial as it does not involve any elaborative procedures, is cheap in manufacturing, and is accessible. Aristolochic acid and its derivatives and monoterpenes such as thujene carene, camphene, and triterpenes, such as lupeol, kaurene-type diterpenes, are the secondary metabolites responsible for biological effects of the *Aristolochia* genus. Likewise, aristolactams-derived alkaloid metabolite and lignan-type phenolic compounds are involved in these functions.

Ethnomedical uses of different parts of *Aristolochia* mostly include the leaves, aerial parts, the whole plant, stems, and roots, listed in Table 9.2. Ethnomedical uses of *Aristolochia* plants are drinking whole plant juice in case of snakebite or application of root paste on the affected area or by giving orally. In case of skin disorders, root powder dried in shade is consumed via the oral route

TABLE 9.2
Different Species of *Aristolochia* and Their Ethnomedical Uses

Species	Plant Part	Treatments	Key References
Aristolochia acuminata Lam.	LV, FT, S, and RT	Anti-inflammatory, antimalarial, health tonic, muscle relaxant, abdominal pain	Borah et al., 2021
Aristolochia albida Duch.	RT and LV	Larvicide, snake anti-venom, antiparasitic, hepatoprotection, nephroprotection	Usman et al., 2018
Aristolochia argentina Griseb.	WP and AP	Diuretic, antioxidant emmenagogue, antiseptic, insecticide	Lerma-Herrera et al., 2022
Aristolochia bracteolata Lam.	AP, RT, LV, and WP	Cytotoxic, antioxidant, osteogenic differentiation of gingival mesenchymal stem cells, antidiabetic, cancer treatment, snakebite, lung inflammation, dysentery, scorpion stings, flu, vomiting, pneumonia treatment of malaria, headache, fever	Taha et al., 2019; Sivakkumar et al., 2019
Aristolochia clematitis L.	AP and RZ	Antioxidant, antibacterial, antifungal	Abbouyi et al., 2016
Aristolochia elegans Mast.	RT and LV	Antitoxin, antiviral, antifungal, scorpion anti-venom, antibacterial	Meela et al., 2019

(*Continued*)

TABLE 9.2 (*Continued*)
Different Species of *Aristolochia* and Their Ethnomedical Uses

Species	Plant Part	Treatments	Key References
Aristolochia fordiana Hemsl.	WP and S	Antibacterial, neuroprotective effect, antioxidant, and cytotoxic	Xiao-dan et al., 2019
Aristolochia indica L.	LV, WP, RT, L, and FR	Snakebites, antibacterial, dryness of tongue, blood purifier, cholera, menstrual irregularities, watery eyes, swelling in leg, pulmonary problems, gangrene, abdomen burning	Padhy, 2021
Aristolochia longa L	RT, LV, and AP	Antibacterial, antioxidant, and cytotoxic	Idrissi et al., 2020
Aristolochia macroura Gomes	LV AP	Cytotoxic and antioxidant	Abd El-Hafeez et al., 2018
Aristolochia Ringens Vahl.	AP and RT	Antibacterial, antifungal, cytotoxic, antidiabetic, antioxidant, antidyslipidemic	Mazadu et al., 2018
Aristolochia saccata Wall.	LV, RT, S, and T	Curing of wounds, diarrhoea, dysentery, haemorrhage, body pain, tonsil, jaundice	Borah et al., 2021
Aristolochia tagala Cham.	RT, LV, and WP	Stomach pain, skin disease, flatulent, dyspepsia, diarrhoea, vomiting, headache, antimalarial, stimulate the menstrual flow, snakebite, gynaecological disorders	Rajani et al., 2020
Aristolochia tribolata L.	S	Antinociceptive, anti-inflammatory	Quintans et al., 2017

Note: FT = fruits, *AP* = aerial parts, *FR* = fresh root, *LV* = leaves, *WP* = whole plant, *RZ* = rhizomes, *RT* = roots, *SB* = stem bark, *T* = tuber, *S* = stems.

for 48 days (Lerma-Herrera et al., 2022). For fever, leaves are prescribed for chewing, and for headache, the paste is applied on the forehead. For abdominal pain and in malarial treatment, the roots of the plants are used for making a decoction and consumed. Following are the different species of *Aristolochia* and their ethnomedical uses (Lerma-Herrera et al., 2022).

9.2 CONCLUSION

Medicinal plants are an abundant source of bioactive compounds that have disease-preventing properties, such as anticancer, anti-inflammatory, antioxidant, and antimicrobial properties. They are in demand in developed and developing countries and being used as alternative medicines as they have long-lasting effects, have less side effects, are easily available, and are less costly. For a long time, the people of North-East India have regarded the genus *Aristolochia* as a plant with significant medicinal value. However, in recent years, its potential in fighting illnesses and diseases has gained recognition on a global scale. Despite having multiple uses, *Aristolochia* was used as a traditional medicine for treating malaria. The entire *Aristolochia* plant is thought to provide a variety of therapeutic benefits in Indian traditional medicine. Deeper studies in the genus *Aristolochia* can help unveil its potential as a possible plant-based medicine, thereby advancing public health on a worldwide scale.

REFERENCES

Abbouyi, A.E., Maliki, S.E., Filali-Ansari, N. & Khyari, S.E. 2016. Antibacterial and antifungal activities of rhizomes extract from *Aristolochia clematitis*. *International Journal of Pharmaceutical, Chemical, and Biological Sciences* 6: 72–75.

Abd El-Hafeez, A.A., Khalifa, H.O., Elgawish, R.A., Shouman, S.A., Abd El-Twab, M.H. & Kawamoto, S. 2018. *Melilotus indicus* extract induces apoptosis in hepatocellular carcinoma cells via a mechanism involving mitochondria-mediated pathways. *Cytotechnology* 70: 831–842.

Babu, M., Ashok, K. & Kuberan. 2021. Medicinal importance and phytochemical constituents of *Aristolochia* plant: A short review. *International Journal of Zoological Investigations* 7(2): 863–865.

Bartha, G.S., Tóth, G., Horváth, P., Kiss, E., Papp, N. & Kerényi, M. 2019. Analysis of aristolochlic acids and evaluation of antibacterial activity of *Aristolochia clematitis* L. *Biologia Futura* 70: 323–329.

Borah, D., Tangjang, S., Das, A.P., Upadhaya, A. & Mipun, P. 2020. Assessment of non-timber forest products (NTFPs) in Behali Reserve Forest, Assam, Northeast India. *Ethnobotany Research and Applications* 19(0): 1–15.

Borah, P.J., Borah, D., Das, U., Das, T.J. & Sarma, R. 2021. A review on ethnopharmacological utility, traditional knowledge and phytochemistry of *Aristolochia* species in Assam, India. *Notulae Scientia Biologicae* 13(3): 11027.

Das, S.T., Latha, R. & Agastian, P. 2016. Evaluation of *Aristolochia Bracteolata* Linn. for antimicrobial activity, α-glucosidase inhibition, and its phytochemical constituents. *Asian Journal of Pharmaceutical and Clinical Research* 9: 102–107.

El Omari, N., Sayah, K., Fettach, S., El Blidi, O., Bouyahya, A., Faouzi, M.E.A. & Barkiyou, M. 2019. Evaluation of in vitro antioxidant and antidiabetic activities of *Aristolochia longa* extracts. *Evidence-Based Complementary and Alternative Medicine*: 1–9.

Hadem, K.L., & Sen, A. 2018. In vitro free radical scavenging activity of *Aristolochia tagala*. *International Journal of Current Pharmaceutical Research* 10(3): 40–42.

Husen, A. 2021. *Traditional Herbal Therapy for the Human Immune System*. CRC Press, Boca Raton, FL. https://doi.org/10.1201/9781003137955

Husen, A. 2022a. *Herbs, Shrubs and Trees of Potential Medicinal Benefits*. CRC Press, Boca Raton, FL. https://doi.org/10.1201/9781003205067

Husen, A. 2022b. *Exploring Poisonous Plants: Medicinal Values, Toxicity Response and Other Potential Uses*. CRC Press, Boca Raton, FL. https://doi.org/10.1201/b23017

Husen, A. 2023. *Antidiabetic Medicinal Plants and Herbal Treatments*. CRC Press, Boca Raton, FL.

Idrissi, A.Y., Khouchlaa, A., Bouyahya, A., Bakri, Y. & Tijane, Y.B.M. 2020. Phytochemical characterization, in vitro antioxidant, cytotoxic, and antibacterial effects of *Aristolochia longa* L. *Biointerface Research in Applied Chemistry* 11: 8129–8140.

Kumar, S.J.U., Chaitanya, K.M.J., Semotiuk, A.J. & Krishna, V. 2019. Indigenous knowledge of medicinal plants used by ethnic communities of South India. *Ethnobotany Research and Applications* 18(4): 1–112.

Lerma-Herrera, M.A., Beiza-Granados, L., Ochoa-Zarzosa, A., López-Meza, J.E., Navarro-Santos, P., Herrera-Bucio, R., Aviña-Verduzco, J. & García-Gutiérrez, H.A. 2022. Biological activities of organic extracts of the genus *Aristolochia*: A review from 2005 to 2021. *Molecules* 27: 3937.

Mariyammal, V., Sathiageetha, V., Amalraj, S. & Shailendra, S. 2023. Chemical profiling of *Aristolochia tagala* Cham. leaf extracts by GC-MS analysis and evaluation of its antibacterial activity. *Journal of the Indian Chemical Society* 100(1).

Mathew, L.S, Mtewa, A.G, Ajayi, C., Deyno, S., Weisheit, A., Tolo, CU., Deng, A.L. & Ogwang, P.E. 2020. Ethnopharmacology, pharmacology and phytochemistry of *Aristolochia Bracteolata* Lam: A review of an antimalarial plant. *The Eastern Africa Social Science Research Review* 1(2): 22–28.

Mazadu, E.A., Misau, M.S. & Gwallameji, L.B. 2018. Phytochemical screening and antimicrobial activity of some medicinal trees grown in Bauchi state, north eastern, Nigeria. *International Journal of Pharmacognosy and Phytochemical Research* 7: 3503–3507.

Meela, M.M., Mdee, L.K., Masoko, P. & Eloff, J.N. 2019. Acetone leaf extracts of seven invasive weeds have promising activity against eight important plant fungal pathogens. *The South African Journal of Botany* 121: 442–446.

Nath, S., Gosh, N. & Mundhra, A. 2022. Genetic diversity assessment and biotechnological aspects in Aristolochia spp. Applied *Microbiology and Biotechnology* 106(19–20): 1–16.

Oladeji, O. 2016. The characteristics and roles of medicinal plants: Some important medicinal plants in Nigeria. *Natural Products: An Indian Journal* 12(3): 102.

Padhy, G.K. 2021. A review of *Aristolochia indica*: Ethnomedicinal uses, phytochemistry, pharmacological and toxicological effects. *Current Traditional Medicine*7: 372–386.

Paul, S., Devi, N. & Sarma, G.C. 2011. Medicinal plants of subankhata reserve forest under dhansiri forest division, Manas Biosphere Reserve in Assam, India. *Pleione* 5(2): 286–291.

Quintans, J.S.S., Alves, R.S., Santos, D.A., Serafini, M.R., Alves, P.B., Costa, E.V., Zengin, G., Quintans-Júnior, L.J. & Guimarães, A.G. 2017. Antinociceptive effect of *Aristolochia trilobata* stem essential oil and 6-methyl-5-hepten-2yl acetate, its main compound, in rodents. *Zeitschrift fur Naturforschung. C, Journal of Biosciences* 72: 93–97.

Rahman, S. & Husen, A. 2021. Potential role of medicinal plants in the cure of liver and kidney diseases. In: *Non-Timber Forest Products* (Eds. Husen, A., Bachheti, R.K. & Bachheti, A.). Springer International Publishing AG, Cham, pp. 229–254. https://doi.org/10.1007/978-3-030-73077-2_10

Rahman, S. & Husen, A. 2022. *Asparagus racemosus* (Shatavari) and *Dioscorea villosa* (Wild yams). In: *Herbs, Shrubs, and Trees of Potential Medicinal Benefits* (Ed. Husen, A.). CRC Press, Boca Raton, FL.

Rahman, S. & Husen, A. 2023a. *Trigonella foenum-graecum* L.: A useful medicinal plant for diabetes and other diseases. In: *Antidiabetic Medicinal Plants and Herbal Treatments* (Ed. Husen, A.). CRC Press, Boca Raton, FL.

Rahman, S. & Husen, A. 2023b. Role of Green Chiretta [*Andaphis paniculata* (Burm. F.)] nees in diabetes and other ailments. In: *Antidiabetic Medicinal Plants and Herbal Treatments* (Ed. Husen, A.). CRC Press, Boca Raton, FL.

Rajani, M.B., Raviraja, S.G. & Pooja, D.A. 2020. A review on medicinal uses, pharmacology and phytochemistry of *Aristolochia tagala* Cham. An endangered medicinal plant. *Journal of Pharmacognosy and Phytochemistry* 9(4): 580–583.

Rao, R.R. 2019. Medicinal plants of India: Diversity, conservation and bioprospection-concerns and strategies for 21st century. *TTPP* 1.

Sarmah, R. & Saikia, A. 2014. Non-timber forest products: Diversity and used pattern at Majuli the Brahmaputra River Island of Assam, India. *Journal of Botanical Sciences* 3(3): 41–43.

Saxena, M., Saxena, J., Nema, R., Singh, D. & Gupta, A. 2013. Phytochemistry of medicinal plants. *Journal of Pharmacognosy and Phytochemistry* 1: 168–182.

Sivakkumar, S., Iyswarya, S., Juliet, L. & Ganapathy, G. 2019. A review on ingredients of anti-diabetic siddha preparation *Naaval Kottai Mathirai. International Journal of Pharmaceutical Sciences Review and Research* 55: 69–76.

Sivaraj, D., Shanmugam, S., Rajan, M., Sasidharan, S.P., Sathyanarayanan, S., Muniyandi, K., Thangaraj, P. & de Souza Araújo, A.A. 2018. Evaluation of *Aristolochia indica* L. and *Piper nigrum* L. methanol extract against centipede *Scolopendra moristans* L. using Wistar albino rats and screening of bioactive compounds by high pressure liquid chromatography: A polyherbal formulation. *Biomedicine & Pharmacotherapy* 97: 1603–1612.

Taha, M., Parveen, B., Osman, B., Abdoon, I.H., Mohamed, M.S., Osman, W.J.A. & Ahmad, S. 2019. In vitro profiling of plants used in sudanese traditional medicine for antioxidant and anti-breast cancer activities. *Annals of Phytomedicine International Journal* 8: 119–126.

Taid, T.C., Rajkhowa, R.C. & Kalita, J.C. 2014. A study on the medicinal plants used by the local traditional healers of Dhemaji district, Assam, India for curing reproductive health related disorders. *Advances in Applied Science Research* 5(1): 296–301.

Teron, R. 2019. Cross-cultural ethnobotanical exploration of diversity and utilization of medicinal plants in Karbi Anglong district, Assam, Northeast India. *NeBIO* 10: 35–46.

Usman, H.S., Sallau, A.B., Salihu, A. & Nok, A.J. 2018. Larvicidal assessment of fractions of *Aristolochia albida* rhizome on culex quinquefasciatus. *Tropical Journal of Natural Product Research* 2: 227–234.

Vineeta, V., Shukla, G., Bhat, J.A. & Chakravarty, S. 2022. Species richness and folk therapeutic uses of ethnomedicinal plants in West Bengal, India – A meta-analysis. *Phytomedicine Plus* 2(1).

WHO. 2023. *Malaria*, World Health Organization. Geneva, Switzerland (https://www.who.int/news-room/fact-sheets/detail/malaria)

Xiao-dan, S., Yang, G., Ying-Xin, X., Peng-Xiang, L. & Xiang, X. 2019. Chemical composition and biological activities of the essential oil from *Aristolochia fordiana* Hemsl. *Records of Natural Products* 13: 346–354.

10 Antimalarial Response, Traditional and Other Potential Uses of *Aspidosperma* Genera

Surjit Sen, Shuvadip Mondal, Manjula Rai, Mangal Chandra Biswas, Rupa Sanyal, and Abhijit Dey

10.1 INTRODUCTION

The most prevalent parasitic infection in the tropical and subtropical regions is represented by malaria (Mitaine-Offer et al., 2002). A report in 2020, pronounced by the World Health Organization, stated that around 409,000 died from malaria in 2019 (WHO, 2020; Milliken et al., 2021). However, there is a significant decrease in new malaria cases that has been observed from 2019 to 2020, but it is still endemic in almost 18 countries in South America and the Caribbean, where nearly 139 million people are documented to have been exposed (WHO, 2020; Milliken et al., 2021). Five different protozoan species under the genus *Plasmodium* have been recorded to cause malaria in humans (*Plasmodium falciparum*, *P. malariae*, *P. ovale*, *P. vivax*, and *P. knowlesi*), where *P. falciparum* and *P. vivax* cause most of the malarial infections, while *P. falciparum* is severely fatal and found to be predominant in Africa (Schlitzer, 2008; Ashok et al., 2013; Uzor, 2020).

For decades, malaria has been treated with quinine, chloroquine, amodiaquine, mefloquine, and artemisinin derivatives (Uzor, 2020). Among these derivatives, chloroquine (4-aminoquinoline) – a synthetic derivative from quinine – is still in use for its cheapness, effectiveness, and less toxicity to the human body (Uzor, 2020). But nowadays, this drug has been restricted in the treatment of malaria as the parasite has become resistant to the drug (Wiesner et al., 2003; Uzor, 2020). To overcome this situation, WHO recommends an alternative therapy for malaria, which is artemisinin-based combination therapy, or ACT, in order to check resistance (Chierrito et al., 2014; Uzor, 2020). However, the limited accessibility to ACT makes an indication to discover a new and cheap drug to control this life-threatening disease (Chierrito et al., 2014). Plant-derived natural drugs to treat various diseases are an age-old practice in Asian and African countries (Husen, 2021, 2022a, 2022b, 2023; Rahman and Husen, 2021, 2022, 2023a, 2023b) but now have come to common in Western World also (Krettli et al., 2009; Willcox et al., 2011; Ceravolo et al., 2018). In the large regions of the Amazon basin, including Brazil, plants are still in use to treat malaria, as these plants possess potent sources of compounds to develop new antimalarial drugs (Coutinho et al., 2013). Among the various plant species used as medicine in Latin American countries, a member from the family Apocynaceae has proved to be active against malarial parasite: the genus *Aspidosperma* (Garavito et al., 2006; Lima et al., 2015; Ceravolo et al., 2018). *Aspidosperma* spp. are vastly grown in Brazilian caatinga and used for different medicinal purposes (Coutinho et al., 2013; Rocha et al., 2019). Due to their high habitat concentration in Amazon basins, they are popularly known as carapanaúba, pau-pereiro, guatambus, and quina (Rocha et al., 2019). Figure 10.1 presents few species of the genus *Aspidosperma* alongside some useful plant parts and obtained phytochemical(s).

Studies have shown that various species of *Aspidosperma* have acted against malarial parasite. The stem bark of *Aspidosperma nitidum* is used to treat fever and malaria in the regions of Amazon and even remains active in low concentrations against *Plasmodium falciparum* (Chierrito

 DOI: 10.1201/9781003378396-10

FIGURE 10.1 (a) *Aspidosperma macrocarpon*: habit; (b) *Aspidosperma tomentosum*: habit; (c) *Aspidosperma marcgravianum*: habit; (d) *Aspidosperma cylindrocarpon*: seeds; € *Aspidosperma melanocalyx*: flowers; (f) *Aspidosperma quebracho-blanco*: the unusual seed pods; (g) *Aspidosperma quebracho-blanco*: flowers; (h) *Aspidosperma spruceanum*: flowers; (i) *Aspidosperma melanocalyx*: pods; (j) *Aspidosperma quebracho-blanco*: pods; (k) *Aspidosperma quebracho-blanco*: pods and branch; (l) (+)-aspidospermidine structural formula.

Source: a, b, e, f, h, i, k: Creative Commons Attribution 2.0; c, g, j: Creative Commons Attribution – Share Alike 3.0; d: Creative Commons Attribution – Share Alike 4.0; l: Public domain. All retrieved from https://commons.wikimedia.org/.

et al., 2014; Ceravolo et al., 2018). Another promising species to act against malaria is *A. olivaceum* (Dolabela et al., 2012; Chierrito et al., 2014). It has been reported that *A. excelsum* is used to treat malaria by the infusion of bark (Coutinho et al., 2013). Therefore, the aim of our present study is to analyze the antimalarial activity of different species of *Aspidosperma* and to find out future prospects.

10.2 DISTRIBUTION AND MORPHOLOGICAL CHARACTERS OF *ASPIDOSPERMA* SPP.

Aspidosperma Mart. and Zucc. belong to the family Apocynaceae, which is one of the ten largest angiosperm families (Koch et al., 2013; de Paula et al., 2014). It is also the largest genus under the tribe Aspidopsermeae Miers and belongs to the grade Rauvolfioid (Endress et al., 2014; Pereira et al., 2018). The genus comprises a neotropical distribution, found from Mexico to Argentina, except Chile (Morales and Zamora, 2017; Scudeler et al., 2018; Pereira et al., 2018). Reports state that 56 species of *Aspidosperma* are found to occur in Brazil, of which 23 are listed as endemic (Morales and Zamora, 2017; Pereira et al., 2018). *Aspidosperma* spp. occur in almost all Brazilian phytogeographical areas (Pereira et al., 2018). *Aspidosperma* is known to be one of the most outstanding genera in Brazil due to its vast ecological, medicinal, and economic interest, chiefly as a valuable source of timber (de Paula et al., 2014; Pereira et al., 2018). Most of the species under *Aspidosperma* are morphologically

TABLE 10.1
Distinctive Morphological Characters of *Aspidosperma* spp. Showing Antimalarial Activity

Species of Aspidosperma	Distinctive Morphological Characters					Reference
	Branch	Leaves	Inflorescence	Flower	Fruit	
A. tambopatense	–	Leaves with flattened secondary veins on both surfaces	–	Flowers with deltoid corolla lobes	Follicles sessile or substipitate, with an inconspicuous midrib	de Paula et al., 2014; Morales and Zamora, 2017; Pereira et al., 2018; Scudeler et al., 2018
A. ulei	–	Leaves pubescent on the abaxial surface and 3× longer than wide, yellow on the abaxial surface	–	Flowers up to 3 mm long, tomentose externally	Follicles pubescent, yellow	
A. cuspa	–	Leaves with reticulodromous venation	Inflorescence leaf-opposed	–	–	
A. schultesii	Branches with white latex	–	–	Flowers more than 13 mm long, corolla lobes more than 8 mm long, ovary tomentose; seed nuclei with radial lines	–	
A. desmanthum	Branches with red latex	Leaves glabrous on the abaxial surface	–	Flowers up to 11 mm long, corolla lobes up to 5 mm long, ovary glabrous; seed nuclei without radial lines	Follicles brown	
A. steinbachii	–	Leaves brown on the abaxial surface, tertiary veins inconspicuous on the adaxial surface	–	Flowers more than 5.2 mm long	Follicles blackish-green, dolabriform or suborbicular	
A. album	–	Leaves white on the abaxial surface, tertiary veins conspicuous on the adaxial surface and dark green on the adaxial surface	–	Flowers with corolla lobes ca. 2.5 mm long	Follicles brown	
A. spruceanum	–	Leaves white on the abaxial surface, tertiary veins conspicuous on the adaxial surface, and olive green or brown on the adaxial surface	–	Flowers with corolla lobes ca. 4 mm long	Follicles yellow	
A. sandwithianum	–	Leaves 2.4–3.8 × longer than wide; leaf petioles ca. 2.2–4 cm long, tomentose	–	Flowers with corolla tube tomentose below the insertion of the anthers internally and with salverform corolla, lobes more than 2.5 mm long	–	

A. auriculatum	–	Leaf margins with deep revolute base and with eucamptodromous venation	–	–	–
A. cylindrocarpon	Branches sparsely lenticellated	Leaf petioles more than 2 cm long	–	–	Follicles smooth, botuliform, inflated with conspicuous lenticels
A. macrocarpon	–	Leaves with prominent secondary veins on the abaxial surface and pubescent on the abaxial surface	–	Flowers with glabrous ovary	Follicles more than 11 cm long
A. carapanauba	–	Leaves with flattened or immersed secondary veins on the abaxial surface	–	Flowers with tomentose ovary	Follicles up to 9 cm long and usually revolute, velutinous
A. rigidum	–	Leaves flattened, pubescent on the abaxial surface	–	Flowers with glabrous ovary, salverform corolla, glabrous externally, lobes equal to or larger than the tube (1–1.6×)	Follicles suborbicular, smooth
A. nitidum	–	–	Inflorescence axillary	Flowers with corolla ca. 6 mm long, lobes up to 0.5 mm long; flowers with tubular corolla, tomentose externally, lobes smaller than the tube (0.2–0.5×)	Follicles dolabriform, verrucose, or spinescent
A. excelsum	–	Leaves coriaceous, 1.7–2 × longer than wide	Inflorescence terminal	Flowers with calyx lobes equal; follicles spinescent and with corolla ca. 7.5 mm long, lobes more than 1 mm long	–
A. marcgravianum	–	Leaves chartaceous, 2.8–3 × longer than wide	Inflorescence terminal	Flowers with calyx lobes unequal and with corolla ca. 7.5 mm long, lobes more than 1 mm long	Follicles verrucose

similar, with minute differences. The contrasting morphological characters of *Aspidosperma* spp. showing antimalarial activity are presented in Table 10.1.

10.3 ANTIMALARIAL ACTIVITY OF *ASPIDOSPERMA* SPP.

Malaria is a serious and sometimes fatal disease that affects millions of people worldwide, particularly in developing countries. The parasites causing the disease are mainly *Plasmodium falciparum* (De Paula et al., 2014; Negi et al., 2014). Several studies have investigated the antimalarial properties of different *Aspidosperma* species. One of the most studied species is *Aspidosperma quebracho-blanco*, also known as 'white quebracho'; the extracts from the bark of this tree have potent antimalarial activity against both chloroquine-sensitive and chloroquine-resistant strains of *Plasmodium falciparum*, the parasite that causes malaria (Sampietro et al., 2021). Other species of *Aspidosperma*, such as *Aspidosperma nitidum*, have also been investigated for their antimalarial properties. The bark extract of *Aspidosperma nitidum* had significant antimalarial activity against *Plasmodium berghei*, a parasite that infects rodents and often used as a model for testing antimalarial drugs (Brígido et al., 2021). *Aspidosperma cylindrocarpon*, known as 'quebra-pedra' in Brazil, has been used by people to treat malaria and other fevers (Chaves et al., 2017). Similarly, *Aspidosperma excelsum*, known as 'tahuari' in Peru, has been used by indigenous communities in the Amazon basin to treat fever, malaria, and other diseases (do Nascimento et al., 2019). These species have shown some level of antimalarial activity in various studies (Table 10.2).

10.4 ANTIMALARIAL MECHANISM OF ACTION OF VARIOUS *ASPIDOSPERMA* SPECIES

The antimalarial mode of action of *Aspidosperma* extracts and their compounds is not yet fully understood, but several studies have suggested potential mechanisms of action. One proposed mechanism is related to the alkaloids present in *Aspidosperma* species, which can interfere with heme metabolism in the parasite. *Heme* is a toxic by-product of haemoglobin digestion in the parasite, and *Aspidosperma* alkaloids like elliptocine and olivacine can inhibit the formation of hemozoin crystals, which are necessary for the detoxification of heme. This can lead to an accumulation of toxic heme and subsequent parasite death (Rocha E Silva et al., 2012). Another proposed mechanism involves the disruption of the parasite's calcium homeostasis by *Aspidosperma* sp.–derived alkaloid compound. Calcium signalling is crucial for many cellular processes in the parasite, including the regulation of gene expression and cell cycle progression. *Aspidosperma* compounds may interfere with calcium-dependent processes like ROS generation, calcium channel opening, and disrupt the parasite's normal functioning, leading to its death (Zebelo and Maffei, 2015). Some studies have suggested that *Aspidosperma* extracts and their compounds may have immunomodulatory effects, enhancing the host's immune response against the parasite. Some *Aspidosperma* compounds have been needed to study in more detail for their antimalarial properties. For example, the indole alkaloid aspidospermine has shown cytotoxicity against HepG2 cell line by doing cellular damage (Coatti et al., 2016). Another indole alkaloid, quebrachamine, has also been shown to exhibit antimalarial activity *in vitro*, as well as in vivo, in a mouse model of malaria. Other compounds from *Aspidosperma* species have been reported to exhibit antimalarial activity, such as *Aspidosperma quebracho* tannis (8%) consumption, which reduces the load of parasite in the body maybe by showing activity against different stages of the parasite life cycle, including the blood stage and the liver stage (Copani et al., 2013).

10.5 TRADITIONAL USES

Other than against malaria, *Aspidosperma* spp. are used in various folk medicines (Aquino et al., 2015). Folk utilizations are in the treatments of dysentery, appendicitis, wounds, dyspnoea, asthma,

TABLE 10.2
Various *Aspidosperma* Species Showing Antimalarial Activity

Species Name	Parts Used	Against Malarial Parasite	Antimalarial Compounds	References
A. vargasii	Trunk bark	*P. falciparum* (*in vitro*) and *P. berghei* (in vivo)	Ellipticine	De Paula et al., 2014; Pan et al., 2018
A. quebracho-blanco	Leaves, trunk bark	*P. falciparum* (*in vitro*)	Alkaloids aspidospermine, quebrachamine, and 3-epi-aspidospermine	De Paula et al., 2014
A. tomentosum	Trunk wood, seeds, fruits, and leaves	*P. falciparum* (*in vitro*)	Alkaloid tomentosine	De Paula et al., 2014
A. excelsum	Trunk bark	*P. falciparum*	Alkaloids tahuacamine, demethylaspidospermine, aspidocarpine, and ochrolifuanine	De Paula et al., 2014; do Nascimento et al., 2018, 2019
A. desmanthum	Trunk bark	*P. falciparum* (*in vitro*)	Alkaloids desmanthine and 3,4-methylenedioxyaspidospermidine and aspidocarpine	De Paula et al., 2014; Pan et al., 2018
A. cylindrocarpon	Trunk wood and stem bark	*P. falciparum* (*in vitro*)	Alkaloid quebrachidine, N1-cinnamoyl-12-demethoxycylindrocarine (methyl (2β,12β,19α)-1-[(2E)-3-phenylprop-2-enoyl] aspidospermidin-21-oate)	Chaves et al., 2017; De Paula et al., 2014; Dolabela et al., 2012
A. megacarpon	Root bark	*P. falciparum* (*in vitro*)	Alkaloids megacarpine and 3-oxo-megacarpine	De Paula et al., 2014
A. olivaceum	Trunk bark, trunk wood, and leaves	*P. falciparum* (*in vitro*)	Monoterpenoid indole alkaloids (MIAs) aspidoscarpine, uleine, apparicine, and N-methyl-tetrahydrolivacine and olivacine	Chierrito et al., 2014; De Paula et al., 2014; Dolabela et al., 2012
A. macrocarpon	Root bark	*P. falciparum* (*in vitro*)	Ethanolic extract	de Mesquita et al., 2007
A. spruceanum	Trunk bark, trunk wood, and leaves	*P. falciparum*	Ethanolic, dichloromethane, and chloroform extract	Pan et al., 2018
A. polyneuron	Trunk bark	*P. cathemerium* and *P. falciparum* (*in vitro* and in vivo)	Alkaloids polyneuridine and N-methylpolyneuridine	De Paula et al., 2014
A. megalocarpon	Trunk bark	*P. falciparum* (*in vitro*)	Fendlerine, aspidoalbine, and aspidolimidine	De Paula et al., 2014
A. nitidum	Trunk bark	*P. falciparum* (*in vitro*)	Wayer extract	De Paula et al., 2014
A. parvifolium	Trunk bark, stem bark	*P. falciparum*	Uleine	De Paula et al., 2014; Dolabela et al., 2012
A. ramiflorum	Trunk wood, trunk bark, and leaves	*P. falciparum* (*in vitro*) and *P. berghei* (in vivo)	Alkaloids ramiflorine and ramifloridine	Aguiar et al., 2015; De Paula et al., 2014; Dolabela et al., 2012
A. rigidum	Trunk bark	*P. falciparum* (*in vitro*)	Ethanol extract	De Paula et al., 2014
A. spruceanum	Leaves, trunk wood, and trunk bark	*P. falciparum* (*in vitro*)	Dichloromethane and ethanol	De Paula et al., 2014; Dolabela et al., 2012
A. marcgravianum	Leaves	*P. falciparum*	Tetrahydrousambaresine	De Paula et al., 2014
A. ulei	Leaves, trunk bark, trunk wood, root wood, and root bark	*P. falciparum* (*in vitro*)	**Indole alkaloids**: 20-epi-dasycarpidone, 3,4,5,6-tetradehydro-β-yohimbine, 20(E)-nor-subincanadine E, 19E-hunteracine, 12-hydroxy-N-acetyl-21(N)-dehydroplumeran-18-oic acid	De Paula et al., 2014; Torres et al., 2013

scabies, stomach ache, cough, constipation, boils, rheumatism, toothache, urinary tract infection, and dermatitis (Aquino et al., 2015). *A. pyrifolium* is a typical species in South American folk medicine, commonly used to treat dermatitis and gastritis due to its anti-inflammatory activity (Almeida et al., 2018; Andrade et al., 2019). The Shipiko-Conibo Indian tribes from Peru used *A. excelsum* to treat hepatitis (Rocha et al., 2019). Other tribes in Peru exploited it for the treatments of bronchitis, as vasodilator, antiseptic, antimicrobial, and wound healing (Oliveira et al., 2009; Rocha et al., 2019). *A. excelsum* is popularly used in Brazil in the treatment of diabetes, in blood pressure control, and as contraceptive (Bittencourt et al., 2014; Rocha et al., 2019). Investigations show that this genus generally is not recommended for pregnant women as it can show a positive teratogenic effect and is a potent abortifacient (Bourdy et al., 2000; Agra et al., 2007; Martínez and Barboza, 2010; Aquino et al., 2015). Ethnobotanical studies show that the most promising species of *Aspidosperma* is *A. nitidum*, which is thoroughly utilized by different indigenous people of Amazon basins (Machado et al., 2021).

10.6 CONCLUSION

The study suggests that *Asidosperma* species have promising antimalarial activity, and further studies are needed to fully explore their potential as a source of new antimalarial drugs. The main components of the tree are alkaloid in nature; although they are naturally occurring, their safety and efficacy in human trials make a question mark for their conventional use. The exact mechanism by which *Asidosperma* species exert their antimalarial activity is not yet fully understood. Antimalarial alkaloid compounds like ellipticine, vindoline, and aspidospermine are well studied. There is a broad research area of *Aspidosperma* species–derived antimalarial compounds, which requires the interest of young researcher for the establishment of traditional to medicinal use of this important genera.

REFERENCES

Agra, M. F., Baracho, G. S., Nurit, K., Basílio, I. J., & Coelho, V. P. (2007). Medicinal and poisonous diversity of the flora of "Cariri Paraibano", Brazil. *Journal of Ethnopharmacology*, *111*(2), 383–395. https://doi.org/10.1016/j.jep.2006.12.007

Aguiar, A., Cunha, A., Ceravolo, I. P., Gonçalves, R. A. C., Oliveira, A. J. B., & Krettli, A. U. (2015). Aspidosperma (Apocynaceae) plant cytotoxicity and activity towards malaria parasites. Part II: Experimental studies with Aspidosperma ramiflorum in vivo and in vitro. *SciELO Brasil*. www.scielo.br/j/mioc/a/vh5RKMwsF4x7bCXMSVykqCp/abstract/?lang=en

Almeida, V. L., Silva, C. G., Silva, A. F., Campana, P. R. V., Foubert, K., Lopes, J. C. D., & Pierts, L. (2018). Aspidosperma species: A review of their chemistry and biological activities. *Journal of Ethnopharmacology*, *231*, 125–140.

Andrade, J. C., Santo, W. M., Nóbrega, F. P., Almeida, L. F., Fernandes, F. H., Santana, C. P., Véras, G., & Dantas, A. C. (2019). Thermal characterization of Aspidosperma pyrifolium Mart. plant drugs. *Journal of Thermal Analysis and Calorimetry*, *138*(5), 3747–3756. https://doi.org/10.1007/s10973-019-08073-3

Aquino, P. G. V., de Aquino, T. M., Alexandre-Moreira, M. S., Santos, B. V. de O., Santana, A. E. G., & de Araujo, J. X. (2015). Aspidosperma terpenoid alkaloids – Biosynthetic origin, chemical synthesis and importance. In: *Phytochemicals – Isolation, Characterisation and Role in Human Health*, pp. 163–184. In Tech. https://doi.org/10.5772/59758

Ashok, P., Ganguly, S., & Murugesan, S. (2013). Review on *in-vitro* anti-malarial activity of natural β-carboline alkaloid. *Mini-Reviews in Medicinal Chemistry*, *13*(12), 1778–1791. https://doi.org/10.2174/1389557511313120008.

Bittencourt, B. L. G., Lima, P. G. C., & Barros, F. B. (2014). Trade and use of plants and animals of importance magical/religious and medicinal in market of Guamá, city of Belém, state of Pará. *Review FSA*, *11*, 96–158.

Bourdy, G., DeWalt, S. J., Chávez de Michel, L. R., Roca, A., Deharo, E., Muñoz, V., Balderrama, L., Quenevo, C., & Gimenez, A. (2000). Medicinal plants uses of the Tacana, an Amazonian Bolivian ethnic group. *Journal of Ethnopharmacology*, *70*(2), 87–109. https://doi.org/10.1016/s0378-8741(99)00158-0

Brígido, H. P. C., Varela, E. L. P., Gomes, A. R. Q., Bastos, M. L. C., de Oliveira Feitosa, A., do Rosário Marinho, A. M., Carneiro, L. A., Coelho-Ferreira, M. R., Dolabela, M. F., & Percário, S. (2021). Evaluation of acute and subacute toxicity of ethanolic extract and fraction of alkaloids from bark of Aspidosperma nitidum in mice. *Scientific Reports*, *11*(1), 1–14. https://doi.org/10.1038/s41598-021-97637-1

Ceravolo, I. P., Zani, C. L., Figueiredo, F. J. B., Kohlhoff, M., Santana, A. E. G., & Krettli, A. U. (2018). *Aspidosperma pyrifolium*, a medicinal plant from the Brazilian caatinga, displays a high antiplasmodial activity and low cytotoxicity. *Malaria Journal*, *17*(1). https://doi.org/10.1186/s12936-018-2568-y

Chaves, O. A., Teixeira, F. S. M., Guimarães, H. A., Braz-Filho, R., Vieira, I. J. C., Sant'Anna, R. C. M., Netto-Ferreira, J. C., Cesarin-Sobrinho, D., & Ferreira, A. B. B. (2017). Studies of the interaction between BSA and a plumeran indole alkaloid isolated from the stem bark of *Aspidosperma cylindrocarpon* (Apocynaceae). *Journal of the Brazilian Chemical Society*, *28*(7), 1229–1236. https://doi.org/10.21577/0103-5053.20160285

Chierrito, T. P., Aguiar, A. C., De Andrade, I. M., Ceravolo, I. P., Gonçalves, R. A., De Oliveira, A. J., & Krettli, A. U. (2014). Anti-malarial activity of indole alkaloids isolated from Aspidosperma olivaceum. *Malaria Journal*, *13*(1), 1–10. https://doi.org/10.1186/1475-2875-13-142/FIGURES/8

Coatti, G. C., Marcarini, J. C., Sartori, D., Fidelis, Q. C., Ferreira, D. T., & Mantovani, M. S. (2016). Cytotoxicity, genotoxicity and mechanism of action (via gene expression analysis) of the indole alkaloid aspidospermine (antiparasitic) extracted from Aspidosperma polyneuron in HepG2 cells. *Cytotechnology*, *68*(4), 1161–1170. https://doi.org/10.1007/S10616-015-9874-9/FIGURES/4

Copani, G., Hall, J. O., Miller, J., Priolo, A., & Villalba, J. J. (2013). Plant secondary compounds as complementary resources: Are they always complementary? *Oecologia*, *172*(4), 1041–1049. https://doi.org/10.1007/S00442-012-2551-1/FIGURES/3

Coutinho, J. P., Aguiar, A. C., dos Santos, P. A., Lima, J. C., Rocha, M. G., Zani, C. L., Alves, T. M., Santana, A. E., Pereira, M. M., & Krettli, A. U. (2013). *Aspidosperma* (Apocynaceae) plant cytotoxicity and activity towards malaria parasites. Part I: *Aspidosperma nitidum* (Benth) used as a remedy to treat fever and malaria in the Amazon. *Memórias do Instituto Oswaldo Cruz*, *108*(8). https://doi.org/10.1590/0074-0276130246

de Mesquita, M. L., Grellier, P., Mambu, L., de Paula, J. E., & Espindola, L. S. (2007). In vitro antiplasmodial activity of Brazilian Cerrado plants used as traditional remedies. *Journal of Ethnopharmacology*, *110*(1), 165–170. https://doi.org/10.1016/J.JEP.2006.09.015

De Paula, R. C., Dolabela, M. F., & De Oliveira, A. B. (2014). Aspidosperma species as sources of antimalarials. Part III. A review of traditional use and antimalarial activity. *Planta Medica*, *80*(5), 378–386. https://doi.org/10.1055/S-0034-1368168

Dolabela, M. F., Oliveira, S. G., Peres, J. M., Nascimento, J. M. S., Póvoa, M. M., & Oliveira, A. B. (2012). In vitro antimalarial activity of six Aspidosperma species from the state of Minas Gerais (Brazil). *Anais Da Academia Brasileira de Ciências*, *84*(4), 899–910. https://doi.org/10.1590/S0001-37652012000400005

do Nascimento, M. S., Pina, N. di P. V., da Silva, A. S. B., Gomes, L. F. dos S., de Vasconcellos, F., Brandão, G. C., do Nascimento, M. F. A., de Oliveira, A. B., & Barbosa, W. L. R. (2018). In vitro antiplasmodial activity and identification, using tandem LC-MS, of alkaloids from Aspidosperma excelsum, a plant used to treat malaria in Amazonia. *Journal of Ethnopharmacology*, *228*, 99–109. https://doi.org/10.1016/J.JEP.2018.09.012

do Nascimento, M. S., Pina, N. di P. V., da Silva, A. S. B., Gomes, L. F. dos S., de Vasconcellos, F., Brandão, G. C., do Nascimento, M. F. A., de Oliveira, A. B., & Barbosa, W. L. R. (2019). In vitro antiplasmodial activity and identification, using tandem LC-MS, of alkaloids from Aspidosperma excelsum, a plant used to treat malaria in Amazonia. *Journal of Ethnopharmacology*, *228*, 99–109. https://doi.org/10.1016/J.JEP.2018.09.012

Endress, M. E., Liede-Schumann, S., & Meve, U. (2014). An updated classification for Apocynaceae. *Phytotaxa*, *159*(3). https://doi.org/10.11646/phytotaxa.159.3.2

Garavito, G., Rincón, J., Arteaga, L., Hata, Y., Bourdy, G., Gimenez, A., Pinzón, R., & Deharo, E. (2006). Antimalarial activity of some Colombian medicinal plants. *Journal of Ethnopharmacology*, *107*(3), 460–462. https://doi.org/10.1016/j.jep.2006.03.033

Husen, A. (2021). *Traditional Herbal Therapy for the Human Immune System*. CRC Press, Boca Raton, FL. https://doi.org/10.1201/9781003137955

Husen, A. (2022a). *Herbs, Shrubs and Trees of Potential Medicinal Benefits*. CRC Press, Boca Raton, FL. https://doi.org/10.1201/9781003205067

Husen, A. (2022b). *Exploring Poisonous Plants: Medicinal Values, Toxicity Response and Other Potential Uses*. CRC Press, Boca Raton, FL. https://doi.org/10.1201/b23017

Husen, A. (2023). *Antidiabetic Medicinal Plants and Herbal Treatments*. CRC Press, Boca Raton, FL. https://doi.org/10.1201/b23347

Koch, I., Rapini, A., Kinoshita, L. S., Simões, A. O., & Spina, A. P. (2013). Apocynaceae. Lista de Espécies da Flora do Brasil. *Jardim Botânico do Rio de Janeiro*. http://floradobrasil.jbrj.gov.br/jabot/floradobrasil/FB26640

Krettli, A. U., Adebayo, J. O., & Krettli, L. G. (2009). Testing of natural products and synthetic molecules aiming at new antimalarials. *Curr Drug Targets*, *10*(3). https://doi.org/10.2174/138945009787581203

Lima, R. B., Rocha e Silva, L. F., Melo, M. R., Costa, J. S., Picanço, N. S., Lima, E. S., Vasconcellos, M. C., Boleti, A. P., Santos, J. M., Amorim, R. C., Chaves, F. C., Coutinho, J. P., Tadei, W. P., Krettli, A. U., & Pohlit, A. M. (2015). In vitro and in vivo anti-malarial activity of plants from the Brazilian Amazon. *Malaria Journal*, *14*, 508. https://doi.org/10.1186/s12936-015-0999-2

Machado, I. R., Mendes, K. R., da Silva Moço Tavares, M. K., Arêdes, A., & de Giorge Cerqueira, H. (2021). Ethnobotanical and pharmacological study of *Aspidosperma nitidum* benth (Apocynaceae) – a review. *Biodiversidade Brasileira*, *11*(3). https://doi.org/10.37002/biobrasil.v12i1.1716

Martínez, G. J., & Barboza, G. E. (2010). Natural pharmacopoeia used in traditional Toba medicine for the treatment of parasitosis and skin disorders (Central Chaco, Argentina). *Journal of Ethnopharmacology*, *132*(1), 86–100. https://doi.org/10.1016/j.jep.2010.07.049

Milliken, W., Walker, B. E., Howes, M. R., Forest, F., & Lughadha, E. N. (2021). Plants used traditionally as antimalarials in Latin America: Mining the tree of life for potential new medicines. *Journal of Ethnopharmacology*, *279*. https://doi.org/10.1016/j.jep.2021.114221

Mitaine-Offer, A. C., Sauvain, M., Valentin, A., Callapa, J., Mallié, M., & Zèches-Hanrot, M. (2002). Antiplasmodial activity of *Aspidosperma* indole alkaloids. *Phytomedicine*, *9*(2), 142–145. https://doi.org/10.1078/0944-7113-00094.

Morales, J. F., & Zamora, N. A. (2017). A synopsis of *Aspidosperma* (Apocynaceae) in Mexico and Central America with a taxonomic clarification of *Aspidosperma cruentum* and a new cryptic species. *Phytoneuron*, *68*, 1–13.

Negi, A., Gupta, A., & A. Hamid, A. (2014). Combating malaria with plant molecules: A brief update. *Current Medicinal Chemistry*, *21*(4), 458–500.

Oliveira, V., Freitas, M., Mathias, L., Braz-Filho, R., & Vieira, I. (2009). Atividade biológica e alcalóides indólicos do gênero Aspidosperma (Apocynaceae): uma revisão. *Revista Brasileira de Plantas Medicinais*, *11*(1), 92–99. https://doi.org/10.1590/S1516-05722009000100015

Pan, W. H., Xu, X. Y., Shi, N., Tsang, S. W., & Zhang, H. J. (2018). Antimalarial activity of plant metabolites. *International Journal of Molecular Sciences*, *19*(5), 1382. https://doi.org/10.3390/IJMS19051382

Pereira, A. S. deS., Castello, A. C. D., Simões, A. O., & Koch, I. (2018). Reestablishment, new records, and a key for the species of *Aspidosperma* (Apocynaceae) from the Brazilian Amazon. *Acta Botanica Brasilica*, *33*(1), 1–20. https://doi.org/10.1590/0102-33062018abb0168

Rahman, S., & Husen, A. (2021). Potential role of medicinal plants in the cure of liver and kidney diseases. In: *Non-Timber Forest Products* (Eds. Husen, A., Bachheti, R. K., Bachheti, A.). Springer International Publishing AG, Cham, pp. 229–254. https://doi.org/10.1007/978-3-030-73077-2_10

Rahman, S., & Husen, A. (2022). *Asparagus racemosus* (Shatavari) and *Dioscorea villosa* (Wild yams). In: *Herbs, Shrubs, and Trees of Potential Medicinal Benefits* (Ed. Husen, A.). CRC Press, Boca Raton, FL.

Rahman, S., & Husen, A. (2023a). *Trigonella foenum-graecum* L.: A useful medicinal plant for diabetes and other diseases. In: *Antidiabetic Medicinal Plants and Herbal Treatments* (Ed. Husen, A.). CRC Press, Boca Raton, FL.

Rahman, S., & Husen, A. (2023b). Role of Green Chiretta [*Andaphis paniculata* (Burm. F.)] nees in diabetes and other ailments. In: *Antidiabetic Medicinal Plants and Herbal Treatments* (Ed. Husen, A.). CRC Press, Boca Raton, FL.

Rocha, N. A., Barbosaa, D. B., Raposoa, F. J., Latalizaa, A. A. B., do Amaranteb, C. B., & Raposoa, N. R. B. (2019). Assessment of different biological capacities of aspidosperma excelsum benth. *Journal of Multidisciplinary Engineering Science and Technology*, *6*(1), 9384–9390.

Rocha E Silva, L. F., Montoia, A., Amorim, R. C. N., Melo, M. R., Henrique, M. C., Nunomura, S. M., Costa, M. R. F., Andrade Neto, V. F., Costa, D. S., Dantas, G., Lavrado, J., Moreira, R., Paulo, A., Pinto, A. C., Tadei, W. P., Zacardi, R. S., Eberlin, M. N., & Pohlit, A. M. (2012). Comparative in vitro and in vivo antimalarial activity of the indole alkaloids ellipticine, olivacine, cryptolepine and a synthetic cryptolepine analog. *Phytomedicine*, *20*(1), 71–76. https://doi.org/10.1016/J.PHYMED.2012.09.008

Sampietro, D. A., del Huerto Sánchez Matías, M., Vattuone, M. A., & Catalán, C. A. N. (2021). Aspidosperma quebracho-blanco Schltdl. In: *Medicinal and Aromatic Plants of South America*, Vol. 2 (pp. 75–83). https://doi.org/10.1007/978-3-030-62818-5_4

Schlitzer, M. (2008). Antimalarial Drugs – what is in use and what is in the pipeline. *Arch Pharm (Weinheim)*, *341*(3), 149–163. https://doi.org/10.1002/ardp.200700184

Scudeler, A. L., Castello, A. C. D., Pereira, A. S. S., & Koch, I. (2018). A new species of *Aspidosperma* (Apocynaceae) from the Brazilian *Cerrado*. *Phytotaxa*, *333*(1), 117–123. https://doi.org/10.11646/phytotaxa.333.1.9

Torres, Z. dos S., Silveira, E., Silva, L. F. R. E., Lima, E. S., de Vasconcellos, M. C., de Andrade Uchoa, D. E., Filho, R. B., & Pohlit, A. M. (2013). Chemical composition of *Aspidosperma ulei* markgr. and antiplasmodial activity of selected indole alkaloids. *Molecules*, *18*, 6281–6297. https://doi.org/10.3390/molecules18066281

Uzor, P. F. (2020). Alkaloids from plants with antimalarial activity: A review of recent studies. *Evidence-Based Complementary and Alternative Medicine*, Article ID 8749083. https://doi.org/10.1155/2020/8749083

WHO. (2020). *World Malaria Report 2020*. World Health Organisation, Geneva.

Wiesner, J., Ortmann, R., Jomaa, H., & Schlitzer, M. (2003). New antimalarial drugs. *Angewandte Chemie International Edition*, *42*(43), 5274–5293. https://doi.org/10.1002/anie.200200569

Willcox, M., Benoit-Vical, F., Fowler D., Bourdy, G., Burford, G., Giani, S., Graziose, R., Houghton, P., Randrianarivelojosia, M., & Rasoanaivo, P. (2011). Do ethnobotanical and laboratory data predict clinical safety and efficacy of anti-malarial plants? *Malaria Journal*, *1*(1). https://doi.org/10.1186/1475-2875-10-S1-S7

Zebelo, S. A., & Maffei, M. E. (2015). Role of early signalling events in plant – insect interactions. *Journal of Experimental Botany*, *66*(2), 435–448. https://doi.org/10.1093/JXB/ERU480

11 Antimalarial Response, Traditional and Other Potential Uses of *Cinchona* Genera

Esmail Khosropour, Sahar Ahangari, and Leila Hakimi

11.1 HERBAL MEDICINE

Herbal medicine products (HMPs) are extensively utilized all throughout the world. The World Health Organization reports a number of difficulties with the use of HMPs, including regulatory status, evaluations of safety and efficacy, quality control, safety monitoring, and a lack of awareness of herbal medicine among national drug regulatory agencies (Quan et al., 2020). It is exceedingly difficult to evaluate and regulate herbal treatments scientifically because of the variety of herbal remedies and the ways in which they are used across national boundaries. As a result, while herbal medications are sometimes considered dietary supplements and are not allowed to make therapeutic claims, they are often subject to strict production regulations. The Directives 2001/83/EC and 2004/24/EC, for instance, in Europe categorize herbal medicines into two groups: 'well-established herbal medicinal products' and 'traditional herbal medicinal products'. For the first category, it must be shown by credible bibliographic evidence that the herbal medicinal substance has a long history of successful medical usage, with acceptable levels of safety and efficacy (Chinou et al., 2014). A minimum of ten years of safe and effective use of the herbal medicine product is required. For the second category, herbal medicine producers must show that the traditional herbal medicinal product's safety and efficacy are based on long-standing medical usage: 30 years around the globe, at least 15 of which were spent in the European Union (Heinrich et al., 2023). Contrarily, in the United States, herbal products are classified as dietary supplements under the Dietary Supplement Health and Education Act of 1994, a product category that is exempt from pre-approval. As a result, it is the manufacturer's responsibility to ensure that an herbal product is safe before it is sold, and the Food and Drug Administration can only enforce regulations against any harmful herbal product once it has hit the market. The ineffective standardization and poor quality control in the official herbal sector, in addition to the usage of medicinal plants that are intrinsically poisonous, are the main causes of the toxicity issues with herbal medications (Ekor, 2014; Adusei-Mensah et al., 2019). The quality of the raw ingredients used in the creation of herbal medicines does, in fact, have a direct impact on their safety and effectiveness. In turn, environmental factors (such as light exposure, water accessibility, processing temperature, harvesting season, method, and method of harvesting, drying, packing, transportation, and storage of raw herbal material, etc.) and genetic (genetic) factors determine the quality of the source materials. As a result, it might be difficult to conduct quality checks on the raw materials used to make herbal medicines, especially since numerous countries lack the operational equipment required to apply good manufacturing standards (Alamgir and Alamgir, 2017). The seasonal fluctuations in the quantity of the medicinal plant parts and consequent possible quantitative changes in the composition and/or ratios of bioactive substances make quality control of herbal medicines more challenging (Adusei-Mensah et al., 2019). Additionally, there is a dearth of study on complete herbal mixes, in part because undifferentiated mixtures of natural substances cannot be approved as drugs. Typically, the natural product is removed from its

DOI: 10.1201/9781003378396-11

source, concentrated, separated, and purified until only one physiologically active molecule remains (Lahlou, 2013). It is a common practice for researchers to study medicinal plants in an effort to identify the specific chemical compound that is responsible for the therapeutic effect (Alamgir and Alamgir, 2017; Husen, 2021, 2022a, 2022b, 2023; Rahman and Husen, 2021, 2022, 2023a, 2023b). As a result of the isolation procedure, biological activity may be lost or diminished since it may be the result of the fusion of multiple molecules. In fact, it is already well established that phytocomplexes, which are essentially complex mixes of several substances found in herbal medicines, can have more potent effects than single compounds (Carmona and Pereira, 2013; Heinrich et al., 2023). The effectiveness of the latter often decreases when the extract is further refined into more isolated fractions or single components, according to data acquired from experiments comparing the activities of whole plant extracts with pure preparations. Consequently, the diverse composition of herbal medications is one of its benefits (Najmi et al., 2022). Research is still needed, but potential explanations include synergy, improved bioavailability, cumulative effects, or simply the addictive characteristics of the ingredients.

11.2 MALARIA

Malaria is a fatal and contagious illness that is known to claim millions of lives each year around the world, particularly in Africa (Talapko et al., 2019). The WHO predicts that 212 million cases of malaria occurred in 2015, resulting in 429,000 deaths globally. Around 90% of malaria cases and 92% of malaria-related fatalities occurred in sub-Saharan Africa in 2015 (Thellier et al., 2020). The World Health Organization (WHO) revealed in 2018 that statistics for the years 2015 to 2017 showed no appreciable improvement in lowering malaria cases globally during this time (World Health Organization, 2018). The first time malarial parasites were found in a patient's blood occurred on November 6, 1880, by French army physician Charles Louis Alphonse Laveran. In 1907, he received a Nobel Prize. In 1947, there were 330 million people in the world, and it was estimated that 75 million of them had malaria, which caused 0.8 million deaths annually (Noronha et al., 2020). *Plasmodium falciparum*, *P. vivax*, *P. ovale*, and *P. malariae* are the four *Plasmodium* species that may cause malaria in humans. The majority of malaria-related morbidity and death are caused by *P. falciparum* (Arévalo-Herrera, 2015). Humanity is at risk from resistance to the most widely used antimalarial medication class, artemisinin. Drug resistance against *Plasmodium* species is a growing issue, and it appears that finding and developing new lead compounds to treat this condition is urgently needed (Gupta et al., 2016). According to Bero et al. (2009), ethnopharmacology is one of the greatest places to look for a plant chemical that may be utilized as a model or a starting point for developing an antimalarial drug. The practice of traditional medicine continues to be crucial to health care. According to Sharma et al. (2022), almost 80% of the world's population still relies on traditional plant-based healthcare items for their basic health care. The majority of currently available antimalarial medications have their roots in natural ingredients. Quinine, artemisinin, and many other compounds have already been isolated from herbal remedies (Kingston and Cassera, 2022). Because plants contain a chemical agent repertoire with several valences, nature is a rich source of antimalarial compounds (Noronha et al., 2020).

The *Plasmodium* species, which is responsible for malaria, is a member of the prosperous phylum Apicomplexa. This phylum's members are all obligate intracellular protozoan parasites. Many human generations have been afflicted by plasmodia since the mid-Pleistocene. Humans and African apes share a tight link with the malaria pathogen. In order to properly complete its life cycle, the malaria parasite needs a host and a vector (Hajkazemian et al., 2021). The parasite that causes malaria must penetrate the host's erythrocytes in order to complete its life cycle. An infected *Anopheles* mosquito bite, which results in the release of sporozoites, causes the infection to start. Then, these sporozoites advance on the liver and enter the hepatocytes. The sporozoites in the hepatocytes go through schizogony and produce schizonts. Merozoites are discharged into the circulation when the hepatocytes ultimately burst. After entering the uninfected erythrocytes, the merozoites divide once more via schizogony, creating 8–32 additional merozoites. Merozoites return to the uninfected erythrocytes after the rupture of the infected erythrocyte (Frischknecht, 2023). During the parasite's blood stage, clinical symptoms might be seen. A persistent exposure to re-infection can lead to acquired immunity (Antonelli et al., 2020). The proteins HRP-1, HRP-2, EMP1, and EMP2 are found on the surface of infected erythrocytes (Shooraj et al., 2022). Merozoite invasion into erythrocytes depends on four main events: contact between any part of the parasite and the host cell surface, reorientation of the parasite so that its apical prominence makes contact with the host cell, irreversible attachment, and the formation of an electron-dense junction between the apical prominence of the parasite and the host cell. Finally, the parasite invades the host cell and forms a parasitophorous vacuole. The whole image of the invasion process, however, cannot be given. Since this parasite produces a wide range of antigens at each stage of its life cycle, its complicated life cycle provides a significant barrier to the creation of a vaccine to combat it. In order to create an effective vaccination, it is therefore necessary to comprehend the dynamics of genetic variety (Frischknecht, 2023).

Numerous antimalarial medications are becoming ineffective due to the growing issue of drug resistance, which raises the risk of morbidity and death (Newton et al., 2016). The antimalarials that are now administered to patients have already started to lose their efficacy. As a result, it makes sense to search for a new antimalarial with an innovative mode of action to tackle this fatal illness. Drug pressue is the main cause of antimalarial drug resistance. Drug resistance is spreading widely, making it increasingly difficult to manage malaria (Carrasquilla et al., 2022). An ideal malaria vaccine would function by triggering the immune system to eradicate all parasites. However, it would be challenging to implement this kind of security. In order to prevent the parasite from infecting additional healthy erythrocytes, several vaccine scientists have concentrated their efforts on developing such a vaccine (Quagliata et al., 2023).

11.2.1 Quinolone

Quinolone was discovered in the 17th century. Since the 1600s, quinine, a substance found in cinchona trees, has been used to treat malaria. The term 'Jesuits' bark', 'cardinal's bark', or 'sacred bark' was used at that time (Yadav et al., 2023). Mefloquine and other man-made quinoline medicines have been the primary antimalarials in use for the past 50 years (Foley and Tilley, 1998). By making the feeding vacuole enlarge, increasing the cell's granularity, and ultimately triggering cell lyses, quinine kills the malaria parasite. The parasite's development of quinine resistance, however, took too long (Balogun et al., 2020).

11.2.2 Chloroquine

Chloroquine has a toxic impact on the parasite's digestive vacuole, preventing it from detoxifying the hematin (a product generated during the proteolysis of haemoglobin), which ultimately leads to the parasite's mortality (Herraiz et al., 2019). The mutation inside *P. falciparum* chloroquine resistance (PfCRT), which is studied to have happened at position 76 (K76T), is one among the causes of chloroquine resistance (Djimde et al., 2001). Because of its superior clinical performance, constrained host toxicity, ease of administration, and straightforward, affordable production, chloroquine has long been the preferred medication. However, due to extensive usage in recent years, the malarial parasite has developed resistance, diminishing the effectiveness of the drug (White, 2004). The dihydrofolate reductase (PfDHFR) and the dihydropteroate synthase (PfDHPS) in the folate biosynthetic pathway are the principal targets of the sulfadoxine-pyrimethamine medication. Codons 51, 59, 108, 164, and 581 within pfdhps and 437, 540, and 581 within pfdhfr, the genes that encode these enzymes, are among those that have undergone mutations (Amor et al., 2012; Turkiewicz et al., 2020).

11.2.3 Artemisinin

In many regions of the world, particularly South-East Asia, artemisinin resistance has also developed. Artemisinin is a modern and extensively used antimalarial medicine that is used as a first-line therapy for *P. falciparum* (Uwimana et al., 2020). The primary pharmacophore of this medication is artemisinin, a 1,2,4-trioxane tetracyclic with an endoperoxide. This medication is made from the plant sweet wormwood (Artemisia annua). It has been demonstrated that the artemisinin-resistant Kelch protein, which is found on *P. falciparum*'s chromosome 13 (K13 propeller), is subject to mutation. Studies have suggested that secondary loci may possibly play a role in the same processes. Regarding the way in which artemisinin works, the discussion is still ongoing. There has mostly been attention in two acts (Lyu et al., 2021). Firstly, iron from haemoglobin interacts with a unique bridge in artemisinin. As a result, the detoxification process is repeatedly disrupted, and ROSs are produced, attacking the protein of the parasites. It could also target the mitochondria directly. The likelihood of resistance is lowered by artemisinin's brief half-life. The parasite is now resistant to the medicine, nevertheless, as a result of recent heavy artemisinin usage (Zhang et al., 2021).

11.3 MALARIA AND *CINCHONA*

Quinine and its associated alkaloids, the initial treatment for malaria, are derived from the cinchona tree's bark. The earliest known medicinally useful naturally occurring phytochemical is quinine (Dubey and Singh, 2021). Dr Jaime Jaramillo Arango, a former dean of the National Faculty of Medicine in Bogota, made a strong argument in his book *The Conquest of Malaria* (1950) that native Peruvians were aware of this bark's medicinal advantages against periodic fevers even before the Spaniards arrived in the Andes (Oliveros, 2017). Father Burtolome Tafur, who was thought to be the most malaria-prone person at that time, is said to have carried some bark to a religious gathering in Rome at the beginning of the 16th century (Boumediene, 2020). It became well-liked because it was successful in treating malarial fever. The bark was subsequently frequently given as 'Jesuit bark' or 'fever bark' to cure malarial fever by another Jesuit priest stationed in the Andes and brought to the Holy City. Up until two French chemists identified and refined quinine alkaloid as the most effective compound to treat malaria in the following 200 years, the bark had widespread appeal (Nair, 2021). Therefore, throughout this time, huge natural populations of cinchona trees were regularly uprooted and exploited in the Andes, and tons of the bark were sent to other regions of the world. As it was known to hold rich resources, not only of many taxa of cinchona, but also of other valuable sources of herbal products and pharmaceuticals, it brought the high rainfall-receiving Andean mountain ranges, primarily in Peru, Bolivia, Colombia, and Ecuador, into the focus of explorers, herbalists, and botanists (Nair and Nair, 2021). Ipecac root was also another source of a brand-new drug that was found in the area and used to treat severe diarrhoea. Due to the fact that antimalarial drugs are likely to be scarce and that local governments frequently impose restrictions on their collection and export during the latter part of the 18th century, significant investments were made in the development of cinchona plantations in distant locations. *Plasmodium* is the world's most deadly parasite, killing a significant portion of the human population every year. As a result, profitable pharmaceutical plantation industries were established in Indonesia and India, and later, Western Europe saw a rise in its production capacity. The most significant alkaloid found in cinchona bark is quinine. It is used to prevent and cure malaria in the form of its salts, including sulphate, bisulphate, hydrochloride, and dihydrochloride (Canales et al., 2020). Quinine hydrochloride was administered as an intravenous injection slowly in cases of malignant malaria, which was later discovered to be resistant to synthetic antimalarials. Quinine hydrochloride is also utilized as a sclerosing agent in the treatment of internal haemorrhoids and varicose veins when combined with other substances. Quinine sulphate, which has long been used in European formulations to treat night cramps, has mostly been abandoned because of rare issues that can lead to thrombocytopenia (Dedeoglu et al., 2015). Cinchona bark possesses bitter and astringent tonic qualities. As a result, quinine hydrochloride and quinine sulphate are frequently employed in soft drinks, alcoholic bitters, and liquors as bitter ingredients. Ground cinchona bark was primarily utilized in France as a raw ingredient in the creation of aperitifs and restorative liquors. Quinine is bacteriostatic and extremely effective against protozoa *in vitro*. Additionally, it stops yeast from fermenting. Another compound derived from cinchona bark, quinidine, has been shown to be particularly effective in the treatment of cardiac arrhythmias and is used in its sulphate form as a cardiac depressant. Quinine can be catalytically converted into quinidine, making it a primary input for the creation of quinidine (Dedeoglu et al., 2015).

11.4 HISTORY OF *CINCHONA* GENUS

The first Countess of *Cinchona*, the wife of the Viceroy of Peru, became ill in 1638, and her illness is credited with inspiring the widely believed mythology that European immigrants discovered cinchona bark. As soon as the governor of Loja, Ecuador, learned about it, he dispatched a package of 'quina-quina' bark, which totally healed the woman. The Countess is rumoured to have generously dispersed the remedy upon her return to Spain; as a result, the bark powder was initially also known as 'countessa powder'. The first Countess of *Cinchona* passed away in Spain three years before her husband was named Viceroy of Peru (Nair, 2021). The second Countess never visited Spain and lived a rather healthy life up until the time of her death in Colombia. This famous idea was further debunked by Keeble in 1977. According to him, the first documented usage of cinchona bark as a malaria preventative in Europe took place around the year 1630. It then found its way to England in 1655 after entering the continent's trade in 1643. It acquired official recognition after being included to the pharmacopoeia in 1677 (Nair and Nair, 2021). Although the fever bark tree was given the name *Cinchona offcinalis* by the renowned botanist Carolus Linnaeus in 1747, the name of the Countess de Cinchon continues to be used today. There are now 23 different and verified species in the genus *Cinchona*, which range in size from huge shrubs to tiny trees and are found on the slopes of the Andes rain forests at elevations between 800 and 3,000 metres above mean sea level (MSL). Only a few species, notably *Cinchona calisaya*, *C. ledgeriana*, *C. offcinalis*, and *C. pubescens* (*C. succirubra*), are abundant in quinine and cinchonine alkaloids, according to chemical investigations of these taxa, including their numerous natural hybrids. In bark (of twig, trunk, and root), quinine concentrations vary from 2 to 8%. Prior to the Second World War, Java in Indonesia manufactured and exported cinchona bark and its alkaloids for a value ranging from 16 to 22 million UK sterling pounds yearly, accounting for about 90% of the world's supply. The typical Indian export of quinine salts was valued roughly Rs 10 million (about US$1 = Rs 46 at current exchange rates) annually at that time, making India the second-most significant supplier with a yearly production of 25 tons. While this was going on, the Belgians increased output beyond that of India by building sizable plantations in Zaire, now known as the Democratic Republic of the Congo, in Africa. Being more abundant in quinine, Zaire secured its position next to Indonesia in commerce and made significant profits during the Second World War, when quinine supplies were scarce. A crude crystalline combination of alkaloids was extracted from the bark by Bernardino Gomes of Portugal, and in 1820, French scientists Pettetier and Caventon separated and purified quinine crystals from the bark. Cinchonidine was discovered in 1847, whereas quinidine was found in 1833 (Nair and Nair, 2021).

11.5 CHARACTERISTICS OF *CINCHONA*

The *Cinchona* species is made up of tiny evergreen shrubs and trees that may reach a height of 20 m and a diameter of 15 to 30 cm (Figure 11.1). They are mostly found on the eastern and northern slopes of the Andes, ranging from 19° south to 10° north, and spreading on the opposite side of the equator (Oran et al., 2022). In Bolivia, Colombia, Ecuador, and Peru, *Cinchona* species may be

FIGURE 11.1 The images of full tree and bark parts of *Cinchona* sp.

found growing at 800 m in the pre-mountain tract and up to 3,000 m above MSL. Further north in Venezuela, *Cinchona pubescens* is found. Due to the possibility of interspecific forms existing in the natural population, taxonomists have historically assigned this genus to different numbers of species (30, 40–65). Anderson (1995), however, identified just 23 genuine species after thoroughly studying the Andean Rubiaceae (Nair and Nair, 2021). According to Andersson (1995), the cladistic examination of structural data made up of 48 characters. In order to make the genus monophyletic in origin, he reassigned a few species to related genera, erected *Cinchonopsis* as a new genus, and *Cinchona amazonica* as a new taxon. He came to assume that *Cinchona* belongs to the four subendemic genera despite the Andes having low levels of generic endemism due to high levels of species endemism (reported at about 59% in the tropical Andes). Later developed species may have emerged by adaption to mountain environments since the progenitors most likely lived in low forests of the Andes (Andersson, 1995). The leaves of these plants have an entire border and range in shape from elliptic to oval. The petite, pink, cream, or brown flowers have a short-lobed calyx that is connected at the base of the cinchona species, which has a number of different species. The tubular, four spreading-lobed corollas are an attractive feature. A lovely scent is released by the flowers. The fruit is shaped like an oblong capsule and has 40–50 winged seeds. South America's tropical rainforests are home to some of the most well-known and significant industrial enterprises. In all cultivated species, including the four species raised in India, the fundamental chromosomal number is (x) 17 (Mathews and Philips, 1979). There are several polyploid species (2n = 68). Because they are tougher and generally produce more quinine than either of their parents, several hybrids, including 'Hybrida' and 'Robusta', have proven helpful in the plantations. The four species previously mentioned provide the majority of the commercial supply of bark, despite the fact that many other *Cinchona* species have been investigated for their producing potential in the plantation sector. Due to its high quinine content, *C. ledgeriana* and *C. offcinalis* are the species of choice for the plantation sector (Nair and Nair, 2021).

11.6 DESCRIPTION OF SOME SPECIES OF *CINCHONA*

11.6.1 *Cinchona Calisaya* Wedd (Yellow Bark)

The lower regions of Bolivia and south-east Peru are home to this huge bushy tree with a straight trunk, and 400–1,000 m above sea level is preferred for its habitat. It has thick oblong to lanceolate leaves with smooth surfaces. The inflorescence is a huge panicle with many light-pink flowers on it. The oblong-shaped capsules have a length of 8 to 17 mm. The bark is thick (2–5 mm), brownish, and contains a few transverse fissures that are starting to peel off in spots. The quinine concentration of the trunk bark ranges from 0.78% to 5.57%, making up 3.89–7.24% of the total alkaloids (Nair and Nair, 2021). Under planting, the species exhibit a varied quinine content. The species exhibit varying levels of quinine in plantations. It was first introduced for use in commerce in Sikkim in north-eastern India and Moyar Valley in the Nilgiris in Tamil Nadu but was subsequently supplanted by the more reliable source of quinine, *C. ledgeriana* (Nair and Nair, 2021).

11.6.2 *Cinchona Ledgeriana* Moons (Yellow Bark)

It is a fragile, heavily branched, quickly growing tree that, when fully grown, reaches heights of 6–16 m. In the slopes of Darjeeling, the species flourish between 1,000 and 1,900 m above mean sea level. With a little bend in the axil, the leaves are elliptic, acuminate, and light green. They bloom from May to October and are tiny and light yellow. The capsule is 15 to 19 mm and is oval-lanceolate (Nair and Nair, 2021). Although the fissures are more frequent and shallower than those of *C. calisaya*, the bark is thick (2–5 mm) and comparable in appearance. In the root and trunk, which are 10 to 12 years old, quinine makes about 5.4% and 1.98% of the average total alkaloid content. It is the most abundant source of quinine, with some trees having as much as 14% of the mineral. In plantations, 136 species have been found according to a diversity analysis of the source species (Chatterjee, 1982). The species, which is the parent stock of important plantations in India, is quite particular about the environmental conditions it may grow in. A hybrid known as 'Hybrida', created from *C. pubescens* and *C. ledgeriana*, is preferred in less-optimal circumstances. Despite producing strong growth, the hybrid has a decreased quinine content. The output of bark is 6–6.25 tons/ha, and it contains 2–2.5% quinine (Nair, 2021).

11.6.3 *Cinchona Offcinalis* Linn. (Pale or Crown Bark)

A weak, struggling, narrow, slow-growing tree with dark-green foliage that is 6–10 m tall enjoys a colder temperature at a height of 120–200 m above mean sea level in the Nilgiri hills. Small, smooth, ovate-lanceolate leaves with reddish petioles are 4–10 cm in length. Flowers are 1.4–1.6 cm long, deep pink to pinkish in hue, and produced in tiny terminal panicles. The capsule is 1.5–2 cm length and ovoid-oblong. Around 1.5 mm thick, the bark is rough and brown on the outside and yellow on the inside. The outside surface is covered in many transverse fissures with recurved edges. The typical yield of the trunk bark is just 1.5–2% in India, compared to the global average of 4–6% total alkaloids, up to 50% of which include quinine. Thus, plantations favour the 'Robusta' hybrid of *C. offcinalis* and *C. pubescens*. This robust tree can withstand a broad variety of temperatures and elevations (1,200–3,000 m above mean sea level). The trunk bark contains 2% quinine on average (Nair, 2021).

11.6.4 *Cinchona Pubescens* Vahl. (syn. *Cinchona Succirubra* Pav. ex Klotzsch) (Red Bark)

This sturdy tree has a straight trunk and is native to Ecuador and Peru. It grows quickly as well. It prefers elevations between 1,200 and 2,000 m above mean sea level. It is interesting that the species can survive under both drought and high humidity. The leaves are 30–40 cm long, thin, light green, and elliptical in shape. The upper surface of the 1–2 cm long rose-pink flowers' corolla is white with pink stripes, and they bloom throughout the year. The capsule is 2–3 cm in length and is oblong. The bark is drab brown, 2–6 mm thick, and has a few transverse fissures in addition to well-defined longitudinal wrinkles. It has a high cinchonine concentration but a low quinine level (Nair and Nair, 2021). There are 7.21%, 6.09%, and 4% total alkaloids in the bark of the root, trunk, and twigs, respectively, of which 0.76–1.42%, 1.1–1.74%, and 0.8–1.16% are quinine. The bark contains quinamine, a brand-new alkaloid. Due to its hardiness and rapid growth at lower elevations in Indonesia, the species is utilized as a root stock for the grafting of *C. ledgeriana* (Nair and Nair, 2021).

11.7 CHEMISTRY OF *CINCHONA*

Cinchona bark contains 6–10% of all quinoline alkaloids. The primary components of the alkaloids are quinine, quinidine, cinchonine, and cinchonidine, as well as 30 additional minor bases that are connected to quinine. These alkaloids have quinoline and quinuclidine rings with a vinyl group connected to them. The bark also contains polyphenols, an essential oil; flavonoids, colouring matter (up to 10%); and alkaloids (Gurung and De, 2017). The alkaloids mostly occur as salts of quinine and cinchotanic acids, and the relative quantities of these alkaloids vary amongst various species. Since these alkaloids are produced throughout the sap's descent, their percentage concentration is lowest in the twigs, highest in the trunk bark, and highest in the root bark (Hoffmann Frackenpohl, 2004). The collar region (30–45 cm in length, towards the base) of these has the highest concentration of quinine. Depending on the species, trees' alkaloid content rises between 1 and 12 years of age. Quinine is the primary alkaloid in these plantation species: *C. ledgeriana*, *C. calisaya*, and *C. offcinalis*. Out of its greater (6.0–8.2%) total alkaloids, *C. pubescens* possesses a higher percentage of cinchonine (3.3–3.4%). *C. kartamanahs* have the greatest cinchonidine concentration in their bark at ages 12 and beyond, while 'Hybrida' has the highest cinchonidine level at ages 6 to 8 (3.7–5%) (Nair and Nair, 2021). Due to the bark's high (3–4%) ash content, the source species may be determined quantitatively. Crown bark had the highest ash level. Quinidine (melting point 173.5°C) and quinine (melting point 174.4–175°C) are stereoisomers that both contain a methoxy group (Song, 2009). As a result, quinidine is a quinine dextrorotatory isomer. When exposed to filtered UV light, they exhibit blue fluorescence with oxygenated acids, such as sulfuric acid. Because they lack a methoxy group and do not melt at temperatures over 264°C or 204.5°C, cinchonine and cinchonidine do not exhibit fluorescent properties. Quinine sulphate is the isolated form of quinine from the total alkaloids. It is a tasteless, odourless white crystalline substance with a strong bitterness. It only partially dissolves in water, but it readily dissolves in organic solvents. In addition to quinine salts, *C. febrifuge* (powder), which is made up of a variety of residual alkaloids from quinine extraction, is sold. Similar to this, a typical combination of quinine, cinchonidine, is offered under the brand name 'Totaquinine' and is likewise recommended for the treatment of malaria (Nair, 2021).

Quinidine is found in small quantities (0.2%) in the majority of cinchona barks, although it is considerably more abundant in *C. calisaya* and reaches its highest concentration in *C. tayansis*. It is made commercially by chemically converting quinine by oxidation, and its proportion of commercial conversion relies on parameters like the oxidizing agent, catalysts, and reducing agents used, among other chemistry of cinchona (Song, 2009). For the simultaneous detection of four alkaloids in the bark, a high-performance thin-layer chromatography (HPTLC)/densitometry approach is available that is quick, precise, and affordable. Up to 1% of the world's alkaloids are found in leaves, with younger ones having more. *C. ledgeriana* leaves were examined, and it was discovered that there were four monomeric indole alkaloids (including quinomine, aricine, and 3-epi quinine) and seven quasidimeric indole alkaloids. According to an analysis of *C. ledgeriana* leaves from trees grown in Kenya and Zaire, *C. pubescens* leaves from trees grown in Thailand, and 'Hybrida' leaves, all the aforementioned species' alkaloid contents were extremely low (cf. bark), and none of the quinoline group's alkaloids were present to warrant commercial extraction (Joshi and Singh, 2021).

TABLE 11.1
Antimalarial Properties of *Cinchona* Genus

Species	Results	References
Cinchona calisaya	Quinine and cinchonidine derived from bark have a significant role in antimalarial properties.	Maldonado et al. (2017)
Cinchona calisaya	Quinine from cinchona bark had a good potential in treating and preventing malaria.	Gachelin et al. (2017)
Cinchona calisaya, Cinchona succirubra	At the greatest dose examined, only bark exhibited antimalarial action, lowering parasitaemia and death. Alkaloids were found, but not quinine, according to a phytochemical examination.	Andrade-Neto et al. (2003)
Cinchona sp.	Mutation in the *Plasmodium falciparum* CRT protein determines the stereospecific activity of antimalarial cinchona alkaloids.	Griffin et al. (2012)

11.8 ANTIMALARIAL PROPERTIES OF *CINCHONA*

Since quinine was discovered to be an effective treatment for malaria in the 17th century, species of the genus *Cinchona* (Rubiaceae) have been utilized in traditional medicine and as a source of quinine (Griffin et al., 2012). Little is known about the chemical variety between and within species of the antimalarial alkaloids found in the bark, despite the fact that they have been the only treatment for malaria for over 350 years. Table 11.1 shows some antimalarial works for *cinchona*. For example, Maldonado et al. (2017) represented that quinine and cinchonidine derived from bark have a significant role in treating malaria.

11.9 TRADITIONAL USES OF *CINCHONA*

Natural products and traditional medicines provide a number of distinguishing benefits, including extensive clinical experience, a broad variety of unique chemical structures and biological processes, and rational and reasonable drug development (Yuan et al., 2016). It is now obvious that traditional Chinese medicine (TCM) and contemporary medicine are becoming convergent (Zhang et al., 2020). Through the use of conventional medications and natural poisons, medicine and natural goods have been tightly connected for many years. Ethnobotanicals are important in the study of pharmacology, drug synthesis, drug development, and even serve as models for active pharmaceutical compounds (Süntar, 2020). The following are the main benefits of utilizing conventional medicine: it is natural, safe, and therefore more widely available; it reduces the negative effects of synthetic drugs; it enables individualized health care; and it ensures better public access to health information. To enhance health care, herbal remedies are especially employed as an alternative to illnesses that pose a serious threat to life. When modern medicine is ineffective in treating a condition, traditional medicine is extremely important. As the foundation of contemporary medicine, plants and other natural resources play a significant role in the manufacture of pharmaceutical goods today (Süntar, 2020). Secondary metabolites, or compounds with antimalarial action, come in a variety of forms. Cinchona have been used for treating different types of diseases, especially for malaria in traditional medicine. The bark of these plants is eminent in controlling a wide range of diseases. The principal alkaloids, quinine and quinidine, from the South American cinchona bark are discussed along with their discovery and usage. Since being brought to Europe in the 17th century, cinchona has been used in colonization and warfare, as well as to heal emperors and commoners. In those 400 years, the use of cinchona alkaloids for medicinal purposes has progressed from the extraction of bark through chemical synthesis and supervised clinical trials (Dedeoglu et al., 2015). Due in significant part to their toxicity, the usage of quinine and quinidine has currently decreased. Quinine, on the other hand, made a limited comeback in the treatment of a number of cardiac and neurological diseases, and it is still given in settings with low resources, for severe malaria, and for pregnant women (Gelfman, 2021).

11.10 OTHER POTENTIAL USES

Table 11.2 lists some of the *Cinchona* species' medicinal potential, including some of the following.

11.10.1 Anticancer Potential

Quinine reduces cell growth in a dose- and time-dependent way and affects cancer cells' ability to undergo apoptosis. Anticancer drugs, including bleomycin, cisplatin, anthracyclines, and radiation, which are now used to treat cancer, have been shown to enhance the formation of intracellular ROS. Quinine's effect causes the normal morphological changes associated with apoptosis, including apoptotic signals, loss of apoptotic bodies and adhesion, cell contraction, membrane blistering, chromatid condensation, and nuclear fragmentation. In the future, quinine might be a successful anticancer medicine due to the high apoptotic activity it exhibits in cancer.

11.10.2 Antioxidant Activity

Cinchona is a powerful antioxidant since it contains phenolic compounds. The biological features of phenols include inhibition of lipid peroxidation, anti-HIV, anti-virus, and antitumour effects (Malakar et al., 2018).

11.10.3 Antimicrobial Activity

Some helpful bacteria can also be found in plants. The disc diffusion technique has a range of 8–18 mm for the antibacterial action of cinchona alkaloids on *Staphylococcus aureus.* Its antibacterial effectiveness rises with the amount of cinchona alkaloids present. *Cinchona* combats microbes that are damaging to the human body. Herpes and infections brought on by *P. falciparum* are treated with the bark of the cinchona tree (Al-Harbi et al., 2022).

11.10.4 Anti-Obesity

A potent alkaloid found in cinchona bark is cinchoquine. Although it can treat obesity, its primary uses are antimalaria initiatives. Cinchoquina is superior to other chemical compounds derived from other plants in fighting obesity, according to Jung's 2012 experiments. Chincinnovine demonstrated a larger decrease of body weight than EGCG and curcumin. A helpful dietary phytochemical to stop obesity and inflammation of the fat is pentachloroquine (Jung et al., 2012).

11.11 CONCLUSIONS

Woody medicinal plants are important to human health because they may be used to make herbal medicines from a variety of components, including leaves, fruits, and bark. *Cinchona* is a genus of medicinal trees with a lot of advantages for treating a variety of diseases. Due to its principal chemical components, quinine, quinidine, cinchonine, and cinchonidine, these trees are effective in treating cancer, diabetic disorders, malaria, and microbial infections. Consequently, it is a potential practice to cultivate various species of cinchona in order to substitute chemical medications internationally, particularly in areas where the availability of chemical pharmaceuticals is limited.

TABLE 11.2
Some Therapeutic Potential of *Cinchona* Genus

Property	Compounds	Results	References
Antiviral therapeutic	Quinidine	Antiviral therapeutic applications showed that the COVID-19 pneumonia was significantly influenced by bark extracts of Cinchona calisaya.	Inklebarger et al. (2020)
Antiviral	Quinine	Quinine's effectiveness in promoting antiviral genes to lower dengue virus replication. The development of novel pharmacological therapy for dengue virus may benefit from the antiviral activity of quinine.	Malakar et al. (2018)
Antiplasmodial	Quinine	Heparin on the nanocomposite's surface may dramatically lessen cytotoxicity, biotoxicity, and acute toxicity. Additionally, nonstabilized Fe3O4 particles display substantial aggregation; however, the addition of heparin can act as a stabilizing agent. These non-toxic, biocompatible nanocomposites have a lot of potential for antiplasmodial medication delivery.	Khanmohammadi et al. (2022)
Antidiabetic	Saponins flavonoids, quinine, cinchonine, quinidine	Aqueous extracts of *Vitex doniana* leaves and *Cinchona calisaya* bark have antidiabetic activity in alloxan-induced diabetic rats.	Ezekwesili et al. (2012)
Anticancer	Quinine, flavonoids, phenolics	The functional stabilization of bioactive substances like alkaloids, quinine, flavonoids, phenolics, etc. into the iron oxide, providing bioavailability and internalization of cinchona metabolites intracellularly, may explain the enhanced cytotoxic potential of the *Cinchona officianlis* stem bark extract–loaded iron oxide nanoparticles compared to free *Cinchona officianlis* extract.	Al-Harbi et al. (2022)
Anti-obesity	Cinchonine	Cinchonine dramatically reversed the downregulation of major adipogenic genes, WNT10b, and galanin-mediated signalling molecules that high-fat-diet-(HFD)-induced in the epididymal adipose tissues of mice. Cinchonine inhibited the toll-like receptor-2-(TLR) and TLR4-mediated signalling cascades in the mouse adipose tissue, which decreased the HFD-induced elevation of proinflammatory cytokines. Eating cinchonine may help prevent obesity due to its effects on adipogenesis and inflammation.	Jung et al. (2012)
Antimicrobial activity	Quinine, cinchonine, quinidine	Cinchona alkaloids proved to be an excellent scaffold for further optimization of novel gram-negative bacteria.	Ramić et al. (2021)

REFERENCES

Adusei-Mensah, F., Haaranen, A., Kauhanen, J., Tikkanen-Kaukanen, C., Henneh, I. T., Ganu, D., . . . & Ekor, M. (2019). Post-market safety and efficacy surveillance of herbal medicinal products from users' perspective: A qualitative semi-structured interview study in Kumasi, Ghana. *International Journal of Pharmaceutics and Pharmacology*, *3*, 1–11.

Alamgir, A. N. M., & Alamgir, A. N. M. (2017). Herbal drugs: Their collection, preservation, and preparation; evaluation, quality control, and standardization of herbal drugs. In: *Therapeutic Use of Medicinal Plants and Their Extracts: Volume 1: Pharmacognosy*, Springer, Cham, vol. 73, pp. 453–495.

Al-Harbi, L. N., Al-Shammari, G. M., Subash-Babu, P., Mohammed, M. A., Alkreadees, R. A., & Yagoub, A. E. A. (2022). *Cinchona officinalis* phytochemicals-loaded iron oxide nanoparticles induce cytotoxicity and stimulate apoptosis in MCF-7 human breast cancer cells. *Nanomaterials*, *12*(19), 3393.

Amor, A., Toro, C., Fernández-Martínez, A., Baquero, M., Benito, A., & Berzosa, P. (2012). Molecular markers in *Plasmodium falciparum* linked to resistance to anti-malarial drugs in samples imported from Africa over an eight-year period (2002–2010): Impact of the introduction of artemisinin combination therapy. *Malaria Journal*, *11*, 1–5.

Andersson, L. (1995). Tribes and genera of the Cinchoneae complex (Rubiaceae). *Annals of the Missouri Botanical Garden*, 409–427.

Andrade-Neto, V. F., Brandão, M. G. L., Stehmann, J. R., Oliveira, L. A., & Krettli, A. U. (2003). Antimalarial activity of Cinchona-like plants used to treat fever and malaria in Brazil. *Journal of Ethnopharmacology*, *87*(2–3), 253–256.

Antonelli, L. R., Junqueira, C., Vinetz, J. M., Golenbock, D. T., Ferreira, M. U., & Gazzinelli, R. T. (2020). The immunology of *Plasmodium vivax* malaria. *Immunological Reviews*, *293*(1), 163–189.

Arévalo-Herrera, M., Lopez-Perez, M., Medina, L., Moreno, A., Gutierrez, J. B., & Herrera, S. (2015). Clinical profile of *Plasmodium falciparum* and *Plasmodium vivax* infections in low and unstable malaria transmission settings of Colombia. *Malaria Journal*, *14*, 1–11.

Balogun, T. A., Omoboyowa, D. A., & Saibu, O. A. (2020). In silico anti-malaria activity of quinolone compounds against *Plasmodium falciparum* dihydrofolate reductase (pfDHFR). *International Journal of Biochemistry Research & Review*, *29*(8), 10–17.

Bero, J., Ganfon, H., Jonville, M. C., Frédérich, M., Gbaguidi, F., DeMol, P., . . . & Quetin-Leclercq, J. (2009). In vitro antiplasmodial activity of plants used in Benin in traditional medicine to treat malaria. *Journal of Ethnopharmacology*, *122*(3), 439–444.

Boumediene, S. (2020). 11. Jesuit recipes, Jesuit receipts: The Society of Jesus and the introduction of exotic materia medica into Europe. *Cultural Worlds of the Jesuits in Colonial Latin America*, 229.

Canales, N. A., Hansen, T. N. G., Cornett, C., Walker, K., Driver, F., Antonelli, A., . . . & Rønsted, N. (2020). Historical chemical annotations of Cinchona bark collections are comparable to results from current day high-pressure liquid chromatography technologies. *Journal of Ethnopharmacology*, *249*, 112375.

Carmona, F., & Pereira, A. M. S. (2013). Herbal medicines: Old and new concepts, truths and misunderstandings. *Revitas Brasileira de Farmacognosia*, *23*(2), 379–385.

Carrasquilla, M., Drammeh, N. F., Rawat, M., Sanderson, T., Zenonos, Z., Rayner, J. C., & Lee, M. C. (2022). Barcoding genetically distinct *Plasmodium falciparum* strains for comparative assessment of fitness and antimalarial drug resistance. *Mbio*, *13*(5), e0093722.

Chatterjee, S. K. (1982). Cultivation of cinchona in Darjeeling hills, West Bengal. *Cultivation and Utilization of Medicinal Plants*, *2*, 222–229.

Chinou, I., Knoess, W., & Calapai, G. (2014). Regulation of herbal medicinal products in the EU: An up-to-date scientific review. *Phytochemistry Reviews*, *13*, 539–545.

Dedeoglu, B., Catak, S., Yildirim, A., Bolm, C., & Aviyente, V. (2015). Cinchona alkaloid catalyzed asymmetric desymmetrization of meso-cyclic anhydrides: The origins of stereoselectivity. *ChemCatChem*, *7*(24), 4173–4179.

Djimdé, A., Doumbo, O. K., Cortese, J. F., Kayentao, K., Doumbo, S., Diourté, Y., . . . & Plowe, C. V. (2001). A molecular marker for chloroquine-resistant falciparum malaria. *New England Journal of Medicine*, *344*(4), 257–263.

Dubey, A., & Singh, Y. (2021). Medicinal properties of cinchona alkaloids-A brief review. *Asian Journal of Pharmaceutical Sciences*, *11*(3), 2231–5659.

Ekor, M. (2014). The growing use of herbal medicines: Issues relating to adverse reactions and challenges in monitoring safety. *Frontiers in Pharmacology*, *4*, 177.

Ezekwesili, C. N., Ogbunugafor, H. A., & Ezekwesili-Ofili, J. O. (2012). Anti-diabetic activity of aqueous extracts of Vitex doniana leaves and Cinchona calisaya bark in alloxan – induced diabetic rats. *International Journal of Tropical Disease & Health*, *2*(4), 290–300.

Foley, M., & Tilley, L. (1998). Quinoline antimalarials: Mechanisms of action and resistance and prospects for new agents. *Pharmacology & Therapeutics*, *79*(1), 55–87.

Frischknecht, F. (2023). Life cycle of malaria-causing parasites. In: *Malaria: Deadly Parasites, Exciting Research and No Vaccination*. Springer Fachmedien Wiesbaden, Wiesbaden, pp. 9–18.

Gachelin, G., Garner, P., Ferroni, E., Tröhler, U., & Chalmers, I. (2017). Evaluating cinchona bark and quinine for treating and preventing malaria. *Journal of the Royal Society of Medicine*, *110*(1), 31–40.

Gelfman, D. M. (2021). Reflections on quinine and its importance in dermatology today. *Clinics in Dermatology*, *39*(5), 900–903.

Griffin, C. E., Hoke, J. M., Samarakoon, U., Duan, J., Mu, J., Ferdig, M. T., . . . & Cooper, R. A. (2012). Mutation in the *Plasmodium falciparum* CRT protein determines the stereospecific activity of antimalarial cinchona alkaloids. *Antimicrobial Agents and Chemotherapy*, *56*(10), 5356–5364.

Gupta, D. K., Patra, A. T., Zhu, L., Gupta, A. P., & Bozdech, Z. (2016). DNA damage regulation and its role in drug-related phenotypes in the malaria parasites. *Scientific Reports*, *6*(1), 23603.

Gurung, P., & De, P. (2017). Spectrum of biological properties of cinchona alkaloids: A brief review. *Journal of Pharmacognosy and Phytochemistry*, *6*(4), 162–166.

Hajkazemian, M., Bossé, C., Mozūraitis, R., & Emami, S. N. (2021). Battleground midgut: The cost to the mosquito for hosting the malaria parasite. *Biology of the Cell*, *113*(2), 79–94.

Heinrich, M., Sharma, S. K., Suetterle, U., & Bhamra, S. K. (2023). Herbal medicine use in the UK and Germany and pharmacy practice-A commentary. *Research in Social and Administrative Pharmacy*, *19*(3), 535–540.

Herraiz, T., Guillén, H., González-Peña, D., & Arán, V. J. (2019). Antimalarial quinoline drugs inhibit β-hematin and increase free hemin catalyzing peroxidative reactions and inhibition of cysteine proteases. *Scientific Reports*, *9*(1), 1–16.

Hoffmann, H. M. R., & Frackenpohl, J. (2004). Recent advances in cinchona alkaloid chemistry. *European Journal of Organic Chemistry*, *2004*(21), 4293–4312.

Husen, A. (2021). *Traditional Herbal Therapy for the Human Immune System*. CRC Press, Boca Raton, FL. https://doi.org/10.1201/9781003137955

Husen, A. (2022a). *Herbs, Shrubs and Trees of Potential Medicinal Benefits*. CRC Press, Boca Raton, FL. https://doi.org/10.1201/9781003205067

Husen, A. (2022b). *Exploring Poisonous Plants: Medicinal Values, Toxicity Response and Other Potential Uses*. CRC Press, Boca Raton, FL. https://doi.org/10.1201/b23017

Husen, A. (2023). *Antidiabetic Medicinal Plants and Herbal Treatments*. CRC Press, Boca Raton, FL.

Inklebarger, J., Gyer, G., Galanis, N., Michael, J., & Adel, G. (2020). Cinchona bark for the treatment of Covid-19 pneumonia: A modern review of the potential anti-viral therapeutic applications of an old treatment. *International Journal of Medical Science and Clinical Invention*, *7*(5), 4795–4801.

Joshi, H., & Singh, V. K. (2022). Cinchona derivatives as bifunctional H-bonding organocatalysts in asymmetric vinylogous conjugate addition reactions. *Asian Journal of Organic Chemistry*, *11*(2), e202100053.

Jung, S. A., Choi, M., Kim, S., Yu, R., & Park, T. (2012). Cinchonine prevents high-fat-diet-induced obesity through downregulation of adipogenesis and adipose inflammation. *PPAR Research*, *2012*.

Khanmohammadi, A., Sadighian, S., & Ramazani, A. (2022). Anti-plasmodial effects of quinine-loaded magnetic nanocomposite coated with heparin. *International Journal of Pharmaceutics*, *628*, 122260.

Kingston, D. G. I., & Cassera, M. B. (2022). Antimalarial natural products. In: *Antimalarial Natural Products. Progress in the Chemistry of Organic Natural Products* (Eds. Kinghorn, A. D., Falk, H., Gibbons, S., Asakawa, Y., Liu, J. K., Dirsch, V. M.), vol. 117. Springer, Cham, Switzerland.

Lahlou, M. (2013). The success of natural products in drug discovery. *Pharmacology & Pharmacy*, *4*(3A), 17–31.

Lyu, H. N., Ma, N., Meng, Y., Zhang, X., Wong, Y. K., Xu, C., . . . & Wang, J. (2021). Study towards improving artemisinin-based combination therapies. *Natural Product Reports*, *38*(7), 1243–1250.

Malakar, S., Sreelatha, L., Dechtawewat, T., Noisakran, S., Yenchitsomanus, P. T., Chu, J. J. H., & Limjindaporn, T. (2018). Drug repurposing of quinine as antiviral against dengue virus infection. *Virus Research*, *255*, 171–178.

Maldonado, C., Barnes, C. J., Cornett, C., Holmfred, E., Hansen, S. H., Persson, C., . . . & Rønsted, N. (2017). Phylogeny predicts the quantity of antimalarial alkaloids within the iconic yellow Cinchona bark (Rubiaceae: Cinchona calisaya). *Frontiers in Plant Science*, *8*, 391.

Mathews, P. M., & Philips, K. O. (1979). Studies on south Indian Rubiaceae. Pt II. Cytology of cinchona linn. *Nucleus (India)*, *22*(2), 125–128.

Nair, K.P. (2021). Cinchona (Cinchona sp.). In: *Tree Crops*. Springer, Cham. pp 129–151, https://doi.org/10.1007/978-3-030-62140-7_4

Najmi, A., Javed, S. A., Al Bratty, M., & Alhazmi, H. A. (2022). Modern approaches in the discovery and development of plant-based natural products and their analogues as potential therapeutic agents. *Molecules*, *27*(2), 349.

Newton, P. N., Caillet, C., & Guerin, P. J. (2016). A link between poor quality antimalarials and malaria drug resistance? *Expert Review of Anti-Infective Therapy*, *14*(6), 531–533.

Noronha, M., Pawar, V., Prajapati, A., & Subramanian, R. B. (2020). A literature review on traditional herbal medicines for malaria. *South African Journal of Botany*, *128*, 292–303.

Oliveros, M. E. O. (2017). *In Search of Precious Bark: Circulation of Quina and Colonial Medical Culture in the New Kingdom of Granada at the End of the Eighteenth Century*. McGill University, Canada.

Oran, S. A., Althaher, A. R., & Mubarak, M. S. (2022). *Cinchona officinalis* (Cinchona Tree) and Corylus avellana (Common Hazel). In: *Herbs, Shrubs, and Trees of Potential Medicinal Benefits*. (Ed. Husen A) CRC Press 6000 Broken Sound Parkway NW, Suite 300, Boca Raton, FL 33487-2742, pp. 377–394.

Quagliata, M., Papini, A. M., & Rovero, P. (2023). Malaria vaccines. *Expert Opinion on Therapeutic Patents*, *33*, 169–178.

Quan, N. V., Dang Xuan, T., & Teschke, R. (2020). Potential hepatotoxins found in herbal medicinal products: A systematic review. *International Journal of Molecular Sciences*, *21*(14), 5011.

Rahman, S., & Husen, A. (2021). Potential role of medicinal plants in the cure of liver and kidney diseases. In: *Non-Timber Forest Products* (Eds. Husen, A., Bachheti, R. K., Bachheti, A.). Springer International Publishing AG, Cham, pp. 229–254. https://doi.org/10.1007/978-3-030-73077-2_10

Rahman, S., & Husen, A. (2022). *Asparagus racemosus* (Shatavari) and *Dioscorea villosa* (Wild yams), In: *Herbs, Shrubs, and Trees of Potential Medicinal Benefits* (Ed. Husen, A.). CRC Press, Boca Raton, FL.

Rahman, S., & Husen, A. (2023a). *Trigonella foenum-graecum* L.: A useful medicinal plant for diabetes and other diseases. In: *Antidiabetic Medicinal Plants and Herbal Treatments* (Ed. Husen, A.). CRC Press, Boca Raton, FL.

Rahman, S., & Husen, A. (2023b). Role of Green Chiretta [*Andaphis paniculata* (Burm. F.)] nees in diabetes and other ailments. In: *Antidiabetic Medicinal Plants and Herbal Treatments* (Ed. Husen, A.). CRC Press, Boca Raton, FL.

Ramić, A., Skočibušić, M., Odžak, R., Čipak Gašparović, A., Milković, L., Mikelić, A., . . . & Hrenar, T. (2021). Antimicrobial activity of quasi-enantiomeric cinchona alkaloid derivatives and prediction model developed by machine learning. *Antibiotics*, *10*(6), 659.

Sharma, P. K., Singh, A., & Sharma, N. K. (2022). A socio-ecological critique on India's local health traditions amidst rising incidence of global pandemics. *Journal of Herbal Medicine*, 100578.

Shooraj, M., Mousavi Zade, M. A. S., & Mahdavi, S. A. (2022). A review on herbal treatments of malaria. *Tabari Biomedical Student Research Journal*, *4*(4), 28–35.

Song, C. E. (2009). *An Overview of Cinchona Alkaloids in Chemistry*; Wiley-VCH Verlag GmbH & Co. KGaA, Germany, pp. 1–10.

Süntar, I. (2020). Importance of ethnopharmacological studies in drug discovery: Role of medicinal plants. *Phytochemistry Reviews*, *19*(5), 1199–1209.

Talapko, J., Škrlec, I., Alebić, T., Jukić, M., & Včev, A. (2019). Malaria: The past and the present. *Microorganisms*, *7*(6), 179.

Thellier, M., Simard, F., Musset, L., Cot, M., Velut, G., Kendjo, E., & Pradines, B. (2020). Changes in malaria epidemiology in France and worldwide, 2000–2015. *Medecine et Maladies Infectieuses*, *50*(2), 99–112.

Turkiewicz, A., Manko, E., Sutherland, C. J., Diez Benavente, E., Campino, S., & Clark, T. G. (2020). Genetic diversity of the *Plasmodium falciparum* GTP-cyclohydrolase 1, dihydrofolate reductase and dihydropteroate synthetase genes reveals new insights into sulfadoxine-pyrimethamine antimalarial drug resistance. *PLoS Genetics*, *16*(12), e1009268.

Uwimana, A., Legrand, E., Stokes, B. H., Ndikumana, J. L. M., Warsame, M., Umulisa, N., . . . & Menard, D. (2020). Emergence and clonal expansion of in vitro artemisinin-resistant *Plasmodium falciparum* kelch13 R561H mutant parasites in Rwanda. *Nature Medicine*, *26*(10), 1602–1608.

White, N. J. (2004). Antimalarial drug resistance. *The Journal of Clinical Investigation*, *113*(8), 1084–1092.

World Health Organization. (2018). *First Annual Country Surveillance Meeting to Facilitate Malaria Elimination in the Greater Mekong Subregion, Vientiane, Lao People's Democratic Republic, 16 November 2018: Meeting Report*. World Health Organization, Philippines.

Yadav, P., Sharma, R., Khan, S., Patel, J., & Patel, R. (2023). Development and estimation of an oro-dispersible anti-malarial dosage form for pediatric patient. *World Journal of Biology Pharmacy and Health Science*, 13(1), 250–257.

Yuan, H., Ma, Q., Ye, L., & Piao, G. (2016). The traditional medicine and modern medicine from natural products. *Molecules*, *21*(5), 559.

Zhang, D. H., Wu, K. L., Zhang, X., Deng, S. Q., & Peng, B. (2020). In silico screening of Chinese herbal medicines with the potential to directly inhibit 2019 novel coronavirus. *Journal of Integrative Medicine*, *18*(2), 152–158.

Zhang, Q., Yi, H., Yao, H., Lu, L., He, G., Wu, M., . . . & Deng, X. (2021). Artemisinin derivatives inhibit non-small cell lung cancer cells through induction of ROS-dependent apoptosis/ferroptosis. *Journal of Cancer*, *12*(13), 4075.

12 Antimalarial Response, Traditional and Other Potential Uses of *Croton* Genera

Rudra Narayan Sahoo, Sachin Mahawar, and Atul Arya

12.1 INTRODUCTION

Nature is the epicentre of all ecosystem functions. All living organisms, including humans, interact with each other and evolve to produce a set of practices that are later firmly established as culture. For ages, plants have been a significant player in this association; it is deeply rooted in the indigenous knowledge of traditional therapeutic systems (Jaiswal, 2019; Sahoo et al., 2022; Husen, 2021, 2022a, 2022b, 2023; Rahman & Husen, 2021, 2022, 2023a, 2023b). In the developing world, about 80% of the population use folk remedies because they are easily accessible and cheaper than modern medicine and healthcare systems, are more accepted because of their socio-cultural and spiritual beliefs, and are used as alternative medicine (Oyebode et al., 2016; WHO, 2019; Arya et al., 2021).

Malaria is an acute delirious illness that is widespread in Africa, South America, and Asia, with predicted 247 million cases and 619,000 deaths globally as per most recent world malaria report of the WHO in 2021 (WHO, 2022). The key impediments to the worldwide malaria eradication initiative have been recognized as economic (Barat et al., 2004; Njau et al., 2006; Mills et al., 2008), geographical (Tobgay et al., 2011; Kaehler et al., 2019), and health system concerns (Chuma et al., 2010; Rao et al., 2013; Suswardany et al., 2015). Additionally, studies have reported frequently restricted healthcare access in low- to middle-income nations, particularly in rural and isolated areas (Peters et al., 2008; Dewi et al., 2023). Traditional medicine (TM) and traditional healers (TH) are significant and frequently utilized parts of healthcare-seeking and treatment for many people in poor to middle-income Asian and Pacific nations, as well as worldwide (Suswardany et al., 2015; Arya et al., 2021), because of their fewer side effects, cost-effectiveness, and affordability, encouraging herbal-based drug discovery (Chakraborty, 2018; Mohammadi et al., 2020). These practices, treatments, and providers are culturally indigenous and have historically functioned predominantly outside the state-funded healthcare system as well as beyond the practices and curriculum of the mainstream medical profession to treat both acute and chronic diseases, including malaria (Adams & Andrews, 2012; Suswardany et al., 2015; Sahoo et al., 2022). Several cohort studies, such as Debas et al. (2006); Okeke et al. (2006); Sawalha (2007); Assan et al. (2009); Picking et al. (2011); Okoronkwo et al. (2014); WHO (2019); and Arya et al. (2021), have validated the prevalence of TM/TH in low- to middle-income countries and, in some scenarios, the only healthcare accessible in these communities.

Historically, plants, such as *Cinchona* spp. (Rubiaceae) and *Artemisia* spp. (Asteraceae), have contributed to or inspired the development of antimalarial medications, raising the possibility that they will produce further malaria therapies (Milliken, 1997; Achan et al., 2011). Similarly, ethnobotanical/ethnopharmacological investigations corroborated the wide use of the genus *Croton*, of the Euphorbiaceae family, encompassing approximately 1,300 species found in the tropical and subtropical parts across the world, and several species have been extensively reported as medicinal

 DOI: 10.1201/9781003378396-12

herbs in Asia, Africa, and South America (Salatino et al., 2007). The genus is popular for treating a wide array of ailments, such as cancer, diabetes, malaria, fever, dysentery, digestive problems, constipation, external wounds, hypercholesterolaemia, hypertension, inflammation, intestinal worms, pain, tuberculosis, ulcers, and weight loss (Salatino et al., 2007; Ndunda, 2014; Rampa et al., 2022) (Figure 12.1). Furthermore, the genus is characterized by several potent therapeutic secondary metabolites, primarily diterpenoids, such as clerodane, kaurane, labdane, pimarane, phorbol esters, and trachylobane (Block et al., 2002); active alkaloids (Amaral & Barnes, 1998; Milanowski et al., 2002); aromatic oil constituents (Oliveira et al., 2001; Lopes et al., 2003); and red latex reported in case of some species (Sandoval et al., 2002; Risco et al., 2003). This chapter aims to provide an understanding of the antimalarial response, their effective traditional usage, and their other potential uses by various indigenous communities of *Croton* genera as a prospective antimalarial drug.

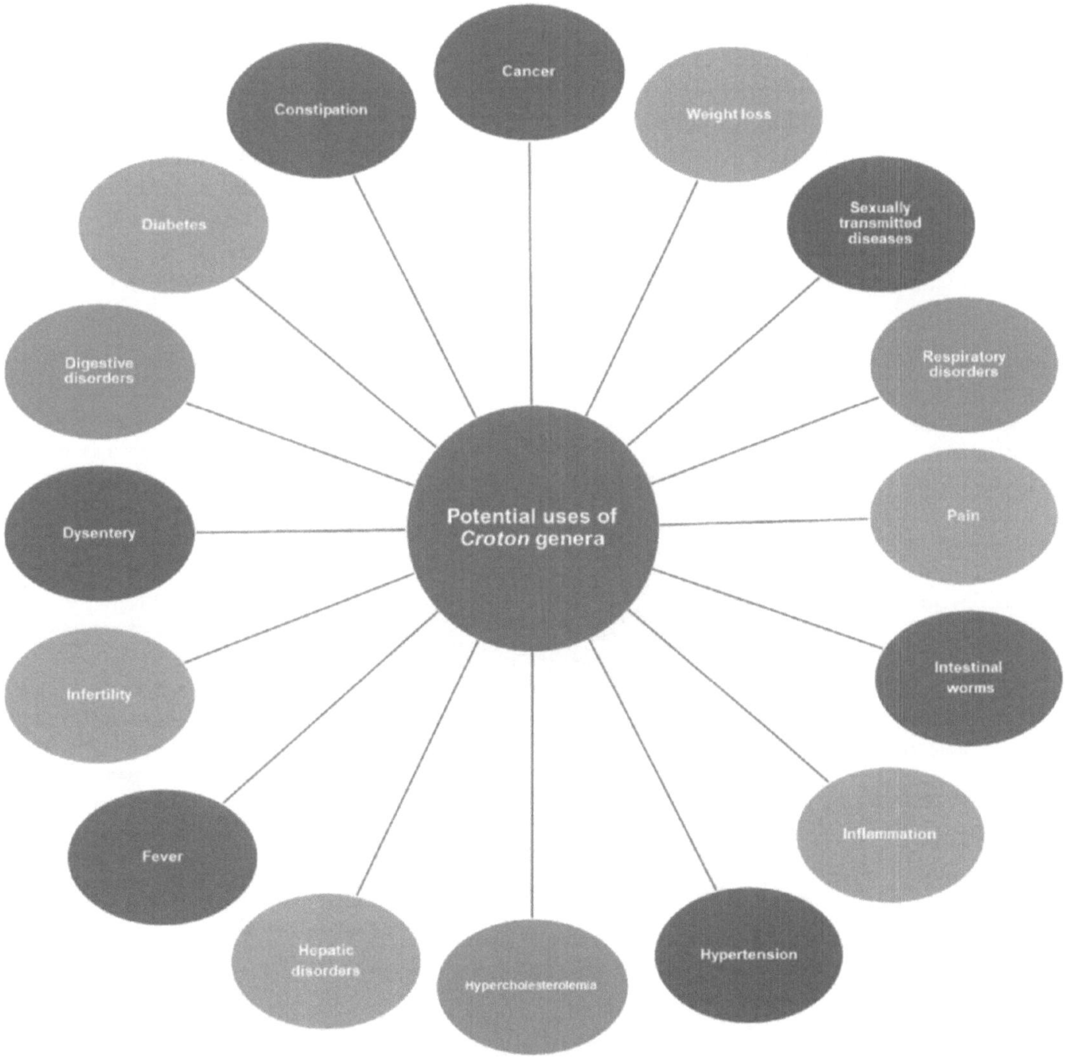

FIGURE 12.1 Outline of other potential uses of *Croton* genera.

12.2 *CROTON* SPECIES: ANTIMALARIAL RESPONSE, TRADITIONAL AND OTHER POTENTIAL USES

The effectiveness of the present treatments is threatened by parasite resistance to antimalarial medications and mosquito resistance to insecticides, potentially leading to a rise in worldwide malaria mortality (Obey et al., 2018). However, many *Croton* species have demonstrated encouraging antimalarial efficacy in diverse *in vitro* and *in vivo* experiments (Okokon et al., 2005; Boyom et al., 2009; Bantie et al., 2014). Several levels of efficiency have been documented in the literature, and a large variety of metabolites produced from these various species have been intensively examined for their antimalarial activity (Ferreira et al., 2008; Kaur et al., 2009; Oliveira et al., 2009; Bantie et al., 2014). This section describes the specific *Croton* species used to cure malaria (Figure 12.2), along with their customary and prospective alternative uses.

***Croton argyratus* Blume**, also known as 'Hujan panas', 'Hamba raja', 'silver croton', 'Melokan, Semangkok', 'Cheret budak', and 'Semelit mayor', is a shrub that is extensively spread throughout South-East Asia, including Malaysia, Thailand, Vietnam, Indonesia, Philippines, Burma, and Bali (Burkill, 1966; Suárez et al., 2006; N. I. M. Ali et al., 2012). The plant includes a variety of secondary metabolites, including cardiac glycosides, flavonoids, phenolics, terpenoids, steroids, and tannins (N. I. M. Ali et al., 2012). The antiplasmodial action of *C. argyratus* Blume's leaves against the *Plasmodium falciparum* D10 strain results in a 60% suppression of the parasite's development, making it a well-liked traditional treatment for malaria fever in Malaysia (Noor Rain et al., 2007). Moreover, the tree's root is historically used in Jambi, Indonesia, to treat malaria, fever, diarrhoea, and afterbirth cure (Noor Rain et al., 2007; Saraswaty et al., 2012), while an infusion of the roots is used to treat thrush and ulcers (Burkill, 1966; Salleh et al., 2022). The cytotoxicity of the methanolic extract of the leaves/twigs, roots, and stem bark against human lung cancer cell lines and Madin-darby bovine kidney (MDBK) cells was

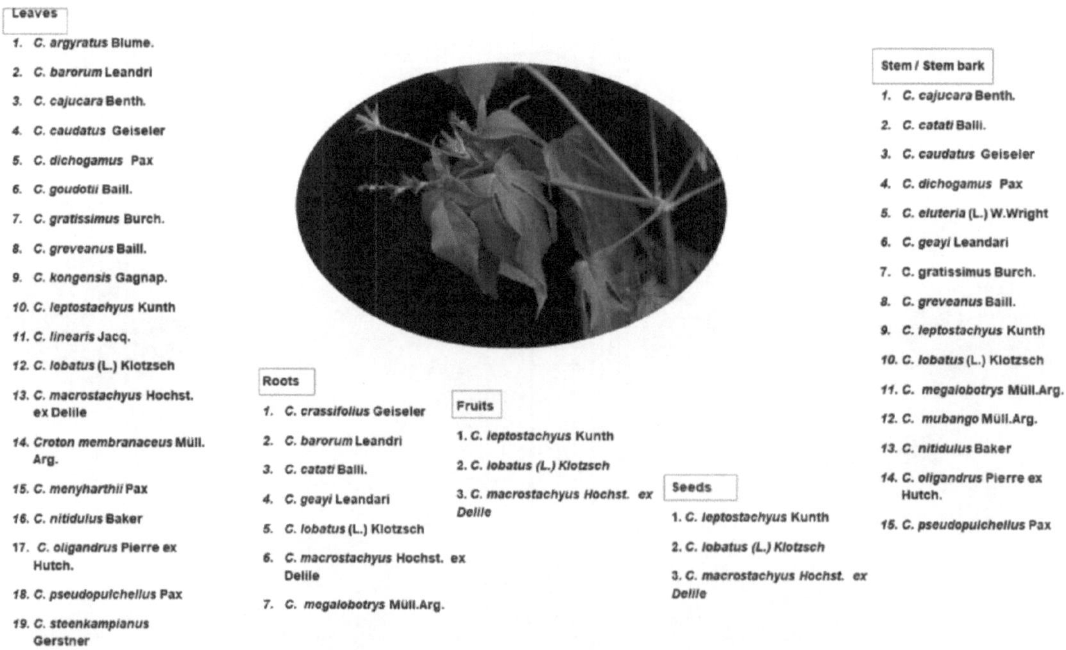

FIGURE 12.2 Overview of different species of *Croton* genera with their parts used for treatment of malaria. [Image used to represent *Croton* genera – *Croton lobatus* L.]

Source: Morales (2014).

discovered (Horgen et al., 2001; Noor Rain et al., 2007). The stem's methanolic and acetone extracts also showed antibacterial action against certain bacterial populations. Furthermore, the leaves of *C. argyratus* Blume showed antioxidant reduction capability in the DPPH assay, with an IC50 value of 1.45 mg/ml, as well as an excellent value for total antioxidant capacity, indicating that leaves extract of the species might be employed as an analgesic (Noor Rain et al., 2007).

***Croton barorum* Leandri**, a plant species native to Madagascar, is known locally as 'Somorombohitse' by Malagasy populations (Fatiany et al., 2016). The plant includes a variety of diterpenoid chemicals that have medicinal effects for a variety of diseases (Hanson, 2011). The leaves and stem bark of *C. barorum* Leandri are well recognized for their antimalarial effects and have long been used to cure malaria in Madagascar (Fatiany et al., 2016). Moreover, the plant's leaves, stems, and roots have been used to treat asthma, breast cancer, diarrhoea, fever, hypertension and contain anticancer, antibacterial, and antiproliferative properties (Rasoanaivo et al., 1992; Schmelzer et al., 2008; Rakotonandrasana et al., 2010; Hanson, 2011; Kuete et al., 2013; Kuete & Efferth, 2014; Fatiany et al., 2016; Maroyi, 2019).

***Croton cajucara* Benth.**, often known as 'Sacaca' in Brazil, is a traditional medicinal plant that grows in South America along the Amazon River (Maciel et al., 1998; Salatino et al., 2007). Its leaves and stem bark are high in phytochemical components, such as flavonoids, phenolics, terpenoids (sesquiterpenes, mono-, di-, and triterpenoids), and volatile oils, giving it medicinal promise for treating a variety of health issues in the local region (Salatino et al., 2007). The plant possesses antimalarial properties against *Plasmodium falciparum* and *Plasmodium berghei* species (Lima et al., 2015), and its leaves and stem bark are used in tea or tablets in Brazil to treat gastrointestinal tract difficulties, high blood cholesterol, hyperglycaemia, and hepatic disorders (Campos et al., 2002; Grassi-Kassisse et al., 2010; Salatino et al., 2007). Additionally, the bark volatile oil has been demonstrated to cure stomach ulcers (Hiruma-Lima et al., 2000; Salatino et al., 2007; Hiruma-Lima et al., 2010), and the bark water extract can aid in weight reduction and boost adipocyte sensitivity to adrenaline and isoprenaline (Salatino et al., 2007; Grassi-Kassisse et al., 2010). The volatile oil from the leaves contains linalool, which possesses antileishmanial properties (Salatino et al., 2007; Grassi-Kassisse et al., 2010). In mice models, active chemicals from the plant, such as trans-crotonin, trans-dehydrocrotonin, and acetyl aleuritolic acid, exhibited hypolipidemic and hypoglycaemic characteristics (Maciel et al., 2000; Campos et al., 2002). Moreover, trans-dehydrocrotonin from the plant's aerial portions has anti-oestrogen and anticancer action in the case of mice models, while linalool from the bark's volatile oil has antibacterial and antifungal properties (Grynberg et al., 1999; Alviano et al., 2005; Salatino et al., 2007).

***Croton catati* Baill.** is a tropical tree endemic to Madagascar that is utilized for a variety of therapeutic applications (Schmelzer et al., 2008). The decoction of the root and stem bark is taken orally for the treatment of malaria (Schmelzer et al., 2008; Maroyi, 2019). In addition, the leaves, roots, and stem bark of *C. catati* Baill. are used in various formulations, such as decoctions or infusions, for the treatment of coughs, disinfectants, dyspnoea, euphoria, and stomach pain (Schmelzer et al., 2008; Maroyi, 2019).

***Croton caudatus* Geiseler**, popularly known as 'Tukul Takal' in Malaysia, is a shrub or tree that is widely distributed in tropical regions of South-East Asia (Dasiman et al., 2020). Phytochemical screening of the plant reveals that its leaves contain alkaloids, flavonoids, cyanogenic glycosides, and phenolic compounds, which provide its medicinal value (Nath et al., 2013b). In Malaysia, specifically, the plant's leaves are used in the treatment of malarial fever (Nath et al., 2013b; Dasiman et al., 2020). *C. caudatus* Geiseler is also used in the treatment of fever, convulsions, pain relief, stomach problems, and rheumatic arthritis (Nath et al., 2013b). The Saikot area of Manipur has reported the use of *C. caudatus* Geiseler in cancer treatment (Nath et al., 2013b). In West Malaysia, the Orang Asli tribe used the boiled infusion of the root of *C. caudatus* Geiseler as a source of nutrient (Dasiman et al., 2020). The leaf extract of the plant also exhibits antidiabetic, antioxidant, antitumour, and cytotoxic activity (Nath, et al., 2013b; Dasiman et al., 2020).

***Croton crassifolius* Geiseler**, known colloquially as 'Pang khi' and 'Phang khi noi', is a monecious undershrub distributed across Asia, especially in South and South-East China, Thailand, Laos, and Vietnam (Zhang et al., 2014). The therapeutic characteristics of *C. crassifolius* Geiseler's roots have been employed in traditional medicine in China, Thailand, and Vietnam (Zhang et al., 2014; Huang et al., 2015). Sesquiterpenoids and diterpenoids found in the roots give the plant its therapeutic effects (Boonyaratavej et al., 1988; Zhang et al., 2014, 2015). The plant roots are historically used to treat malaria in Laos in the form of a decoction of roots with sugar in 2 L of water (Elliott et al., 2020). In Chinese folk medicine, the roots are used to cure rheumatism, sore throats, jaundice, snakebites, stomach ache, joint pain, and pharyngitis (Zhang et al., 2014; Huang et al., 2015; Flora of China, 2023). Thai indigenous peoples also utilize the herb to cure cancer (Boonyaratavej et al., 1988; Zhang et al., 2014; Huang et al., 2015).

***Croton dichogamus* Pax**, also known as 'Kulelwat', 'Kelerwe', 'Mwalula', 'Olokididigai', 'Rachar', 'Muthia', 'Mhande', and 'Maffuka', is a plant species utilized for traditional medicinal purposes in many communities across the world (Tropical Plants Database, 2012). The species has been recorded in Ethiopia, Kenya, Mozambique, Madagascar, Rwanda, Somalia, and Tanzania (Hedberg et al., 1983; Salatino et al., 2007; Tropical Plants Database, 2012; Palhares et al., 2015; Mutie et al., 2020). Phytochemical tests on the plant's crude extract from the leaves and stem bark indicated the presence of a variety of secondary metabolites, including flavonoids, alkaloids, saponins, tannins, terpenoids, and steroids (Johns et al., 1999; Aldhaher et al., 2016, 2017; Matara et al., 2021). These substances are responsible for the plant's historic usage in treating digestive disorders through inhalation of smoke from burnt leaves, functioning as an antipyretic agent, and as a nutritional supplement in milk (Hedberg et al., 1983; Aldhaher, 2016; Kigen et al., 2019; Matara et al., 2021). The milled roots of *C. dichogamus* Pax are used with porridge in Tanzania to treat tuberculosis and respiratory issues (Hedberg et al., 1983; Matara et al., 2021). Moreover, the plant is used by the East African Batemi and Massai tribes to produce soups for cardiovascular diseases associated with excessive cholesterol (Johns et al., 1999; Matara et al., 2021). The Samburu community in the Loitoktok region of Kenya uses the plant to cure gonorrhoea and arthritis (Aldhaher et al., 2017; Matara et al., 2021). The root decoction is used as a drink to cure impotence and infertility, while the stem bark, leaves, and infusion are used to treat back pain, malaria, chest problems, coughs, and stomach ache (Mutie et al., 2020; Matara et al., 2021). In Ethiopia, the roots of *C. dichogamus* Pax are used as a paste to increase female fertility (Sori et al., 2004; Matara et al., 2021). Furthermore, a number of studies, such as Hedberg et al. (1983), Pascaline et al. (2011), Mutie et al. (2020), and Omara (2020), have reported that several African cultures used the aerial parts of the plant to treat malaria, and that the essential oil isolated from the aerial parts has antimalarial activity against *Anopheles gambie*, which is commonly seen in Lake Basin malaria.

***Croton eluteria* (L.) W. Wright**, often known as 'Cascarilla', is a tropical shrub native to the West Indies and Northern South America. This plant is rich in diterpenoids, which have the potential to cure a variety of human ailments (Salatino et al., 2007). *C. eluteria*'s bitter bark has been used to cure malaria, and it has also been claimed to have hypotensive, narcotic, balsamic, stomachic, and tonic characteristics, making it useful in the treatment of bronchitis, diarrhoea, and dysentery (Vigor et al., 2001; Salatino et al., 2007). Also, the bitter bark extract is frequently utilized to increase gastric output, demonstrating its versatility with a wide range of medicinal applications (Appendino et al., 2003).

***Croton geayi* Leandri** is known by the vernacular name 'Pisopisovavy' among the Malagasy/local communities of Madagascar (Fatiany et al., 2016). The plant is rich in various kinds of diterpenoid compounds and essential oils, which give it therapeutic value in the treatment of various diseases (Palazzino et al., 1998; Radulović et al., 2006; Fatiany et al., 2016). The infusion of stem and root bark of the plant is well-known for its antimalarial properties in Madagascar, where it is used to treat malaria (Fatiany et al., 2016; Maroyi, 2019). In addition to its use in malaria treatment, *C. geayi* Leandri leaves, stems, bark, and roots are also used to treat various health ailments, such as cough, fever, flu, constipation, asthma, stomach disorders, wounds healing, injury, hypertension, antibacterial (Palazzino et al., 1998; Andriamparany et al., 2014; Fatiany et al., 2016; Maroyi, 2019).

***Croton goudotii* Baill.** is a medicinal plant native to Madagascar (Rasoanaivo et al., 1992). The plant's leaves and stem bark are used to treat a variety of health issues (Rasoanaivo et al., 1992; Maroyi, 2019). Phytochemical studies have validated the presence of crotobarin and crotogoudin, two diterpenoids found in the plant, responsible for its medicinal properties (Rakotonandrasana et al., 2010; Maroyi, 2019). The infusions of the leaves and stem bark, as well as inhalation of the leaves, have been shown to be useful in the treatment of *P. falciparum* malaria at Madagascar (Rasoanaivo et al., 1992; Schmelzer et al., 2008; Maroyi, 2019). It is also used to treat gonorrhoea, blennorrhoea, and cough and acts as an aphrodisiac (Maroyi, 2019). The leaves and bark infusions, as well as their decoctions used in baths and lotions, are used as aphrodisiac and blennorrhoea treatments, respectively (Rakotonandrasana et al., 2010; Maroyi, 2019). Similarly, cough is treated using a decoction and infusion of the leaf, stem, and root bark, and in the treatment of gonorrhoea, an infusion of the stem bark and leaves is given orally (Schmelzer et al., 2008). Additionally, the plant's cytotoxic activity against the murine lymphocytic leukaemia p388 cell line was assessed, revealing a powerful cytotoxic impact with 100% suppression of the murine p388 cell line (Rakotonandrasana et al., 2010). The existence of the isolated chemicals crotobarin and crotogoudin, which demonstrated high cytotoxic effects with IC50 values of 0.13 0.01 μg/mL, suggests that the plant may be effective in the treatment of cancer (Rakotonandrasana et al., 2010; Maroyi, 2019).

***Croton greveanus* Baill.** is known by the vernacular name 'Andriambolafotsy' among the Malagasy/local communities of Madagascar (Fatiany et al., 2016). The plant is rich in various kinds of essential oils, which give it therapeutic value in the treatment of various diseases (Fatiany et al., 2016). The decoction of stem and root bark of the plant is well-known for its antimalarial properties in Madagascar, where it is used to treat malaria (Fatiany et al., 2016; Maroyi, 2019). *C. greveanus* Baill. aerial parts, bark, and root are also used to treat various health ailments, such as asthma, cough, gonorrhoea, hypertension, insect repellent, pain, and stomach disorder (Fatiany et al., 2016; Maroyi, 2019).

***Croton gratissimus* Burch.**, often known as 'lavender croton' or 'lavender fever berry', is a medicinal plant (Kerry, 2018; Magwilu et al., 2022). In several African cultures, the plant is known locally as 'Gunukia', 'Mukena', and 'Mukinduri' (Moremi et al., 2021). Because of the presence of many phytoconstituents, such as flavonoids, phenolics, polyphenolics, terpenoids, saponins, tannins, steroids, cardenolides, essential oils, carotenoids, indoles, organosulphur, and glycosides, its leaves, stem, and roots are widely utilized for therapeutic reasons in Africa (Salatino et al., 2007; Moremi et al., 2021; Magwilu et al., 2022). *C. gratissimus* Burch. possesses potent antimalarial activity, with both ethanol and dichloromethane root and leaf extracts effective against *Plasmodium falciparum* and *Plasmodium berghei* (Clarkson et al., 2004; Magwilu et al., 2022). Additionally, an infusion of the stem bark has also been used to treat malaria in Nigeria (Mulholland et al., 2010). The plant is also used against various other disorders in Nigeria, such as stomach and intestinal diseases, uterine illnesses, pleurisy, diabetes, dysentery, diarrhoea, and convulsions (Mulholland et al., 2010; Magwilu et al., 2022). Further, the decoction of the leaves has been utilized as an antibacterial and antihypertensive agent in Benin, while the methanolic leaf extract is used to treat HIV in South Africa (Vuuren & Naidoo, 2010; Mthethwa et al., 2014; Moremi et al., 2021). *C. gratissimus* Burch. also exhibits analgesic, antiprotozoal, antitrypanosomal, antifungal, antipyretic, anti-inflammatory, insecticidal, anticancer, and antiulcer properties (Magwilu et al., 2022).

***Croton kongensis* Gagnap**, commonly known as 'Plao Noi' or 'Plao Ngeon' in Thailand, is a shrub or tree that is commonly utilized in Thai medicine, owing to its potential health advantages (Thongtan et al., 2003; Salatino et al., 2007). The leaves of *C. kongensis* Gagnap are high in diterpenoid 'secokauranes', which have demonstrated promising antimalarial efficacy against *P. falciparum* K1 multidrug-resistant strain, with an IC50 of 1.0–2.8 μg/ml (Thongtan et al., 2003; Salatino et al., 2007). Moreover, these leaves show antibacterial activities at a minimum inhibitory concentration of 6.25–25.08 μg/ml and demonstrate cytotoxicity against malaria parasites. *C. kongensis* Gagnap is also used in medicine to treat dysmenorrhoea (Thongtan et al., 2003; Salatino et al., 2007).

***Croton leptostachyus* Kunth**, usually known as 'Mosquerito' or 'Mosquero', is a plant species found in Antioquia, Tolima, Boyacá, and Cundinamarca (García Barriga, 1992; Garavito et al., 2006). Peralta et al. (2015) reported that the aerial portions of the plant are high in flavonoids, polyphenols, terpenes, alkaloids, tannins, steroids, saponins, iridoids, phenylpropanoids, quinones, and coumarins, which contribute to its therapeutic characteristics. *C. leptostachyus* Kunth aerial portions, in particular, have historically been utilized for their febrifuge and antimalaria effects (García Barriga, 1992). The ethanolic extract of *C. leptostachyus* Kunth aerial parts also exhibits *in vitro* antimalarial activity against *P. falciparum*–resistant strains, with an IC50 of 2.10.2 μg/ml (Garavito et al., 2006).

***Croton linearis* Jacq**. is an aromatic evergreen shrub native to the Bahamas and Florida and broadly distributed in Caribbean countries such as Jamaica and eastern regions of Cuba (Díaz et al., 2018, 2022). The phytochemical screening of *C. linearis* Jacq. leaves showed the presence of alkaloids (reticuline, laudanidine and 8,14-dihydrosalutaridine), flavonoids, and essential oil, which provide its therapeutical values (Díaz et al., 2018, 2022). Again, in the study by Díaz et al. (2022), three isolated alkaloids (reticuline, laudanidine, and 8,14-dihydrosalutaridine) from the ethanolic extract of the leaves of *C. linearis* Jacq. were reported to have moderate antiplasmodial activity against the chloroquine-resistant strain *P. falciparum* K1, with IC50 values of 46.8 ± 0.6, 17.7 ± 0.6, and 16.0 ± 0.5 μM, respectively. Further, an ethnobotanical study of *C. linearis* Jacq. reported its use as tea to relieve menstrual pain, childbirth problems, rheumatism, symptoms of cold and fever, and that it exhibits antipyretic, antileishmanial, antitrypanosomal, insecticidal, sedative properties for circulation (Mitchell & Ahmad, 2006; Austin, 2012; Roig y Mesa & Academia de Ciencias de Cuba, 1974; Díaz et al., 2018, 2019, 2022).

***Croton lobatus* (L.) Klotzsch**, commonly known as 'lobed croton', 'Yerba krabo', and 'Cateycillo', is a herbaceous annual plant native to Ethiopia, Eritrea, Senegal, the Caribbean, South America, and the Arabian Peninsula (Ndunda, 2014). Because of the plant's favourable phytochemical composition, which includes alkaloids, sterols, sesquiterpenes, diterpenes, triterpenes, polyphenols, and one nitrogen molecule, aerial portions and roots have been employed in traditional medicine practices across the world (Attioua et al., 2012). The terpenoid molecules betulinic acid and geranylgeraniol, isolated chemicals, in particular, from the leaves and stem bark, have shown considerable antimalarial efficacy against the *P. falciparum* K1 strain (Attioua et al., 2007). Apart from its antimalarial qualities, *C. lobatus* leaves are also used in traditional medicine through various formulations to treat a variety of ailments and disorders, including pregnancy issues, dysentery, rheumatic pain, in Ethiopia, Eritrea, Senegal, Caribbean countries, South America, and the Arabian Peninsula (Neuwinger, 1996, 2000, 2004; B. K. Attioua et al., 2012; Ndunda, 2014). Similarly, in Senegal, the leaves are combined with the seed of 'fufusuf bigger' to treat whooping cough, convulsions, and oral infections, while juice from fresh leaves is used to treat eye issues and unconsciousness (Neuwinger, 1996, 2000, 2004). The leaves are used in Togo as maceration to form a lotion to improve female fertility, or decocted to use as a purgative (Schmelzer & Gurib-Fakim, 2013). The leaves and roots are utilized as an aphrodisiac, whereas the leaves of *C. lobatus* (L.) Klotzsch and *Hildegardia barteri* (Mast.) Kosterm. are used as an antihypertensive medicine (Neuwinger, 1996, 2000, 2004; Ndunda, 2014).

***Croton macrostachyus* Hochst. Ex Delile**, commonly known as 'broad-leaved croton' in English, is a tropical African deciduous shrub or tree found from Guinea through Ethiopia and Somalia, and south to Angola, Madagascar, and Mozambique (Maroyi, 2019). Wush, Masincho, Kitundu, Tebesuet, Taambuk, Dogoma, and Bisana are some of its indigenous names in Africa (Getahun, 1976; Meresa et al., 2019). This species may be found in a variety of environments, including forest edges along rivers and lakes, forested regions, grassland woods, and roadsides (Bantie et al., 2014). Secondary metabolites found in the plant include flavonoids, phenolic compounds, alkaloids, tannins, terpenoids, saponins, phlobatannins, free anthraquinones, phytosterols, and withanoides (Bantie et al., 2014; Meresa et al., 2019). Moreover, in traditional African medicine, the plant's leaves, stem bark, and roots have been used to treat malaria in several African nations, and in vivo tests have validated the crude extract's antiplasmodial effectiveness against *Plasmodium berghei* in mice. *C. macrostachyus* Hochst. Ex Delile has various prospective and historic applications in addition to being an antimalarial agent (Bantie et al., 2014; Ndunda, 2014; Mekonnen, 2015; Maroyi, 2017; Meresa et al., 2019). In Ethiopian medicine, for example, the leaves are used to cure malaria, tuberculosis, constipation, skin cancer, epilepsy, and tetanus (Mesfin et al., 2009; Behailu, 2010; Regassa, 2018; Meresa et al., 2019). The seeds, bark, and roots are used to treat diabetes mellitus, wounds healing, tapeworm infection, syphilis, rabies, venereal disease, gastrointestinal tract problems, and sexually transmitted infections (Desisa, 2001; Giday et al., 2007; Yibralign, 2007; Africa & Ababa, 2007; Regassa, 2018; Schmelzer & Gurib-Fakim, 2013; Ndunda, 2014).

***Croton membranaceus* Müll. Arg.** is a tropical shrub commonly found in Ghana and Nigeria, and it has been traditionally used in these regions due to the presence of various compounds, including coumarin, diterpene, glutarimide alkaloid, glutarimide peptide, labdane diterpene, and phytosterols, which are responsible for its therapeutic properties (Maroyi, 2018). Bayor in (2007) evaluated the *in vitro* antiplasmodial activity of six isolated alkaloid compounds from *C. membranaceus* Müll. Arg. on the *P. falciparum* strain 3D7. Results indicated that β-sitosterol, DL-threitol, and β-sitosterol-3-D-glucoside exhibit limited antiplasmodial activity, while gomojoside H, julocrotine, and crotomembranafuran showed weak activity with an IC50 range of 43.6 to 46.1 μg/ml (Bayor, 2007; Maroyi, 2018). Ethnobotanical studies revealed the extract from leaves, stem bark, and roots of *C. membranaceus* Müll. Arg. has been used to treat various diseases; for example, in Ghana, the root and leaf extracts are used to treat benign prostatic hyperplasia (BPH) and urinary retention, as well as diabetes and measles, and the essential oil from the stem bark is used to treat cough, diarrhoea, fever, flatulence, and nausea (Asare et al., 2011). In Nigeria, the root and leaf extracts are used to improve digestion (Maroyi, 2018). Additionally, several scientific studies on *C. membranaceus* Müll. Arg. highlighted numerous *in vitro* activities in the plant, such as antibacterial, cytotoxic, antifungal, antioxidant, antiproliferative, and inhibitory effects on prostate growth (Maroyi, 2018).

***Croton megalobotrys* Müll. Arg.**, also known as *C. gubouga* S. Moore (WFO, 2023), is a medium-sized, densely leafed tree with characteristic name 'megalobotrys', meaning 'big bunch of grapes' (Palmer & Pitman, 1972; Maroyi, 2017). *Croton megalobotrys* is commonly found in semi-deciduous woodland areas and at low altitudes in various countries, such as Zimbabwe, South Africa, Namibia, Botswana, Zambia, and Malawi (Tropical Plants Database, 2019). In almost 62.5% of the countries where the tree is indigenous, it is referred to as 'fever-berry'. Some of the common names that are used to refer to this species include 'fever bark', 'fever-berry cotton', 'river fever-berry', and 'fever tree' (Maroyi, 2017). The seeds and bark of the tree are believed to be used as an effective natural medicine for fever and malaria (Coates Palgrave et al., 2002; Maroyi, 2017).

According to a study reported by Maberly in (1899), the seeds and bark of *C. megalobotrys*, alongside (*Papaver somniferum* L.) opium, are used as an effective treatment for malaria. Moreover, bark or root decoction is used orally as a malaria cure in Zimbabwe (Nyazema, 1984), while bark decoction is also consumed orally as a malaria herbal medication in Mozambique (Watt & Breyer-Brandwijk, 1962) and Namibia (Chinsembu, 2015). Additionally, *C. megalobotrys* seeds, bark, and root decoction is used orally as a purgative (Nyazema, 1984; Maroyi, 2018) or as a herbal remedy for fever or malaria (Palmer & Pitman, 1972; Maroyi, 2017), and berries are used as poultices for skin infections or muscle pains or discomfort (Bolten, 1998; Selowa et al., 2010). Little is known about the chemical constituents of the species aside from the 4-hydroxyhydroxy acid isolated from the bark by Goodsonand Clewer (1919) and the cinnamic acid derivative (E)-tetratriacontyl-3-(4-hydroxy-3-methoxyphenyl)-2-propenate derived from the chloroform fraction of stem extract of *C. megalobotrys* by Abosi and Majinda (2015). Again, Abosi and Majinda (2015) in the study investigated on the antiplasmodial properties of different extracts from different parts of *C. megalobotrys* in white albino NMRI mice infected with 1 × 107 erythrocytes parasitized with *Plasmodium berghei* (ANKA). The study showed that the stem bark extract showed a significant effect in the initial stage of infection, reducing it by 74.5% and showing a residual inhibitory effect of 86.9% (Abosi & Majinda, 2015). In addition, the hexane fraction of the stem bark extract was shown to be effective *in vitro* against chloroquine-sensitive (D6) and chloroquine-resistant (W2) *Plasmodium falciparum* strains, with IC50 values of 1.74 ± 0.47 μg/ml and 3.78 ± 1.03 μg/ml (Abosi & Majinda, 2015).

In addition to antimalarial activities, the bark and seeds of the plant have also been used as prophylaxis in Southern Africa (Coates Palgrave et al., 2002); the infusion of bark or root is administered orally as an aborticide (Nyazema, 1984), the decoction of seed, root, and bark is taken orally as a laxative (Hedberg et al., 1983); and berries are applied as dermatogen for muscle pains and skin infections, specifically in Botswana and South Africa (Bolten, 1998; Selowa et al., 2010), while in Malawi and Zimbabwe, it is used as an herbal remedy for pain in the alimentary canal (Gelfland et al., 1985). Again, in Malawi, the bark decoction is used in case of wounds or fresh cuts, and the leaf and root decoction is taken orally for headaches (Pullanikkatil et al., 2016), while in Zimbabwe, bark and roots in powdered form are consumed orally against female infertility (Gelfland et al., 1985). Furthermore, the decoction of bark, leaf, roots, or seed is administered orally against sexually transmitted diseases (STDs), such as gonorrhoea in Zambia (Ndubani & Höjer, 1999; Chinsembu, 2015), and other venereal diseases in South Africa (Mabogo, 1999). The study reported by Mabogo (1999) further elucidated the seed decoction of *C. megalobotrys*, effective for blood purification and kidney-related disorders. Another study by Tietjen et al. (2016) reported the therapeutic use of the plant in treatment of various diseases, such as acquired immunodeficiency syndrome (AIDS), diarrhoea, and stomach- and other digestive-related problems. Several other studies have also emphasized the alternative uses of *C. megalobotrys* as a source of fish poison in Botswana (Parratt, 1996), Mozambique (Watt & Breyer-Brandwijk, 1962), and South Africa (Watt & Breyer-Brandwijk, 1962; Neuwinger, 2004), again, as source of pesticide, timber, and ornamental plants (Mutakela, 2001; Neelo et al., 2015).

***Croton menyharthii* Pax** is a widely widespread shrub in tropical Africa, otherwise known as 'rough-leaved croton', and locally in Kenya as 'Muyama' or 'Mualikaji', especially in the Tana River region (Kaingu, 2020). The plant has been used medicinally, and its different components, including leaves, twigs, bark, and roots, have been proven to be beneficial in treating a variety of health issues (Flora of Zimbabwe, 2010; Kaingu, 2020). *C. menyharthii* Pax phytochemical screening revealed that the plant has a high concentration of secondary metabolites, such as phenolic chemicals, alkaloids, saponins, cardiac glycosides, and tannins, which provide it therapeutic qualities (Kaingu, 2020). Research findings on the plant's antimalarial activity have indicated that a dichloromethane–methanol (1:1) extract of the leaves has an IC50 of 1.7 g/ml against the *P. falciparum* K10 resistant strain, highlighting its promise as a source of natural chemicals for creating novel malaria medicines (Clarkson et al., 2004; Pillay et al., 2008). In addition to antimalarial characteristics, *C. menyharthii* Pax leaf extract has been shown to have antibacterial, anti-inflammatory, and antidiabetic qualities (Aderogba et al., 2013). In Somalia, the plant's fresh or dried roots and root bark are used to treat dysmenorrhoea, influenza, and intestinal blockage (Schmelzer et al., 2010; Kaingu, 2020).

***Croton mubango* Müll.Arg.**, known vernacularly as 'Kabujimutshi (Tshiluba)', 'Mulemba (Songye)' in Lubumbashi region, and 'Onganga' in the Congolese dialects of Otetela or Djongo language of Democratic Republic of the Congo (Mesia et al., 2005; Mbayo et al., 2016), is a balletic shrub or tree with a thick, spreading canopy that can get up to 17 m tall and trunk up to 30 cm in diameter (Isyaka, 2020). It is prevalent in Angola, Central African Republic, Gabon, and Democratic Republic of the Congo (Schmelzer, 2007). The plant parts are used for the treatment of a wide variety of ailments; however, very limited is known about the phytochemistry of *C. mubango* (Isyaka, 2020). The stem bark of the plant is used traditionally as an aqueous decoction for antimalarial remedies due to the characteristic presence of several phytochemicals, such as alkaloids, tannins, saponins, terpenes, or steroids (Mesia et al., 2005). The antimalarial activities were further validated through both *in vitro* study against cultures of chloroquine-sensitive *Plasmodium falciparum* and *in vivo* study in mice infected with ANKA strain of *Plasmodium berghei* (Mesia et al., 2005). Further, the study by Mesia et al. (2005) corroborated the most active crude extracts *in vitro* to be aqueous extract with IC_{50} = 53.2 μg/ml, while methanolic and dichloromethane extracts of *C. mubango* exhibited median inhibitory concentrations with IC_{50} value of < 1 μg/ml. Furthermore, the *in vivo* study validated that a daily oral dosage of 200 mg/kg for all the petroleum-ether, ethyl-acetate, residual water-soluble fractions, and dichloromethane extracts from *C. mubango* exhibited 80% chemosuppression of the parasitaemias by day 4 (Mesia et al., 2005). In addition to the antimalarial activities, the plant is used traditionally in Democratic Republic of the Congo as a purgative, a febrifuge, and against abdominal pain, blennorrhoea, cancer, dysentery, fever, gastritis, inflammation, skin eruptions, splenomegaly, wounds, toothache, and tuberculosis (Tangenyi, 1996; Mesia et al., 2005; Isyaka, 2020).

***Croton nitidulus* Baker** is a monoecious shrub or tree of central and eastern Madagascar, where people use its alkaloid-rich leafy branches and stem bark in different formulation to cure certain diseases (Schmelzer et al., 2008). In Madagascar, the stem bark decoction is taken orally in the treatment of malaria. Despite being used in malaria, *C. nitidulus* Baker stem bark is also used in cough, leafy branches as insect and lice repellent, and leafy branches applied as steam bath to clear the nose (Schmelzer et al., 2008; Maroyi, 2019).

***Croton oligandrus* Pierre ex Hutch**. Is a widespread tree species in western and central Africa, known locally as 'Ebin' in Cameroon (Guetchueng Tamdem, 2019). Further, Guetchueng Tamdem (2019) reported the phytochemical screening of the plant's leaves and stem bark with the presence of di- and triterpenes, stigmasterol, and essential oil (linalool), all of which have shown varied medicinal capabilities for the treatment of various health conditions. Again, Guetchueng Tamdem (2019) isolated diterpenoid compounds from dichloromethane and n-hexane extracts of *C. oligandrus* Pierre ex Hutch. That exhibit powerful antimalarial and chemopreventive activities. The stem bark of this plant is used in Cameroon traditional medicine to cure splenomegaly and pneumonia and employed in ritual rites (Jiofack et al., 2009; Guetchueng Tamdem, 2019). In Gabon, the decoction of the stem bark is used to alleviate anaemia and colic (Mpondo Mpondo et al., 2018; Guetchueng Tamdem, 2019). Moreover, the powdered plant is applied topically to cure scabies, and the powder is inhaled to treat nasal tumours (Betti, 2013; Guetchueng Tamdem, 2019).

***Croton penduliflorus* Hutch**., also known as 'Nyamaradza' in Ghana, is a plant species widespread from Sierra Leone eastward to Nigeria, Central African Republic, and Gabon (Schmelzer et al., 2008; Laryea & Borquaye, 2019). According to Ndunda (2014), the plant's aerial portions and roots contain flavonoids, tannins, and glycoside chemicals that have potential medicinal benefits in the treatment of many ailments in the African subcontinent. The ethanolic extract of *C. penduliflorus* Hutch. Leaves is found to have *in vitro* antimalarial activity against three strains of *P. falciparum*, namely, chloroquine-sensitive, chloroquine-resistant, and CAM06 field isolate, with IC50 values of 5.37 0.18 μg/ml, 14.03 17.04 μg/ml, and 14.66 2.02 μg/ml, respectively (Laryea & Borquaye, 2019). Several parts of the *C. penduliflorus* Hutch. Plant have been identified as potentially medicinal, including the roots, seeds, stem bark, and leaf infusion. Furthermore, Adesogan (1981), Anika and Shetty (1983), and Ndunda (2014) reported the traditional therapeutic use of the roots, seeds, stem bark, and seed extract as purgative acting against stomach aches, labour pains, migraines, and impotence, while the leaf infusion is used to treat menstruation abnormalities and to cure fever in Ghana and uterine tumours and stomach ailments in Nigeria.

***Croton pseudopulchellus* Pax**, commonly known as 'small lavender fever-berry', is a shrub that grows up to 4 m tall and is spread in Angola, Ethiopia, Kenya, Mali, Mozambique, Nigeria, Somalia, South Africa, and Zimbabwe (Schmidt et al., 2002; Balkwill et al., 2004; Isyaka, 2020). The plant parts such as leaves, stem, and roots have been used extensively for the treatment of different ailments (Isyaka, 2020). The plant is reported to possess seven kaurane diterpenoids, a labdane, three sesquiterpenoids, triterpenoids, stigmasterol (Langat et al., 2012), while the leaves and the roots of *C. pseudopulchellus* contain the toxalbumin crotin (Githens, 1948; Langat et al., 2012). Further, the chloroform extract of the stem bark of *C. pseudopulchellus* has been found to exhibit 82% inhibition against PfUP1, a chloroquine-resistant strain of the malarial parasite *Plasmodium falciparum*, validating the antimalarial activity of the plant. In addition to the antimalarial activity, the plant's leaves are used in Tanzania against syphilitic ulcers and for curing of chest infections (von Bruchhausen et al., 1999; Isyaka, 2020). Asthma is treated with the root decoction, while colds are

TABLE 12.1
Antimalarial and Other Activities of Different Species of *Croton* Genera

S. No.	Scientific Name	Common Name	IUCN Status	Parts Used	Secondary Metabolites	Mode of Administration in Malaria	Other Potential and Traditional Uses	References
1.	*Croton argyratus* Blume.	Hujan panas, Hamba raja, silver croton, Melokan, Semangkok, Cheret budak, Semelit mayor	Least concern	Leaves, stem, and roots	Flavonoids, phenolics, terpenoids, steroids, tannins, reducing sugar, and cardiac glycosides	Decoction of the grind leaves	Diarrhoea, afterbirth cure, ulcer, cytotoxic, analgesic, antioxidant, anticancerous, antibacterial	(Suárez et al., 2006; Noor Rain et al., 2007; N. M. Ali, 2012; Saraswaty et al., 2012; Salleh et al., 2022)
2.	*Croton barorum* Leandri	Somorombohitse	Endangered	Leaves, stem, and roots	Diterpenoids,	Stem and root bark infusion taken orally	Asthma, breast cancer, diarrhoea, fever, hypertensions, insect repellent, leukaemia, lice repellent, stomach problems, anticancer, antimicrobial, and antiproliferative activity	(Rasoanaivo et al., 1992; Schmelzer et al., 2008; Rakotonandrasana et al., 2010; Hanson, 2011; Kuete et al., 2013; Kuete & Efferth, 2014; Fatiany et al., 2016; Maroyi, 2019)
3.	*Croton cajucara* Benth.	Sacaca	Least concern	Leaves and stem bark	Flavonoids, phenolics, terpenoids (sesquiterpenes, mono-, di-, and triterpenoids), volatile oils	Leaves and stem bark infusion	Diabetes, liver disorder, digestive disorders, weight loss, gastric ulcer, high blood cholesterol, antibacterial, antifungal, antileishmanial, anti-oestrogen, and anticancer activity	(Grynberg et al., 1999; Hiruma-Lima et al., 2000; Maciel et al., 2000; Campos et al., 2002; Alviano et al., 2005; Salatino et al., 2007; Grassi-Kassisse et al., 2010; Hiruma-Lima et al., 2010; Lima et al., 2015)
4.	*C. catati* Baill	Not Available (NA)	Least concern	Leaves, stem bark, and roots	Not Available (NA)	Root and stem bark taken orally	Cough, disinfectant, dyspnoea, euphoria, stomach pain	(Schmelzer et al., 2008; Maroyi, 2019)
5.	*Croton caudatus* Geiseler	Tukul Takal	Not evaluated	Leaves and roots	Flavonoids, cyanogenetic glycosides, alkaloids, and phenolic compounds	Decoction of leaves	Fever, convulsions, pain relief, stomach problems, rheumatic arthritis, nutritional supplement, anticancer, antidiabetic, antifertility, antioxidant, cytotoxic activity	(Nath, et al., 2013a; Dasiman et al., 2020)

6.	*Croton crassifolius* Geiseler	Pang khi, Phang khi noi	Not evaluated	Roots	Sesquiterpenoids and bicyclic diterpenoids	Decoction of root with sugar in water	Jaundice, rheumatoid arthritis, sternalgia, joint pain, pharyngitis, snakebites, stomach ache, and anticancer	(Boonyaratavej et al., 1988; Zhang et al., 2014, 2015; Huang et al., 2015; Elliott et al., 2020; Flora of China, 2023)
7.	*Croton dichogamus* Pax	Kulelwet, Kelerwe, Mwalula, Olokididigai, Rachar, Muthima, Mhande, Maffuka	Least concern	All parts	Flavonoids, terpenoids saponins, alkaloids, steroids, and tannins	Direct inhalation of smoke of dry leaves, infusion/ decoction of leaves/ stem/bark	Tuberculosis, fever, chest pain, stomach ache, arthritis, respiratory tract infections, infertility, syphilis, gonorrhea, and antipyretic	(Hedberg et al., 1983; Johns et al., 1999; Sori et al., 2004; Pascaline et al., 2011; Muthee et al., 2011; A. Aldhaher et al., 2017; Kigen et al., 2019; Mutie et al., 2020; Omara, 2020; Matara et al., 2021)
8.	*Croton eluteria* (L.) W.Wright	Cascarilla	Least concern	Stem bark	Diterpenoids	Decoction of stem bark	Bronchitis, fever, digestive disorder, hypertension, dysentery, diarrhoea	(Vigor et al., 2001; Appendino et al., 2003; Salatino et al., 2007)
9.	*Croton geayi* Leandri	Pisopisovavy	Not evaluated	Leaves, stem, bark, and roots	Diterpenoids and essential oil	Stem and root bark infusion taken orally	Cough, fever, flu, constipation, asthma, stomach disorders, wound healing, injury, hypertension, antibacterial	(Palazzino et al., 1998; Andriamparany et al., 2014; Fatiany et al., 2016; Maroyi, 2019)
10.	*Croton goudotii* Baill.	Not available (NA)	Least concern	Leaves and stem bark	Diterpenoids (crotobarin, crotogoudin)	Decoction or inhalation of the leaves	Blennorrhoea, cough, aphrodisiac, gonorrhea and anticancer	(Rasoanaivo et al., 1992; Schmelzer et al., 2008; Rakotonandrasana et al., 2010; Maroyi, 2019)
11.	*Croton greveanus* Baill.	Andriambolafotsy	Least concern	Aerial parts, bark, and roots	Essential oil	Stem and root bark decoction taken orally	Asthma, cough, gonorrhea, hypertension, insect repellent, pain, stomach disorder	(Schmelzer et al., 2008; Fatiany et al., 2016; Maroyi, 2019)

(*Continued*)

TABLE 12.1 (*Continued*)
Antimalarial and Other Activities of Different Species of *Croton* Genera

S. No.	Scientific Name	Common Name	IUCN Status	Parts Used	Secondary Metabolites	Mode of Administration in Malaria	Other Potential and Traditional Uses	References
12.	*Croton gratissimus* Burch.	Gunukia, Mukena, Mukinduri	Least concern	Leaves and stem bark and roots	Flavonoids, phenolics, polyphenolics, terpenoids, saponins, tannins, steroids, cardenolides, essential oils, carotenoids, indoles, organosulphur, and glycosides	Infusion of stem bark, leaves decoction	Fever, dysentery, diarrhoea, convulsions, constipation, wound healing, arthritis, urinary tract infections, antimicrobial and antihypertensive, anti-HIV/AIDS, analgesic, antiprotozoal, antitrypanosomial, antifungal, antimalarial, antiviral, antidiabetic, antipyretic, antihypersensitive, anti inflammatory, insecticidal, anticancer, and antiulcer	(Clarkson et al., 2004; Salatino et al., 2007; Mulholland et al., 2010; Robert et al., 2010; van Vuuren & Naidoo, 2010; Mthethwa et al., 2014; Abdalaziz et al., 2016; Kumar et al., 2017; Moremi et al., 2021; Magwilu et al., 2022)
13.	*Croton kongensis* Gagnap.	Plao Ngeon, Plao Noi	Not evaluated	Leaves	Diterpenoids (Secokauranes)	Dichloromethane extract of leaves	Dysmenorrhoea, antileishmanial, antimycobacterial, cytotoxic	(Thongtan et al., 2003; Salatino et al., 2007)
14.	*Croton leptostachyus* Kunth	Mosquerito, Mosquero	Not evaluated	Aerial parts	Flavonoids, polyphenols, terpenes, alkaloids, tannins, steroids, saponins, iridoids, phenylpropanoids, quinones, and coumarins	Ethanolic extract of aerial parts	Anti-inflammatory and antioxidant	(García Barriga, 1992; Garavito et al., 2006; Peralta et al., 2015)
15.	*Croton linearis* Jacq.	Not available (NA)	Not evaluated	Leaves	Alkaloids (reticuline, laudanidine, and 8,14-dihydrosalutaridine), flavonoids, essential oil	Ethanolic extract of leaves	Menstrual pain, childbirth pain, rheumatism, cold, fever, antipyretic, insecticidal, sedative for circulation, antileishmanial, antitrypanosomal activity	(Roig y Mesa & Academia de Ciencias de Cuba, 1974; Mitchell & Ahmad, 2006; Austin, 2012; Díaz et al., 2018, 2019, 2022)

16.	*Croton lobatus* (L.) Klotzsch	Lobed croton, Yerba Krabo, Cateycillo	Not evaluated	Aerial parts and roots	Alkaloids, sterols, sesquiterpenes, diterpenes, triterpenes, polyphenol, and one nitrogen compound	Methanolic extract of aerial parts and root	Pregnancy troubles, dysentery, purgative, unconsciousness, eye disease, convulsions, mouth infections, rheumatic pain, female sterility, whooping cough, antihypertensive, antispasmodic, antihypersensitive	(Neuwinger, 1996, 2000, 2004; B. Attioua et al., 2007; Attioua et al., 2012; Schmelzer & Gurib-Fakim, 2013; Ndunda, 2014)
17.	*Croton macrostachyus* Hochst. ex Delile	Broad-leaved croton, Wush, Masincho, Kitundu, Tebesuet, Taambuk, Dogoma, Bisana	Least concern	Aerial parts and roots	Flavonoids, phenolic compounds, alkaloids, tannins, terpenoids, saponins, phlobatannins, free anthraquinones, phytosterols, and withanoides	Decoction of the leaves, root, and fruits	Constipation, tuberculosis, skin cancer, epilepsy, tetanus, tapeworm infection, syphilis, diabetes mellitus, wound healing, rabies, venereal disease, GI tract problems, and sexually transmitted infections	(Getahun, 1976; Desisa, 2001; Africa & Ababa, 2007; Giday et al., 2007; Yibralign, 2007; Bantie et al., 2014; Mesfin et al., 2009; Behailu, 2010; Schmelzer & Gurib-Fakim, 2013; Ndunda, 2014; Mekonnen, 2015; Maroyi, 2017; Regassa, 2018; Maroyi, 2019; Meresa et al., 2019)
18.	*Croton membranaceus* Müll.Arg.	Not available (NA)	Not evaluated	Leaves, stem bark, and roots	Coumarin, diterpene, glutarimide alkaloid, glutarimide peptide, labdane diterpene, and phytosterol	Ethanolic extract of leaves	Benign prostatic hyperplasia (BPH), urinary retention problem, cough, diarrhoea, fever, diabetes, nausea, digestive problems, flatulence, measles, antibacterial, cytotoxic, antifungal, antioxidant, antiproliferative, prostate growth inhibitory effects	(Bayor, 2007; Asare et al., 2011; Maroyi, 2018)

(Continued)

TABLE 12.1 (*Continued*)
Antimalarial and Other Activities of Different Species of *Croton* Genera

S. No.	Scientific Name	Common Name	IUCN Status	Parts Used	Secondary Metabolites	Mode of Administration in Malaria	Other Potential and Traditional Uses	References
19.	*Croton megalobotrys* Müll.Arg.	Large fever berry, Koorsbessie, Motsibi	Least concern	Leaves, stem bark, fruits, seeds, and roots	Alkaloid, cinnamic acid derivatives	Decoction of mixture of bark, root, and seed or in combination with opium	Purgative, abortifacient, HIV/AIDS, muscle pain, sexually transmitted infections, abortion, skin infections, female infertility, radical scavenging activity, antibacterial, antiplasmodial	(Ndunda, 2014; Maroyi, 2017; Moremi et al., 2021)
20.	*Croton menyharthii* Pax	Rough-leaved croton, Muyama, Mualikaji	Least concern	Leaves, twigs, bark, and roots	Phenolic compounds, alkaloids, saponins, cardiac glycosides, and tannins	Dichloromethane–methanol (1:1) extract of leaves	Dysmenorrheal, anti-influenza and intestinal obstruction, antimicrobial activity, anti-inflammatory activity, antidiabetic activity	(Clarkson et al., 2004; Pillay et al., 2008; Schmelzer et al., 2010; Aderogba et al., 2013; Kaingu, 2020)
21.	*Croton mubango* Müll.Arg.	Kabujimutshi (Tshiluba), Mulemba (Songye), and Onganga	Not evaluated	Leaves, bark, fruits, and roots	Flavonoids, alkaloids, tannins, saponins, steroids, terpenes, triterpenes, and reducing sugars	Bark decoction/infusion taken orally	Rheumatism, female sterility, asthma, paralysis, sleeping sickness, toothache, tuberculosis, fever, hernia, abdominal pain, diarrhoea, purgative, vermifuge, Dysentery, intestinal worms, and as purgative	(Ndunda, 2014; Moremi et al., 2021)
22.	*Croton nitidulus* Baker	NA	Least concern	Leafy branches and stem bark	Alkaloids	Stem bark decoction taken orally	Cough, insect repellent, lice repellent, nose clearance	(Schmelzer et al., 2008; Maroyi, 2019)
23.	*Croton oligandrus* Pierre ex Hutch.	Ebin	Least concern	Leaves and stem bark	Di- and triterpenes, stigmasterol, essential oil (linalool)	N-Hexane/Dichloromethane (DCM) extract of stem bark	Pneumonia, splenomegaly, anaemia, colic, nasal tumour, scabies, chemoprotective, and used in ritual practices	(Jiofack et al., 2009; Betti, 2013; Mpondo Mpondo et al., 2018; Guetchueng Tamdem, 2019)

24.	*Croton penduliflorus* Hutch.	Nyamaradza	Least concern	Leaves, twigs, stem bark, seeds, and roots	Flavonoids, tannins, glycosides	Ethanolic extract of leaves	Fever, stomach ache, purgative, body pain, headache, impotence, labour pain, menstrual disorders, uterine tumours	(Ndunda, 2014; Laryea & Borquaye, 2019)
25.	*Croton pseudopulchellus* Pax	Small lavender croton	Not evaluated	Leaves, stem bark, and roots	Terpenoids, stigmasterol, phytosterols, essential oil	Chloroform extract of stem bark	Syphilitic ulcers, chest infections, tuberculosis, anthrax, asthma, colds, viral and tissue infections, antimicrobial, anti-inflammatory	(Pillay et al., 2008; Ndunda, 2014; Moremi et al., 2021)
26.	*Croton steenkampianus* Gerstner	Marsh fever berry, Tonga croton	Not evaluated	Leaves	Flavonoids, diterpenes, triterpenes	Ethanolic extract of leaves	Antibacterial and antioxidant	(Prozesky, 2008; Schmelzer et al., 2008; Adelekan, 2009; Ndunda, 2014)

treated by snorting the powdered roots of the plant (Schmidt et al., 2002; Isyaka, 2020). Moreover, *C. pseudopulchellus* is utilized to treat coughs and viral infections (Prozesky, 2008; Isyaka, 2020).

***Croton steenkampianus* Gerstner** is a shrub that is frequently known as 'marsh fever berry' and 'Tonga croton' in Southern Africa, Tanzania, and Mozambique (Ndunda, 2014). The presence of different flavonoids, diterpenes, and triterpenes chemicals in its leaves has been discovered, indicating the plant's potential therapeutic significance, particularly, *in vitro* antimalarial potential of *C. steenkampianus* Gerstner, in which a novel diterpenoid component was extracted from the plant's ethanolic extract and tested against the *P. falciparum*–resistant strain (Prozesky, 2008). The study's findings revealed that the isolated chemical had considerable antiplasmodial action, suggesting its potential as an antimalarial drug (Prozesky, 2008; Adelekan, 2009). *C. steenkampianus* Gerstner further exhibits potential benefits, such as in fresh leaves vapor inhalation, which has been suggested as a viable remedy for relieving bodily pain (Schmelzer et al., 2008). Additionally, it has also been identified as having antibacterial and antioxidant properties (Prozesky, 2008; Adelekan, 2009), indicating its wider therapeutic usefulness. The antimalarial activities of different species of *Croton* genera, with common name, IUCN status, parts used, secondary metabolites, mode of administration, and their other potential and traditional uses are enumerated in Table 12.1.

12.3 CONCLUSION AND FUTURE PERSPECTIVES

In general, both the scientific and traditional studies of *Croton* genera in malaria-endemic areas are validated to possess promising antimalarial activities. Several compounds isolated from different parts of the plant exhibited potent antimalarial activities against different strains of *Plasmodium* sp., the dreadful causative agent of the malaria. In addition to its antimalarial activities, the species have shown potential for the treatment of a wide array of ailments, such as cancer, digestive disorders, inflammatory skin infections, respiratory diseases, tuberculosis, HIV/AIDS, etc. However, at present, there is limited systematic data pertaining to phytochemistry, pharmacognosy, clinical trials, and organ toxicity of the plant species extracts and fractions. Extensive research into the phytochemistry of *Croton* genera using reliable analytical techniques is needed to ensure identification, characterization of novel compounds that could be potential therapeutic sources, paving way for formulation of new drugs and therapies for treatment of malaria and other diseases.

REFERENCES

Abdalaziz, M. N., Ali, A., & Kabbashi, A. S. (2016). In vitro antioxidant activity and phytochemical screening of *Croton zambesicus*. *Journal of Pharmacognosy and Phytochemistry*, *5*(6), 12–16.

Abosi, A. O., & Majinda, R. R. T. (2015). Anti-plasmodial and radical scavenging activities of Croton megalobotrys. *Research Journal of Pharmacognosy and Phytochemistry*, *3*, 73.

Achan, J., Talisuna, A. O., Erhart, A., Yeka, A., Tibenderana, J. K., Baliraine, F. N., Rosenthal, P. J., & D'Alessandro, U. (2011). Quinine, an old anti-malarial drug in a modern world: Role in the treatment of malaria. *Malaria Journal*, *10*(1), 144. https://doi.org/10.1186/1475-2875-10-144

Adams, J., & Andrews, G. J. (Eds.). (2012). *Traditional, Complementary and Integrative Medicine: An International Reader*. Palgrave Macmillan.

Adelekan, A. M. (2009). *In Vitro Medicinal Properties of Novel Compounds from Croton Steenkampianus* [Thesis, University of Pretoria]. https://repository.up.ac.za/handle/2263/24934

Aderogba, M. A., Ndhlala, A. R., Rengasamy, K. R. R., & Van Staden, J. (2013). Antimicrobial and selected in vitro enzyme inhibitory effects of leaf extracts, flavonols and indole alkaloids isolated from Croton menyharthii. *Molecules (Basel, Switzerland)*, *18*(10), 12633–12644. https://doi.org/10.3390/molecules181012633

Adesogan, E. K. (1981). The structure of penduliflaworosin, a new furanoid diterpene from Croton penduliflorus. *Journal of the Chemical Society, Perkin Transactions*, *1*, 1151. https://doi.org/10.1039/p19810001151

Africa, F., & Ababa, A. (2007). *The Key Steps in Establishing Participatory Forest Management. A Field Manual to Guide Practitioners in Ethiopia*. Addis Ababa (Ethiopia) Farm Africa. https://scholar.google.com/scholar_lookup?title=The+key+steps+in+establishing+participatory+forest+management.+a+field

+manual+to+guide+practitioners+in+Ethiopia&author=Farm+Africa%2C+Addis+Ababa+%28Ethiopia%29%0A++++eng&publication_year=2007

Aldhaher, A. H. (2016). *Phytochemical Investigations of Members of the Meliaceae, Euphorbiaceae and Amaryllidaceae Families*. University of Surrey. https://openresearch.surrey.ac.uk/esploro/outputs/doctoral/Phytochemical-investigations-of-members-of-the-Meliaceae-Euphorbiaceae-and-Amaryllidaceae-families/99514027902346

Aldhaher, A. H., Langat, M., Ndunda, B., Chirchir, D., Midiwo, J. O., Njue, A., Schwikkard, S., Carew, M., & Mulholland, D. (2017). Diterpenoids from the roots of Croton dichogamus Pax. *Phytochemistry*, *144*, 1–8. https://doi.org/10.1016/j.phytochem.2017.08.014

Aldhaher, A. H., Langat, M., Schwikkard, S., Carew, M., & Mulholland, D. (2016). New terpenoids from Croton dichogamus Pax. *Planta Medica*, *81*(S01), S1–S381. https://doi.org/10.1055/s-0036-1596327

Ali, N. I. M. (2012). Phytochemical screening, antioxidant and analgesic activities of Croton argyratus ethanolic extracts. *Journal of Medicinal Plants Research*, *6*(21). https://doi.org/10.5897/JMPR11.743

Ali, N. I. M., Annegowda, H. V., Mansor, S. M., Ismail, S., Ramanathan, S., & Mordi, M. N. (2012). Phytochemical screening, antioxidant and analgesic activities of Croton argyratus ethanolic extracts. *Journal of Medicinal Plants Research*, *6*(21), 3724–3731. https://doi.org/10.5897/JMPR11.743

Alviano, W. S., Mendonca-Filho, R. R., Alviano, D. S., Bizzo, H. R., Souto-Padron, T., Rodrigues, M. L., Bolognese, A. M., Alviano, C. S., & Souza, M. M. G. (2005). Antimicrobial activity of Croton cajucara Benth linalool-rich essential oil on artificial biofilms and planktonic microorganisms. *Oral Microbiology and Immunology*, *20*(2), 101–105. https://doi.org/10.1111/j.1399-302X.2004.00201.x

Amaral, A. C. F., & Barnes, R. A. (1998). A tetrahydroprotoberberine alkaloid from *Croton hemiargyreus*. *Phytochemistry*, *47*(7), 1445–1447. https://scholar.google.com/scholar_lookup?title=tetrahydroprotoberberine+alkaloid+from+Croton+hemiargyreus&author=Amaral%2C+A.C.F.&publication_year=1998

Andriamparany, J. N., Brinkmann, K., Jeannoda, V., & Buerkert, A. (2014). Effects of socio-economic household characteristics on traditional knowledge and usage of wild yams and medicinal plants in the Mahafaly region of south-western Madagascar. *Journal of Ethnobiology and Ethnomedicine*, *10*(1), 82. https://doi.org/10.1186/1746-4269-10-82

Anika, S. M., & Shetty, S. N. (1983). Investigations on *Croton penduliflorus* Hutch.: II. A study on the mechanism of the hypotensive activity in pentobarbital-anesthetized dogs. *International Journal of Crude Drug Research*, *21*(2), 59–65. https://doi.org/10.3109/13880208309070611

Appendino, G., Borrelli, F., Capasso, R., Campagnuolo, C., Fattorusso, E., Petrucci, F., & Taglialatela-Scafati, O. (2003). Minor diterpenoids from cascarilla (Croton eluteria Bennet) and evaluation of the cascarilla extract and cascarillin effects on gastric acid secretion. *Journal of Agricultural and Food Chemistry*, *51*(24), 6970–6974. https://doi.org/10.1021/jf034780h

Arya, A., Kumar, S., Paul, R., Suryavanshi, A., Kain, D., & Sahoo, R. N. (2021). Ethnopharmacological survey of indigenous medicinal plants of Palampur, Himachal Pradesh in north-western Himalaya, India. *Advances in Traditional Medicine* 23, 169–212, https://doi.org/10.1007/s13596-021-00607-1

Asare, G. A., Sittie, A., Bugyei, K., Gyan, B. A., Adjei, S., Addo, P., Wiredu, E. K., Nyarko, A. K., Otu-Nyarko, L. S., & Adjei, D. N. (2011). Acute toxicity studies of Croton membranaceus root extract. *Journal of Ethnopharmacology*, *134*(3), 938–943. https://doi.org/10.1016/j.jep.2011.02.004

Assan, J. K., Assan, S. K., Assan, N., & Smith, L. (2009). Health inequality in resource poor environments and the pursuit of the MDGs: Traditional versus modern healthcare in rural Indonesia. *Journal of Health Management*, *11*(1), 93–108. https://doi.org/10.1177/097206340901100107

Attioua, B. K., Harisolo, R., Boti, J. B., Adiko, V. A., Tonzibo, F. Z., & Djakoure, L. A. (2012). Isolation and identification of alkaloids from Croton lobatus. *International Journal of Pharmaceutical Sciences Review and Research*, *13*(2). www.researchgate.net/publication/274074043_ISOLATION_AND_IDENTIFICATION_OF_ALKALOIDS_FROM_CROTON_LOBATUS

Attioua, B. K., Weniger, B., & Chabert, P. (2007). Antiplasmodial activity of constituents isolated from *Croton lobatus*. *Pharmaceutical Biology*, *45*(4), 263–266. https://doi.org/10.1080/13880200701214607

Austin, D. F. (2012). Bush medicine of The Bahamas. A cross-cultural perspective from San Salvador island, including pharmacology and oral histories. *Economic Botany*, *66*(2), 214–226. https://doi.org/10.1007/s12231-012-9200-9

Balkwill, K., Porter, B., & Van Gogh, J. (2004). *Sappi Tree Spotting: Lifer List*. Jacana Media.

Bantie, L., Assefa, S., Teklehaimanot, T., & Engidawork, E. (2014). In vivo antimalarial activity of the crude leaf extract and solvent fractions of Croton macrostachyus Hocsht. (Euphorbiaceae) against Plasmodium berghei in mice. *BMC Complementary and Alternative Medicine*, *14*(1), 79. https://doi.org/10.1186/1472-6882-14-79

Barat, L. M., Palmer, N., Basu, S., Worrall, E., Hanson, K., & Mills, A. (2004). Do malaria control interventions reach the poor? A view through the equity lens. *The American Journal of Tropical Medicine and Hygiene*, *71*(2 Suppl), 174–178.

Bayor, M. T. (2007). *The Anticancer and Other Bioacttvity Investigations on the Extract and Some Compounds of Croton Membranaceus [Euphorbiaceae]* [PhD Thesis, Kwame Nkrumah University of Science and Technology].

Behailu, E. (2010). *Ethnobotanical Study of Traditional Medicinal Plants of Goma Wereda, Jima Zone of Oromia Region, Ethiopia* [Thesis, Addis Ababa University]. http://etd.aau.edu.et/handle/123456789/11339

Betti, J. (2013). An ethnobotanical and floristical study of medicinal plants among the Baka pygmies in the periphery of the Ipassa-Biosphere Reserve, Gabon. *European Journal of Medicinal Plants*, *3*(2), 174–205. https://doi.org/10.9734/EJMP/2013/2550

Block, S., Stévigny, C., De Pauw-Gillet, M.-C., De Hoffmann, E., Llabrès, G., Adjakidjé, V., & Quetin-Leclercq, J. (2002). Ent-Trachyloban-3β-ol, a new cytotoxic diterpene from *Croton zambesicus*. *Planta Medica*, *68*(7), 647–649. https://doi.org/10.1055/s-2002-32903

Bolten, C. (1998). *Healing Knowledge and Cultural Practices in a Modern Tswana Village*. Williams College.

Boonyaratavej, S., Roengsumran, S., Bates, R. B., Ortega, R. B., & Petsom, A. (1988). X-ray structure of cyperenoic acid from Croton crassifolius. *Journal of Natural Products*, *51*(4), 769–770. https://doi.org/10.1021/np50058a019

Boyom, F. F., Kemgne, E. M., Tepongning, R., Ngouana, V., Mbacham, W. F., Tsamo, E., Zollo, P. H. A., Gut, J., & Rosenthal, P. J. (2009). Antiplasmodial activity of extracts from seven medicinal plants used in malaria treatment in Cameroon. *Journal of Ethnopharmacology*, *123*(3), 483–488. https://doi.org/10.1016/j.jep.2009.03.008

Burkill, I. H. (1966). Dictionary of Economic Products of the Malay Peninsula; Ministry of Agriculture and Cooperatives Malaysia: Kuala Lumpur, Malaysia, 2: 2344–2345.

Campos, A. R., Albuquerque, F. A. A., Rao, V. S. N., Maciel, M. A. M., & Pinto, A. C. (2002). Investigations on the antinociceptive activity of crude extracts from Croton cajucara leaves in mice. *Fitoterapia*, *73*(2), 116–120. https://doi.org/10.1016/S0367-326X(02)00004-7

Chakraborty, P. (2018). Herbal genomics as tools for dissecting new metabolic pathways of unexplored medicinal plants and drug discovery. *Biochimie Open*, *6*, 9–16. https://doi.org/10.1016/j.biopen.2017.12.003

Chinsembu, K. C. (2015). Plants as antimalarial agents in Sub-Saharan Africa. *Acta Tropica*, *152*, 32–48.

Chuma, J., Okungu, V., & Molyneux, C. (2010). Barriers to prompt and effective malaria treatment among the poorest population in Kenya. *Malaria Journal*, *9*(1), 144. https://doi.org/10.1186/1475-2875-9-144

Clarkson, C., Maharaj, V. J., Crouch, N. R., Grace, O. M., Pillay, P., Matsabisa, M. G., Bhagwandin, N., Smith, P. J., & Folb, P. I. (2004). In vitro antiplasmodial activity of medicinal plants native to or naturalised in South Africa. *Journal of Ethnopharmacology*, *92*(2–3), 177–191. https://doi.org/10.1016/j.jep.2004.02.011

Coates Palgrave, K., Drummond, R. B., Moll, E. J., & Coates Palgrave, M. (2002). *Trees of Southern Africa* (3rd ed., New ed./revised and updated by Meg Coates Palgrave). Struik Publishers.

Croton crassifolius in Flora of China @ efloras.org. (2023). *Efloras.org*. http://www.efloras.org/florataxon.aspx?flora_id=2&taxon_id=200012551

Dasiman, R., Malek, M. A., Bahari, E. A., Zakaria, F. N., Hashim, N. K., Samsuddin, A., Daud, K. A., Sham, F., & Amom, Z. (2020). Effect of Croton caudatus Geiseler aqueous root extract on reproductive and biochemical parameters in male wistar rats. *The Indonesian Biomedical Journal*, *12*(3), Article 3. https://doi.org/10.18585/inabj.v12i3.1186

Debas, H. T., Laxminarayan, R., & Straus, S. E. (2006). Complementary and alternative medicine. In D. T. Jamison, J. G. Breman, A. R. Measham, G. Alleyne, M. Claeson, D. B. Evans, P. Jha, A. Mills, & P. Musgrove (eds.), *Disease Control Priorities in Developing Countries* (2nd ed.). The International Bank for Reconstruction and Development/The World Bank. http://www.ncbi.nlm.nih.gov/books/NBK11796/

Desisa, D. (2001). *A Preliminary Economic Evaluation of Medicinal Plants in Ethiopia: Trade, Volume and Price*. https://scholar.google.com/scholar_lookup?title=A+preliminary+economic+evaluation+of+medicinal+plants+in+Ethiopia%3A+trade%2C+volume+and+price&author=Desalegn+Desisa+%28The+National+Herbarium%2C+Addis+Abeba+%28Ethiopia%29%29&publication_year=2001

Dewi, S.P., Kasim, R., Sutarsa, I.N., Hunter, A., & Dykgraaf, S.H. (2023). Effects of climate-related risks and extreme events on health outcomes and health utilization of primary care in rural and remote areas: A scoping review. *Family Practice*, 40, 486–497.

Díaz, J. G., Arranz, J. C. E., Batista, D. da G. J., Fidalgo, L. M., Acosta, J. de la V., de Macedo, M. B., & Cos, P. (2018). Antileishmanial potentialities of *Croton linearis* leaf essential oil. *Natural Product Communications*, *13*(5), 1934578X1801300. https://doi.org/10.1177/1934578X1801300527

Díaz, J. G., Tuenter, E., Escalona Arranz, J. C., Llauradó Maury, G., Cos, P., & Pieters, L. (2019). Antimicrobial activity of leaf extracts and isolated constituents of Croton linearis. *Journal of Ethnopharmacology*, *236*, 250–257. https://doi.org/10.1016/j.jep.2019.01.049

Díaz, J. G., Tuenter, E., Escalona Arranz, J. C., Llauradó Maury, G., Cos, P., & Pieters, L. (2022). Antiplasmodial activity of alkaloids from Croton linearis leaves. *Experimental Parasitology*, *236–237*, 108254. https://doi.org/10.1016/j.exppara.2022.108254

Elliott, E., Chassagne, F., Aubouy, A., Deharo, E., Souvanasy, O., Sythamala, P., Sydara, K., Lamxay, V., Manithip, C., Torres, J. A., & Bourdy, G. (2020). Forest fevers: Traditional treatment of malaria in the southern lowlands of Laos. *Journal of Ethnopharmacology*, *249*, 112187. https://doi.org/10.1016/j.jep.2019.112187

Fatiany, P., Ruphin, Baholy, R., Sylver, S., Aubin, R., Oscar, Y., Mahamoud, A., Fienena, F., Raymond, Rakotoniriana, H., Amélie, R., Ruphin, R., Baholy, S., Sylver, R., Aubin, Y., Oscar, A., Mahamoud, Raymond, S., Marcelin, H., & Ngbolua, K.-T.-N. (2016). GC-FID and GC/MS analyses and Antimicrobial activity of Croton greveanus, C. borarium and C. geayi (Euphorbiaceae) essential oils from Madagascar. *Journal of Pharmacognosy and Phytochemistry*, *5*, 188–197.

Ferreira, A., Balla, J., Jeney, V., Balla, G., & Soares, M. P. (2008). A central role for free heme in the pathogenesis of severe malaria: The missing link? *Journal of Molecular Medicine*, *86*(10), 1097–1111. https://doi.org/10.1007/s00109-008-0368-5.

Flora of China. (2023). @ *efloras.org*. (Accessed on November 11, 2023). http://www.efloras.org/florataxon.aspx?flora_id=2&taxon_id=200012551

Flora of Zimbabwe. (2010). www.zimbabweflora.co.zw/speciesdata/species.php?species_id=134800

Garavito, G., Rincón, J., Arteaga, L., Hata, Y., Bourdy, G., Gimenez, A., Pinzón, R., & Deharo, E. (2006). Antimalarial activity of some Colombian medicinal plants. *Journal of Ethnopharmacology*, *107*(3), 460–462. https://doi.org/10.1016/j.jep.2006.03.033

García Barriga, H. (1992). *Flora medicinal de Colombia: Botánica médica* (2nd ed.). Tercer mundo.

Gelfland, M., Mavi, S., Drummond, R. B., & Ndemera, B. (1985). The traditional medical practitioner in Zimbabwe: His principles of practice and pharmacopoeia. In *The Traditional Medical Practitioner in Zimbabwe: His Principles of Practice and Pharmacopoeia*. Mambo Press.

Getahun, A. (1976). *Some Common Medicinal and Poisonpous Plants Used in Ethiopian Folk Medicine*. Addis Ababa University.

Giday, M., Teklehaymanot, T., Animut, A., & Mekonnen, Y. (2007). Medicinal plants of the Shinasha, Agew-awi and Amhara peoples in northwest Ethiopia. *Journal of Ethnopharmacology*, *110*(3), 516–525. https://doi.org/10.1016/j.jep.2006.10.011

Githens, T. S. (1948). *Drug Plants of Africa. African Handbook No. 8*. www.cabdirect.org/cabdirect/abstract/19511603536

Goodson, J. A., & Clewer, H. W. B. (1919). LXXVIII. – Examination of the bark of Croton gubouga. Isolation of 4-hydroxyhygric acid. *Journal of the Chemical Society, Transactions*, *115*, 923–933.

Grassi-Kassisse, D. M., Wolf-Nunes, V., Miotto, A. M., Farias-Silva, E., Brito, A. R. M. S., Nunes, D. S., & Spadari-Bratfisch, R. C. (2010). Sensitivity to β-adrenoceptor agonists of adipocytes from rats treated with an aqueous extract of Croton cajucara Benth. *Journal of Pharmacy and Pharmacology*, *55*(2), 253–257. https://doi.org/10.1211/002235702595

Grynberg, N. F., Echevarria, A., Lima, J. E., Pamplona, S. S., Pinto, A. C., & Maciel, M. A. (1999). Anti-tumour activity of two 19-nor-clerodane diterpenes, trans-dehydrocrotonin and trans-crotonin, from Croton cajucara. *Planta Medica*, *65*(8), 687–689. https://doi.org/10.1055/s-1999-14042

Guetchueng Tamdem, S. (2019). *Phytochemical, Chemopreventive and Antimalarial Activity Evaluation of Five Selected Medicinal Plants from the Cameroonian Flora*. Liverpool John Moores University. http://researchonline.ljmu.ac.uk/10347/

Hanson, J. R. (2011). Diterpenoids of terrestrial origin. *Natural Product Reports*, *28*(10), 1755–1772. https://doi.org/10.1039/C1NP90021H

Hedberg, I., Hedberg, O., Madati, P. J., Mshigeni, K. E., Mshiu, E. N., & Samuelsson, G. (1983). Inventory of plants used in traditional medicine in Tanzania. II. Plants of the families Dilleniaceae – Opiliaceae. *Journal of Ethnopharmacology*, *9*(1), 105–127. https://doi.org/10.1016/0378-8741(83)90030-2

Hiruma-Lima, C. A., Gracioso, J. S., Nunes, D. S., & Brito, A. S. (2010). Effects of an essential oil from the bark of Croton cajucara Benth. On experimental gastric ulcer models in rats and mice. *Journal of Pharmacy and Pharmacology*, *51*(3), 341–346. https://doi.org/10.1211/0022357991772349

Hiruma-Lima, C. A., Gracioso, J. S., Rodriguez, J. A., Haun, M., Nunes, D. S., & Souza Brito, A. R. M. (2000). Gastroprotective effect of essential oil from Croton cajucara Benth. (Euphorbiaceae). *Journal of Ethnopharmacology*, *69*(3), 229–234. https://doi.org/10.1016/S0378-8741(99)00127-0

Horgen, F. D., Edrada, R. A., de los Reyes, G., Agcaoili, F., Madulid, D. A., Wongpanich, V., Angerhofer, C. K., Pezzuto, J. M., Soejarto, D. D., & Farnsworth, N. R. (2001). Biological screening of rain forest plot trees from Palawan Island (Philippines). *Phytomedicine: International Journal of Phytotherapy and Phytopharmacology*, *8*(1), 71–81. https://doi.org/10.1078/0944-7113-00019

Huang, W., Wang, J., Liang, Y., Ge, W., Wang, G., Li, Y., & Yin Chung, H. (2015). Potent anti-angiogenic component in Croton crassifolius and its mechanism of action. *Journal of Ethnopharmacology*, *175*, 185–191. https://doi.org/10.1016/j.jep.2015.09.021

Husen, A. (2021). *Traditional Herbal Therapy for the Human Immune System*. CRC Press. https://doi.org/10.1201/9781003137955

Husen, A. (2022a). *Herbs, Shrubs and Trees of Potential Medicinal Benefits*. CRC Press. https://doi.org/10.1201/9781003205067

Husen, A. (2022b). *Exploring Poisonous Plants: Medicinal Values, Toxicity Response and Other Potential Uses*. CRC Press. https://doi.org/10.1201/b23017

Husen, A. (2023). *Antidiabetic Medicinal Plants and Herbal Treatments*. CRC Press.

Isyaka, S. M. (2020). *The Chemistry of African Croton Species* [Thesis, University of Surrey]. https://doi.org/10.15126/THESIS.00853360

Jaiswal, A. (2019). Indigenous knowledge system and traditional medicinal practitioner. *Anthropology and Ethnology Open Access Journal*, *2*(1). https://doi.org/10.23880/AEOAJ-16000119

Jiofack, T., Ayissi, L., Fokunang, C., Guedje, N., & Kemeuze, V. (2009). Ethnobotany and phytomedicine of the upper Nyong valley forest in Cameroon. *African Journal of Pharmacy and Pharmacology*, *3*(4), 144–150.

Johns, T., Mahunnah, R. L., Sanaya, P., Chapman, L., & Ticktin, T. (1999). Saponins and phenolic content in plant dietary additives of a traditional subsistence community, the Batemi of Ngorongoro District, Tanzania. *Journal of Ethnopharmacology*, *66*(1), 1–10. https://doi.org/10.1016/s0378-8741(98)00179-2

Kaehler, N., Adhikari, B., Cheah, P. Y., von Seidlein, L., Day, N. P. J., Paris, D. H., Tanner, M., & Pell, C. (2019). Prospects and strategies for malaria elimination in the Greater Mekong sub-region: A qualitative study. *Malaria Journal*, *18*(1), 203. https://doi.org/10.1186/s12936-019-2835-6

Kaingu, C. K. (2020). Acute and sub-acute oral toxicity and phytochemical profile of Croton menyharthii plant from Tana River County Kenya. *Discovery Phytomedicine*, *7*(3). https://doi.org/10.15562/phytomedicine.2020.135Kaur, K., Jain, M., Kaur, T., & Jain, R. (2009). Antimalarials from nature. *Bioorganic & Medicinal Chemistry*, *17*(9), 3229–3256. https://doi.org/10.1016/j.bmc.2009.02.050

Kerry, N. (2018, February 15). Croton gratissimus. *CJM Tree Growers*. https://cjmgrowers.co.za/croton-gratissimus/

Kigen, G., Kamuren, Z., Njiru, E., Wanjohi, B., & Kipkore, W. (2019). Ethnomedical survey of the plants used by traditional healers in Narok County, Kenya. *Evidence-Based Complementary and Alternative Medicine*, *2019*, 1–8. https://doi.org/10.1155/2019/8976937

Kuete, V., & Efferth, T. (2014). Contribution of African flora in a global fight against cancer. In V. Kuete & T. Efferth (eds.), *Biodiversity, Natural Products and Cancer Treatment* (pp. 289–305). World Scientific. https://doi.org/10.1142/9789814583510_0008

Kuete, V., Viertel, K., & Efferth, T. (2013). Antiproliferative potential of African medicinal plants. In *Medicinal Plant Research in Africa* (pp. 711–724). Elsevier. https://doi.org/10.1016/B978-0-12-405927-6.00018-7

Kumar, P., Kumar, R., Rastogi, M. K., & Murti, K. (2017). Exploration of antidiabetic and hypolipidemic activity of roots of *Croton zambesicus*. *American Journal of Pharmacology and Toxicology*, *12*(1), 1–6. https://doi.org/10.3844/ajptsp.2017.1.6

Langat, M. K., Crouch, N. R., Pohjala, L., Tammela, P., Smith, P. J., & Mulholland, D. A. (2012). Ent-kauren-19-oic acid derivatives from the stem bark of Croton pseudopulchellus Pax. *Phytochemistry Letters*, *5*(3), 414–418. https://doi.org/10.1016/j.phytol.2012.03.002

Laryea, M. K., & Borquaye, L. S. (2019). Antimalarial efficacy and toxicological assessment of extracts of some Ghanaian medicinal plants. *Journal of Parasitology Research*, *2019*, 1–9. https://doi.org/10.1155/2019/1630405

Lima, R. B. S., Rocha e Silva, L. F., Melo, M. R. S., Costa, J. S., Picanço, N. S., Lima, E. S., Vasconcellos, M. C., Boleti, A. P. A., Santos, J. M. P., Amorim, R. C. N., Chaves, F. C. M., Coutinho, J. P., Tadei, W. P., Krettli, A.

U., & Pohlit, A. M. (2015). In vitro and in vivo anti-malarial activity of plants from the Brazilian Amazon. *Malaria Journal*, *14*, 508. https://doi.org/10.1186/s12936-015-0999-2

Lopes, D., Bizzo, H. R., Sá Sobrinho, A. F., & Pereira, M. V. (2003). Essential oil from leaves of Croton sacaquinha Benth. *Journal of Essential Oil Research*, *15*(1), 48–49. https://doi.org/10.1080/10412905.2003.9712263

Maberly, J. (1899). The pharmacology of some South African plants. *Lancet*, *157*, 874.

Mabogo, D. E. N. (1999). *The Ethnobotany of the Vhavenda* [Dissertation, University of Pretoria]. https://repository.up.ac.za/handle/2263/28881

Maciel, M. A. M., Pinto, A. C., Arruda, A. C., Pamplona, S. G. S. R., Vanderlinde, F. A., Lapa, A. J., Echevarria, A., Grynberg, N. F., Côlus, I. M. S., Farias, R. A. F., Luna Costa, A. M., & Rao, V. S. N. (2000). Ethnopharmacology, phytochemistry and pharmacology: A successful combination in the study of Croton cajucara. *Journal of Ethnopharmacology*, *70*(1), 41–55. https://doi.org/10.1016/S0378-8741(99)00159-2

Maciel, M. A. M., Pinto, A. C., Brabo, S. N., & Da Silva, M. N. (1998). Terpenoids from Croton cajucara. *Phytochemistry*, *49*(3), 823–828. https://doi.org/10.1016/S0031-9422(98)00163-0

Magwilu, K. D., Nguta, J. M., Mapenay, I., & Matara, D. (2022). Phylogeny, phytomedicines, phytochemistry, pharmacological properties, and toxicity of Croton gratissimus Burch (Euphorbiaceae). *Advances in Pharmacological and Pharmaceutical Sciences*, *2022*, 1238270. https://doi.org/10.1155/2022/1238270

Maroyi, A. (2017). Ethnopharmacological uses, phytochemistry, and pharmacological properties of *Croton macrostachyus* Hochst. Ex Delile: A comprehensive review. *Evidence-Based Complementary and Alternative Medicine*, *2017*, 1–17. https://doi.org/10.1155/2017/1694671

Maroyi, A. (2018). Review of Croton membranaceus Mûll. Arg.: Phytochemical, pharmacological and toxicological perspective. *Journal of Complementary Medicine Research*, *9*(2), 24. https://doi.org/10.5455/jcmr.20180718123349

Maroyi, A. (2019). Croton species in Madagascar: Their ethnomedicinal uses, phytochemistry and biological activities. *Asian Journal of Agriculture and Biology*, *7*(2), 279–288.

Matara, D. N., Nguta, J. M., Musila, F. M., & Mapenay, I. (2021). Phytochemical analysis and investigation of the antimicrobial and cytotoxic activities of Croton dichogamus Pax crude root extracts. *Evidence-Based Complementary and Alternative Medicine: ECAM*, *2021*, 2699269. https://doi.org/10.1155/2021/2699269

Mbayo, M., Kalonda, M., Tshisand, T., Kisimba, K., Mulamba, M., Muya, K., Sangwa, K., Kalubandika, G., Maseho, M., Bakari, S., Mpiana, P. T., Kahumba, B., & Lumbu, S. (2016). Contribution to ethnobotanical knowledge of some Euphorbiaceae used in traditional medicine in Lubumbashi and its surroundings (DRC). *Journal of Advanced Botany and Zoology*, *4*, 1–16. https://doi.org/10.15297/JALS.V4I2.01

Mekonnen, L. B. (2015). In vivo antimalarial activity of the crude root and fruit extracts of Croton macrostachyus (Euphorbiaceae) against Plasmodium berghei in mice. *Journal of Traditional and Complementary Medicine*, *5*(3), 168–173. https://doi.org/10.1016/j.jtcme.2014.07.002

Meresa, A., Ashebir, R., Gemechu, W., Basha, H., Arega, E., Woldkidan, S., Tadele, A., Teka, F., & Abebe, A. (2019). Ethno medicinal uses, phytochemistry and anti-malarial effect of Croton macrostachyus (Bisana): A review. *Journal of Medicinal Plants Studies*, *7*, 79–88.

Mesfin, F., Demissew, S., & Teklehaymanot, T. (2009). An ethnobotanical study of medicinal plants in Wonago Woreda, SNNPR, Ethiopia. *Journal of Ethnobiology and Ethnomedicine*, *5*(1), 28. https://doi.org/10.1186/1746-4269-5-28

Mesia, G. K., Tona, G. L., Penge, O., Lusakibanza, M., Nanga, T. M., Cimanga, R. K., Apers, S., Van Miert, S., Totte, J., Pieters, L., & Vlietinck, A. J. (2005). Antimalarial activities and toxicities of three plants used as traditional remedies for malaria in the Democratic Republic of Congo: *Croton mubango, Nauclea pobeguinii* and *Pyrenacantha staudtii*. *Annals of Tropical Medicine & Parasitology*, *99*(4), 345–357. https://doi.org/10.1179/136485905X36325

Milanowski, D. J., Winter, R. E. K., Elvin-Lewis, M. P. F., & Lewis, W. H. (2002). Geographic distribution of three alkaloid chemotypes of Croton lechleri. *Journal of Natural Products*, *65*(6), 814–819. https://doi.org/10.1021/np000270v

Milliken, W. (1997). Traditional anti-malarial medicine in Roraima, Brazil. *Economic Botany*, *51*(3), 212–237.

Mills, A., Lubell, Y., & Hanson, K. (2008). Malaria eradication: The economic, financial and institutional challenge. *Malaria Journal*, *7*(S1), S11. https://doi.org/10.1186/1475-2875-7-S1-S11

Mitchell, S. A., & Ahmad, M. H. (2006). A review of medicinal plant research at the University of the West Indies, Jamaica, 1948–2001. *The West Indian Medical Journal*, *55*(4), 243–269. https://doi.org/10.1590/s0043-31442006000400008

Mohammadi, S., Jafari, B., Asgharian, P., Martorell, M., & Sharifi-Rad, J. (2020). Medicinal plants used in the treatment of Malaria: A key emphasis to Artemisia, Cinchona, Cryptolepis, and Tabebuia genera. *Phytotherapy Research*, *34*(7), 1556–1569. https://doi.org/10.1002/ptr.6628

Morales, C. G. (2014). *Croton Lobatus L.* (collected in Mexico). US National Herbarium, Department of Botany, NMNH, Smithsonian Institution, in collaboration with Herbario Nacional de México (MEXU), Instituto de Biologia. Occurrence dataset https://doi.org/10.15468/hnhrg3 accessed via GBIF.org on 2023–03–14. www.gbif.org/occurrence/1320015012

Moremi, M. P., Makolo, F., Viljoen, A. M., & Kamatou, G. P. (2021). A review of biological activities and phytochemistry of six ethnomedicinally important South African Croton species. *Journal of Ethnopharmacology*, *280*, 114416. https://doi.org/10.1016/j.jep.2021.114416

Mpondo Mpondo, E., Vandi, D., Ngouondjou Foze, T., Mvogo Ottou, P. B., Embolo Enyegue, E., Yinyang, J., & Dibong, S. D. (2018). Contribution des populations des villages du centre Cameroun aux traitements traditionnels des affections des voies respiratoires. *Journal of Animal & Plant Sciences*, *32*(3).

Mthethwa, N. S., Oyedeji, B. A., Obi, L. C., & Aiyegoro, O. A. (2014). Anti-staphylococcal, anti-HIV and cytotoxicity studies of four South African medicinal plants and isolation of bioactive compounds from Cassine transvaalensis (Burtt. Davy) codd. *BMC Complementary and Alternative Medicine*, *14*(1), 512. https://doi.org/10.1186/1472-6882-14-512

Mulholland, D. A., Langat, M. K., Crouch, N. R., Coley, H. M., Mutambi, E. M., & Nuzillard, J.-M. (2010). Cembranolides from the stem bark of the southern African medicinal plant, Croton gratissimus (Euphorbiaceae). *Phytochemistry*, *71*(11–12), 1381–1386. https://doi.org/10.1016/j.phytochem.2010.05.014

Mutakela, P. S. (2001). State of forest genetic resources in Botswana. Prepared for the SADC regional workshop on forest and tree genetic resources, Arusha, Tanzania, 5–9 June 2000. *Forest Genetic Resources Working Papers, Working Paper FGR/11E. Forest Resources Development Service, Forest Resources Division*. FAO.

Muthee, J. K., Gakuya, D. W., Mbaria, J. M., Kareru, P. G., Mulei, C. M., & Njonge, F. K. (2011). Ethnobotanical study of anthelmintic and other medicinal plants traditionally used in Loitoktok district of Kenya. *Journal of Ethnopharmacology*, *135*(1), 15–21. https://doi.org/10.1016/j.jep.2011.02.005

Mutie, F. M., Gao, L.-L., Kathambi, V., Rono, P. C., Musili, P. M., Ngugi, G., Hu, G.-W., & Wang, Q.-F. (2020). An ethnobotanical survey of a dryland botanical garden and its environs in Kenya: The Mutomo hill plant sanctuary. *Evidence-Based Complementary and Alternative Medicine: ECAM*, *2020*, 1543831. https://doi.org/10.1155/2020/1543831

Nath, R., De, B., & Choudhury, M. D. (2013a). Phytochemical study of Croton caudatus Geisel by TLC profiling. *Biosciences Biotechnology Research Asia*, *10*(1), 445–449. https://doi.org/10.13005/bbra/1151

Nath, R., Roy, S., De, B., & Duttachoudhury, M. (2013b). Anticancer and antioxidant activity of Croton: A review. *International Journal of Pharmacy and Pharmaceutical Sciences*, *5*, 63–70.

Ndubani, P., & Höjer, B. (1999). Traditional healers and the treatment of sexually transmitted illnesses in rural Zambia. *Journal of Ethnopharmacology*, *67*(1), 15–25.

Ndunda, B. (2014). *Phytochemistry and Bioactivity Investigations of Three Kenyan Croton Species* [Thesis, University of Nairobi]. http://erepository.uonbi.ac.ke/handle/11295/76387

Neelo, J., Kashe, K., Teketay, D., & Masamba, W. (2015). Ethnobotanical survey of woody plants in Shorobe and Xobe villages, northwest region of Botswana. *Ethnobotany Research and Applications*, *14*, 367–379.

Neuwinger, H. D. (1996). *African Ethnobotany. Poisons and Drugs. Chemistry, Pharmacology, Toxicology*. Chapman and Hall Ltd. https://scholar.google.com/scholar_lookup?title=African+ethnobotany.+Poisons+and+drugs.+Chemistry%2C+pharmacology%2C+toxicology&author=Neuwinger%2C+H.D.+%28Correspondence+address%3A+Hauptstrasse+190%2C+68789+St.+Leon-Rot+%28Germany%29%29&publication_year=1996

Neuwinger, H. D. (2000). African traditional medicine: A dictionary of plant use and applications. With supplement: search system for diseases. *African Traditional Medicine: A Dictionary of Plant Use and Applications. With Supplement: Search System for Diseases*. www.cabdirect.org/cabdirect/abstract/20026790056

Neuwinger, H. D. (2004). Plants used for poison fishing in tropical Africa. *Toxicon*, *44*(4), 417–430. https://doi.org/10.1016/j.toxicon.2004.05.014

Njau, J. D., Goodman, C., Kachur, S. P., Palmer, N., Khatib, R. A., Abdulla, S., Mills, A., & Bloland, P. (2006). Fever treatment and household wealth: The challenge posed for rolling out combination therapy for malaria. *Tropical Medicine and International Health*, *11*(3), 299–313. https://doi.org/10.1111/j.1365-3156.2006.01569.x

Noor Rain, A., Khozirah, S., Mohd Ridzuan, M. A. R., Ong, B. K., Rohaya, C., Rosilawati, M., Hamdino, I., Badrul, A., & Zakiah, I. (2007). Antiplasmodial properties of some Malaysian medicinal plants. *Tropical Biomedicine*, *24*(1), 29–35.

Nyazema, N. Z. (1984). Poisoning due to traditional remedies. *Central African Journal of Medicine*, *30*(5), 80–83.

Obey, J. K., Ngeiywa, M. M., Kiprono, P., Omar, S., von Wright, A., Kauhanen, J., & Tikkanen-Kaukanen, C. (2018). Antimalarial activity of *Croton macrostachyus* stem bark extracts against *Plasmodium berghei In Vivo*. *Journal of Pathogens*, *2018*, 1–6. https://doi.org/10.1155/2018/2393854

Okeke, T. A., Okafor, H. U., & Uzochukwu, B. S. C. (2006). Traditional healers in Nigeria: perception of cause, treatment and referral practices for severe malaria. *Journal of Biosocial Science*, *38*(4), 491–500. https://doi.org/10.1017/S002193200502660X

Okokon, J., Iyadi, K., & Effiong, C. (2005). Effect of sub chronic administration of ethanolic leaf extract of *Croton zambesicus* on hematological parameters of rats. *Nigerian Journal of Physiological Sciences*, *19*(1). https://doi.org/10.4314/njps.v19i1.32629

Okoronkwo, I., Onyia-pat, J., Okpala, P., Agbo, M.-A., & Ndu, A. (2014). Patterns of complementary and alternative medicine use, perceived benefits, and adverse effects among adult users in Enugu Urban, Southeast Nigeria. *Evidence-Based Complementary and Alternative Medicine*, *2014*, 1–6. https://doi.org/10.1155/2014/239372

Oliveira, A. B., Dolabela, M. F., Braga, F. C., Jácome, R. L. R. P., Varotti, F. P., & Póvoa, M. M. (2009). Plant-derived antimalarial agents: New leads and efficient phythomedicines. Part I. Alkaloids. *Anais Da Academia Brasileira de Ciências*, *81*(4), 715–740. https://doi.org/10.1590/S0001-37652009000400011

Oliveira, A. C., Leal-Cardoso, J. H., Santos, C. F., Morais, S. M., & Coelho-de-Souza, A. N. (2001). Antinociceptive effects of the essential oil of Croton zehntneri in mice. *Brazilian Journal of Medical and Biological Research*, *34*(11), 1471–1474. https://doi.org/10.1590/S0100-879X2001001100016

Omara, T. (2020). Antimalarial plants used across Kenyan communities. *Evidence-Based Complementary and Alternative Medicine*, *2020*, 1–31. https://doi.org/10.1155/2020/4538602

Oyebode, O., Kandala, N.-B., Chilton, P. J., & Lilford, R. J. (2016). Use of traditional medicine in middle-income countries: A WHO-SAGE study. *Health Policy and Planning*, *31*(8), 984–991. https://doi.org/10.1093/heapol/czw022

Palazzino, G., Federici, E., Rasoanaivo, P., Galeffi, C., & Delle Monache, F. (1998). 3, 4-seco diterpenes of Croton geayi. *ChemInform*, *29*(12).

Palhares, R. M., Gonçalves Drummond, M., Dos Santos Alves Figueiredo Brasil, B., Pereira Cosenza, G., das Graças Lins Brandão, M., & Oliveira, G. (2015). Medicinal plants recommended by the world health organization: DNA barcode identification associated with chemical analyses guarantees their quality. *PLoS ONE*, *10*(5), e0127866. https://doi.org/10.1371/journal.pone.0127866

Palmer, E., & Pitman, N. (1972). *Trees of Southern Africa: Covering All Known Indigenous Species in the Republic of South Africa, South-West Africa, Botswana, Lesotho & Swaziland* (Volumes 1 & 2., p. 972). A. A. Balkema.

Parratt, N. T. (1996). The potential of non-timber forest products of Botswana. *Botswana Notes & Records*, *28*(1), 203–218.

Pascaline, J., Charles, M., & George, O. (2011). An inventory of medicinal plants that the people of Nandi use to treat malaria. *Journal of Animal & Plant Sciences*, *9*(3).

Peralta, W. F. S., Giraldo, L. F. O., Jaramillo, C. C. P., Arango, W. M., & Arteaga, J. J. M. (2015). Phytochemical screening, free radical scavenging and anti-inflammatory activity of Croton leptostachyus Kunth leaf extracts. *Asian Journal of Pharmaceutical and Clinical Research*, 190–194.

Peters, D. H., Garg, A., Bloom, G., Walker, D. G., Brieger, W. R., & Hafizur Rahman, M. (2008). Poverty and access to health care in developing countries. *Annals of the New York Academy of Sciences*, *1136*(1), 161–171. https://doi.org/10.1196/annals.1425.011

Picking, D., Younger, N., Mitchell, S., & Delgoda, R. (2011). The prevalence of herbal medicine home use and concomitant use with pharmaceutical medicines in Jamaica. *Journal of Ethnopharmacology*, *137*(1), 305–311. https://doi.org/10.1016/j.jep.2011.05.025

Pillay, P., Maharaj, V. J., & Smith, P. J. (2008). Investigating South African plants as a source of new antimalarial drugs. *Journal of Ethnopharmacology*, *119*(3), 438–454. https://doi.org/10.1016/j.jep.2008.07.003

Prozesky, E. A. (2008). *Antiplasmodial- and Chloroquine Resistance Reversal Properties of a New Diterpene from Croton Steenkampianus* [Thesis, University of Pretoria]. https://repository.up.ac.za/handle/2263/27220

Pullanikkatil, D., Palamuleni, L. G., & Ruhiiga, T. M. (2016). Land use/land cover change and implications for ecosystems services in the Likangala River Catchment, Malawi. *Physics and Chemistry of the Earth, Parts A/B/C*, *93*, 96–103.

Radulović, N., Mananjarasoa, E., Harinantenaina, L., & Yoshinori, A. (2006). Essential oil composition of four Croton species from Madagascar and their chemotaxonomy. *Biochemical Systematics and Ecology*, *34*(8), 648–653. https://doi.org/10.1016/j.bse.2006.02.005

Rahman, S., & Husen, A. (2021). Potential role of medicinal plants in the cure of liver and kidney diseases. In A. Husen, R. K. Bachheti & A. Bachheti (eds.), *Non-Timber Forest Products*. Springer International Publishing AG, pp. 229–254. https://doi.org/10.1007/978-3-030-73077-2_10

Rahman, S., & Husen, A. (2022). *Asparagus racemosus* (Shatavari) and *Dioscorea villosa* (Wild yams). In A. Husen (ed.), *Herbs, Shrubs, and Trees of Potential Medicinal Benefits*. CRC Press.

Rahman, S., & Husen, A. (2023a). *Trigonella foenum-graecum* L.: A useful medicinal plant for diabetes and other diseases. In A. Husen (ed.), *Antidiabetic Medicinal Plants and Herbal Treatments*. CRC Press.

Rahman, S., & Husen, A. (2023b). Role of Green Chiretta [*Andaphis paniculata* (Burm. F.)] nees in diabetes and other ailments. In A. Husen (ed.), *Antidiabetic Medicinal Plants and Herbal Treatments*. CRC Press.

Rakotonandrasana, O. L., Raharinjato, F. H., Rajaonarivelo, M., Dumontet, V., Martin, M.-T., Bignon, J., & Rasoanaivo, P. (2010). Cytotoxic 3,4- *seco* -atisane diterpenoids from *Croton barorum* and *Croton goudotii*. *Journal of Natural Products*, *73*(10), 1730–1733. https://doi.org/10.1021/np1005086

Rampa, K. M., Van De Venter, M., Koekemoer, T. C., Swanepoel, B., Venables, L., Hattingh, A. C., Viljoen, A. M., & Kamatou, G. P. (2022). Exploring four South African Croton species for potential anti-inflammatory properties: In vitro activity and toxicity risk assessment. *Journal of Ethnopharmacology*, *282*, 114596. https://doi.org/10.1016/j.jep.2021.114596

Rao, V. B., Schellenberg, D., & Ghani, A. C. (2013). Overcoming health systems barriers to successful malaria treatment. *Trends in Parasitology*, *29*(4), 164–180. https://doi.org/10.1016/j.pt.2013.01.005

Rasoanaivo, P., Petitjean, A., Ratsimamanga-Urverg, S., & Rakoto-Ratsimamanga, A. (1992). Medicinal plants used to treat malaria in Madagascar. *Journal of Ethnopharmacology*, *37*(2), 117–127. https://doi.org/10.1016/0378-8741(92)90070-8

Regassa, R. (2018). Assessment of indigenous knowledge of medicinal plant practice and mode of service delivery in Hawassa city, southern Ethiopia. *Journal of Medicinal Plant Research*, *7*. https://doi.org/10.5897/JMPR012.1126

Risco, E., Ghia, F., Vila, R., Iglesias, J., Alvarez, E., & Cañigueral, S. (2003). Immunomodulatory activity and chemical characterisation of sangre de drago (dragon's blood) from Croton lechleri. *Planta Medica*, *69*(9), 785–794. https://doi.org/10.1055/s-2003-43208

Robert, S., Baccelli, C., Devel, P., Dogné, J.-M., & Quetin-Leclercq, J. (2010). Effects of leaf extracts from Croton zambesicus Müell. Arg. On hemostasis. *Journal of Ethnopharmacology*, *128*(3), 641–648. https://doi.org/10.1016/j.jep.2010.02.007

Roig y Mesa, J. T., & Academia de Ciencias de Cuba, L. H. I. de A. spa. (1974). *Plantas medicinales, aromaticas o venenosas de Cuba – 2. Ed.* La Habana (Cuba) Ciencia y Tecnica, Inst. del Libro. https://scholar.google.com/scholar_lookup?title=Plantas+medicinales%2C+aromaticas+o+venenosas+de+Cuba++-+2.+ed.&author=Roig+y+Mesa%2C+J.T.&publication_year=1974

Sahoo, R. N., Kumar, S., Suryavanshi, A., Kain, D., Arya, A., & Chaudhry, B. (2022). Venerated and medicinal aspects of plants used in India: An ethnobotanical review. *Journal of Drug Research in Ayurvedic Sciences*, *6*(3), 128–140. https://doi.org/10.4103/jdras.jdras_15_21

Salatino, A., Salatino, M. L. F., & Negri, G. (2007). Traditional uses, chemistry and pharmacology of Croton species (Euphorbiaceae). *Journal of the Brazilian Chemical Society*, *18*(1), 11–33. https://doi.org/10.1590/S0103-50532007000100002

Salleh, W. M. N. H. W., Nafiah, M. A., Khamis, S., & Jauri, M. H. (2022). Chemical composition of the essential oil of Croton argyratus. *Chemistry of Natural Compounds*, *58*(3), 556–557. https://doi.org/10.1007/s10600-022-03736-4

Sandoval, M., Okuhama, N. N., Clark, M., Angeles, F. M., Lao, J., Bustamante, S., & Miller, M. J. S. (2002). Sangre de grado Croton palanostigma induces apoptosis in human gastrointestinal cancer cells. *Journal of Ethnopharmacology*, *80*(2–3), 121–129. https://doi.org/10.1016/s0378-8741(02)00013-2

Saraswaty, V., Setiyanto, H., & Nurhajati, J. (2012). Antibacterial activity from Croton argyratus stem bark extract. *International Journal of PharmTech Research*, *4*, 190–193.

Sawalha, A. F. (2007). Complementary and Alternative Medicine (CAM) in Palestine: Use and safety implications. *The Journal of Alternative and Complementary Medicine*, *13*(2), 263–270. https://doi.org/10.1089/acm.2006.6280

Schmelzer, G. H., & Gurib-Fakim, A. (2013). Plant resources of tropical Africa 11(2): Medicinal plants 2. *Plant Resources of Tropical Africa*, *11*(2): Medicinal Plants 2. www.cabdirect.org/cabdirect/abstract/20163198829

Schmelzer, G. H., Achigan-Dako, E., & Bosch, C. (eds.). (2010). *Medicinal Plants of Tropical Africa. Conclusions and Recommendations Based on PROTA* (Vol. 11). 'Medicinal plants'. PROTA Foundation.

Schmelzer, G. H., Gurib-Fakim, A., Arroo, R., Bosch, C. H., Ruijter, A. de, Simmonds, M. S. J., Lemmens, R. H. M. J., & Oyen, L. P. A. (2008). *Plant Resources of Tropical Africa 11(1): Medicinal Plants 1*. PROTA Foundation [etc.]. https://research.wur.nl/en/publications/plant-resources-of-tropical-africa-111-medicinal-plants-1

Schmidt, E., Lotter, M., & McCleland, W. (2002). *Trees and Shrubs of Mpumalanga and Kruger National Park*. Jacana Media.

Selowa, S. C., Shai, L. J., Masoko, P., Mokgotho, M. P., & Magano, S. R. (2010). Antibacterial activity of extracts of three croton species collected in Mpumalanga region in South Africa. *African Journal of Traditional, Complementary and Alternative Medicines*, *7*(2). https://doi.org/10.4314/ajtcam.v7i2.50861

Sori, T., Bekana, M., Adugna, G., & Kelbessa, E. (2004). Medicinal plants in the ethnoveterinary practices of Borana pastoralists, Southern Ethiopia. *International Journal of Applied Research in Veterinary Medicine*, *2*(3).

Suárez, A. I., Blanco, Z., Compagnone, R. S., Salazar-Bookaman, M. M., Zapata, V., & Alvarado, C. (2006). Anti-inflammatory activity of Croton cuneatus aqueous extract. *Journal of Ethnopharmacology*, *105*(1–2), 99–101. https://doi.org/10.1016/j.jep.2005.10.006

Suswardany, D. L., Sibbritt, D. W., Supardi, S., Chang, S., & Adams, J. (2015). A critical review of traditional medicine and traditional healer use for malaria and among people in malaria-endemic areas: Contemporary research in low to middle-income Asia-Pacific countries. *Malaria Journal*, *14*(1), 98. https://doi.org/10.1186/s12936-015-0593-7

Tangenyi, O. (1996). *Contribution á l'étude des plantes médicinales utilisées contre le Paludisme au Sankuru* [MSc Thesis, Patrice Emery Lumumba Methodist University, Wembo-Nyama].

Thongtan, J., Kittakoop, P., Ruangrungsi, N., Saenboonrueng, J., & Thebtaranonth, Y. (2003). New antimycobacterial and antimalarial 8,9-secokaurane diterpenes from *Croton k ongensis*. *Journal of Natural Products*, *66*(6), 868–870. https://doi.org/10.1021/np030067a

Tietjen, I., Gatonye, T., Ngwenya, B. N., Namushe, A., Simonambanga, S., Muzila, M., Mwimanzi, P., Xiao, J., Fedida, D., & Brumme, Z. L. (2016). Croton megalobotrys Müll Arg. and Vitex doniana (Sweet): Traditional medicinal plants in a three-step treatment regimen that inhibit in vitro replication of HIV-1. *Journal of Ethnopharmacology*, *191*, 331–340.

Tobgay, T., Torres, C. E., & Na-Bangchang, K. (2011). Malaria prevention and control in Bhutan: Successes and challenges. *Acta Tropica*, *117*(3), 225–228. https://doi.org/10.1016/j.actatropica.2010.11.008

Tropical Plants Database. (2012). https://tropical.theferns.info/viewtropical.php?id=Croton+dichogamus

Tropical Plants Database. (2019). Ken Fern. Tropical.Theferns.Info. tropical.theferns.info/viewtropical.php?id=Croton+megalobotrys

van Vuuren, S. F., & Naidoo, D. (2010). An antimicrobial investigation of plants used traditionally in southern Africa to treat sexually transmitted infections. *Journal of Ethnopharmacology*, *130*(3), 552–558. https://doi.org/10.1016/j.jep.2010.05.045

Vigor, C., Fabre, N., Fourasté, I., & Moulis, C. (2001). Three clerodane diterpenoids from Croton eluteria Bennett. *Phytochemistry*, *57*(8), 1209–1212. https://doi.org/10.1016/s0031-9422(01)00183-2

von Bruchhausen, F., Ebel, S., & Hackenthal, E. (1999). *Hagers Handbuch der Pharmazeutischen Praxis: Stoffe AK*. Springer.

Watt, J. M., & Breyer-Brandwijk, M. G. (1962). *The Medicinal and Poisonous Plants of Southern and Eastern Africa Being an Account of Their Medicinal and Other Uses, Chemical Composition, Pharmacological Effects and Toxicology in Man and Animal*. E. & S. Livingstone.

WFO. (n.d.). *Croton Megalobotrys Müll.Arg*. Published on the Internet. Retrieved March 13, 2023, from www.worldfloraonline.org/taxon/wfo-0000931663

World Health Organization. (2019). *WHO Global Report on Traditional and Complementary Medicine*. World Health Organization. https://iris.who.int/handle/10665/312342

World Health Organization. (2022). *World Malaria Report 2022*. World Health Organization. https://www.who.int/teams/global-malaria-programme/reports/world-malaria-report-2022. Accessed on February 12, 2023.

Yibralign, Z. (2007). *Phytochemical Investigation on the Stem Bark of Croton Macrostachyus (Bisana)* [Thesis, Addis Ababa Universty]. http://etd.aau.edu.et/handle/123456789/4189

Zhang, Z.-X., Li, H.-H., Fan, G.-X., Li, Z.-Y., Dong, L.-L., Li, H.-Y., & Fei, D.-Q. (2015). A Novel norclerodane diterpenoid from the roots of *Croton crassifolius*. *Natural Product Communications*, *10*(11), 1934578X1501001. https://doi.org/10.1177/1934578X1501001128

Zhang, Z.-X., Li, H.-H., Qi, F.-M., Xiong, H.-Y., Dong, L.-L., Fan, G.-X., & Fei, D.-Q. (2014). A new halimane diterpenoid from Croton crassifolius. *Bulletin of the Korean Chemical Society*, *35*(5), 1556–1558. https://doi.org/10.5012/BKCS.2014.35.5.1556

13 Antimalarial Response, Traditional and Other Potential Uses of *Cryptolepis* Genera

Patrick Valere Tsouh Fokou, Vincent Ngouana, Raoul Kemzeu, Aubin Kamche Youbi, and Issakou Bakarnga-Via

13.1 INTRODUCTION

The orchid family is home to Apocynaceae, a dicotyledonous plant with sophisticated entomophilic pollination mechanisms. Categorization in the family is influenced by the pollination mechanism's morphology, discernible across five subfamilies. Periplocoididae serves as the root subfamily connecting the earliest Lauborfiodyceae and Vulcanomataceae to the advanced Sechamonoideae and Asclepiadaceae. *Cryptolepis*, with 30 species, is the second-largest genus in the Periplocoideae subfamily, which is mainly made up of monotypic or dimorphic genera. *Cryptolepis* species are generally thought of as tropical and subtropical climbers, and among the better-known species are *Cryptolepis apiculata* K. Schum., *Cryptolepis sanguinolenta* (Lindl.) Schltr. (African taxa), *Cryptolepis buchananii* Roem and Schult., *Cryptolepis grandiflora* Wight, and *Cryptolepis sinensis* (Lour.). Occupying various habitats from moist forests to stony semi-deserts, the genus in the subfamily spreads extensively, making it the most widely distributed. Shrubs are the common species of *Cryptolepis* found in arid regions with less foliage, slender inflorescences, and scanty blooming. Its significant radiation began 16 million years ago, while its appearance dates to 21 million years ago. Despite its widespread distribution, it has not evolved significantly, possibly due to the lack of succulence. Three SE Asian species are nested among African taxa and are likely to have arrived in SE Asia via long-distance dispersal (Joubert et al., 2016; Joubert and Venter, 2008; Venter et al., 2006).

Cryptolepis is a plant that has been used in traditional medicine for centuries. In recent years, *Cryptolepis* has gained attention for its potential as an alternative treatment for various health conditions. For thousands of years, Africans have employed the medicinal herb *Cryptolepis sanguinolenta* to treat ailments like diabetes, hypertension, malaria, Lyme disease, respiratory diseases, hepatitis B viral infection, liver damage, and diarrhoea (Adeleye et al., 2021; Borquaye et al., 2020; Luo et al., 1998). *Cryptolepis* is almost exclusively used as an antimicrobial against parasitic, bacterial, amoebic, and fungal infections. *Cryptolepis sanguinolenta* has proved to contain antiparasitic activities against *Plasmodium falciparum* and is effective in treating malaria, especially when used in combination with other antimalarial drugs. Its root bark bioactive alkaloids, viz, cryptolepine, 11-hydroxycryptolepine, and quindoline, were found to have more *in vitro* antiplasmodial action than chloroquine. In addition to its potential as an alternative treatment for malaria, *Cryptolepis sanguinolenta* acts as an antimicrobial against parasitic, bacterial, amoebic, fungal, and viral infections and as anticancer agents (Borquaye et al., 2020; Zofou et al., 2013). Several alkaloids, including cryptolepine, have been isolated from *Cryptolepis* species, and these compounds and the plant extracts possess broad-spectrum antipathogenic activity. One of the most well-known derived compounds of *Cryptolepis sanguinolenta* is cryptolepine, which is one of the active principles involved in the reported pharmacological properties, such as antimalarial, anticancer, and antidiabetic properties (Borquaye et al., 2020; Luo et al., 1998; Zhang et al., 2021; Zofou et al., 2013).

 DOI: 10.1201/9781003378396-13

In this chapter, we will explore the pharmacology of *Cryptolepis*, its benefits, especially in malaria treatments, and its potential risks.

13.2 BOTANICAL DESCRIPTION AND DISTRIBUTION

The Periplocoideae is a relatively small subfamily of the Apocynaceae, with 34 genera and about 190 species. *Cryptolepis* R. Br. is the second-largest genus in the Periplocoideae, with 30 species (Joubert et al., 2016; Joubert and Venter, 2008). *Cryptolepis* has the following characteristics: A shrub or vine with woody stems. The leaves are bluish on the dorsal side. The cymes are terminal, axillary, or extra-axillary, with stalks. The calyx has 5–10 basal glands. The buds are cylindrical and pointed. The corolla has a stalk and a short cylindrical or bell-shaped tube with overlapping lobes on the right side. The corolla lobes are linear or ovate, fibrous, and inserted near the centre of the corolla tube. The filament is wide at the bottom and narrow at the top. The anthers are fused and attached to the stigma. Pollen tetrads are solitary within each anther cell, and pollen carriers are spatulate and erect. The head of the stigma is broad conical. The follicles are paired, widely fanned out, and narrowly lanceolate in outline (WFO, 2023).

Cryptolepis R. Br. grows in the Old World, in southern Europe, in Africa (both sub-Saharan Africa and North Africa) and Madagascar, southern Yemen, the Socotra Archipelago, and South Asia from India–Pakistan to southern China, Taiwan, the Philippines, and Indonesia and northern Australia (Joubert, 2013; Joubert et al., 2016). *Cryptolepis* is a genus of plants found in tropical forests and savannahs, of which 13 species are adapted to dry environments. The majority of *Cryptolepis* species are concentrated in four centres of diversity along the east coast of Africa and Socotra (Joubert, 2013). The genus *Cryptolepis* includes *Cryptolepis brazzei*, *C. deciduas*, *C. oblongifolia*, *C. nigritana*, *C. triangularis*, and *C. sanguinolenta*. *Cryptolepis sanguinolenta* is common to West Africa. It is also known as *Pergularia sanguinolenta* (Ameyaw et al., 2007; Barku et al., 2012). With the exception of *C. sanguinolenta*, which is widely distributed in West African coastal countries, and *C. purpureus*, which occurs in Cameroon, all African *Cryptolepis* species are found near the east coast. *C. sanguinolenta* is found in countries ranging from Senegal to Nigeria, as well as Gambia, Guinea-Bissau, Guinea, Ghana, Ivory Coast, and Sierra Leone. Cameroon and Sudan have not reported the species, but it has been reported from Zaire, Angola (Malange and Cazengo districts), and there is evidence of an outbreak in Uganda in East Africa. *C. obtusa* is the most common species in South-East Africa, with populations found in South Africa (Transvaal), Mozambique, and Tanzania. Two more species have been identified in South Africa. *C. cryptolepioides* is from Transvaal, and *C. capensis* is from Natal and Transkei. *C. sinensis* subsp. Africana has only been found in Kenya. *C. hypoglauca* has been discovered in Tanzania and Kenya, as has *C. apiculata* in these two countries and Zimbabwe. *C. pendulina* lives in Mozambique, and *C. newii* lives in Tanzania/Kilimanjaro Kenya's mountains (Paulo and Houghton, 2003). Macro- and micromorphological studies have demonstrated that many features, such as growth habits, leaf shape and size, leaf epidermal features, leaf veins, inflorescence structure, flower structure, and surface features of the seed coat, have been identified at the species level. Molecular phylogenetic analysis indicated that this genus is paraphyletic and that the genus definition should be extended to include *Parquetina* as a synonym in order to establish a monophyletic genus. Several vegetative and reproductive traits exhibit high levels of homogeneity and morphological plasticity, and *C. oblongifolia* is the most common species due to its high resistance to disturbance (Joubert, 2013).

13.3 PHYTOCHEMICAL CONSTITUENTS

Only three *Cryptolepis* species have had their phytochemistry investigated. Cardenolides and pseudo-alkaloids (nicotinoyl glycosides) are the main secondary metabolites found in the leaves and roots of *C. buchananii*, and cardenolides were also found in the roots *of C. apiculate. C. sanguinolenta* roots were discovered to be high in bioactive indole alkaloids (Paulo and Houghton,

2003). Phytochemical analysis of *C. sanguinolenta* has shown the presence of a variety of secondary metabolites: alkaloids, tannins, and flavones, reducing sugars, polyuronides, alkaloids, and anthocyanosides, triterpenes in addition (Osafo et al., 2017).

The major alkaloid in *C. sanguinolenta* root is the indoloquinoline cryptolepine. Extensive research on cryptolepine has shown its anti-inflammatory, hypotensive, antithrombotic, antidiabetic, antibacterial, antiplasmodial, antipyretic, and renovascular vasodilatory effects (Osafo et al., 2017). Other alkaloids reported from the plant *C. sanguinolenta* include neocryptolepine, quindoline, cryptospirolepine, cryptomisrine, 11-isopropylcryptolepine, cryptolepinone, biscryptolepine, hydroxycryptolepine, cryptoheptine, cryptoquindoline, and isocryptolepine (Figure 13.1) (Osafo et al., 2017).

13.4 TRADITIONAL AND OTHER POTENTIAL USES

Cryptolepis is native to tropical regions of Africa and has been used for centuries in traditional medicine to treat various ailments. One of the most well-known uses of *Cryptolepis* is its antimalarial properties. In recent years, however, researchers have discovered other potential uses for this plant. In addition to its use as an antimalarial agent, *Cryptolepis* has been used in traditional medicine to treat a variety of other ailments. It has been used to treat fever, urinary tract infections, septicaemia, stomach and intestinal disorders, tuberculosis, hepatitis, wounds, fever, hypertension, inflammation, pyrexia, malaria, diarrhoea, and respiratory infections. It has also been used as a diuretic and to treat skin conditions such as eczema and psoriasis (Osafo et al., 2017; Zhang et al., 2021). The traditional uses of *Cryptolepis* are still being studied, and researchers are working to determine the efficacy of these treatments. While some traditional uses of the plant have been validated by scientific research, others have not.

13.5 PHARMACOLOGICAL STUDIES

Among all *Cryptolepis* species, *Cryptolepis sanguinolenta* is the most studied and has been shown in preclinical studies to have anti-inflammatory, antimicrobial, anti-amoebic, anticancer, and antimalarial properties. These properties make it a useful treatment for a variety of infections. *Cryptolepis*

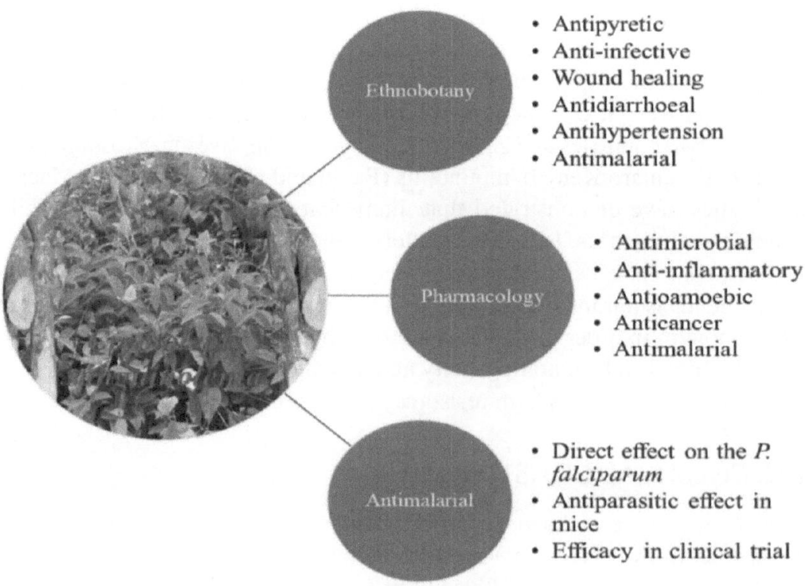

FIGURE 13.1 Several potential applications for *Cryptolepis sanguinolenta* ethnopharmacology.

sanguinolenta effectively treats neurodegenerative diseases, such as Alzheimer's and Parkinson's. In addition, researchers have discovered that *Cryptolepis* contains compounds that have been shown to protect neurons from damage and improve cognitive function (Olajide and Sarker, 2020). Recently, *Cryptolepis sanguinolenta* herbal product extracts support for the treatment of tick-borne infections such as Lyme diseases and *Bartonella* infections, *Bartonella henselae*, which showed high activity respectively against *Borrelia burgdorferi* and *Bartonella henselae* (Feng et al., 2019; Ma et al., 2020). The whole herb *Cryptolepis sanguinolenta* was effective against *Babesia* spp. and had the strongest inhibitory activity against *B. duncani in vitro* with no regrowth visible after six days of subculture at doses of 2×, 4×, and 8× IC_{50} values. Cryptolepine, one important active constituent, showed significant activity against *B. duncani* and further suggests that cryptolepine may be a promising compound to treat human babesiosis. In addition to the significant activity against multiple *Babesia* spp. *In vitro*, Cryptolepine inhibited *B. microti* in mice at a dose of 5 mg/kg (Feng et al., 2019; Zhang et al., 2021).

Cryptolepine has a great potential in the treatment and prevention of several diseases, such as malaria, intestinal disorders, cancer, urinary tract infections, rheumatism, and upper respiratory tract infections. In fact, cryptolepine exhibits a wide range of pharmacological or biological properties, including antimalarial, antimicrobial, antibacterial, antifungal, antihyperglycaemic, anticancer, antidiabetic, anti-inflammatory, hepatoprotective, hypotensive, and antipyretic activities, presynaptic α-adreno receptor blocking action, antimuscarinic, and anticholinesterase activities (Grellier et al., 1996; Olajide and Sarker, 2020; Osafo et al., 2017; Tudu et al., 2023). The antineurodegenerative effects of cryptolepine are mediated by its ability to inhibit enzymes such as acetylcholinesterase and butyrylcholinesterase. However, there is still some research to be carried out on cryptolepine analogues so that they may be more suitable for human clinical trials in the future (Grellier et al., 1996; Olajide and Sarker, 2020; Osafo et al., 2017; Tudu et al., 2023). The alkaloids in *Cryptolepis* have been shown to inhibit the growth of cancer cells and induce apoptosis (cell death) in cancer cells (Olajide and Sarker, 2020).

13.6 ANTIMALARIAL RESPONSE

Cryptolepis sanguinolenta has been used in traditional medicine to treat malaria for centuries. Majority of antimalarial herbal preparations in Ghana are made up of 47.8% of *C. sanguinolenta* (Osafo et al., 2017). Preliminary preclinical and clinical data indicate that *Cryptolepis sanguinolenta* and cryptolepine can have systemic effects *in vivo* and therefore are candidates for clinical development. Two open-label trials using different formulations of *Cryptolepis sanguinolenta* showed significant efficacy without observed toxicity in the treatment of patients with uncomplicated malaria. The pharmacokinetics of cryptolepine revealed that it has low bioavailability, a moderate half-life (4.5h), and extensive distribution (Feng et al., 2019; Zhang et al., 2021). The plant contains several alkaloids, including indole alkaloids, that have been shown to be effective against the malaria parasite (Figure 13.2). One of these alkaloids, cryptolepine, has been extensively studied and has shown promising results in clinical trials (Osafo et al., 2017). Different extracts and alkaloids from the leaves and root bark of *Cryptolepis sanguinolenta* showed *in vitro* activity against various strains of *Plasmodium falciparum* with varying degrees of resistance to chloroquine: *Plasmodium falciparum* D6 (chloroquine-sensitive strain), *Plasmodium falciparum* K1 (multidrug-resistant strain) and T996 (chloroquine-sensitive clone), CQ-sensitive strain F32/Tanzania and the CQ-resistant strains FcB1/Colombia and FcR3/Gambia, and W2 (chloroquine-resistant strains). Extracts displayed 80–90% inhibition of *P. falciparum* growth, with the aqueous extracts being the most potent *in vitro* with IC_{50} ranging from 1 to 2 µg/ml. Cryptolepine remained the most active alkaloid tested, with IC_{50} values (0.23 µM to K1; 0.059 µM to T996). The indolobenzazepine alkaloid cryptoheptine was the second most active, with IC_{50} values of 0.134–0.8 µM (K1) and 1.2 µM (T996). Cryptolepinoic acid showed no significant activity, while its ethyl ester derivative was active against *P. falciparum* K1 (IC_{50} = 3.7 µM). Cryptolepine and its hydrochloride,

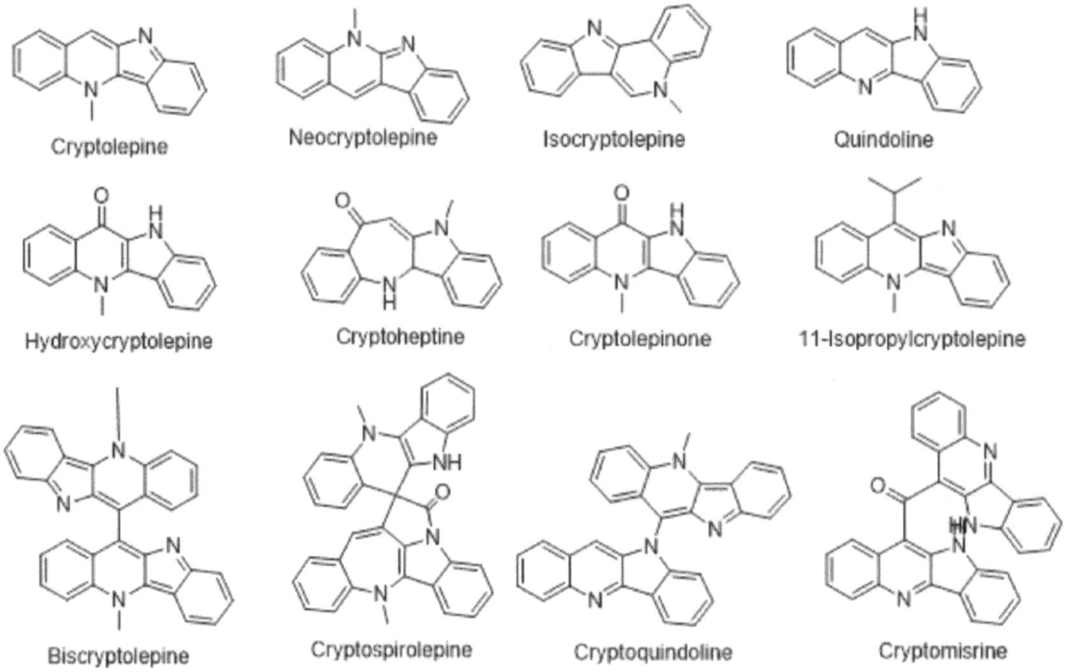

FIGURE 13.2 Alkaloids isolated from *Cryptolepis* genus.

11-hydroxycryptolepine, and neocryptolepine showed a strong antiplasmodial activity against *P. falciparum* chloroquine-resistant strains with IC_{50} values in the same range, 0.2–0.6 μM (Grellier et al., 1996; Osafo et al., 2017; Zhang et al., 2021).

In vivo tests on infected mice showed that crytpolepine, when tested as its hydrochloride, exhibited a significant chemosuppressive effect against *Plasmodium berghei yoelii* and *Plasmodium berghei berghei*, while cryptolepine had the same effect against *P. berghei yoelii* only; 11-hydroxycryptolepine and quindoline did not show activity in this *in vivo* test system. Like 9-aminoacridine, this compound appears to intercalate with DNA, and this may explain the high degree of antimalarial activity demonstrated *in vitro*. The bioactivity of cryptolepine is believed to contribute to DNA intercalating activity targeting non-alternating CC sites, cell morphology change, and topoisomerase II–inhibiting effects. *In vivo*, cryptolepine given orally at 50 mg/kg/day to mice infected with *Plasmodium berghei* was found to have moderate antimalarial activity to suppress parasitaemia by 80%. In a four-day suppression test, there was no significant reduction in parasitaemia in *P. berghei*–infected mice treated subcutaneously with cryptolepine (7–113 mg/kg/day) when compared with untreated controls. However, additional *in vivo* experiments in the infected mice against *P. berghei* and *Plasmodium vinckei petteri* were performed using the classical four-day suppressive test, and the results indicated higher suppression percentage against *P. vinckei petteri*. Even with significant variability in the response to a single dose observed for both parasites in a single group of mice in three independent *in vivo* experiments, the results of this study strongly indicate the plants' inhibitory effect on *Plasmodium* species. Evidence provided by other independent studies indicates the antiplasmodial *in vivo* (mice, ED_{50} = 10 mg/kg i.p.) properties of the *C. sanguinolenta*–derived alkaloid cryptolepine. A nanoformulation of cryptolepine also exhibited a more favourable pharmacokinetic profile and a stronger antiplasmodial effect compared to the unaltered form of cryptolepine (Grellier et al., 1996; Osafo et al., 2017; Zhang et al., 2021). Overall, cryptolepine works by inhibiting the growth of the malaria parasite and preventing it from replicating.

In a clinical trial, the plant's root powder was formulated as teabags and prescribed to 44 patients. The results showed at least 50% reduction in parasitaemia rate in all the patients after 7 days; however, two cases of recurrence have been reported after 25 days (Bugyei et al., 2010). A controlled clinical trial compared an aqueous root extract of *Cryptolepis sanguinolenta* with chloroquine in patients with *P. falciparum* malaria. Twelve patients were treated with *Cryptolepis* extract (25 mg/kg body weight three times daily), and ten were given chloroquine (25 mg/kg over 3 days). Parasite clearance time was only one day longer than with chloroquine (3.3 vs 2.2 days), and symptom clearance was faster (36 vs 48 hours). Fewer side effects were reported by patients taking the extract than by those on chloroquine. No significant changes were detected in the blood and urine samples analyzed. No cases of recrudescence were reported in any of the participants in the follow-up 28-day period. A larger trial in 46 patients confirmed a parasite clearance time of 82.3 h and a fever clearance time of 25.4 h (Willcox, 2011).

13.7 SAFETY

Cryptolepis sanguinolenta has potential health benefits, but there are potential risks associated with it, such as toxicity data, which suggests it is safe at levels under 500 mg/kg of body weight. One of the main risks associated with *Cryptolepis sanguinolenta* is its potential to interact with other medications. One of the main risks of *Cryptolepis sanguinolenta* is its interaction with other drugs or herbs. *Cryptolepis sanguinolenta* contains alkaloids such as cryptolepine, which can affect the metabolism of certain drugs by inhibiting or inducing cytochrome P450 enzymes in the liver. This can result in increased or decreased levels of drugs in the blood, leading to adverse effects or reduced efficacy. Therefore, *Cryptolepis sanguinolenta* should be used with caution if you are taking any prescription or over-the-counter medications, especially anticoagulants, anticonvulsants, antidepressants, antimalarials, antibiotics, or antidiabetics. It is important to consult with a healthcare provider before using *Cryptolepis sanguinolenta*, especially if you are taking any medications. Another risk of *Cryptolepis* is its potential toxicity to the liver and kidneys. Cryptolepis has been shown to cause hepatotoxicity and nephrotoxicity in animal studies at high doses or prolonged use. Even if there is no evidence that *Cryptolepis sanguinolenta* can cause liver damage in humans, it is important to use caution when using high doses of *Cryptolepis sanguinolenta*. Therefore, *Cryptolepis* should be avoided if you have any pre-existing liver or kidney disease or impairment. A third risk of *Cryptolepis* is its possible allergic reaction. *Cryptolepis* belongs to the Apocynaceae family of plants, which includes some common allergens, such as oleander and periwinkle. Some people may be sensitive or allergic to *Cryptolepis* or its constituents and experience symptoms such as rash, itching, swelling, hives, difficulty breathing, or anaphylaxis. Therefore, *Cryptolepis* should be tested on a small area of skin before using it internally or externally (Bugyei et al., 2010; Osafo et al., 2017; Tudu et al., 2023; Zhang et al., 2021).

13.8 CONCLUSION AND FUTURE REMARKS

In conclusion, *Cryptolepis*, especially *C. sanguignolia*, is a plant with a long history of use in traditional medicine. While some traditional uses of the plant have been validated by scientific research, others are still being studied. Its antimalarial properties have been extensively studied and confirmed with clinical trials. Additionally, researchers have discovered other potential uses for *Cryptolepis sanguinolenta*, including its use in treating cancer, diabetes, hypertension, and neurodegenerative and infectious diseases. Overall, *Cryptolepis sanguinolenta* is a powerful herb that can offer many benefits for treating infections and other conditions. However, it also carries some risks that should not be ignored. *Cryptolepis* should be used with caution and under professional guidance, especially if you are taking other drugs or herbs, have liver or kidney problems, or have a history of allergies.

REFERENCES

Adeleye, O.A., Femi-Oyewo, M.N., Bamiro, O.A., Bakre, L.G., Alabi, A., Ashidi, J.S., Balogun-Agbaje, O.A., Hassan, O.M., Fakoya, G., 2021. Ethnomedicinal herbs in African traditional medicine with potential activity for the prevention, treatment, and management of coronavirus disease 2019. *Future Journal of Pharmaceutical Sciences* 7, 72.

Ameyaw, Y., Akotoye, H., Boahen, Y.O., Eyiah-bediako, F., 2007. Assessment of variation in Cryptolepis sanguinolenta (Lindl.) Schter., an antimalaria plant of the family periploceae using principal component analysis. *Asian Journal of Chemistry* 19, 1809.

Barku, V.Y.A., Opoku-Boahen, Y., Dzotsi, E., 2012. Isolation and pharmacological activities of alkaloids from Cryptolepis sanguinolenta (Lindl) schlt. *International Research Journal of Biochemistry and Bioinformatics* 2, 56–61.

Borquaye, L.S., Gasu, E.N., Ampomah, G.B., Kyei, L.K., Amarh, M.A., Mensah, C.N., Nartey, D., Commodore, M., Adomako, A.K., Acheampong, P., Mensah, J.O., Mormor, D.B., Aboagye, C.I., 2020. Alkaloids from Cryptolepis sanguinolenta as potential inhibitors of SARS-CoV-2 viral proteins: An in silico study. *BioMed Research International* 2020, 5324560.

Bugyei, K.A., Boye, G.L., Addy, M.E., 2010. Clinical efficacy of a tea-bag formulation of cryptolepis sanguinolenta root in the treatment of acute uncomplicated falciparum malaria. *Ghana Medical Journal* 44, 3–9.

Feng, J., Leone, J., Schweig, S., Zhang, Y., 2019. Evaluation of natural and botanical medicines for activity against growing and non-growing forms of B. burgdorferi. *bioRxiv*, 652057.

Grellier, P., Ramiaramanana, L., Millerioux, V., Deharo, E., Schrével, J., Frappier, F., Trigalo, F., Bodo, B., Pousset, J.-L., 1996. Antimalarial activity of Cryptolepine and isocryptolepine, alkaloids isolated from Cryptolepis sanguinolenta. *Phytotherapy Research* 10, 317–321.

Joubert, L., 2013. A taxonomic study of the genus cryptolepis (periplocoideae: apocynaceae). Ph.D. Thesis, Plant Sciences, University of the Free State, South Africa. Available online http://hdl.handle.net/11660/1811

Joubert, L., Klak, C., Venter, A.M., Venter, H.J.T., Bruyns, P.V., 2016. A widespread radiation in the Periplocoideae (Apocynaceae): The case of Cryptolepis. *Taxon* 65, 487–501.

Joubert, L., Venter, A.M., 2008. Taxonomic value of floral morphology in the genus Cryptolepis (Apocynaceae, Periplocoideae) in southern Africa. *South African Journal of Botany* 74, 369.

Luo, J., Fort, D.M., Carlson, T.J., Noamesi, B.K., nii-Amon-Kotei, D., King, S.R., Tsai, J., Quan, J., Hobensack, C., Lapresca, P., Waldeck, N., Mendez, C.D., Jolad, S.D., Bierer, D.E., Reaven, G.M., 1998. Cryptolepis sanguinolenta: An ethnobotanical approach to drug discovery and the isolation of a potentially useful new antihyperglycaemic agent. *Diabetic Medicine* 15, 367–374.

Ma, X., Leone, J., Schweig, S., Zhang, Y., 2020. Botanical medicines with activity against stationary phase Bartonella henselae. *bioRxiv*, 2020.2008.2019.256768.

Olajide, O.A., Sarker, S.D., 2020. Alzheimer's disease: Natural products as inhibitors of neuroinflammation. *Inflammopharmacology* 28, 1439–1455.

Osafo, N., Mensah, K.B., Yeboah, O.K., 2017. Phytochemical and pharmacological review of Cryptolepis sanguinolenta (Lindl.) schlechter. *Advances in Pharmacological Sciences* 2017, 3026370.

Paulo, A., Houghton, P.J., 2003. Chemotaxonomic analysis of the genus Cryptolepis. *Biochemical Systematics and Ecology* 31, 155–166.

Tudu, C.K., Bandyopadhyay, A., Kumar, M., Radha, Das, T., Nandy, S., Ghorai, M., Gopalakrishnan, A.V., Prоćków, J., Dey, A., 2023. Unravelling the pharmacological properties of cryptolepine and its derivatives: A mini-review insight. *Naunyn-Schmiedeberg's Archives of Pharmacology* 396, 229–238.

Venter, H.J.T., Thulin, M., Verhoeven, R.L., 2006. Two new species of Cryptolepis (Apocynaceae: Periplocoideae) from Somalia, north-east Africa. *South African Journal of Botany* 72, 139–143.

WFO, 2023. *Cryptolepis R.Br.* Available online: http://www.worldfloraonline.org/taxon/wfo-4000009895. Accessed 06 Nov 2023.

Willcox, M., 2011. Improved traditional phytomedicines in current use for the clinical treatment of malaria. *Planta Medica* 77, 662–671.

Zhang, Y., Alvarez-Manzo, H., Leone, J., Schweig, S., Zhang, Y., 2021. Botanical Medicines Cryptolepis sanguinolenta, Artemisia annua, Scutellaria baicalensis, Polygonum cuspidatum, and Alchornea cordifolia Demonstrate Inhibitory Activity Against Babesia duncani. *Frontiers in Cellular and Infection Microbiology* 11.

Zofou, D., Kuete, V., Titanji, V.P.K., 2013. 17 – Antimalarial and other antiprotozoal products from African medicinal plants. In: Kuete, V. (Ed.), *Medicinal Plant Research in Africa*. Elsevier, Oxford, pp. 661–709.

14 Antimalarial Response, Traditional and Other Potential Uses of *Momordica* Genera

Nayomi Deshani De Silva, Keddagoda Gamage Piyumi Wasana, and Anoja Priyadarshani Attanayake

ABBREVIATIONS

GSK3β: glycogen synthase kinase-3β
IC_{50}: concentration (of extracts) at 50% inhibition
NADH: nicotinamide adenine dinucleotide

14.1 INTRODUCTION

The genus *Momordica* (family: Cucurbitaceae) is native to the Paleotropics. The generic name '*momordica*' comes from the Latin word 'mordeo', with the meaning 'to bite', which probably refers to the bitten appearance of serrated edges of the leaves and grooved margins of the seeds (Bharathi and John, 2013; Jia et al., 2017). Although the exact origin of this genus is ambiguous, most research has stated that Eastern Asia (possibly Eastern India or Southern China) is the centre of the *Momordica* genus (Nagarani et al., 2014). In contrast, analysis of dated biogeographic data suggested that this genus originated in tropical Africa. However, several species of *Momordica* genera are widely distributed around the tropical and subtropical regions, where they flourish freely in the wild (Schaefer and Renner, 2010). The taxonomic treatment of the genus *Momordica* is quite extensive. The main diagnostic characteristics, such as the presence of conspicuous floral bracts, the entire petal, pendulous, calyx cup, scales on the corolla, sculptured seeds, muricate fruits, and viny habit, have been considered throughout taxonomic treatment to define and distinguish the genus (Bharathi and John, 2013). *Momordica* belongs to the subtribe Thalidianthinae Pax., tribe Joliffieae Schrad., and subfamily Cucurbitoideae of Cucurbitaceae (Jeffrey, 1980; Bharathi and John, 2013). However, taxonomic confusion exists due to the extensive use of common names. Nine names are reported as synonyms of *Momordica*. They are *Amordica* Neck., *Calpidosicyos* Harms., *Dimorphochlamys* Hook.f., *Eulenburgia* Pax., *Muricia* Lour., *Nevrosperma* Raf., *Raphanistrocarpus* (Balli.) Pax., *Raphanocarpus* Hook.f., and Zucca Comm. Ex. Juss. (Govaerts et al., 2021; RBGK, 1753).

Momordica is a monophyletic genus that encompasses more than 80 species, while 14 species have been botanically described by tropical Africans and 12 species by Asians (Schaefer and Renner, 2010; Ramalhete et al., 2022). Even though a total number of 50 species are listed as accepted species under *Momordica* genera, 6 species, such as *Momordica balsamina* L., *Momordica charantia* L., *Momordica cochinchinensis* L. Spreng., *Momordica cymbalaria* Hook.f., *Momordica dioica* Roxb. Ex Willd., and *Momordica foetida* Schumach., are the most documented species (Behera et al., 2011; Nagarani et al., 2014; Muronga et al., 2021).

DOI: 10.1201/9781003378396-14

14.2 BOTANICAL DESCRIPTION

Momordica species belong to the annual crops category and therefore can be considered perennial crops based on their performance during the season (Muronga et al., 2021). The climber consists of either glabrous or pubescent simple leaves, lobed leaves, or pedately foliate leaves with 3–5 units, and unbranched tendrils (Figure 14.1). The colour of monoecious or dioecious flowers of *Momordica* species varies as off-white, cream, or yellow. There are five free petals in the corolla with entire margins. Each base of the petals has 1–3 incurved scales. The male flowers are in the form of solitary, or in some cases, the florets are loosely arranged as a pseudo-raceme with short stalks, while the female flowers are always axile and solitary and are protected by rudimentary or conspicuous bracts (Behera et al., 2011). The pedicel of the male flower can be either minute or long while with a saucer or cupular-shaped short receptacle tube. The calyx consists of scarious lobes with full margins and an adnate base. Stamens have very short filaments that are inserted into the opening of the receptacle tube (Ramalhete et al., 2022).

The calyx of the female flower has several similarities with that of the male flower. Fusiform and oblong-shaped ovary, which is ribbed, watery or soft, and fleshy, is present with numerous ovules. Stigma is horizontal and consists of three lobes. Sterile stamens are absent. The medium- to big-sized ovoid ellipsoid to fusiform and fleshy fruits of *Momordica* genera are ornamented with soft spines, splits, warts, and tubercles from the base into three valves. The raw fruits of *Momordica* genera are emerald green, and it becomes orange-red when it is fully ripe. The seeds of the fruit are enclosed with orange-red sarcotesta (Nagarani et al., 2014).

14.3 DISTRIBUTION

The genera *Momordica* could be widely found in tropical and subtropical regions of the world, while it is native to the Paleotropics, including Angola, Burkina, South-East China, Gabon, India, Kenya, Nigeria, Pakistan, Philippines, Queensland, Sri Lanka, Swaziland, Taiwan, Thailand, Vietnam, and Zimbabwe. In fact, at first, the species were introduced to Brazil, Argentina north-east, Colombia, Jamaica, Korea, Nex Mexico, Venezuela, Oman, Pennsylvania, Paraguay, and Cuba, and extinct in Socotra (Kattukunnel and Antony, 2004; Nagarani et al., 2014). The plants of the genera are cultivated in areas with an average rainfall of 400 mm during mild and frost-free winters. A temperature of 18°C is required for emergence and germination, while 24°C to 27°C is required for the vegetative and flowering stages. Furthermore, crops adapt well to slit loam soil and well-drained sandy loam. Indeed, the species are well-grown in soil rich in organic matter and with an optimal soil pH ranging from 4.3 to 8.0. In fact, plants are cultivated throughout the tropics, particularly in China, India, and countries of Central and South American countries of East Africa (Tan et al., 2008; Muronga et al., 2021).

14.4 ANTIMALARIAL RESPONSES OF *MOMORDICA* GENERA

Since time immemorial, the *Momordica* species have been used in the indigenous medical system for the management of several diseases, including malaria, diabetes mellitus, cancers, cardiovascular diseases, obesity, cholera, diarrhoea, blood diseases, ulcers, bronchitis, colic worms, and diseases of the liver and spleen (Waako et al., 2005; Dans et al., 2007; Lii et al., 2009; Rupachandra and Sarada, 2013; Ahmad et al., 2016). Among the reported diseases, malaria has become a major concern, as it is one of the critical causes of mortality worldwide. Malaria folklore drugs originated from antimalarial compounds like quinine and artemisinin. With the emergence of chloroquine-resistant *Plasmodium falciparum*, attention toward alternative antimalarials has become prominent. In this context, several investigators have paid attention to species of *Momordica* genera, including *M. charantia* and *M. balsamina* (Graz et al., 2011). Generally, Africans do not use *M. balsamina* alone, and they practiced *M. balsamina* extensively as a component in a mixture of different herbs to treat malaria fever. They used *M. balsamina* with sesamum and limeum in a ratio of 8:1:1 (Benoit-Vical et al., 2006).

FIGURE 14.1 Fresh leaves, flowers, and raw fruit of *Momordica charantia* L. in a natural habitat.

Furthermore, dietary supplements using *M. charantia* have been introduced in the form of a capsule in United States (Ma et al., 2012). It is well-known that *M. balsamina* is commonly used to treat malaria in the form of decoction or infusion. Both the root and the leaves in hot water are also used to brew to make tea as a remedy for malaria (Benoit-Vical et al., 2006; Ahmad et al., 2016).

Despite the wide utilization of *Momordica* species in the indigenous medicine system to manage malaria, it is important to scientifically assess its overall quality. From this point on, the eminence of *Momordica* species is described in-depth to introduce promising antimalarial agents. In parallel to that, traditional and other potential uses of *Momordica* species are focused on. Utilization of herbal medicine is on the rise for basic healthcare needs since they are affordable, easily accessible, and culturally acceptable (Van Wyk and Prinsloo, 2020). However, there may be impromptu applications that could be toxic due to overdosing and inappropriate preparation. Therefore, it is vital to assess the quality in terms of the safety and efficacy of the plant-based requirements through preclinical and clinical studies.

14.4.1 Preclinical Studies

Several preclinical studies have been reported to date on the antimalarial activity of *Momordica* species. In general, antimalarial activity is classified according to the criteria as IC_{50} (concentration at 50% inhibition) $\leq 1 \mu M$: excellent; $1 < IC_{50} \leq 10 \mu M$: good; $10 < IC_{50} \leq 30 \mu M$: moderate; and $IC_{50} > 30 \mu M$: weak/inactive (Kamaraj et al., 2012). Adia et al. (2016) evaluated *in vitro* antiplasmodial activity of the ethyl acetate, methanol, and water extracts of *M. foetida* leaf. Micro-Mark III test was used to assess the response against chloroquine-sensitive (NF54) and chloroquine-resistant (FCR3) *P. falciparum*. The study results revealed that the aqueous and ethyl acetate extracts exert higher antiplasmodial activity against the FCR3 strain while promising activity in the NF54 strain than that of the methanol extract (Adia et al., 2016).

An ethnopharmacological survey carried out by Kamaraj et al. (2012) has identified medicinal plants which are used to manage malaria-like fever. Based on the documented herbs that were mentioned by traditional healers and herbalists, most cited 23 herbs were selected for antiplasmodial and cytotoxic activities, and among them, *M. charantia* was identified as one of the herbs with potent antimalarial activity. Furthermore, in the study, the ethyl acetate leaf extract of *M. charantia* was assayed for antimalarial activity against chloroquine-sensitive 3D7 and chloroquine-resistant INDO strains of *P. falciparum* (Kamaraj et al., 2012). In addition, cytotoxicity was assessed against HeLa cells using an MTT assay. As a result, the ethyl acetate extract of *M. charantia* was able to report promising antimalarial activity with an IC_{50} value of 17.5 ± 0.5 μg/mL, and analysis against the chloroquine-resist strain INDO of *P. falciparum* showed an IC_{50} of 9 ± 0.4 μg/mL. It was nontoxic to the mammalian HwLa cell line, as revealed by the MTT assay (Kamaraj et al., 2012). Remarkable antimalarial activities were reported for the ethyl acetate extract of *M. charantia* due to the presence of a high concentration of lipophilic secondary metabolites, such as cardiac glycosides, saponins, alkaloids, and flavonoids (Muronga et al., 2021). Fractionation of the methanol extract of *M. balsamina* (aerial part) was used to isolate new triterpenoids and evaluate their antimalarial activities by Ramalhete et al. (2010). The results revealed that balsaminoside A and karavilagenin E had the highest antimalarial activities against both *P. falciparum* chloroquine-sensitive strain 3D7 and the chloroquine-resistant clone Dd2 (Ramalhete et al., 2010).

As evident in several studies, species in *Momordica* genera, particularly *M. charantia*, *M. balsamina*, and *M. foetida*, are extensively being evaluated for their antimalarial activity through *in vivo* studies (Waako et al., 2005; Abdillah et al., 2015; Christy et al., 2016; Ali et al., 2022). The antimalarial activity of the ethanol extract of *M. charantia* was evaluated by Abdillah et al. (2015) using a classic four-day suppressive test against *Plasmodium berghei*. In the study, Swiss male mice were inoculated with infected red blood cells intraperitoneally at the beginning, and the extract was introduced for three consecutive days. Inhibition of parasite growth by the extract was determined after five days and demonstrated that the ethanol extract of *M. charantia* is a potent antimalarial agent ($Ed_{50} \leq$ 101–250 mg/kg body weight/day) (Abdillah et al., 2015).

Most medicinal plants are rich in primary and secondary metabolites with anti-inflammatory potency. To date, researchers have highlighted the anti-inflammatory conditions of medicinal plants that could offer an attractive strategy against malaria. In this regard, the antimalarial effect of the aqueous fruit extract of *M. charantia* was used to explicate the mechanism of action involving glycogen synthase kinase-3β (GSK3β) inhibition. In summary, administration of the aqueous fruit extract of *M. charantia* to *P. berghei* NK 65–infected mice (50, 75, and 100 mg/kg body weight) for four consecutive days was able to inhibit parasite development in a dose-dependent manner. Furthermore, they have proven that the antimalarial effect of *M. charantia* involved in cytokine modulation is mediated through the inhibition of GSK3β in *P. berghei*–infected mice (Ali et al., 2022). In the same study, toxicity assessment revealed that the aqueous fruit extract of *M. charantia* is nontoxic against Vero cells. Based on their findings, they have proven that *M. charantia* is a potent immunomodulator for treating malarial infection (Ali et al., 2022).

In recent years, the synthesis of nanoparticles using plant extracts has achieved much attention to suppressing the toxic effect of chemical agents and targeted drug delivery. To date, several studies have been carried out on the management of malaria using photosynthetic nanoparticles. For example, the antimalarial activity of the synthesized TiO_2 nanoparticles using *M. charantia* leaf extract was evaluated via an *in vitro* study (Gandhi et al., 2018). Vector control plays a vital role in the prevention and reduction of the transmission of malaria. Larval stages of vector populations are also used to prevent malaria. In this context, the results of Gandhi et al. (2018) revealed that the TiO_2 nanoparticles using the aqueous leaf extract of *M. charantia* are stable over time and strongly reduce the population of malaria vectors in low doses (Gandhi et al., 2018).

14.4.2 Clinical Studies

It is vital to clinically evaluate the efficacy and safety of traditional medicine to be validated and to establish reliable quality-controlled herbal preparations. According to the WHO guidelines for the antimalarial screening of drugs, there are four stages. Clinical testing is described as the fourth stage, and it is mandatory to screen active compounds, establish quality, and perform a comparative analysis of analogues before stage 4 (Mojab, 2012). Preclinical studies on *Momordica* species made available information about their safety and efficacy and therefore provided hints to initiate clinical studies. However, none of the clinical information on antimalarial responses of the species belonging to *Momordica* genera has been reported so far. This situation has become further complicated due to the immune status of the patient. Malaria is one of the reasons for most deaths occurring in Africa and in children under the age of five years (Wright, 2005). Therefore, there is a straightforward issue in the clinical demonstration of an antimalarial drug. In addition, practical and ethical difficulties may affect to restrict the clinical trials on antimalarial activity (Wright, 2005).

14.4.3 *In Silico* Studies

In recent decades, *in silico* molecular modelling has become a crucial part of the drug discovery process since it expedites the screening of drug leads from natural medicinal plants. Several approaches have been taken to discover novel drug lead molecules with antimalarial activity (A Barmade et al., 2015; Mamudha and Sunilson, 2021; Alzain et al., 2022). *In silico* approaches can facilitate antimalarial drug discovery via various available databases of chemical compounds and proteins in *Plasmodium.* Drug development could fail during clinical trials due to the toxicity and poorly exerted pharmacokinetic properties, but *in silico* modelling facilitates the adjustment of the candidate's pharmacodynamics or pharmacokinetic properties of the candidate (A Barmade et al., 2015). Malaria could be transmitted via a bite of infected female *Anopheles* mosquitoes. An evaluation of *Momordica tuberosa* (Roxb) against *Anopheles* was carried out using molecular coupling studies by Mamudha and Sunilson (2021). They aimed to evaluate *in silico* larvicidal potent compounds against *Anopheles* and to analyze the mechanism of inhibition of *Anopheles* angiotensin-converting enzyme. The result revealed that the selected compounds, including limonene dioxide, neophytadiene, and palmitic acid bind to the binding sites of the angiotensin-converting enzyme with better docking binding energy and, thereby, design a safe inhibitor against angiotensin-converting enzyme. Hence, they have proven that these compounds are lethal for larvae development. (Mamudha and Sunilson, 2021).

14.4.4 Active Phytoconstituents of *Momordica* Genera

The growing amount of evidence establishes the value of isolated biological active compounds for the synthesis of novel drug leads. Numerous phytochemical and pharmacological studies have reported that several extracts of *Momordica* species have huge potential as antimalarial, antidiabetic, antioxidant, anticancer, anti-inflammatory, antifungal, antiproliferative, antiallergic, and anthelmintic agents (Siboto et al., 2018; Fang et al., 2019; Elangovan et al., 2021; Hassan et al., 2022; Nogbou et al., 2022; Silva et al., 2022). However, these phytochemical and pharmacological studies have been limited to a few species in *Momordica* genera over about 80 species. Various phytochemicals and secondary metabolites have been isolated and identified from different parts of the plant species, such as leaves, aerial, fruit, seeds, and roots, as drug lead molecules for the management of several diseases, including malaria. In general, the presence of polyphenols, alkaloids, glycosides, steroids, flavonoids, saponins, and terpenoids was proven in the *Momordica* species by phytochemical screening and has been implicated in the treatment of various diseases. Froelich et al. (2007) carried out a phytochemical analysis on the leaf extract of *M. foetida*, and as a result, six compounds were isolated. Of them, only eriodictyol-7-*O*-β-D-glucopyranoside (Figure 14.2) showed antimalarial activity (Froelich et al., 2007).

In several attempts to identify the phytochemical constituents of aerial parts of *M. balsamina*, the presence of antimalarial compounds was reported (Ramalhete et al., 2009; Ramalhete et al.,

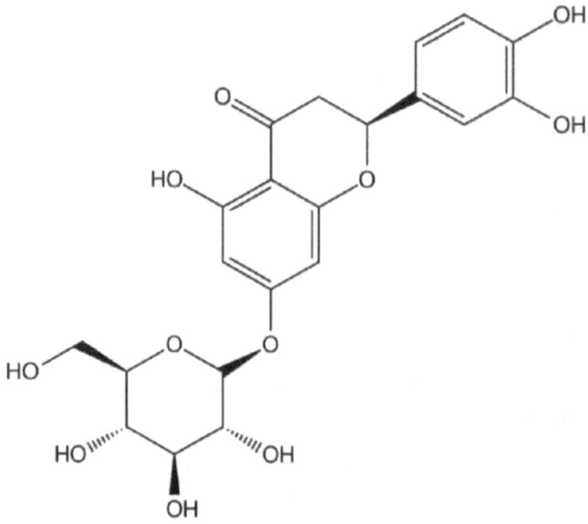

FIGURE 14.2 Structure of eriodictyol-7-*O*-β-D-glucopyranoside.

2010; Ramalhete et al., 2011). An effort was taken to isolate cucurbitane-type triterpenoids, such as balsaminoside A, balsaminol B, balsaminol C, and balsaminol F, from the aerial part of *M. balsamina*. Balsaminol F was acylated, producing two new triesters, called triacetylbalsaminol F and tribenzoylbalsaminol F. Furthermore, glycosylated cucurbitacin, such as kaguaglycoside A, karavilagenin C, karavilagenin E, karavoate A, B, C, D, M, and N, were isolated from the aerial part of *M. balsamina*, and their antimalarial activity was proven. The molecular structures of the aforementioned compounds are shown in Figure 14.3 (Ramalhete et al., 2010; Ramalhete et al., 2011).

Accordingly, a few antimalarial compounds have been isolated from several extracts of *Momordica* genera, and lack of efficacy and safety data has restricted their wider clinical applications. Given this, it is vital to isolate antimalarial active compounds and assess their efficacy and safety via preclinical and clinical studies.

14.5 CELLULAR MECHANISMS FOR ANTIMALARIAL ACTIVITY

Various compounds present in plant extracts are beneficial in regulating the mechanisms of action to manage various ailments, including malaria. Prevailing antimalarial drugs have been designed to control metabolic differences between malarial parasites with their host, including nucleic acid metabolism, oxidative stress, heme detoxification, and fatty acid biosynthesis (Saxena et al., 2019). In general, there are some suggested mechanisms for isolating antimalarial compounds from plants. Indole alkaloids are capable of inhibiting hemozoin polymerization in parasites (Onyeibor et al., 2005; Adebayo and Krettli, 2011). Lactate dehydrogenase in the malaria parasite *P. falciparum* comprises many unique amino acid residues that define substrate and cofactor binding sites. This uniqueness of *P. falciparum* lactate dehydrogenase could be an attractive target for the development of novel antimalarials. Lactate dehydrogenase in parasite evince high specificity for pyruvic acid since it is more susceptible to pyruvate reduction with high catalytic efficiency. Despite several amino acid alterations, the parasite lactate dehydrogenase exhibits cofactor specificity to the human isoforms in L-lactate oxidation, suggesting the same environment of cofactor binding. Inhibitors with antimalarial activity such as chloroquine bind in or near the cofactor binding site of *P. falciparum* lactate dehydrogenase. Here, chloroquine acts as a competitive inhibitor. Similarly, gossypol, an isolated disesquiterpene from the seeds of *Gossypium* species, was shown to be a competitive

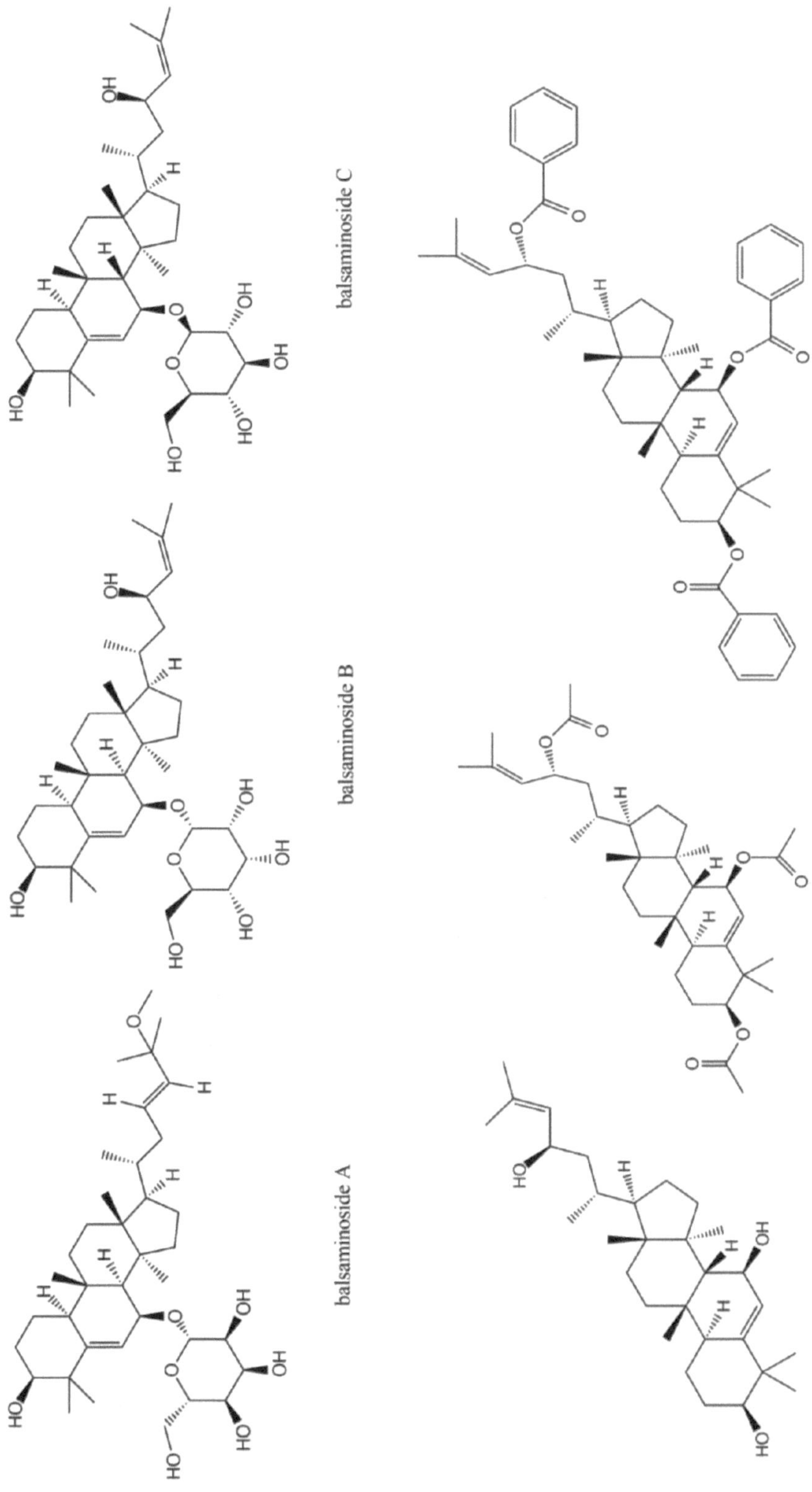

FIGURE 14.3 Molecular structures of antimalarial active compounds.

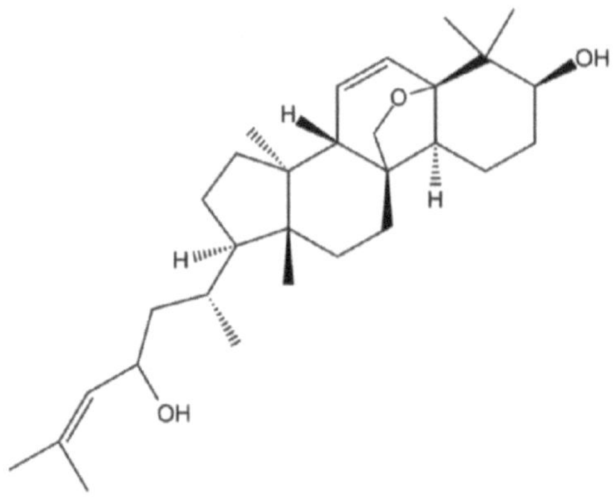

karavilagenin E

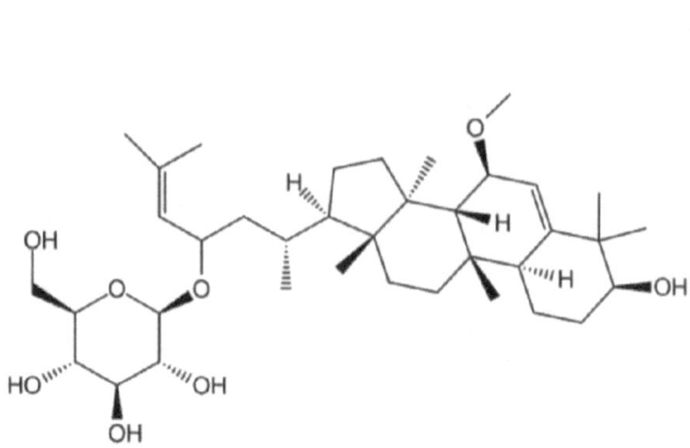

kaguaglycoside A

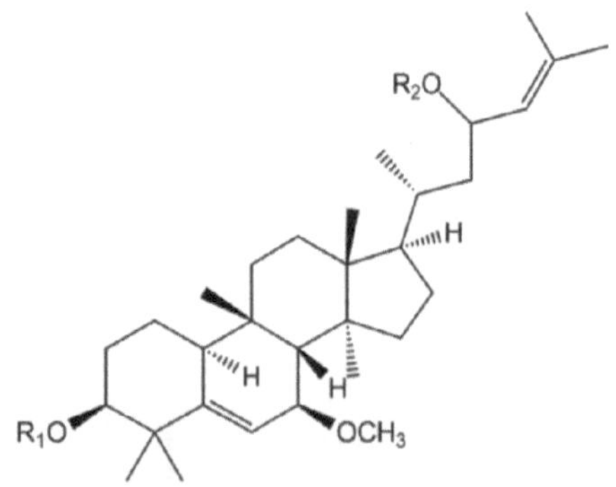

R_1= H; R_2 = H - karavilagenin C
R_1 = H; R_2 = $COCH_3$ - karavoate A
R_1 = R_2 = $COCH_3$ - karavoate B
R_1 = H; R_2 = $COCH_2CH_3$ - karavoate C
R_1 = R_2 = $COCH_2CH_3$ - karavoate D
R_1 = H; R_2 = *p*-methoxyBz - karavoate M
R_1 = R_2 = *p*-methoxyBz - karavoate N

Bz =

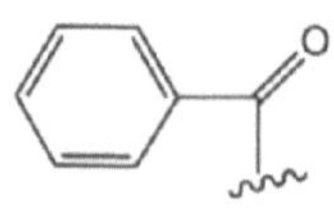

FIGURE 14.3 (Continued)

inhibitor of the binding of nicotinamide adenine dinucleotide (NADH). Alteration of the produced structure leads to the inhibition of dehydrogenase (Gomez et al., 1997; Saxena et al., 2019). This phenomenon avoids the energy generation of malarial parasites through glycolysis. Herein, the said mechanism could be applied to *Momordica* species that are rich in natural polyphenol-derived compounds.

The functional ingredients in *M. charantia* that involve anthelmintic action are arachidonic acid metabolism, β-tubulin binding, improved calcium permeability, oxidative phosphorylation, and acetylcholinesterase inhibitors (Saeed et al., 2018). Saponin can inhibit acetylcholinesterase and cause death via paralysis. In addition, saponin can affect the parasite permeability of the cell membrane and lead to disintegration. The inhibition of larval growth and inhibition of arachidonic acid metabolism by flavonoids may lead to the degeneration of neurons and ultimately lead to death (Poolperm and Jiraungkoorskul, 2017).

Moreover, Billker et al. (2002) suggested that azadirachtin is much capable of interfering with the formation of mitotic spindles and the assembly of microtubules into complex structures as axonemes in gametes, which leads to inhibiting the formation of mobile microgametes (Billker et al., 2002). The antimalarial mechanism of esters present in *Momordica* species could be the same as mentioned earlier. In addition, a great deal of effort has been made to explore novel mechanisms of antimalarial activity related to inflammation. GSK3 is currently recognized as a target for the regulation of cytokine production, while dysregulation of GSK3 implicates many diseases and inflammation-related conditions. Protein kinase B (Akt) is a critical regulator of cell proliferation and survival (Dai et al., 2012). Administration of *M. charantia* aqueous fruit extract (75 mg/kg body weight) for *P. berghei* NK65–infected mice demonstrated that the aqueous fruit extract could increase phosphorylation of host GSK3β and Akt (Ali et al., 2022). Furthermore, levels of IFN-γ and TNF-γ were reported to decrease, while IL-10 was increased after treatment. According to these facts, the antimalarial effect of the aqueous fruit extract of *M. charantia* could be due to modulating inflammatory cytokines through the inhibition of GSK3β (Ali et al., 2022).

14.6 TRADITIONAL AND OTHER POTENTIAL USES OF *MOMORDICA* SPECIES

In addition to growing widely, *Momordica* species are cultivated around the world because of their traditional and imperative potential value for the management of several health conditions. In the native medicine system, *Momordica* species are used to treat diabetes mellitus, cancer, inflammatory diseases, fungal and viral infections, allergies, blood cholesterol, bilious disorders, cataract, snakebite, jaundice, and haemorrhoids. (Kameswararao et al., 2003; Behera et al., 2011; Rakh et al., 2012; Saeed et al., 2018; Attar and Ghane, 2019). In addition to their medicinal uses, *Momordica* species are widely utilized as crafts, food, veterinary, and ritual agents (Behera et al., 2011). The aqueous leaf extract of *M. foetida* is very prominent among Ugandans as an agent for the treatment of cough, vomiting, stomach pain, and expel worms. Leaf decoction is used to manage tuberculosis (Tabuti et al., 2010). Furthermore, the stem and leaves of *M. foetida* are boiled in water and the steam inhaled during bedtime as a treatment for febrile illnesses (Mukaila et al., 2023). Although the leaves of *M. foetida* exert a bitter taste, people in Gabon, Sudan, Uganda, Tanzania, and Malawi use them as vegetables. The powdered form of the root and leaves is added to burning charcoal to inhale the generated smoke to manage epidemic diseases by Ethiopians (Mukaila et al., 2023).

The entire plant of *M. balsamina* is used as a stomachic, a component in aphrodisiac preparation, and as anthelminthic medicine, to stimulate lactation and treat vomiting together with bile and fever, malignant ulcers, and skin problems (Bharathi and John, 2013). The leaves of *M. balsamina* are used to manage hyperglycaemia, burns, abdominal pains, uterine bleeding, and induce lactation (Bharathi and John, 2013; Ramalhete et al., 2022). Both the root and the fruit of *M. balsamina* are used to induce abortion, to treat stomach cramps, diarrhoea, chapped hands, and haemorrhoids. In all treatments, the whole plant, leaves, roots, and fruits are used in the forms such as infusion, maceration, pulverization, and decoction (Bot et al., 2007). In addition, the leaves and fruits of *M. balsamina* are cooked as a vegetable, sometimes used as an ingredient in sauces and soups in India, Sudan, South Africa, and Cameroon. Pickles are also prepared using the unripe fruit of *M. balsamina* (Bharathi and John, 2013).

In the Western world, the leaves and fruit of *M. charantia* are utilized to prepare tea, beer, or seasonal soups. The root of *M. charantia* can reduce scars in smallpox when the root paste is mixed with milk. The pulp prepared from the leaves of *M. charantia* could be applied to wounds as a haemostasis agent. Administration of a small quantity of *M. charantia* leaves produces a therapeutic effect for anaemic patients. Three drops of *M. charantia* leaf juice in goat milk are expended to reduce chest congestion and phlegm (Joseph and Antony, 2008). Six tablespoons of the chopped leaves of *M. charantia* are boiled with two glasses of water, and the resultant is used to treat diabetes mellitus (Kumar and Bhowmik, 2010).

Among the species in the *Momordica* genera, *M. doica* is believed to be an antihyperglycemic, gastroprotective, hepatoprotective, and antiviral agent. Soft and deseeded fruits are used as a vegetable in the form of chutney and roasted state. Aliphatic compounds in the fruit of *M. doica* act as astringents and appetizers. The whole plant of *M. doica* is important for the treatment of eye diseases and fever. Tubes are used to prepare electuaries or jam-like medicine, which is a worthy treatment for piles, intestinal ulcers, and snakebites (Joseph and Antony, 2008; Hassan et al., 2022).

Traditionally, *M. cochinchinensis* is used as a food-cum-medicine. Aril of cooked ripe fruits with red glutinous rice (called 'xoi gac') is used as a traditional food in Vietnam and Thailand. This special rice is added to its soup to enhance its appearance as a nutritional supplement. This gac fruit offers a wide range of protection against various ailments, such as xerophthalmia, liver and spleen disorders, chest complaints, wounds, abdominal pains, haemorrhoids, swelling, and dysentery (Chuyen et al., 2015). However, the aril of *M. cochinchinensis* fruit exerts strong potential against numerous disorders; nevertheless, other parts of the fruit including peel and pulp are discarded. The seeds of the fruit of *M. cochinchinensis* are used to treat anaemia and arthritis (Chuyen et al., 2015).

One of the *Momordica* species, *M. cymbalaria*, has been identified as an imperative wild crop, especially in the field of traditional medicine. The fruits of *M. cymbalaria* are shown to have antihyperlipidemic and antidiabetic properties (Dhasan et al., 2010). In addition, it is used as an abortifacient and to manage rheumatism, skin diseases, and diarrhoea. The fruits of *M. cymbalaria* are preserved in the form of pickling or sun-dried. The roots are used to heal some gynaecological ailments. The decoction made out of *M. cymbalaria* fruits has been used to treat gastric ulcers (Bharathi and John, 2013). The potential uses of the *Momordica* genera are presented in Table 14.1. However, the reported healing and other properties of the *Momordica* genera are limited to a few species, and evaluation of other potent species in terms of therapeutic efficacy and safety is necessary.

TABLE 14.1
Potential Uses of *Momordica* Species

Plant Species	Part of the Plant	Type of Extract	Ethnic Use	Reference
Momordica balsamina L.	Aerial parts	Methanol	Anticancer	Silva et al., 2022
		95% ethanol	Antibiotic	Coutinho et al., 2010
	Fruits	Methanol	Antibacterial	Nogbou et al., 2022
		Chloroform	Antidiabetic	Kaushik et al., 2017
		Pulp	Inhibitor of HIV-1 replication	Bot et al., 2007
	Leaves	Acetone	Anticancer	Serala et al., 2021
		Methanol	Cardiovascular complications	Ludidi et al., 2019
		Ethyl acetate	Antibacterial	Madureira et al., 2012
Momordica charantia L.	Fruits	Ethanol	Hypoglycaemic, antiproliferative	Hsiao et al., 2013
		Aqueous	Antidiabetic, immunomodulatory	Xu et al., 2015; Sun et al., 2016
		As a juice	Anticancer, antidiabetic	Kaur et al., 2013
	Leaves	Methanol	Anticancer	Li et al., 2012
		80% ethanol	Anticancer	Pitchakarn et al., 2012
	Seeds	75% ethanol	Anticancer	Soundararajan et al., 2012
		80% ethanol	Antiproliferative	Kai et al., 2011
Momordica cochinchinensis (Lour.) Spreng.	Fruits	Aqueous	Anticancer, antioxidant	Tien et al., 2005; Abdulqader et al., 2019; Wimalasiri et al., 2020
	Seeds	Aqueous	Anticancer	Le et al., 2018
		10% ethanol	Anti-inflammatory	Jung et al., 2013
		Ethanol	Antiproliferative	Yu et al., 2017
Momordica cymbalaria Hook. f.	Fruits	Methanol	Antioxidant, antibacterial	Swamy et al., 2015
		Aqueous	Antibacterial	Hemalatha et al., 2021
	Roots	Methanol	Neuroprotective	Koneri et al., 2014
	Seeds	Methanol	Protective effect against hyperglycaemia, hyperlipidaemia	Elangovan et al., 2019

(*Continued*)

TABLE 14.1 (*Continued*)
Potential Uses of *Momordica* Species

Plant Species	Part of the Plant	Type of Extract	Ethnic Use	Reference
Momordica dioica Roxb. Ex Willd.	Fruit	Hydroalcoholic	Antiulcer activity	Vijayakumar et al., 2011
		Protein extract	Antidiabetic, antilipidemic, antioxidant	Poovitha et al., 2017
		70% ethanol	Antidiabetic	Hassan et al., 2022
		Ethanol	Antiulcer	Kumar et al., 2022
	Seed	Methanol	Anticancer, antiallergic, analgesic, antihyperglycemic, antioxidant, antilipidemic peroxidative	Rupachandra and Sarada, 2013; Rao and Mohan, 2017
		Ethanol	Nephroprotective	Jain and Singhai, 2010
Momordica foetida Schumach.	Leaves	Aqueous	Antioxidant	Acquaviva et al., 2013
		Ethanol	Antibacterial	Odeleye and Oyedeji, 2008
		Ethyl acetate	Antiplasmodial	Adia et al., 2016

Due to the potent health benefits of *Momordica* species, it is highly focused on the isolation of its active phytoconstituents (Ramalhete et al., 2022). For instance, *M. charantia* compromises several triterpenoids, including kuguacins A-S, charantin, karavilagenin A-E, momordicine I, II, and III (Matsuda et al., 2007; Chen et al., 2008). Hus et al. (2011) were able to isolate five new terpenoids from the *M. charantia* fruit, which exhibited numerous biological activities, such as antidiabetic, hepatoprotective, cardiovascular, antiparasitic, and anti-inflammatory effects. The isolated compounds were 5β,19-epoxycucurbita-6,22*(E)*,24-triene-3β,19-diol, cucurbita-6,22*(E)*, 24-triene-3β-ol-19,5β-olide, 19-*nor*-cucurbita-5(10),6,8,22,24-pentaen-3β-ol, 3β-hydroxycucurbita-5(10),6,22*(E)*,24-tetraen-19-al, and 19-dimethoxycucurbita-5(10),6,22(*E*),24-tetraen-3β-ol (Hsu et al., 2011). Sterols in herbs are more susceptible to reducing the concentration of low-density lipoprotein cholesterol. The prominent phytosterols isolated from *Momordica* species are decortinol, clerosterol, 5 α,6 α-epoxy-(22*E*,24R)-ergosta8,22-diene-3β,7 α-diol, and 3 β-hydroxy-(22*E*,24R)-ergosta-5,8,22-trien-7-one,22-diene-3β,7α-diol (Liao et al., 2011). Furthermore, momordicoside F_2, kuguacin J, goyaglycoside B, goyaglycoside D, and stigmasterol glucoside compounds extracted from *M. charantia*. Anticancer activity is depicted by isolated compounds isorhamnetin glycoside, isorhamnetin acetyl glycoside from *M. foetia*, and karavilagenin C, balsaminagenin A and B from the herb *M. balsamina* (Muronga et al., 2021; Ramalhete et al., 2022). However, it is worth mentioning that species of *Momordica* have powerful traditional and numerous other uses because of their active phytoconstituents and specific mechanisms.

14.7 CONCLUSION

Momordica species could be considered nutritious vegetables and play a promising role in the management of various diseases, including malaria. This attribute is due to the presence of potent primary and secondary metabolites, which are the key governess of therapeutic potential. Over 80 species of *Momordica* genera, *Momordica balsamina* L., *Momordica charantia* L., *Momordica cochinchinensis* L. Spreng., *Momordica cymbalaria* Hook.f., *Momordica dioica* Roxb. Ex Willd., and *Momordica foetida* Schumach, are the most utilized species in terms of managing numerous diseases, such as malaria, diabetes mellitus, cancers, cardiovascular diseases, etc. This is scientifically proven through the performance of preclinical and clinical studies. However, clinical studies on *Momordica* species remain sparse, and especially, none of the clinical information has been documented concerning antimalarial activity. So far, several active compounds, such as balsaminol

F, balsaminoside A, balsaminol B, balsaminol C, kuguacins A-S, charantin, momordicine I, II, and III. etc., have been isolated from various species in the *Momordica* genera. Further efforts must be directed towards clinical studies on the therapeutic potential of *Momordica* species and thereby the isolation of drug lead molecules.

REFERENCES

Abdillah, S., Tambunan, R.M., Farida, Y., Sandhiutami, N.M.D. & Dewi, R.M. 2015. Phytochemical screening and antimalarial activity of some plants traditionally used in Indonesia. *Asian Pacific Journal of Tropical Disease* 5(6): 454–457.

Abdulqader, A., Ali, F., Ismail, A. & Esa, N.M. 2019. Antioxidant compounds and capacities of Gac (*Momordica cochinchinensis* Spreng) fruits. *Asian Pacific Journal of Tropical Biomedicine* 9(4): 158.

Acquaviva, R., Di Giacomo, C., Vanella, L., Santangelo, R., Sorrenti, V., Barbagallo, I., Genovese, C., Mastrojeni, S., Ragusa, S. & Iauk, L. 2013. Antioxidant activity of extracts of *Momordica foetida* Schumach. et Thonn. *Molecules* 18(3): 3241–3249.

Adebayo, J. & Krettli, A. 2011. Potential antimalarials from Nigerian plants: A review. *Journal of Ethnopharmacology* 133(2): 289–302.

Adia, M.M., Emami, S.N., Byamukama, R., Faye, I. & Borg-Karlson, A.K. 2016. Antiplasmodial activity and phytochemical analysis of extracts from selected Ugandan medicinal plants. *Journal of Ethnopharmacology* 186: 14–19.

Ahmad, N., Hasan, N., Ahmad, Z., Zishan, M. & Zohrameena, S. 2016. *Momordica charantia*: For traditional uses and pharmacological actions. *Journal of Drug Delivery and Therapeutics* 6(2): 40–44.

Ali, M.H., Ibrahim, I., Jasamai, M., Embi, N. & Sidek, H. 2022. Anti-malarial effect of *Momordica charantia* involved modulation of cytokine mediated via GSK3β inhibition in *Plasmodium berghei*-infected Mice. *Jordan Journal of Biological Sciences* 15(3): 523–529.

Alzain, A.A., Ahmed, Z.A.M., Mahadi, M.A. & Elbadwi, F.A. 2022. Identification of novel *Plasmodium falciparum* dihydroorotate dehydrogenase inhibitors for malaria using *in silico* studies. *Scientific African* 16: e01214.

Attar, U. & Ghane, S. 2019. *In vitro* antioxidant, antidiabetic, antiacetylcholine esterase, anticancer activities and RP-HPLC analysis of phenolics from the wild bottle gourd (*Lagenaria siceraria* (Molina) Standl. *South African Journal of Botany* 125: 360–370.

Barmade, M.A., Murumkar, P., Kumar Sharma, M., Shingala, K., Giridhar, R. & Ram Yadav, M. 2015. Discovery of anti-malarial agents through application of *in silico* studies. *Combinatorial Chemistry & High Throughput Screening* 18(2): 151–187.

Behera, T., John, K.J., Bharathi, L. & Karuppaiyan, R. (2011). Momordica. In: Kole, C. (eds.), *Wild Crop Relatives: Genomic and Breeding Resources*: 217–246, Springer, Berlin and Heidelberg.

Benoit-Vical, F., Grellier, P., Abdoulaye, A., Moussa, I., Ousmane, A., Berry, A., Ikhiri, K. & Poupat, C. 2006. *In vitro* and *in vivo* antiplasmodial activity of *Momordica balsamina* alone or in a traditional mixture. *Chemotherapy* 52(6): 288–292.

Bharathi, L.K. & John, K.J. (2013). *Momordica Genus in Asia: An Overview*, Springer, New Delhi.

Billker, O., Shaw, M.K., Jones, I.W., Ley, S.V. & Sinden, R.E. 2002. Azadirachtin disrupts formation of organized microtubule arrays during microgametogenesis of *Plasmodium berghei*. *Journal of Eukaryotic Microbiology* 49(6): 489–497.

Bot, Y., Mgbojikwe, L., Nwosu, C., Abimiku, A., Dadik, J. & Damshak, D. 2007. Screening of the fruit pulp extract of *Momordica balsamina* for anti HIV property. *African Journal of Biotechnology* 6(1): 47–52.

Chen, J.C., Lu, L., Zhang, X.M., Zhou, L., Li, Z.R. & Qiu, M.H. 2008. Eight new cucurbitane glycosides, kuguaglycosides A – H, from the root of *Momordica charantia* L. *Helvetica Chimica Acta* 91(5): 920–929.

Christy, A.O., Mojisola, C.O.C., Taiwo, E.O. & Ola, O.O. 2016. The antimalaria effect of *Momordica charantia* L. and *Mirabilis jalapa* leaf extracts using animal model. *Journal of Medicinal Plants Research* 10(24): 344–350.

Chuyen, H.V., Nguyen, M.H., Roach, P.D., Golding, J.B. & Parks, S.E. 2015. Gac fruit (*Momordica cochinchinensis* Spreng.): A rich source of bioactive compounds and its potential health benefits. *International Journal of Food Science & Technology* 50(3): 567–577.

Coutinho, H.D., Costa, J.G., Falcão-Silva, V.S., Siqueira-Júnior, J.P. & Lima, E.O. 2010. Effect of *Momordica charantia* L. in the resistance to aminoglycosides in methicilin-resistant *Staphylococcus aureus*. *Comparative Immunology, Microbiology and Infectious Diseases* 33(6): 467–471.

Dai, M., Freeman, B., Shikani, H.J., Bruno, F.P., Collado, J.E., Macias, R., Reznik, S.E., Davies, P., Spray, D.C. & Tanowitz, H.B. 2012. Altered regulation of Akt signaling with murine cerebral malaria, effects on long-term neuro-cognitive function, restoration with lithium Treatment. *PLoS ONE* 7(10): e44117.

Dans, A.M.L., Villarruz, M.V.C., Jimeno, C.A., Javelosa, M.A.U., Chua, J., Bautista, R. & Velez, G.G.B. 2007. The effect of *Momordica charantia* capsule preparation on glycemic control in type 2 diabetes mellitus needs further studies. *Journal of Clinical Epidemiology* 60(6): 554–559.

Dhasan, P.B., Jegadeesan, M. & Kavimani, S. 2010. Antiulcer activity of aqueous extract of fruits of *Momordica cymbalaria* Hook f. in Wistar rats. *Pharmacognosy Research* 2(1): 58.

Elangovan, A., Durairaj, S., Subramanian, A., Ramakrishnan, S., Lakshmanan, D.K., Ravichandran, G. & Thilagar, S. 2021. *Momordica cymbalaria* improves reproductive parameters in alloxan-induced male diabetic rats. *3 Biotech* 11: 1–14.

Elangovan, A., Subramanian, A., Durairaj, S., Ramachandran, J., Lakshmanan, D.K., Ravichandran, G., Nambirajan, G. & Thilagar, S. 2019. Antidiabetic and hypolipidemic efficacy of skin and seed extracts of *Momordica cymbalaria* on alloxan induced diabetic model in rats. *Journal of Ethnopharmacology* 241: 111989.

Fang, E.F., Froetscher, L., Scheibye-Knudsen, M., Bohr, V.A., Wong, J.H. Ng, T.B. 2019. Emerging antitumor activities of the bitter melon (*Momordica charantia*). *Current Protein and Peptide Science* 20(3): 296–301.

Froelich, S., Onegi, B., Kakooko, A., Siems, K., Schubert, C. & Jenett-Siems, K. 2007. Plants traditionally used against malaria: Phytochemical and pharmacological investigation of *Momordica foetida. Revista Brasileira de Farmacognosia* 17: 1–17.

Gandhi, P.R., Jayaseelan, C., Kamaraj, C., Rajasree, S.R. & Mary, R.R. 2018. *In vitro* antimalarial activity of synthesized TiO_2 nanoparticles using *Momordica charantia* leaf extract against *Plasmodium falciparum. Journal of Applied Biomedicine* 16(4): 378–386.

Gomez, M.S., Piper, R.C., Hunsaker, L.A., Royer, R.E., Deck, L.M., Makler, M.T. & Vander Jagt, D.L. 1997. Substrate and cofactor specificity and selective inhibition of lactate dehydrogenase from the malarial parasite *P. falciparum. Molecular and Biochemical Parasitology* 90(1): 235–246.

Govaerts, R., Nic Lughadha, E., Black, N., Turner, R. & Paton, A. 2021. The world checklist of vascular plants, a continuously updated resource for exploring global plant diversity. *Scientific Data* 8: 215.

Graz, B., Kitua, A.Y. & Malebo, H.M. 2011. To what extent can traditional medicine contribute a complementary or alternative solution to malaria control programmes. *Malaria Journal* 10(1): 1–7.

Hassan, M.M., Uddin, S., Bhowmik, A., Ashraf, A., Islam, M.M. & Rokeya, B. 2022. Phytochemical screening and antidiabetic effects of fruit rind of *Momordica dioica* Roxb. on streptozocin induced type 2 diabetic rats. *Heliyon* 8(1): e08771.

Hemalatha, K.P.J., Shantakani, S. & Botcha, S. 2021. Green synthesis of silver nanoparticles using aqueous fruit and tuber extracts of *Momordica cymbalaria. Journal of Plant Biochemistry and Biotechnology* 30: 196–204.

Hsiao, P.C., Liaw, C.C., Hwang, S.Y., Cheng, H.L., Zhang, L.J., Shen, C.C., Hsu, F.L. & Kuo, Y.H. 2013. Antiproliferative and hypoglycemic cucurbitane-type glycosides from the fruits of *Momordica charantia. Journal of Agricultural and Food Chemistry* 61(12): 2979–2986.

Hsu, C., Hsieh, C.L., Kuo, Y.H. Huang, C.J. 2011. Isolation and identification of cucurbitane-type triterpenoids with partial agonist/antagonist potential for estrogen receptors from *Momordica charantia. Journal of Agricultural and Food Chemistry* 59(9): 4553–4561.

Jain, A. & Singhai, A. 2010. Effect of *Momordica dioica* Roxb on gentamicin model of acute renal failure. *Natural Product Research* 24(15): 1379–1389.

Jeffrey, C. 1980. A review of the Cucurbitaceae. *Botanical Journal of the Linnean society* 81(3): 233–247.

Jia, S., Shen, M., Zhang, F. & Xie, J. 2017. Recent advances in *Momordica charantia*: Functional components and biological activities. *International Journal of Molecular Sciences* 18(12): 2555.

Joseph, J.K. & Antony, V. 2008. Ethnobotanical investigations in the genus *Momordica* L. in the Southern Western Ghats of India. *Genetic Resources and Crop Evolution* 55: 713–721.

Jung, K., Chin, Y.W., Yoon, K.D., Chae, H.S., Kim, C.Y., Yoo, H. & Kim, J. 2013. Anti-inflammatory properties of a triterpenoidal glycoside from *Momordica cochinchinensis* in LPS-stimulated macrophages. *Immunopharmacology & Immunotoxicology* 35(1): 8–14.

Kai, H., Akamatsu, E., Torii, E., Kodama, H., Yukizaki, C., Sakakibara, Y., Suiko, M., Morishita, K., Kataoka, H. & Matsuno, K. 2011. Inhibition of proliferation by agricultural plant extracts in seven human adult T-cell leukaemia (ATL)-related cell lines. *Journal of Natural Medicines* 65: 651–655.

Kamaraj, C., Kaushik, N.K., Rahuman, A.A., Mohanakrishnan, D., Bagavan, A., Elango, G., Zahir, A.A., Santhoshkumar, T., Marimuthu, S. & Jayaseelan, C. 2012. Antimalarial activities of medicinal plants traditionally used in the villages of Dharmapuri regions of South India. *Journal of Ethnopharmacology* 141(3): 796–802.

Kameswararao, B., Kesavulu, M. & Apparao, C. 2003. Evaluation of antidiabetic effect of *Momordica cymbalaria* fruit in alloxan-diabetic rats. *Fitoterapia* 74(1–2): 7–13.

Kattukunnel, J.J. & Antony, V. 2004. *Momordica sahyadrica* sp. nov. (Cucurbitaceae), an endemic species of Western Ghats of India. *Nordic Journal of Botany* 24(5): 539–542.

Kaur, M., Deep, G., Jain, A.K., Raina, K., Agarwal, C., Wempe, M.F. & Agarwal, R. 2013. Bitter melon juice activates cellular energy sensor AMP-activated protein kinase causing apoptotic death of human pancreatic carcinoma cells. *Carcinogenesis* 34(7): 1585–1592.

Kaushik, U., Aeri, V., Showkat, R.M. & Ali, M. 2017. Cucurbitane-type triterpenoids from the blood glucose-lowering extracts of *Coccinia indica* and *Momordica balsamina* fruits. *Pharmacognosy Magazine* 13(Suppl 1): S115.

Koneri, R.B., Samaddar, S., Simi, S. & Rao, S.T. 2014. Neuroprotective effect of a triterpenoid saponin isolated from *Momordica cymbalaria* Fenzl in diabetic peripheral neuropathy. *Indian Journal of Pharmacology* 46(1): 76.

Kumar, A., Dubey, A. & Singh, R. 2022. Investigation on anti-ulcer activity of *Momordica dioica* fruits in Wistar Rat. *International Journal for Research in Applied Sciences and Biotechnology* 9(1): 105–111.

Kumar, K.S. & Bhowmik, D. 2010. Traditional medicinal uses and therapeutic benefits of *Momordica charantia* Linn. *International Journal of Pharmaceutical Sciences Review and Research* 4(3): 23–28.

Le, A.V., Huynh, T.T., Parks, S.E., Nguyen, M.H. & Roach, P.D. 2018. Bioactive composition, antioxidant activity, and anticancer potential of freeze-dried extracts from defatted Gac (*Momordica cochinchinensis* Spreng) seeds. *Medicines* 5(3): 104.

Li, C.J., Tsang, S.F., Tsai, C.H., Tsai, H.Y., Chyuan, J.H. & Hsu, H.Y. 2012. *Momordica charantia* extract induces apoptosis in human cancer cells through caspase-and mitochondria-dependent pathways. *Evidence-Based Complementary and Alternative Medicine* 2012: 1–11.

Liao, Y.W., Chen, C.R., Hsu, J.L., Cheng, H.L., Shih, W.L., Kuo, Y.H., Huang, T.C. & Chang, C.I. 2011. Sterols from the Stems of *Momordica charantia*. *Journal of the Chinese Chemical Society* 58(7): 893–898.

Lii, C.K., Chen, H.W., Yun, W.T. & Liu, K.L. 2009. Suppressive effects of wild bitter gourd (*Momordica charantia* Linn. var. abbreviata ser.) fruit extracts on inflammatory responses in RAW 264.7 macrophages. *Journal of Ethnopharmacology* 122(2): 227–233.

Ludidi, A., Baloyi, M., Khathi, A., Sibiya, N. & Ngubane, P. 2019. The effects of *Momordica balsamina* methanolic extract on haematological function in streptozotocin-induced diabetic rats: Effects on selected markers. *Biomedicine & Pharmacotherapy* 116: 108925.

Ma, J., Krynitsky, A.J., Grundel, E. & Rader, J.I. 2012. Quantitative determination of cucurbitane-type triterpenes and triterpene glycosides in dietary supplements containing bitter melon (*Momordica charantia*) by HPLC-MS/MS. *Journal of AOAC International* 95(6): 1597–1608.

Madureira, A.M., Ramalhete, C., Mulhovo, S., Duarte, A. & Ferreira, M.J.U. 2012. Antibacterial activity of some African medicinal plants used traditionally against infectious diseases. *Pharmaceutical Biology* 50(4): 481–489.

Mamudha, M. & Sunilson, A. 2021. Extraction, identification and molecular docking studies evaluation of *Momordica tuberosa* (Roxb) against Anopheles. *Indian Journal of Science and Technology* 14(31): 2550–2556.

Matsuda, H., Nakamura, S., Murakami, T. & Yoshikawa, M. 2007. Structures of new cucurbitane-type triterpenes and glycosides, karavilagenins D and E, and karavilosides VI, VII, VIII, IX, X, and XI, from the fruit of *Momordica charantia*. *ChemInform* 38(30). https://doi.org/10.1002/chin.200730164.

Mojab, F. 2012. Antimalarial natural products: A review. *Avicenna Journal of Phytomedicine* 2(2): 52.

Mukaila, Y.O., Ajao, A.A. & Ajao, A.A.N. 2023. A review of the ethnopharmacological significance of *Momordica foetida* Schumach. (Cucurbitaceae: Cucurbitales). *Egyptian Journal of Basic and Applied Sciences* 10(1): 45–54.

Muronga, M., Quispe, C., Tshikhudo, P.P., Msagati, T.A.M., Mudau, F.N., Martorell, M., Salehi, B., Abdull Razis, A.F., Sunusi, U. & Kamal, R.M. 2021. Three selected edible crops of the genus *Momordica* as potential sources of phytochemicals: Biochemical, nutritional, and medicinal values. *Frontiers in Pharmacology* 12: 625546.

Nagarani, G., Abirami, A. & Siddhuraju, P. 2014. Food prospects and nutraceutical attributes of *Momordica* species: A potential tropical bioresources – a review. *Food Science and Human Wellness* 3(3–4): 117–126.

Nogbou, N.D., Mabela, D.R., Matseke, B., Mapfumari, N.S., Mothibe, M.E., Obi, L.C. & Musyoki, A.M. 2022. Antibacterial activities of *Monsonia angustifolia* and *Momordica balsamina* Linn extracts against carbapenem-resistant *Acinetobacter baumannii*. *Plants* 11(18): 2374.

Odeleye, O.M. & Oyedeji, A.O. 2008. Antibacterial activity of crude and fractions of *Momordica foetida* leaf extracts. *International Journal of Biomedical and Pharmaceutical Sciences* 2(2): 75–78.

Onyeibor, O., Croft, S.L., Dodson, H.I., Feiz-Haddad, M., Kendrick, H., Millington, N.J., Parapini, S., Phillips, R.M., Seville, S. & Shnyder, S.D. 2005. Synthesis of some cryptolepine analogues, assessment of their antimalarial and cytotoxic activities, and consideration of their antimalarial mode of action. *Journal of Medicinal Chemistry* 48(7): 2701–2709.

Pitchakarn, P., Ohnuma, S., Pintha, K., Pompimon, W., Ambudkar, S.V. & Limtrakul, P. 2012. Kuguacin J isolated from *Momordica charantia* leaves inhibits P-glycoprotein (ABCB1)-mediated multidrug resistance. *The Journal of Nutritional Biochemistry* 23(1): 76–84.

Poolperm, S. & Jiraungkoorskul, W. 2017. An update review on the anthelmintic activity of bitter gourd, *Momordica charantia*. *Pharmacognosy Reviews* 11(21): 31.

Poovitha, S., Sai, M.S. & Parani, M. 2017. Protein extract from the fruit pulp of *Momordica dioica* shows anti-diabetic, anti-lipidemic and antioxidant activity in diabetic rats. *Journal of Functional Foods* 33: 181–187.

Rakh, M.S., Khedkar, A.N., Aghav, N.N. & Chaudhari, S.R. 2012. Antiallergic and analgesic activity of *Momordica dioica* Roxb. Willd fruit seed. *Asian Pacific Journal of Tropical Biomedicine* 2(1): S192–S196.

Ramalhete, C., Gonçalves, B.M., Barbosa, F., Duarte, N. & Ferreira, M.J.U. 2022. *Momordica balsamina*: Phytochemistry and pharmacological potential of a gifted species. *Phytochemistry Reviews* 21(2): 617–646.

Ramalhete, C., Lopes, D., Molnár, J., Mulhovo, S., Rosário, V.E. & Ferreira, M.J.U. 2011. Karavilagenin C derivatives as antimalarials. *Bioorganic & Medicinal Chemistry* 19(1): 330–338.

Ramalhete, C., Lopes, D., Mulhovo, S., Molnár, J., Rosário, V.E. & Ferreira, M.J.U. 2010. New antimalarials with a triterpenic scaffold from *Momordica balsamina*. *Bioorganic & Medicinal Chemistry* 18(14): 5254–5260.

Ramalhete, C., Mansoor, T.A., Mulhovo, S., Molnár, J. Ferreira, M.J.U. 2009. Cucurbitane-type triterpenoids from the African plant *Momordica balsamina*. *Journal of Natural Products* 72(11): 2009–2013.

Rao, P.S. & Mohan, G.K. 2017. *In vitro* alpha-amylase inhibition and *in vivo* antioxidant potential of *Momordica dioica* seeds in streptozotocin-induced oxidative stress in diabetic rats. *Saudi Journal of Biological Sciences* 24(6): 1262–1267.

Royal Botanic Graderns Kew (RBGK). 1753. *Momordica* L., Plants of the World Online | Kew Science.

Rupachandra, S. & Sarada, D. 2013. Anticancer activity of methanol extract of the seeds of *Borreria hispida* and *Momordica dioica*. *Journal of Pharmacy Research* 6(5): 565–568.

Saeed, F., Afzaal, M., Niaz, B., Arshad, M.U., Tufail, T., Hussain, M.B. & Javed, A. 2018. Bitter melon (*Momordica charantia*): A natural healthy vegetable. *International Journal of Food Properties* 21(1): 1270–1290.

Saxena, S., Durgam, L. & Guruprasad, L. 2019. Multiple e-pharmacophore modelling pooled with high-throughput virtual screening, docking and molecular dynamics simulations to discover potential inhibitors of *Plasmodium falciparum* lactate dehydrogenase (PfLDH). *Journal of Biomolecular Structure and Dynamics* 37(7): 1783–1799.

Schaefer, H. & Renner, S.S. 2010. A three-genome phylogeny of *Momordica* (Cucurbitaceae) suggests seven returns from dioecy to monoecy and recent long-distance dispersal to Asia. *Molecular Phylogenetics and Evolution* 54(2): 553–560.

Serala, K., Steenkamp, P., Mampuru, L., Prince, S., Poopedi, K. & Mbazima, V. 2021. *In vitro* antimetastatic activity of *Momordica balsamina* crude acetone extract in HT-29 human colon cancer cells. *Environmental Toxicology* 36(11): 2196–2205.

Siboto, A., Sibiya, N., Khathi, A. & Ngubane, P. 2018. The effects of *Momordica balsamina* methanolic extract on kidney function in STZ-induced diabetic rats: Effects on selected metabolic markers. *Journal of Diabetes Research* 2018: 1–7.

Silva, C.D., Ramalhete, C., Spengler, G., Mulhovo, S., Molnar, J. & Ferreira, M.J.U. 2022. Triterpenes from *Momordica balsamina* (African pumpkin): ABCB1 inhibition and synergistic interaction with doxorubicin in resistant cancer cells. *Phytochemistry* 203: 113354.

Soundararajan, R., Prabha, P., Rai, U. & Dixit, A. 2012. Antileukemic potential of *Momordica charantia* seed extracts on human myeloid leukemic HL60 cells. *Evidence-Based Complementary and Alternative Medicine* 2012: 1–11.

Sun, Y., Sun, F., Li, J., Wu, M., Fan, X. & Meng, Y. 2016. Mono-PEGylation of alpha-MMC and & MAP30 from *Momordica charantia* L.: Production, identification and anti-tumor activity. *Molecules* 21(11): 1457.

Swamy, M.K., Akhtar, M.S., Mohanty, S.K. & Sinniah, U.R. 2015. Synthesis and characterization of silver nanoparticles using fruit extract of *Momordica cymbalaria* and assessment of their *in vitro* antimicrobial, antioxidant and cytotoxicity activities. *Spectrochimica Acta Part A: Molecular and Biomolecular Spectroscopy* 151: 939–944.

Tabuti, J.R., Kukunda, C.B. & Waako, P.J. 2010. Medicinal plants used by traditional medicine practitioners in the treatment of tuberculosis and related ailments in Uganda. *Journal of Ethnopharmacology* 127(1): 130–136.

Tan, M.J., Ye, J.M., Turner, N., Hohnen-Behrens, C., Ke, C.Q., Tang, C.P., Chen, T., Weiss, H.C., Gesing, E.R. & Rowland, A. 2008. Antidiabetic activities of triterpenoids isolated from bitter melon associated with activation of the AMPK pathway. *Chemistry & Biology* 15(3): 263–273.

Tien, P.G., Kayama, F., Konishi, F., Tamemoto, H., Kasono, K., Hung, N.T.K., Kuroki, M., Ishikawa, S.E., Van, C.N. & Kawakami, M. 2005. Inhibition of tumor growth and angiogenesis by water extract of Gac fruit (*Momordica cochinchinensis* Spreng). *International Journal of Oncology* 26(4): 881–889.

Van Wyk, A.S. & Prinsloo, G. 2020. Health, safety and quality concerns of plant-based traditional medicines and herbal remedies. *South African Journal of Botany* 133: 54–62.

Vijayakumar, M., Eswaran, M.B., Ojha, S., Rao, C.V. & Rawat, A. 2011. Antiulcer activity of hydroalchol extract of *Momordica dioica* roxb. fruit. *Indian Journal of Pharmaceutical Sciences* 73(5): 572.

Waako, P., Gumede, B., Smith, P. & Folb, P. 2005. The *in vitro* and *in vivo* antimalarial activity of *Cardiospermum halicacabum* L. and *Momordica foetida* Schumch. Et Thonn. *Journal of Ethnopharmacology* 99(1): 137–143.

Wimalasiri, D., Dekiwadia, C., Fong, S.Y., Piva, T.J. & Huynh, T. 2020. Anticancer activity of *Momordica cochinchinensis* (red gac) aril and the impact of varietal diversity. *BMC Complementary Medicine and Therapies* 20(1): 1–15.

Wright, C.W. 2005. Plant derived antimalarial agents: New leads and challenges. *Phytochemistry Reviews* 4: 55–61.

Xu, X., Shan, B., Liao, C.H., Xie, J.H., Wen, P.W. & Shi, J.Y. 2015. Anti-diabetic properties of *Momordica charantia* L. polysaccharide in alloxan-induced diabetic mice. *International Journal of Biological Macromolecules* 81: 538–543.

Yu, J.S., Roh, H.S., Lee, S., Jung, K., Baek, K.H. & Kim, K.H. 2017. Antiproliferative effect of *Momordica cochinchinensis* seeds on human lung cancer cells and isolation of the major constituents. *Revista Brasileira de Farmacognosia* 27: 329–333.

15 Antimalarial Response, Traditional and Other Potential Uses of *Piper* Genera

Yashashree Pradhan, Hina Alim, Nimisha Patel, Jamal Akhtar, and Ahmad Ali

15.1 INTRODUCTION

Various plants possess various activities and bioactive compounds which are helpful or harmful to mankind. Some of the plants are used as medicine since the ancient times to treat diseases. Traditional medicines use herbs, shrubs, and many parts of trees to treat diseases like respiratory infections, liver cure, fever, headaches, skin infections, insect bite, snakebite, etc. (Husen 2021, 2022a, 2022b, 2023; Rahman and Husen 2021, 2022, 2023a, 2023b). These traditional medicines are slow-acting. Nowadays, synthetic medicines, which are fast-acting, against diseases are used. Although they are fast-acting, they may cause side effects after their long-term use. To avoid these side effects, the world is looking forward to medicines produced using traditional medicines with increased rate of action (Ali 2020).

Malaria is one of the parasitic diseases in tropical countries. It is caused via the transmission of parasite through a mosquito vector. This parasite is a member of the *Plasmodium* genera, and the main infective species of this genus are *Plasmodium falciparum*, *Plasmodium vivax*, *Plasmodium ovale*, and *Plasmodium malariae*. *Plasmodium* is transmitted by the female *Anopheles* mosquito to humans by the bite of an infected mosquito. *Plasmodium* completes its life cycle in two different organisms, that is, the mosquito vector and the human host (Tuteja 2007). To treat this disease, medicines like quinines are used, which cause side effects like fatigue, vomiting, and nausea. To overcome such effects, new traditional medicines are gaining a lot of interest in the field of therapeutics. Many herbs are being tested for their phytochemicals, which act as antimalarial agents.

Piper genera is one of the plants used to treat malaria in ancient medicine. It contains different species which are used in therapeutics as well as in the food industry. It contains various phytochemicals which show different activities, like antimicrobial, antimalarial, antidiabetic, antioxidant, anticancer, hepatoprotective, neuroprotective, cardioprotective, aphrodisiac, and many more (Durant-Archibold et al. 2018).

15.2 BOTANICAL DESCRIPTION AND DISTRIBUTION

Piper genera is a member of the Piperaceae family, which includes approximately 1,000 species. It mainly contains slender shrubs and climbers (Figure 15.1). Leaves of *Piper* genera plants are elliptic, ovate, round, and cordate. Flowers of these plants are small, sessile, and produced on erect inflorescence. It possesses fleshy fruits. Examples of *Piper* genera are *Piper longum*, *Piper betle*, *Piper nigrum*, etc. (Tebbs 1993).

It is distributed in the tropical and subtropical regions of Asia, America, and Middle Eastern countries (Kumar et al. 2016). Some of the main plants from the *Piper* genera which show antimalarial

DOI: 10.1201/9781003378396-15

FIGURE 15.1 Examples of *Piper* genera: (a) *Piper nigrum*; (b) *Piper longum*; (c) *Piper betle*.

activity are *Piper longum*, *Piper betle*, *Piper nigrum*, *Piper tricuspe*, *Piper cubeba*, *Piper aduncum*, *Piper piedecuestanum*, *Piper chaba*, *Piper hostmannianum*, ***Piper Sarmentosum***, and ***Piper heterophyllum***(Gutierrez et al. 2013). All these plant species are found in the Indian subcontinental region.

15.3 MAJOR PHYTOCHEMICALS

Plants from the *Piper* genera contain many bioactive compounds in their fruits, roots, leaves, essential oils, and extracts. These compounds show many activities, like antimalarial, antimicrobial, antidiarrhoeal, antiparasitic, antidiabetic, anticancer, insecticidal, antiasthmatic, analgesic, antihyperlipidemic, etc. (Gutierrez et al. 2013). Phytochemicals present in *Piper longum*, *Piper betle*, *Piper nigrum*, *Piper tricuspe*, *Piper aduncum*, *Piper piedecuestanum*, *Piper chaba*, *Piper hostmannianum*, *Piper sarmentosum*, and *Piper heterophyllum* which show antimalarial response are as follows (Table 15.1 and Figure 15.2).

TABLE 15.1
Plants and Their Antimalarial Compounds and Extracts

Plant Name	Phytochemicals	References
Piper longum	Piperine	(Yadav et al. 2019)
Piper betle	Methanolic extract	(Al-Adhroey et al. 2010)
Piper nigrum	Piperine	(Jeon et al. 2019)
Piper tricuspe	2′*E*,6′*E* 2-farnesyl hydroquinone	(Vega et al. 2008)
Piper aduncum	Essential oils	(Monzote et al. 2017)
Piper piedecuestanum	Dichloromethane extract, ethyl acetate extracts	(Mesa Vanegas et al. 2018)
Piper chaba	Piperine	(Thiengsusuk et al. 2018)
Piper hostmannianum	Methyllinderatin, linderatone, 2',6'-dihydroxy-4'-methoxydihydrochalcone	(Portet et al. 2007)
Piper peltatum	4-nerolidylcatechol	(Rocha e Silva et al. 2015)
Piper sarmentosum	Ethanolic extract	(Tuntiwachwuttikul et al. 2006)
Piper heterophyllum	Dichloromethane extract	(Flores et al. 2009)

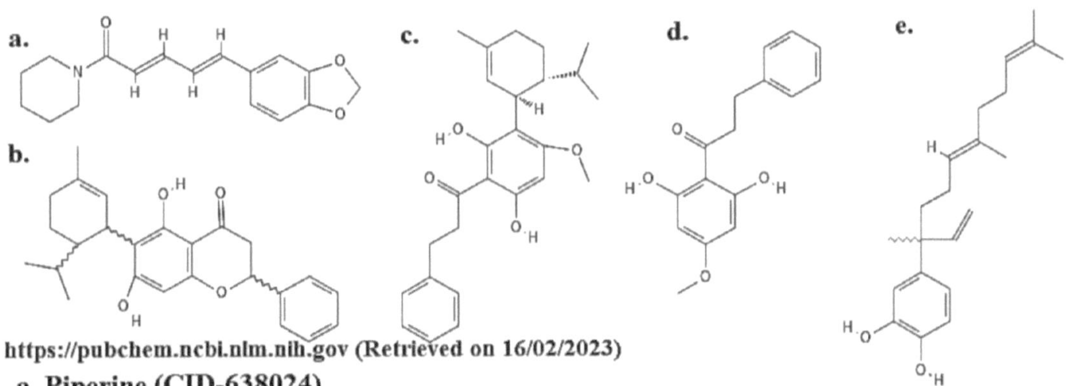

https://pubchem.ncbi.nlm.nih.gov (Retrieved on 16/02/2023)

a. Piperine (CID-638024)
b. Linderatone (CID-14781322)
c. Methyllinderatin (CID-42607684)
d. 2',6'-dihydroxy-4'-methoxydihydrochalcone (CID- 169676)
e. 4-Nerolidylcatechol (CID- 5352089)

FIGURE 15.2 Chemical structures of antimalarial compounds from *Piper* genera.

15.4 ANTIMALARIAL RESPONSE OF *PIPER* GENERA

Piper genera contains a variety of plants which have different phytochemicals acting against malarial parasites. It acts on *Plasmodium* by different pathways and shows different modes of action. *Piper tricuspe* is used to extract phytochemicals which act against *Plasmodium* species. 2′*E*,6′*E* 2-farnesyl hydroquinone shows IC_{50} at the concentration of 1.37μM against *Plasmodium falciparum* late-ring throphozoites (Vega et al. 2008).

Piperine is one of the phytochemicals present in many *Piper* genera plants. It shows activity against malarial parasites. When piperine from *Piper chaba* is isolated and given to mice infected with *P. chabaudi* in combination with curcumine and chloroquine in concentration of 20 mg, 250 mg, and 2.5 mg, respectively, it shows inhibitory activity against *P. chabaudi* late-ring trophozoites

(Neto et al. 2013). Piperine from *Piper longum* shows larvicidal activity against *Anopheles stephensi* LC_{50} *at concentration of 8.969 ppm by inducing apoptosis* (Yadav et al. 2019). Similarly, piperine from *Piper nigrum* also shows larvicidal activity against *Anopheles aegypti* (Jeon et al. 2019).

4-nerolidylcatechol is one of the phytochemicals isolated from *Piper peltatum* which show antimalarial activity by inhibiting protein synthesis in *Plasmodium falciparum*. Derivatives of 4-nerolidylcatechol also show inhibitory effect of 44% against *Plasmodium berghei* NK65 when tested in infected mice with the oral dose of 50 mg/kg body weight/day (Rocha e Silva et al. 2015).

Essential oils present in *Piper aduncum* show antimalarial activity when tested *in vitro*. It showed antimalarial activity against *Plasmodium falciparum* at a concentration of 1.2μg/mL when tested with reference drug (Monzote et al. 2017). *Piper piedecuestanum* shows antiplasmodial activity with IC_{50} of 17.93 μg/mL when tested using dichloromethane extract and ethyl acetate extract (Mesa Vanegas et al. 2018). Methanolic extract of leaves of *Piper betle* shows antiplasmodial activity against *Plasmodium berghei* schizonts (Al-Adhroey et al. 2010).

N-hexane extract of *Piper hostmannianum* was purified, and methyllinderatin, linderatone, 2',6'-dihydroxy-4'-methoxydihydrochalcone are the phytochemicals isolated and tested for antimalarial activity. These compounds showed IC_{50} of 5.64 μM, 10.33 μM, and 12.69 μM, respectively. These values are obtained after *in vitro* studies performed by using *Plasmodium falciparum*. 2',6'-dihydroxy-4'-methoxydihydrochalcone also showed decrease in parasitaemia when tested in vivo against *Plasmodium vinckei petteri* in mice (Portet et al. 2007).

Ethanolic extract of *Piper sarmentosum* showed antimalarial activity against *Plasmodium falciparum* at a concentration of 0.16μg/mL when tested with chloroquine diphosphate as reference drug (Tuntiwachwuttikul et al. 2006).

Dichloromethane extract of leaves of *Piper heterophyllum* shows activity against *Plasmodium falciparum* when tested using chloroquine as a standard drug at a concentration of 0.1μg/mL (Flores et al. 2009).

15.5 TRADITIONAL AND POTENTIAL USES OF *PIPER* GENERA

Plants from *Piper* genera has been used in traditional medicine since the ancient times. These plants show various properties. Though *Piper* genera contain almost 1,000 species, *Piper longum*, *Piper nigrum*, *Piper chaba*, and *Piper betle* are widely documented in Ayurveda and Arabian medicine. According to Ayurveda, *Piper longum* contains rasa-katu (i.e. pungent), veerya-Anushnashita (i.e. slight cold), vipaka-Madhur (i.e. sweet), guna-snigdha (i.e. unctuous), Laghu (i.e. light), tikshna (i.e. sharp), and dosha-calms *vata* and *kapha* (Pradhan et al. 2022). The roots of *P. longum* are used for the management of heart diseases. Fruits are used as an abortifacient, antiasthmatic, stomachic, liver tonic, aphrodisiac, laxative, antidiarrhoeal, antidysenteric, and antibronchitis (Johri and Zutshi 1992). In the treatment of acute and chronic bronchitis, the decoction of root and dried young fruits is used. Formation of liniment using long pepper boiled with ginger, mustard oil, butter milk, and curds is used in case of paralysis (Singh 2018). In Chinese medicine, *P. longum* is used for the treatment of stomach ache, rhinitis, vomiting, and headache (Gajurel et al. 2021). *Piper nigrum* contains properties like rasa-katu (i.e. pungent), veerya-ushna (i.e. hot), vipaka-katu (i.e. pungent), guna-Laghu (i.e. light), tikshna (i.e. sharp), and dosha-calms *vata*, *pitta*, and *kapha* (Deshpande et al. 2018). The root of *Piper nigrum* is used in the form of ghee, powders, enemas, and balms for the treatment of abdominal tumours. Black pepper can be used to treat abdominal fullness, adenitis, cancer, cholera, cold, colic, gravel, and headache. White pepper can be used to treat cholera, malaria, and stomach ache (Chaveerach et al. 2006). For *Piper chaba*, it is rasa-katu (i.e. pungent), tikta (i.e. bitter), veerya-ushna (i.e. hot), vipaka-katu (i.e. pungent), guna-ruksha (i.e. harsh), Laghu i.e. (light), tikshna (i.e. sharp), and dosha-calms *vata* and *kapha* (Deshpande et al. 2018). *Piper chaba* is used in the treatment of asthma, bronchitis, piles, colic pain, dyspepsia, and gastralgia. The aerial part of the plant shows carminative, diuretic, stimulant, expectorant, analgesic, and smooth muscle relaxant properties (Islam et al. 2020). Leaves of *Piper betle* can be used to alleviate kidney inflammation and thirst resulting from diabetes. They can be used as an expectorant effect for coughs, asthma, and bronchitis. The fresh leaves are used as an antiflatulent material in Thailand (Chaveerach et al. 2006). Due to these properties, plants from *Piper* genera are used in traditional medicine for the treatment of appetite loss, indigestion, acidity, anaemia, splenomegaly, pyrexia, and many more (Deshpande et al. 2018).

15.5.1 Antioxidant Activity

3-farnesyl *p*-hydroxy benzoic acid, dictyochromenol, and 2′*E*,6′*E* 2-farnesyl hydroquinone present in *Piper tricuspe* petroleum ether extract shows good antioxidant activity when tested using DPPH and FRAP assays (Vega et al. 2008). Antithyroid drug–induced oxidative stress in rats was reduced by supplementation of piperine in diet. Amount of antioxidant enzymes are also increased with the supplementation of piperine (Gutierrez et al. 2013). Petroleum ether extract of *P. longum* fruits shows reduction in lipid peroxide level as well, as it maintains glutathione content due to its antioxidant activity (Singh 2018). Piperine from *Piper nigrum* decreases lipid peroxidation tested *in vitro* (Damanhouri and Ahmad 2014). Methanolic extract of *Piper heterophyllum* shows antioxidant activity when tested using DPPH assay (Valdivia et al. 2009).

15.5.2 Antimicrobial Activity

Major components like camphor, viridiflorol, and piperitone from *Piper aduncum* essential oils show antimicrobial activity. Methanolic extract of *Piper aduncum* shows activity against *C. albicans* (Monzote et al. 2017). Chloroform extract of the root of the *Piper chaba* shows the presence of Bornyl piperate, piperlonguminine, and piperine, which shows antifungal activity against *Cladosporium cladosporioides* and *Cladosporium sphaerospermum* and antileishmanial activities (Lago et al. 2011). Bornyl piperate, piperlonguminine from root of the *Piper chaba* shows activity against *Aspergillus niger*, *Aspergillus fumigatus*, *Aspergillus flavus*, and *Candida albicans* (Islam et al. 2020). *P. aeruginosa*, *S. albus*, *S. typhi*, *E. coli*, and *B. megaterium* are inhibited by alcoholic extract of *P. longum* (Khan and Siddiqui 2007). It also shows antifungal effect against phytopathogenic fungi, viz, *Puccinia recondita*, *Botrytis cinerea*, *Erysiphe graminis*, *Rhizoctonia solani*, *Pyricularia oryzae*, and *Phytophthora infestans* (Pradhan et al. 2022). *Piper nigrum* shows antimicrobial activity against plant pathogens (Damanhouri and Ahmad 2014). *Piper sarmentosum* ethanolic extract shows activity against *Mycobacterium tuberculosis* and *Candida albicans* (Tuntiwachwuttikul et al. 2006).

15.5.3 Anti-Inflammatory Activity

Piperine from *P. longum* shows anti-inflammatory activity by inhibiting the prostaglandin and leukotrienes COX-1 when tested on carrageenan rat paw oedema (Kumar et al. 2005). Freund's complete adjuvant-induced arthritis in rats is reduced by the use of aqueous extract of *P. longum* (Dhanalakshmi et al. 2016). Piperine acts as an anti-inflammatory agent on early phases of inflammation mediated through histamine and serotonin, but not in later phases of prostaglandin-induced local inflammatory changes (Martha Perez Gutierrez et al. 2013). The expression of interleukin 6 and matrix metallo-proteinase 13 can be inhibited by the piperine present in *Piper nigrum* (Damanhouri and Ahmad 2014).

15.5.4 Hepatoprotective Activity

Piperine from *Piper nigrum* shows inhibition of elevated levels of GOT and GPT enzymes when tested in mice with hepato-toxicity induced by D-galactosamine (Damanhouri and Ahmad 2014). Lipid peroxidation induced by chemicals such as carbon tetrachloride and *tert*-butyl hydroperoxide causing hepatotoxicity is reduced after administration of piperine extracted from *P. longum*. Liver fibrosis is also inhibited by using *P. longum* extracts (Pradhan et al. 2022). Piperine from *Piper chaba* also shows hepatoprotective activity when tested in mice with hepato-toxicity induced by D-galactosamine (Islam et al. 2020).

15.5.5 Anti-Obesity Activity

(2E,4Z,8E)-N-[9-(3,4-methylenedioxyphenyl)-2,4,8-nonatrienoyl]-piperidine, retrofractamide C, pipernonaline, and piperrolein B isolated from chloroform extracts of *Piper nigrum* and *P. longum* show anti-obesity effect (Martha Perez Gutierrez et al. 2013). Administration of methyl piperine isolated from *P. longum* reduces hepatic cholesterol in hypercholesterolemic rats (Pradhan et al. 2022).

15.5.6 Anticancer Activity

Piperine inhibits the adhesion of neutrophils to endothelium, whereas piperlonguminine shows inhibitory activity in alpha-MSH-induced tyrosinase synthesis and reduces carcinogenesis in hamsters. Piperine also shows effect as chemopreventive in animals having lung cancer and reduces metastasis. Piplartine and piperine alkaloid amide show cytotoxic activity against tumour and hence act as an antitumour agent (Pradhan et al. 2022).

15.5.7 Antidiabetic Activity

Piper longum shows antihyperglycaemic effect, which reduces the sugar level when tested in alloxan-induced diabetic rats. Oral administration of ethanolic extract of *P. longum* showed more effect against hyperglycaemia (Pradhan et al. 2022). Oral administration of aqueous extract of *Piper nigrum* to alloxan-induced diabetic rats leads to significant lowering of blood sugar level and reduction in serum lipids (Kaleem et al. 2005). Administration of *Piper betle* extract to the STZ-induced diabetic mice showed antidiabetic activity by lowering blood glucose level (Rekha et al. 2014).

15.5.8 Antihyperlipidemic Activity

Ethanolic extract of *Piper betle* shows antihyperlipidemic activity tested on Wistar rats. Administration of methyl piperine extracted from *P. longum* shows reduction in hepatic cholesterol in hypercholesterolaemic rats (Wu and Bao 1992).

15.5.9 Wound Healing Activity

Wound healing and repair were accelerated by application of ointment formulation containing leaves and stem extract *of Piper betle* (Rekha et al. 2014).

15.5.10 Immunomodulating Activity

The phagocytic activity and macrophage migration index were seen to be elevated when tested by hemagglutination titre and phagocytic index due to the use of *P. longum*. These observations indicate that the long pepper shows immunomodulatory effect (Dhanalakshmi et al. 2016). Piperine present in *P. longum* also increases the WBC count, α-esterase positive cells, as well as bone marrow cellularity in mice (Srivastava and Saxena 2020). *Piper betle* methanolic extract showed immunosuppressive activity at various concentrations as well as reduces antibody titre (Rekha et al. 2014). Increased secretion of Th-1 cytokines (IFN-γ and IL-2) as well as increased macrophage activation and proliferation of T and B cell were observed by the piperine from *Piper nigrum* (Damanhouri and Ahmad 2014). *Piper chaba* shows immunomodulatory effect by increasing NK cell activity and lymphocyte proliferation (Islam et al. 2020).

15.5.11 Antiasthmatic Activity

Piper betle shows the ability to reduce bronchial asthma in guinea pigs (Rekha et al. 2014). Antigen-induced bronchospasm, which can be the cause of asthma, was reduced by the usage of extracts of *P. longum* administered with milk to guinea pigs (Zaveri et al. 2010).

15.5.12 Antifertility Activity

Reduction in reproductive organ weights, circulating levels of oestrogen, fertility as compared to control values were observed in albino rats treated with *Piper betle* ethanolic extract (Rekha et al. 2014). Hexane extract of *P. longum* exhibits anti-implantation activity. It also shows contraceptive activity in males and females without interfering with hormones (Srivastava 2014).

15.5.13 Antidepressant Activity

Piperine from *Piper longum* reversed the corticosterone-induced reduction of brain-derived neuropathic factor, and it also inhibits monoamine oxidase, which suggests the antidepressant activity (Pradhan et al. 2022). Ethanolic extract of *Piper betle* also shows antidepressant activity when tested on mice (Dwivedi and Tripathi 2014). Piperine from *Piper nigrum* produces an antidepressant-like effect in corticosterone-induced model of depression in mice (Damanhouri and Ahmad 2014).

15.5.14 Radioprotective Activity

Hampered glutathione production due to increase in level of glutathione pyruvate transaminase, lipid peroxidation, and alkaline phosphatase is normalized by the use of alcoholic extracts of *P. longum* fruits in radiation-treated animals (Kumar et al. 2011). The extract of *Piper betle* effectively prevented γ-ray-induced lipid peroxidation when tested on rat liver mitochondria (Dwivedi and Tripathi 2014).

15.6 CONCLUSION AND FUTURE REMARKS

Malaria is one of the most common tropical diseases encountered mainly in equatorial regions. It is a parasitic disease caused by *Plasmodium* and spread through mosquito vectors from one to another. *Piper* genera belonging to the Piperaceae family is one of the medicinal plant groups which include approximately 1,000 species in it. These plants contain bioactive compounds, such as alkaloids, flavonoids, saponins, starch, protein, essential oils. *Piper* genera plants possess activities like antioxidant activity, antimicrobial activity, anti-inflammatory activity, hepatoprotective activity, anti-obesity activity, anticancer activity, antidiabetic activity, antihyperlipidemic activity, wound healing activity, immunomodulating activity, antiasthmatic activity, antifertility activity, antidepressant activity, and radioprotective activity. Due to these activities, plants from *Piper* genera can be used as potential medicines against various ailments. Nearly 9 species of *Piper* genera show antimalarial activity. Phytochemicals like 4-nerolidylcatechol, piperine, 2′*E*,6′*E* 2-farnesyl hydroquinone, methyllinderatin, linderatone, 2',6'-dihydroxy-4'-methoxydihydrochalcone, as well as essential oils and extracts like methanolic extract, dichloromethane extract, ethyl acetate extracts from *Piper betle*, *Piper nigrum*, *Piper tricuspe*, *Piper aduncum*, *Piper piedecuestanum*, *Piper chaba*, and *Piper hostmannianum* show antimalarial activity. The major effect of these phytochemicals is inhibition of malarial parasites as well as larvicidal activity against mosquito larvae. Due to this activity, these plants from *Piper* genera can prove to be good medicine against malaria.

REFERENCES

Al-Adhroey, A. H., Nor, Z. M., Al-Mekhlafi, H. M., Amran, A. A. & Mahmud, R. 2010. Antimalarial activity of methanolic leaf extract of *Piper betle* L. *Molecules* 16(1): 107–118.

Ali, A. 2020. Herbs that heal: The philanthropic behaviour of nature. *Annals of Phytomedicine* 9(1): 7–17.

Chaveerach, A., Mokkamul, P., Sudmoon, R. & Tanee, T. (2006). Ethnobotany of the genus Piper (Piperaceae) in Thailand. *Ethnobotany Research and Applications* 4: 223–231.

Damanhouri, Z. A. & Ahmad, A. 2014. A review on therapeutic potential of *Piper nigrum* L. *Black Pepper): The King of Spices. Medical and Aromatic Plants* 3(3): 161.

Deshpande, A., Javalgekar, R. & Ranade, S. 2018. *Dravyaguna Vidnyan*, 4th ed. Proficient Publishing House, Pune, India.

Dhanalakshmi, D., Umamaheswari, S., Balaji, D., Santhanalakshmi, R. & Kavimani, S. (2016). phytochemistry and pharmacology of *piper longum* – a systematic review. *World Journal of Pharmacy and Pharmaceutical Sciences* 6(1): 381–398.

Durant-Archibold, A. A., Santana, A. I. & Gupta, M. P. 2018. Ethnomedical uses and pharmacological activities of most prevalent species of genus Piper in Panama: A review. *Journal of Ethnopharmacology* 217: 63–82.

Dwivedi, V. & Tripathi, S. 2014. Review study on potential activity of *Piper betle. Journal of Pharmacognosy and Phytochemistry* 3(4): 93–98.

Flores, N., Jiménez, I. A., Giménez, A., Ruiz, G., Gutiérrez, D., Bourdy, G. & Bazzocchi, I. L. (2009). Antiparasitic activity of prenylated benzoic acid derivatives from Piper species. *Phytochemistry* 70(5): 621–627.

Gajurel, P., Kashung, S., Nopi, S., Panmei, R. & Singh, B. 2021. Can the Ayurvedic pippali plant (*Piper longum* L.) be a good option for livelihood and socio-economic development for Indian farmers. *Current Science* 120(10): 1567–1572.

Gutierrez, R. M. P., Gonzalez, A. M. N. & Hoyo-Vadillo, C. 2013. Alkaloids from piper: A review of its phytochemistry and pharmacology. *Mini Review in Medicinal Chemistry* 13(2): 163–193.

Husen, A. 2021. *Traditional Herbal Therapy for the Human Immune System.* CRC Press, Boca Raton, FL. https://doi.org/10.1201/9781003137955

Husen, A. 2022a. *Herbs, Shrubs and Trees of Potential Medicinal Benefits.* CRC Press, Boca Raton, FL. https://doi.org/10.1201/9781003205067

Husen, A. 2022b. *Exploring Poisonous Plants: Medicinal Values, Toxicity Response and Other Potential Uses.* CRC Press, Boca Raton, FL. https://doi.org/10.1201/b23017

Husen, A. 2023. *Antidiabetic Medicinal Plants and Herbal Treatments.* CRC Press, Boca Raton, FL. https://doi.org/10.1201/b23347

Islam, M. T., Hasan, J., Snigdha, H. S. H., Ali, E. S., Sharifi-Rad, J., Martorell, M. & Mubarak, M. S. 2020. Chemical profile, traditional uses, and biological activities of *Piper chaba* Hunter: A review. *Journal of Ethnopharmacology* 257: 112853.

Jeon, H., Kim, K., Kim, Y. & Lee, S. 2019. Naturally occurring Piper plant amides potential in agricultural and pharmaceutical industries: Perspectives of piperine and piperlongumine. *Applied Biological Chemistry* 62(1): 1–7.

Johri, R. & Zutshi, U. 1992. An Ayurvedic formulation 'Trikatu' and its constituents. *Journal of Ethnopharmacology* 37(2): 85–91.

Kaleem, M., Sheema, S. H. & Bano, B. 2005. Protective effects of *Piper nigrum* and *Vinca rosea* in alloxan induced diabetic rats. *Indian Journal of Physiology and Pharmacology* 49(1): 65–71.

Khan, M. & Siddiqui, M. 2007. Antimicrobial activity of fruits of *Piper longum. Natural Product Radiance* 6: 111–113.

Kumar, S., Arya, P., Mukherjee, C., Singh, B., Singh, N., Parmar, V., Prasad, A. & Ghosh, B. 2005. Novel aromatic ester from *Piper longum* and its analogues inhibit expression of cell adhesion molecules on endothelial cells. *Biochemistry* 44(48): 15944–15952.

Kumar, S., Kamboj, J., & Sharma, S. 2011. Overview for various aspects of the health benefits of *Piper longum* Linn. fruit. *Journal of Acupuncture and Meridian Studies* 4(2): 134–140.

Kumar, V., Marković, T., Emerald, M., & Dey, A. H. 2016. Composition and dietary importance. In *Encyclopedia of Food and Health*, Eds. Caballero, B., Finglas, P. M., Toldrá, F. Academic Press, Oxford, pp. 332–337.

Lago, J. H., Ito, A. T., Fernandes, C. M., Young, M. C. & Kato, M. J. 2011. Secondary metabolites isolated from *Piper chimonantifolium* and their antifungal activity. *Natural Product Research* 26(8): 770–773.

Martha Perez Gutierrez, R., Maria Neira Gonzalez, A. & Hoyo-Vadillo, C. 2013. Alkaloids from piper: A review of its phytochemistry and pharmacology. *Mini Reviews in Medicinal Chemistry* 13(2): 163–193.

Mesa Vanegas, A. M., Toro Suaza, J. F., Vásquez Cardona, A. M., Blair Trujillo, S., Peláez Jaramillo, C., Diaz-Oltra, S., Angulo-Pachón, C. A. & Carda, M. 2018. Antiplasmodial and cytotoxic activity of *Piper piedecuestanum* Trel. and Yunck. *Research Journal of Medicinal Plants* 12: 57–64.

Monzote, L., Scull, R., Cos, P. & Setzer, W. N. 2017. Essential oil from *Piper aduncum*: Chemical analysis, antimicrobial assessment, and literature review. *Medicines* 4(3): 49.

Neto, Z., Machado, M., Lindeza, A., do Rosário, V., Gazarini, M. L. & Lopes, D. 2013. Treatment of *Plasmodium chabaudi* parasites with curcumin in combination with antimalarial drugs: Drug interactions and implications on the ubiquitin/proteasome system. *Journal of Parasitology Research* 2013: 11.

Portet, B., Fabre, N., Roumy, V., Gornitzka, H., Bourdy, G., Chevalley, S., Sauvain, M., Valentin, A. & Moulis, C. 2007. Activity-guided isolation of antiplasmodial dihydrochalcones and flavanones from *Piper hostmannianum* var. berbicense. *Phytochemistry* 68(9): 1312–1320.

Pradhan, Y., Alim, H., Patel, N., Zahra, K. F., Yiğit, B. M., Khan, J. & Ali, A. 2022. *Piper longum* (Long Pepper or Pipli) and *Tinospora cordifolia* (Giloy or Heart-Leaved Moonseed). In *Herbs, Shrubs, and Trees of Potential Medicinal Benefits*, Eds. Husen, A. CRC Press, Boca Rato, pp. 217–234.

Rekha, V. P. B., Kollipara, M., Gupta, B. R. S. S., Bharath, Y. & Pulicherla, K. K. 2014. A review on *Piper betle* L.: Nature's promising medicinal reservoir. *American Journal of Ethnomedicine* 1(5): 276–289.

Rocha e Silva, L., Nogueira, K., Pinto, A., Katzin, A., Sussmann, R., Muniz, M., Neto, V., Chaves, F., Coutinho, J., Lima, E. & Krettli, A. 2015. *In* vivo antimalarial activity and mechanisms of action of 4-nerolidylcatechol derivatives. *Antimicrobial Agents and Chemotherapy* 59(6): 3271–3280.

Singh, A. 2018. Critical review on various ethonomedicinal and pharmacological aspects of *piper longum* Linn. (long pepper or pippali). *International Journal of Innovative Pharmaceutical Sciences and Research* 6(1): 48–60.

Srivastava, N. & Saxena, V. 2020. A review on scope of immuno-modulatory drugs in Ayurveda for prevention and treatment of Covid-19. *Plant Science Today* 7(3): 417.

Srivastava, P. 2014. Therapeutic potential of *Piper longum* L. for disease management – a review. *International Journal of Pharma Sciences* 4(4): 692–696.

Tebbs, M. C. 1993. Piperaceae. In The families and genera of vascular plants II. Flowering plants. In *Dicotyledons. Magnoliid, Hamamelid and Caryophyllid Families*, Eds. Kubitzki, K., Rohwer, J. G., Bittrich, V. Springer, Berlin, Heidelberg and New York, pp. 516–520.

Thiengsusuk, A., Muhamad, P., Chaijaroenkul, W. & Na-Bangchang, K. 2018. Antimalarial activity of piperine. *Journal of Tropical Medicine* 2018: 1–7.

Tuntiwachwuttikul, P., Phansa, P., Pootaeng-On, Y., & Taylor, W. C. 2006. Chemical constituents of the roots of *Piper sarmentosum. Chemical and Pharmaceutical Bulletin* 54(2): 149–151.

Rahman, S. & Husen, A. 2021. Potential role of medicinal plants in the cure of liver and kidney diseases. In *Non-Timber Forest Products*, Eds. Husen, A., Bachheti, R. K., Bachheti, A. Springer International Publishing AG, Cham, pp. 229–254. https://doi.org/10.1007/978-3-030-73077-2_10

Rahman, S. & Husen, A. 2022. *Asparagus racemosus* (Shatavari) and *Dioscorea villosa* (Wild yams). In: *Herbs, Shrubs, and Trees of Potential Medicinal Benefits*, Ed. Husen, A. CRC Press, Boca Raton, FL.

Rahman, S. & Husen, A. 2023a. *Trigonella foenum-graecum* L.: A useful medicinal plant for diabetes and other diseases. In *Antidiabetic Medicinal Plants and Herbal Treatments*, Ed. Husen, A. CRC Press, Boca Raton, FL.

Rahman, S. & Husen, A. 2023b. Role of Green Chiretta [*Andaphis paniculata* (Burm. F.)] nees in diabetes and other ailments. In *Antidiabetic Medicinal Plants and Herbal Treatments*, Ed. Husen, A. CRC Press, Boca Raton, FL.

Tuteja, R. 2007. Malaria– an overview. *The FEBS Journal* 274(18): 4670–4679.

Valdivia, A. C., Mollinedo, P., Vilaseca, A. & Sterner, O. 2009. Prenylated protocatechuic acid derivatives with anti-oxidant activity from Piper heterophyllum. *Revista Boliviana de Química* 26(2): 83–89.

Vega, A. S., Rojano, B., Blair, S. & Saez, J. 2008. Antimalarials and antioxidants compounds from *Piper tricuspe* (Piperaceae). *Pharmacologyonline* 1: 1–8.

Wu, E. & Z. Bao. 1992. Effects of unsaponificable matter of *Piper longum* oil on cholesterol biosynthesis in experimental hypocholestrolaemic mice, *Honggacayano* 23(4): 197–200.

Yadav, R., Saini, H., Kumar, D., Pasi, S. & Agrawal, V. 2019. Bioengineering of *Piper longum* L. extract mediated silver nanoparticles and their potential biomedical applications. *Materials Science and Engineering: C* 104: 109984.

Zaveri, M., Khandhar, A., Patel, S. & Patel, A. 2010. Chemistry and pharmacology of *Piper longum* L. *International Journal of Pharmaceutical Sciences Review and Research* 5(1): 67–76.

16 Antimalarial Response, Traditional and Other Potential Uses of the Genus *Senna*

Kolagani Chandramohan, Salman Khan, and Azamal Husen

16.1 INTRODUCTION

Ethnobotanical studies focus on traditional knowledge of tribal/local/Adivasi people and development of databases on ethnic knowledge, but also focus on preservation and regeneration of traditional beliefs and maintenance of traditional knowledge. Since ancient days, the preliminary and safest therapeutic approach has been plant-based medicine, which has displayed a significant role in the treatment of different ailments (Husen, 2021, 2022a, 2022b, 2023; Rahman and Husen, 2021, 2022, 2023a, 2023b). About 40,000 to 70,000 medicinal plant species were documented based on their therapeutic potentiality and positioned in developing countries on utilization of herbal medicines (Ekor, 2013; Verpoorte, 2015; WHO, 2016; Oladeji et al., 2019). The recent estimates show that there are about 25,000 active plant-based preparations used in traditional medicine and known to remote communities all over from India, and the records stated that nearly 10,000 medicated formulations are available in the Aboriginal medicinal texts.

Globally, malaria fever or viral fever are alarming to public health problem in recent years. About 3.2 billion people remain at risk of malarial infections. Annually, 149–303 million clinical cases were noticed globally, and it kills more than 400,000 people (Ramírez et al., 2017). Malaria is infected through natural or vector borne in humans by six species from the genus *Plasmodium*: *P. falciparum*, *P. vivax*, *P. ovale*, *P. malariae*, *P. knowlesi*, and recently *P. cynomolgi* (Ta et al., 2014; WHO, 2015). Human beings are being transmitted the parasites by the bite of the *Anopheles* mosquito. Once inside the organism, the parasite causes clinical symptoms that can range from mild to grave or even lead to fatality (Kitua and Malebo, 2004; Campuzano-Zuluaga et al., 2010). Usage of traditional medicine to treat different illnesses was initiated, and the World Health Organization (WHO) recognized that 80% of the world population uses plants as alternative therapy (Bermúdez et al., 2005). Since 1670, a metabolite called quinine extracted from quina plant by French chemists Pierre Pelletier and Joseph Caventou has been used to treat malarial fevers and served as a template for the synthesis of aminochinolines, such as chloroquine, which has malarial resistance (Rieckmann et al., 1971). After this, search of new antimalarial metabolites was stimulated. In 1972, the artemisinin derivatives artesunate, arteether, and artemether have been synthesized from *Artemisia annua* by Chinese scientists, which consist a new pillar of antimalarial therapy (Tang and Eisenbrand, 1992; Hien and White, 1993).

Despite the fewer side effects, medicinal plants are the principal sources of therapeutic substances which could lead to the preparation of novel therapeutic compounds and in drug development (Veeresham, 2012). Several medicinal plants have been investigated to authenticate the pharmacologically bioactive metabolites which cure malarial diseases. The plant species of *Senna* have been used universally in traditional medicines since many years. The active compounds of *Senna* species are used for active therapy in improving human health and treatment of a wide range of diseases and infections, such as diabetes, sexually transmitted diseases (STDs), measles, malaria, inflammation, abdominal pain, typhoid, and gonorrhoea in China, India, Mexico, Brazil,

 DOI: 10.1201/9781003378396-16

Malaysia, Thailand, and Africa (Silva et al., 2008). Finally, the bioactive metabolites from different *Senna* spp. have been explored, and research revealed that over 120 structurally diverse phytochemicals, including piperidine alkaloids, terpenoids, glycosides, tannins, saponins, steroids, flavonoids, anthraquinones, polyphenol, and anthrones (Hennebelle et al., 2009), were investigated in recent years to possess antimalarial, antidiabetic, antimicrobial, antioxidant, anti-inflammatory, analgesic, antitumour, antinociceptive, and anticancer effects (Ibrahim and Islam, 2014). Among them, comprehensive information on traditional uses, phytochemistry, and pharmacological studies of the genus *Senna* and their therapeutic potential on antimalarial response are discussed here.

16.2 ETHNOBOTANICAL AND TRADITIONAL USES OF GENUS *SENNA*

The species of *Senna* have been used in different traditional medicines, like Indian, Chinese, and Tibetan folk medicine and African traditional medicine, with diverse therapeutic activities (Jain et al., 1997; Jafri et al., 1999; Viegas et al., 2004; Bum et al., 2010; Silva et al., 2008; Singh et al., 2013). Most of the plant parts of *Senna* spp. are commonly used in preparation of different herbal crude drugs or herbal medicine for the treatment of various diseases and infections. Generally, herbal medicines can be extracted from the genus *Senna* by the methods maceration, decoction, infusion, exudates, or stem bath and can be used as therapeutic substances (Oladeji et al., 2016, 2020). In Siddha, Ayurveda, and African system of medicine, *Senna alata* and *S. occidentalis* are widely used in the treatment of typhoid, malaria, yellow fever, and cholera (Sansores-Perazaa et al., 2000; Yadav et al., 2010; Singh et al., 2013; Oladeji et al., 2020). In northern Nigeria, the stem, leaf, and root decoctions of *Senna* species are used in the treatment of wound, skin, and respiratory tract infection, burns, diarrhoea, and constipation (Adedayo et al., 2001). In the south-western regions and Egypt, a decoction of leaves serves as abdominal pain, constipation, and toothache relief (Benjamin and Lamikanra, 2008; Leung et al., 2011). In Cameroon, the stem, bark, and leaves of *S. alata* were reported to be used as a remedy for gastroenteritis, hepatitis, and skin infections (Day and Underwood, 1986).

In India, Philippines, and China, decoctions of *S. alata*'s stem, bark, and leaves were found to be effective in treating haemorrhoids, inguinal hernia, syphilis, malarial or parasital infection, and diabetes (Igoli et al., 2005). Recent investigation on extractions of seeds and roots revealed that regulation of uterus disorder and worms was noticed (Makinde et al., 2007; Selvi et al., 2012). The extractions of leaves and flowers are typically used as antifungal agents and laxatives (Farnsworth and Bunyapraphatsara, 1992). In Guatemala, Brazil, and Guinea, the whole plant is currently used in the treatment of flu and malaria (Hennebelle et al., 2009). In China, India, Benin Republic, Ghana, Nigeria, and Togo, the extraction of whole plant is used as a curative for diabetes mellitus (Woradulayapinij et al., 2005; Abo et al., 2008). The leaves of *Senna alata* are extensively used in traditional systems than roots, flowers, and stem bark due to its pharmacological activities. Because of that, it is processed into capsule, pellet, and tea in Nigeria for preventing diseases and maintaining sound health (Figure 16.1c, d, and e).

In various traditional systems, leaf extractions of *Senna angustifolia* are used for the treatment of constipation, loss of appetite, hepatomegaly, splenomegaly, digestive problems, malaria, jaundice, anaemia, and as purgative (Anon, 1966; Tripathi, 1999).

Senna occidentalis is an annual or perennial herb (Figure 16.1f and g) which contains a wide range of chemical compounds, including achrosin, aloe-emodin, emodin, anthraquinones, anthrones, apigenin, aurantiobtusin, campesterol, cassiollin, chryso-obtusin, chrysophanic acid, chrysarobin, chrysophanol, and chrysoeriol. It has been known to possess antibacterial, antifungal, antidiabetic, anti-inflammatory, anticancerous, antimutagenic, and hepatoprotective activity in different traditional systems (Yadav et al., 2010).

The leaves, pods, and flowers of *Cassia fistula* (Figure 16.1a and b) are used as purgative, febrifugal, biliousness, and astringent in traditional systems (Kirtikar and Basu, 2006; Gupta, 2010; Danish et al., 2011; Ruth et al., 2021). Ash from burnt pods is used to cure cough (Gupta, 2010). The

FIGURE 16.1 (a, b). *Cassia fistula*; (c, d, e) *Senna alata*; (f, g) *Senna occidentalis*; (h, i) *Senna tora*; (j, k) *Senna siamea*.

root is useful in fever, heart diseases, retained excretions, and biliousness (Nadkarni, 1976). The pulp of *Cassia fistula* is used as analgesic and as an antipyretic; it is a remedy for malaria and fever. It is also applied in blood poisoning, anthrax, and antidysenteric, leprosy, and antidiabetic, for the removal of abdominal obstruction (Anonymous, 2007; Ruth et al., 2021).

In different traditional systems, the whole plant of *C. obtusifolia* has been used as a laxative and for the treatment of eye infections, diarrhoea, urinary tract infections, gingivitis, fever, and as cough remedy, due to the presence of various chemical constituents like anthraquinones and naphthopyrones: aloe-emodin, emodin, 1,2-dihydroxyanthraquinone, obtusin, chryso-obtusin, aurantio-obtusin, gluco-obtusifolin, gluco-aurantio-obtusin, gluco-chryso-obtusin, chryso-obtusin-2-O-β-d-glucopyranoside, physicon-8-O-β-d-glucoside, obtusifolin, O-methyl-chrysophanol, emodin-1-O-β-gentio-bioside, chrysophanol-1-O-β-gentiobioside, chrysophanol-1-O-β-d-glucopyranosyl-(13)-β-d-glucopyranosyl-(1→6)-β-d-glucopyranoside, physcion-8-O-β-glucoside, 1,3-dihydroxy-8-methylanthraquinone,(Doug hari et al., 2008; Ali et al., 2021).

The seeds of *Senna tora* are used as antiparasitic, antidiarrhoeal, antioxidant, antihepatotoxic, antiplasmodial, and antimutagenic due to the presence of bioactive constituents like anthraquinones, including 1-desmethylaurantion-obtusin, 1-desmethylchryso-obtusin, aurantio-obtusin, chryso-obtusin, obtusin, etc. The methanolic extract of seeds of plant has properties of larvicidal activity against *Aides aegypti* mosquito (Jang et al., 2002) and shows antiplasmodial activity against chloroquine-sensitive *Plasmodium falciparum* (EI-Tahir et al., 1999) (Figure 16.1 H&I).

The extracts of *Senna auriculata* are widely used in traditional medicine for rheumatism, conjunctivitis, and diabetes (Siva and Krishnamurthy, 2005). Pills prepared from ground leaves and fruits have been used for the treatment of leucorrhoea in Eastern Ghats of Andhra Pradesh (Ratnam and Raju, 2005). The tribes of the north-east region use the plants for treating skin disease and as purgative (Baruah and Sharma, 1987). Tribes from Southern India prepare paste from leaves along vinegar, which is applied on skin for various skin diseases, dandruff, bone fracture (Natarajan and Paulsen, 1999; Kshirsagar and Singh, 2000; Jeeva et al., 2007), and scorpion bite (Ratnam and Raju, 2008). It was also found that tenders of leaves are used for treatment of stomachache (Nagnur et al., 2009). Flowers are used to treat venereal diseases and spermatorrhea. The whole plant is used to treat diarrhoea. The Meena community of Rajasthan used the leaf extract of the plant for treating tuberculosis. Roots are useful in urinary discharge and cure tumours, skin disease, and asthma. Powder of bark is used for fixing teeth and decoction for chronic dysentery. Decorticated seeds in fine powder and paste are valued local applications to purulent ophthalmia and conjunctivitis (Latha and Pari, 2003; Thabrew et al., 2004; Duraipandiyan et al., 2006; Rajendran et al., 2008; Meena and Rao, 2010; Meena et al., 2019).

Cassia singueana is well distributed in northern Nigeria, is used traditionally for the relief of pain and body ache of acute malaria attack. The test caused a significant ($P < 0.05$) and dose-dependent reduction in the mean parasitaemia in mice similar to chloroquine (5 mg/kg) (Bulus et al., 2003). The extracts of plant have ethnomedicinal applications that include its usage as an anticonvulsant, in constipation, snakebite, abortive, urinary schistosomiasis, and hernia (Baoua et al., 1976; Ndamba et al., 1984, Chhabra et al., 1987; Audu, 1989; Mutasa et al., 1990). It has shown antinociceptive, antipyretic, anti-inflammatory, and antiplasmodial activities of the methanol extract of the root bark of plant (Bulus et al., 2003).

The leaves, stems, roots, flowers, and seeds of *C. siamea* (Figure 16.1j and k) are therapeutically proven in treatment of malarial diseases. The decoction of the leaves became a part of herbal remedies in African and Asian countries. In Burkina Faso, the decoction of fresh and dried leaves is used to treat malaria and liver disorders (Sanon et al., 2003; Fowler, 2006; Nadembega et al., 2011). It is also used in the treatment of cough, stomach pains, and malaria in regions of Cote d'Ivoire and Sierra Leone and Togo (Mbatchi et al., 2006; Kone and Atindehou, 2008; Koudouvo et al., 2011). Recent investigations revealed that it cures constipation, hypertension, toothache, and as a laxative (Thongsaard et al., 1996; Mbatchi et al., 2006). The decoction of roots is used against fever, constipation, hypertension, insomnia, snakebite, diabetes, sore throat, and malaria

(Owuor et al., 2006; Odason et al., 2007; Allabi et al., 2011). The decoction of stem or stem bark is widely used against malaria and liver disorders in African countries and Malaysia (Sanon et al., 2003; Odugbemi et al., 2007; Asase et al., 2010; Kamatenesi et al., 2011; Adebayo and Krettli, 2011). It also works as a laxative, antidiabetic, purgative, and antiscabies (Kiepe, 1995; Odason and Kolawole, 2007). The decoction of seeds is used in treatment of intestinal worms, snake-and scorpion bites, and laxative (Kiepe, 1995; Kaur et al., 2006; Mbatchi et al., 2006). The decoction of the whole plant is used for the treatment of diabetes, antirheumatic, swellings, purgative, malarial fever, and as laxative in parts of India and China (Hu et al., 2019; Mbatchi et al., 2006; Ali, 2009; Lawal et al., 2010).

Senna spinescens is known to Amazonian regions that grows wild on the slopes of the rivers. The part used for the treatment of malaria is the fresh stem bark. The remedy can be extracted from both juvenile and adult trees, is prepared by decoction, and can be handled by both men and women (Schultes and Raffauf, 1990).

16.3 PHYTOCHEMISTRY, PHARMACOLOGICAL, AND OTHER POTENTIAL IMPORTANCE OF GENUS *SENNA*

Many researchers have reported that different parts of the *Senna tora* contain different constituents. The leaves of the plant contain emodin, tricontain-1-ol, stigmasterol, β-sitosterolβ-D-glucoside, succinic acid, tartaric acid, uridine, quercetin, isoquercitin (Verma et al., 2021; Keneda et al, 1969). Anthraquinones as chrysophanol, physcion, emodin, rhein, obtusifolin, obtusin, chyrso-obtusin, rubrofusarin, aurantio-obtusin, chysophanic acid-9-anthrone have been extracted from leaves and seeds (Meena et al., 2010), and alcoholic extraction of seeds also demonstrated the presence of anthraquinones derivative like 1-desmethylaurantio-obtusin and 1-desmethylchryso-obtusin (Zhu et al., 2008). The gum which is extracted from the seeds of *Cassia tora* contains 70% high polysaccharide with a liner chain of 1, 4-βD-mannopyarnose linked to 1,6-linked β-dgalactopyranose unit. *Cassia tora* leaves were also evaluated to find out proximate nutrient content, amino acid composition, and some special mineral element. The presence of 11.25% of crude protein, 28.00% of crude fibre protein, and 17 amino acids has been evaluated from the extract (Kubmarawa et al., 2011).

Several classes of bioactive metabolites have been extracted from *Senna obtusifolia* L., including anthraquinones, terpenoids, flavonoids, and lipids. The main plant chemicals include anthraquinone, emodin, chrysophanol, physcion, obtusifolin, obtusin, aurantio-obtusin, chryso-obtusin, alaternin, questin, aloe-emodin, gluco-aurantio-obtusin, gluco-obtusifolin, chrysophanol-2-O-tetraglucoside, and chryso-obtusin-2-glucoside. In addition, components like naphthopyrone glycosides, toralactone-9-β-gentiobioside, toralactone gentiobioside, cassiaside, rubrofusarin-6-O-gentiobiosideol, rubrofusarin-6-β-gentiobioside, cassiaside C, cassiaside B2, cassiaside C2, xanthones (1, 8-dihydroxy-3-methoxy-6-methylxanthone, isogentisin, 1,7-dihydroxy-3-methylxanthone, euxanthone, 1,3,6-trihydroxy-8-methylxanthone), triterpenoids, and toralactone have been extracted. Recent investigations revealed that the whole *C. obtusifolia* L. plant (seeds, twigs, leaves, and roots) is rich in free and bound anthraquinones (Ali et al., 2021; Sob et al., 2008; Yun-Choi et al., 1990; Kitanaka et al., 1998; Sung et al., 2004; Kim et al., 2007; Vadivel et al., 2012; Deshpande & Naik, 2016; Jung et al., 2016; Jung et al., 2017).

Preliminary phytochemical screening of *Senna siamea* showed the presence of chromones and its derivatives (chromone alkaloids, chromones glycosides, dihydronaphthalenone compounds, bischromone), polyphenols (anthraquinones, bianthraquinones, anthrone, flavonoids, isoflavonoids, phenolics, tannins), alkaloids, saponins, steroids, carotenoids, antinutrients (oxalate, phytate), reducing sugars, vitamins, minerals, and enzymes (Kamagate et al., 2014).

After recent investigation on screening of phytochemicals in *Senna angustifolia* revealed small quantities of other anthraquinone glycosides, that is, sennosides C and D (Tyler, 1982), rhein-anthrone-8-diglucose, rhein-8-glucoside, rhein-8-diglucoside, aloe-emodine-8-glucoside, and aloe-emodine-anthrone diglucoside have also been found in the leaves and pods (Lemli and Cuveeli, 1972, 1974).

The following phytochemicals are present in extracts of *Cassia fistula* (Ruth et al., 2021). In leaf extracts, anthraquinones like rhein, chrysophanol, and physcion and butyric and formic acids, pectin, tannin, sennosides A and B, oxalic acid, physcion, proanthocyanidin B dimer, apigenin-6,8-di-C-glycoside, myricetin hexoside, apigenin-C-hexoside-opentoside phenolic acids (e.g. coumaric acid derivative and 3,4-di-O-caffeoylquinic acid, chrysophanol) were found. The flower extracts have alkaloids, rhein, leucopelargonidin tetramer, fistulin, kaempferol, heptacosanoic acids, hentriacontanoic, triacontanoic, nonacosanoic, and gibberellic acid, and bark extracts have oxyanthraquinone and flavonol glycosides, dihydroxyanthraquinone, leucocyanidin, fistucacidin (3,4,7,8,4β-pentahydroxyflavon), and lupeol. The extraction of pods resulted in phytochemicals 1,8-dihydroxy-3-anthraquinolone derivative, rhein, fistulin, oxyanthraquinone, tannin, ceryl alcohol, glycosides-sennosides A and B, volatile oil (essential oils), chrysophanol (Figure 16.1), resinous, and waxy derivatives. The roots have phytochemicals oxyanthraquinone and phlobaphenes, tannins, and rhamnetin-3-O-gentiobioside. The seed extracts have phytoconstituents anthraquinone derivatives, carbohydrates, proteins, amino acids, resinous and waxy derivatives, volatile oil, stetculic and malvalic acids, (2′S)-7-hydroxy-2-(2′-hydroxypropyl)-5-methylchromone, 5-hydroxymethylfurfural, Benzyl-2β-O-D-glucopyranosyl-3,6-dimethoxybenzoate, benzyl-2-hydroxy-3,6-dimethoxybenzoate and 5-(2-hydroxyphenoxymethyl) furfural.

The investigation on phytochemical studies on *Senna auriculata* resulted in leaf extracts showing the presence of 3-o-methy-d glucose, alpha-tocopherol-beta-D-mannosidase, n-hexadecanoic acid, resorcinol, octadecenal, and carboxylic acid. The seed extracts were reported to have the presence of fatty acid content, such as palmitic, oleic, and linoleic acids. The ethanolic seed extract showed the presence of benzoic acid, 2-hydroxyl methyl ester, glycine, n-(trifluroacetyl), 1-methybutul ester, cupric acid ethyl ester, resorcinol, water-soluble galactomannan like beta-D-manopyranosyl-1(1–4)-o-beta-D-manopyronosyl (1–4)-o-beta – D-monopyranose. Root extracts showed the presence of anthraquinone glycosides as 1,3-dihydroxy-2-methylanthraquinone, 1,8-dihydroxy-6-methoxy-2-methylanthraquinone-3-o-rutinoside, 8-dihydroxy-2-methylanthraquinone-3-o-rutinoside, flavone glycoside, and two leucoanthocyanins like leucocyanidin-3-o-rhamnopyroside and leucopeonidin-3-o-1-rhamanopyroside. (Meena et al., 2019).

The extracts of *Senna alata* have bioactive compounds, such as kaempferol, luteolin, and aloe emodin isolated from the methanolic extracts, which were assessed against MDR bacteria strains. Aloe emodin exhibited strong MIC activities. Anthraquinone and flavonoids compounds, chrysoeriol-7-O-(2″-O-beta-D-mannopyranosyl)-beta-Dallopyranoside, 3-O-gentiobioside, quercetin, naringenin, and rhamnetin-3-O-(2″-O-beta-D-mannopyranosyl)-beta-D-allopyranoside were isolated from chloroform, ethanol, petroleum ether, methanol, and aqueous leaves extract of *S. alata*. The crude extracts and isolates obtained from purified fractions of *S. alata* flower extract (anthraquinone glycosides, steroids, tannins, and volatile oils) were assessed on selected bacteria isolates (Oladeji et al., 2020).

16.4 ANTIMALARIAL ACTIVITY AND ANTIPLASMODIAL ACTIVITY

The *S. occidentalis* plant extract has a significant antimalarial activity. The ethanolic, dichloromethane, and lyophilized aqueous extracts of *S. occidentalis* root bark were evaluated for their antimalarial activity in vivo, in four-day suppressive assays against *Plasmodium berghei* ANKA in mice (Neamsuvan, 2015). No toxic effect or mortality was observed in mice treated, orally, with any of the extracts as a single dose, of 500 mg/kg body weight, or as the same dose given twice weekly for 4four weeks (to give a total dose of 4 g/kg). At doses of 200 mg/kg, all the ethanolic and dichloromethane extracts produced significant chemo suppression of parasitaemia of >60% for *C. occidentalis* root bark when administered orally. The *S. occidentalis* was active and caused 60% chemosuppression. It is also observed that the lyophilized aqueous extract was less active than the corresponding ethanolic extract (Neamsuvan, 2015). Ethanol and chloroform extract of the *C. occidentalis* leaves have been found to have good antimalarial activity. These extracts produce more than 60% inhibition of the parasite growth *in vitro* at a concentration of 6 µg/ml (Ramstad, 1959; Brown, 2011).

As per procedure described by Saidu et al. (2000), the experimental research was carried out on mice. A donor mouse infected with rodent malaria (*Plasmodium berghei*) was anaesthetized with chloroform and the abdomen opened. Blood was collected through cardiac puncture with a sterile needle and syringe (Zarinject, Nigeria) in such a way that 0.2 ml of the blood contained about 1 × 107 infected red cells. Thirty mice were selected, and inoculums of 0.2 ml were given to each mouse (i.p.). The mice were then grouped into five ($n = 6$) and treated as follows on day 3 (D3). Group 1 received normal saline, groups 2–4 received the extract (50–200 mg/kg, i.p.), while group 5 received CQ (5 mg/kg, i.p.). The treatment continued daily until D7. Thick and thin blood smear firms were collected daily from tail blood, fixed with methanol, stained with 4% Giemsa at pH 7.2 for 45 min and examined microscopically (Nikon YS2-H, Japan). The increase/decrease in parasitaemia, defined as the number of infected and uninfected RBC, were counted in five different fields, and mean survival time (within 30 days) was recorded (Bulus et al., 2003).

Various extracts of leaves, stem bark, and flowers of *S. siamea* were screened for its antimalarial activity (Jun et al., 2012). Most of the activities described were determined *in vitro* on *Plasmodium falciparum* strains. Specified and bio-guided fractionation was also based on this antimalarial test. Activities were assessed on different strains, among which are chloroquine-sensitive (3D7), chloroquine-resistant (W2, FcM29-Cameroon), and multidrug resistant (K1), in order to find effective compounds against resistant malaria. In all studies, alkaloids fraction of the leaves exhibited better antiplasmodial activity than other extracts (Hussain, 1991). This alkaloid's fraction, the chloroform and ethanol extracts of the leaves, showed activity against 3D7 with IC50 value of 0.24, 2.41, and 7.06 µg/ml, respectively. Cassiarin A is the alkaloid compound which has more potential activities. Its activity is similar to that of chloroquine against 3D7 with IC50 value of 0.005 and 0.006 µg/mL, respectively (Wiwied et al., 2009). Among stem bark extracts, chloroform extract (IC50 21±3 µg/mL) was the most interesting, with promising antimalarial activity, followed by ethanol extract (IC50 31±5 µg/mL) and aqueous extract (IC50 > 100 µg/mL) on FcM2930. Against K1, ethyl acetate fraction of this plant part was active with IC50 31±3 µg/mL, and this activity was associated to emodin and lupeol, which displayed similar IC50 value of 5 µg/mL. (Ajaiyeoba and Ashidi, 2008; Gallo and Sarachine, 2009).

The studies on *Senna singueana* have resulted in acute toxicity, and antiplasmodial activity tests on Swiss albino mice indicated that 70% ethanol extract of *S. singueana* leaves is relatively safe to mice and demonstrated antimalarial activity, which may contribute to its traditional use for the management of malaria and related symptoms, such as fever (Hiben et al., 2016).

16.5 CONCLUSION AND FUTURE REMARKS

This review provides a comprehensive information of the traditional significance, phytochemistry, pharmacology, and antimalarial properties of the genus *Senna*. The studies unveiled several phyto-constituents embedded in different species of the genus *Senna* and *Cassia* with considerable therapeutic potentials in ameliorating of various infectious and degenerative diseases in light of antimalarial significant. The phytochemical reports of the genus were focused on specific groups of compounds which were discussed here. The geographical distribution and adaptability of the species to various climates could influence the embedded secondary metabolites produced by plants. The antimicrobial, antidiabetic, antinociceptive, antioxidant, anti-inflammatory, antimalarial, and antipyretic effects have been studied, but more studies into the pharmacological activities are necessitated specifically by applying appropriate models (in vivo). More initiatives should be taken up towards strategizing mechanisms which could help corroborate the pharmacology and toxicology of the species. The potential of genus *Senna* to regulate antiplasmodial, antimicrobial, and antimalarial properties has been scientifically proven in various methods. Hence, isolation and clinical trials should be systematically considered, which could lead to the discovery of novel bioresources, thus quenching the biotechnological demands of scientists in search of safe therapeutic drugs which could regulate malarial diseases as well as current health challenges.

16.6 ACKNOWLEDGEMENTS

The authors are greatly indebted to Shri. Anoop Singh IFS, Director General, Forest Survey of India, Dehradun, and Shri. Chaturbhuja Behera IFS, Regional Director, Forest Survey of India Central Zone, Nagpur, for their support and facilities.

REFERENCES

Abo, K.A., Fred-Jaiyesimi, A.A. & Jaiyesimi, A.E.A. 2008. Ethnobotanical studies of medicinal plants used in the management of diabetes mellitus in South Western Nigeria. *Journal of Ethnopharmacology* 115(1): 67–71.

Adedayo, J.O., Anderson, W.A., Moo-Young, M., Snieckus, V., Patil, P.A. & Kolawole, D.O. 2001. Phytochemistry and antibacterial activity of *Senna alata* flower. *Pharmaceutical Biology* 39(6): 408–412.

Adebayo, J.O. & Krettli, A.U. 2011. Potential antimalarials from Nigerian plants: A review. *Journal of Ethnopharmacology* 133: 289–302.

Adzu, B., Abbah, J., Vongtau, H. & Gamaniel, K. 2003. Studies on the use of *Cassia singueana* in malaria ethnopharmacy. *Journal of Ethnopharmacology* 88(2–3): 261–267.

Ajaiyeoba, E.O. & Ashidi, J.S. 2008. Antiplasmodial compounds from *Cassia siamea* stem bark extract. *Phytotherapy Research* 22: 254–255.

Ali, M.Y., Park, S. & Chang, M. 2021. Phytochemistry, ethnopharmacological uses, biological activities and therapeutic applications of *Cassia obtusifolia* L.: A comprehensive review. *Molecules* 26(20): 6252.

Ali, S. 2009. Determination of chemical composition of *Senna siamea* (Cassia leaves). *Pakistan Journal of Nutrition* 8: 119–121.

Allabi, A.C., Busia, K., Ekanmian, V. & Bakiono, F. 2011. The use of medicinal plants in self-care in the Agonlin region of Benin. *Journal of Ethnopharmacology* 133: 234–243.

Anonymous. 1966. *Pharmacopoeia of India* (2nd ed.). Manager of Government Publications, Delhi, India, pp. 647–649.

Anonymous. 2007. *The Wealth of India, First Supplement Series* (Raw Materials), National Institute of Science Communication and Information Resources, CSIR, New Delhi, Vol 3 (Ca-Ci), pp. 340–342.

Asase, A., Akwetey, G.A. & Achel, D.G. 2010. Ethnopharmacological use of herbal remedies for the treatment of malaria in the Dangme West District of Ghana. *Journal of Ethnopharmacology* 129: 367–376.

Audu, J. 1989. Medicinal herbs and their uses in Bauchi State. *The Nigerian Field* 54: 157–168.

Baoua, M., Fayn, J. & Bassiere, J. 1976. Preliminary phytochemical testing of some medical plants of Niger. *Plantes Medicinales et Phytotherapie* 10: 251–266.

Baruah, P. & Sharma, G.C. 1987. Studies on the medicinal use of plants by the North east tribes. *Journal of Economic & Taxonomic Botany* 11: 71–76.

Benjamin, T.V. & Lamikanra, A. 2008. Investigation of *Cassia alata*, a plant used in Nigeria in the treatment of skin diseases. *Quarterly Journal of Crude Drug Research* 19(2&3): 93–96.

Bermúdez, A., Oliveira, M.M.A. & Velázquez, D. 2005. La investigación etnobotánica sobre plantas medicinales: una revisión de sus objetivos yenfoques actuales. *Interciencia* 30(8): 453–459.

Brown, S.H. 2011. *S. Alata*. The Institute of Food and Agricultural Sciences, University of Florida, Gainesville, FL.

Bum, E.N., Nkantchoua, G.N., Njikam, N., Taiwe, G.S., Hgoupaye, G.T., Pelanken, M.M., Maidawa, F., Rakotonirina, A. & Rakotonirina, S.V. 2010. Anticonvulsant and sedative activity of leaves of *Senna spectabilis* in mice. *International Journal of Pharmacology* 6(2): 123–128.

Campuzano-Zuluaga, G., Alvarez-Sánchez, G., Escobar-Gallo, G.E., Valencia-Zuluaga, L.M., Ríos-Orrego, A.M., Pabón-Vidal, A., Miranda-Arboleda, A.F., Blair-Trujillo, S. & Campuzano-Maya, G. 2010. Design of malaria diagnostic criteria for the Sysmex XE-2100 hematology analyzer. *American Journal of Tropical Medicine and Hygiene* 82(3): 402–411.

Chhabra, S., Mahunnah, R.L.A. & Mushiu, E.N. 1987. Plants used in traditional medicine in eastern Tanzania. I. Pteridophytes and angiosperms (Acanthaceae to Canellaceae). *Journal of Ethnopharmacology* 21: 253–277.

Danish, M., Singh, P., Mishra, G., Srivastava, S., Jha, K.K. & Khosa, R.L. 2011. *Cassia fistula* Linn. (Amulthus) – An important medicinal plant: A review of its traditional uses, phytochemistry and pharmacological properties. *Journal of Natural Product and Plant Resources* 1(1): 101–118.

Day, R.A. & Underwood, A.L. 1986. *Quantitative Analysis*. Prentice-Hall Publication, London.

Deshpande, S.R. & Naik, B.S. 2016. Evaluation of in vitro antimicrobial activity of extracts from *Cassia obtusifolia* L. and *Senna sophera* (L.) Roxb against pathogenic organisms. *Journal of Applied Pharmaceutical Science* 6: 83–85.

Doughari, J.H., El-mahmood, A.M. & Tyoyina, I. 2008. Antimicrobial activity of leaf extracts of *Senna obtusifolia* (L). *African Journal of Pharmacy and Pharmacology* 2: 7–13.

Duraipandiyan, V., Ayyanar, M. & Ignacimuthu, M. 2006. Antimicrobial activity of some ethnomedicinal plants used by Paliyar tribe from Tamil Nadu, India. *Complementary and Alternative Medicine* 6: 35–39.

Ekor, M. 2013. The growing use of herbal medicines: Issues relating to adverse reactions and challenges in monitoring safety. *Frontier of Pharmacology* 4(1–10): 177.

El-Tahir, A., Satti, G.M. & Khalid, S.A. 1999. Antiplasmodial activity of selected Sudanese medicinal plants with emphasis on *Acacia nilotica*. *Phytotherpy Research* 13(6): 474–478.

Farnsworth, N.R. & Bunyapraphatsara, N. 1992. *Thai Medicinal Plants*. Recommended for Primary Health Care System, Medicinal Plant Information Center, Faculty of Pharmacy, Mahidol University, Salaya, Thailand.

Fowler, D.G. 2006. *Traditional Fever Remedies: A List of Zambian Plants*, https://press.uchicago.edu/ucp/books/book/distributed/Z/bo12390424.html

Gallo, B.C. & Sarachine, J. 2009. Biological activities of lupeol. *International Journal of Biomedical and Pharmaceutical Sciences* 3: 46–66.

Gupta, R.K. 2010. *Medicinal & Aromatic Plants* (1st ed.). CBS Publishers & Distributors, Delhi pp. 116–117.

Hennebelle, T., Weniger, B., Joseph, H., Sahpaz, S. & Bailleul, F. 2009. *Senna alata*. *Fitoterapia* 80: 385–393.

Hiben, M.G., Sibhat, G.G., Fanta, B.S., Gebrezgi, H.G. & Tesema, S.B. 2016. Evaluation of *Senna singueana* leaf extract as an alternative or adjuvant therapy for malaria. *Journal of Traditional and Complementary Medicine* 6: 112–117.

Hien, T.T., & White, N.J. 1993. Qinghaosu. *The Lancet* 341(8845): 603–608.

Husen, A. 2021. *Traditional Herbal Therapy for the Human Immune System*. CRC Press, Boca Raton, FL. https://doi.org/10.1201/9781003137955

Husen, A. 2022a. *Herbs, Shrubs and Trees of Potential Medicinal Benefits*. CRC Press, Boca Raton, FL. https://doi.org/10.1201/9781003205067

Husen, A. 2022b. *Exploring Poisonous Plants: Medicinal Values, Toxicity Response and Other Potential Uses*. CRC Press, Boca Raton, FL. https://doi.org/10.1201/b23017

Husen, A. 2023. *Antidiabetic Medicinal Plants and Herbal Treatments*. CRC Press, Boca Raton, FL. https://doi.org/10.1201/b23347

Hussian, H.S. 1991. Plant in Kano. Ethnomedicine: Screening for antimicrobial activity and alkaloids. *Journal of Pharmacology* 29: 51–56.

Ibrahim, M.A., & Islam, M.S. 2014. Anti-diabetic effects of the acetone fraction of *Senna singueana* stem bark in a type 2 diabetes rat model. *Journal of Ethnopharmacology* 153(2): 392–399. doi: 10.1016/j.jep.2014.02.042

Igoli, J., Ogaji, O., Igoli, N. & Tor-Anyiin, T.A. 2005. Traditional medicinal practices among the Igede people of Nigeria (part II). *African Journal of Traditional Complementary and Alternative Medicine* 2: 134–152.

Jafri, M.A., Subhami, M.J., Javed, K. & Singh, S. 1999. Hepatoprotective activity of leaves of *Cassia occidentalis* against paracetamol and ethyl alcohol intoxication in rats. *The Journal of Ethnopharmacology* 66: 355–361.

Jain, S.C., Jain, R., Sharma, R.A. & Capasso, F. 1997. Pharmacological investigation of *Cassia italica*. *The Journal of Ethnopharmacology* 58: 135–142.

Jang, Y.S., Baek, B.R., Yang, Y.C., Kim, M.K. & Lee, H.S. 2002. Larvicidal activity of leguminous seed and grains against *Aedes agypti* and *Culex pipiens* pallens. *Journal of the American Mosquito Control Association* 18(3): 210–213.

Jeeva, G.M., Jeeva, S. & Kingston, C. 2007. Traditional treatment of skin diseases in South Travancore, Southern Peninsular India. *Indian Journal of Traditional Knowledge* 6: 498–501.

Jun, D., Tomoe, H., Yusuke, H., Wiwied, E., Aty, W., Osamu, S., Motoo, S. & Hiroshi, M. 2012. New tricyclic alkaloids, cassiarins G, H, J, and K from leaves of *Cassia siamea*. *Chemical and Pharmaceutical Bulletin* 60: 219–222.

Jung, H.A., Ali, M.Y. & Choi, J.S. 2017. Promising inhibitory effects of anthraquinones, naphthopyrone, and naphthalene glycosides from *Cassia obtusifolia* on α-glucosidase and human protein tyrosine phosphatases 1B. *Molecules* 22(1): 28.

Jung, H.A., Ali, M.Y., Jung, H.J., Jeong, H.O., Chung, H.Y. & Choi, J.S. 2016. Inhibitory activities of major anthraquinones and other constituents from *Cassia obtusifolia* against β-secretase and cholinesterases. *Journal of Ethnopharmacology* 191: 152–160.

Kamagate, M., Koffi, C., Kouame, N.M., Akoubet, A., Yao, N.A.R. & Die-Kakou, H.M. 2014. Ethnobotany, phytochemistry, pharmacology and toxicology profiles of *Cassia siamea* Lam. *Journal of Phytopharmacology* 3: 57–76.

Kamatenesi, M.M., Acipa, A. & Oryem-Origa, H. 2011. Medicinal plants of Otwal and Ngai Sub counties in Oyam District, Northern Uganda. *Journal of Ethnobiology and Ethnomedicine* 7: 1–14.

Kaur, G., Sarwar Alam, M., Jabbar, Zoobi, Javed, Kaleem & Athar, Mohammad. 2006. Evaluation of antioxidant activity of *Cassia siamea* flowers. *Journal of Ethnopharmacology* 108(3): 340–348. https://doi.org/10.1016/j.jep.2006.05.021.

Keneda, M., Morishita, E. & Shibata, S. 1969. Chemical studies on the oriental plant drugs, XXI. The constituents of *Cassia tora* L., A glycoside of rubrofusarin. *Chemical Pharmaceutical Bulletin* 17(3): 458–461.

Kiepe, P. 1995. Effect of *Cassia siamea* hedgerow barriers on soil physical properties. *Geoderma* 66: 113–120.

Kim, D.H., Yoon, B.H., Kim, Y.W., Lee, S., Shin, B.Y., Jung, J.W., Kim, H.J., Lee, Y.S., Choi, J.S., Kim, S.Y., Lee, K.T. & Ryu, J.H. 2007. The seed extract of *Cassia obtusifolia* ameliorates learning and memory impairments induced by scopolamine or transient cerebral hypoperfusion in mice. *Journal of Pharmacological Sciences* 105(1): 82–93.

Kirtikar, K.R. & Basu, B.D. 2006. *Indian Medicinal Plants*. Lalit Mohan Basu Press, Allahabad, India, Vol. 2, pp. 856–860.

Kitanaka, S., Nakayama, T., Shibano, T., Ohkoshi, E. & Takido, M. 1998. Antiallergic agent from natural sources: Structures and inhibitory effect of histamine release of naphthopyrone glycosides from seeds of *Cassia obtusifolia* L. *Chemical and Pharmaceutical Bulletin* 46: 1650–1652.

Kitua, A.Y. & Malebo, H.M. 2004. Malaria control in Africa and the role of traditional medicine. In: *Traditional Medicinal Plants and Malaria*, CRC Press, Boca Raton. pp. 3–18.

Kone, W.M. & Atindehou, K.K. 2008. Ethnobotanical inventory of medicinal plants used in traditional veterinary medicine in Northern Côte d'Ivoire (West Africa). *South African Journal of Botany* 74: 76–84.

Koudouvo, K., Karou, D.S., Kokou, K., Essien, K., Aklikokou, K., Glitho, I.A., Simpore, J., Sanogo, R., Souza, C. & Gbeassor, M. 2011. An ethnobotanical study of antimalarial plants in Togo Maritime Region. *Journal of Ethnopharmacology* 134: 183–190.

Kshirsagar, R.D. & Singh, N.P. 2000. Less known ethnomedicinal use of plants reported by Jeuknuba tribe of Mysore district, South India. *Ethnobotany* 12: 118–122.

Kubmarawa, D., Magomya, A.M., Yebpella, G.G. & Adedayo, S.A. 2011. Nutrient content and amino acid composition of the leaves of *Cassia tora* and *Celtis integrifolia. International Research Journal of Biochemistry and Bioinformatics* 1(9): 222–225.

Latha, M. & Pari, L. 2003. Preventive effects of *Cassia auriculata* L. lowers on brain lipid peroxidation in rats treated with streptozotocin. *Molecular and Cellular Biochemistry* 243: 23–28.

Lawal, I.O., Uzokwe, N.E., Igboanugo, A.B., Adio, A.F., Awosan, E.A., Nwogwugwu, J.O., Faloye, B., Olatunji, B.P. & Adesoga, A.A. 2010. Ethnomedicinal information on collection and identification of some medicinal plants in Research Institutes of South-west Nigeria. *African Journal of Pharmacy and Pharmacology* 4: 1–7.

Lemli, J. & Cuveeli, J. 1972. Study of anthraquinone producing plants: XXII, Detection of anthrones in purgative plants. *Plant Medicinales et Phytotherapie* 6: 271–274.

Lemli, J. & Cuveeli, J. 1974. Investigations of the anthraquinone principles drugs: XXIII. Chromatography of anthrone glycosides in purgative drugs. *Planta Medica* 26: 193–195.

Leung, L., Riutta, T., Kotecha, J. & Rosser, W. 2011. Chronic constipation: An evidence-based review. *Journal of the American Board of Family Medicine* 24(4): 436–451.

Makinde, A., Igoli, J., Amal, L., Shaibu, S. & Garbal, A. 2007. Antimicrobial activity of *Cassia alata. African Journal of Biotechnology* 6: 1509–1510.

Mbatchi, S., Mbatchi, B., Banzouzi, J., Bansimba, T., Nsonde, N.G., Ouamba, J., Berry, A. & Benoit-Vical, F. 2006. In vitro antiplasmodial activity of 18 plants used in Congo Brazzaville traditional medicine. *Journal of Ethnopharmacology* 104: 168–174.

Meena, A.K., Niranjan, U.S., Yadav, A.K. & Singh, B. 2010. A review on *Cassia tora* ethnobotany, phytochemical and pharmacological profile. *Journal of Pharmacy Research* 3(3): 557–560.

Meena, A.K. & Rao, M.M. 2010. Folk herbal medicines used by the Meena community in Rajasthan. *Asian Journal of Traditional Medicines* 5: 19–31.

Meena, V., Baruah, H. & Parveen, R. 2019. *Cassia auriculata.* A heeling herb for all remedy. *Journal of Pharmacognosy & Phytochemistry* 8(3): 4093–4097.

Mutasa, S.L., Khan, M.R. & Jewers, K. 1990. 7-Methyl physcion and cassiamin A from the root bark of *Cassia singueana. Planta Medica* 56: 244–245.

Nadembega, P., Boussim, J.I., Nikiema, J.B., Poli, F. & Antognoni, F. 2011. Medicinal plants in Baskoure, Kourittenga Province, Burkina Faso: An ethnobotanical study. *Journal of Ethnopharmacology* 133(2): 378–395. doi: 10.1016/j.jep.2010.10.010.

Nadkarni, K.M. 1976. *Indian Materia Medica.* Popular Prakashan, Bombay, Vol. 1, pp. 285–286.

Nagnur, S., Channal, G. & Channamma, N. 2009. Indigenous home remedies for common ailments. *Indian Journal of Traditional Knowledge* 8: 577–580.

Natarajan, B. & Paulsen, B.S. 1999. An ethnopharmacological study from the Coimbatore district, Tamil Nadu: Traditional knowledge compared with modern biological science. *Pharmaceutical Biology* 37: 378–380.

Ndamba, J., Nyazema, N., Nakaza, N. & Anderson, C. 1984. Traditional herbal remedies used for the treatment of schistosomiasis in Zimbabwe. *Journal of Ethnopharmacology* 42: 125–132.

Neamsuvan, O. 2015. A survey of herbal formulas for skin diseases from Thailand's three southern border provinces. *Journal of Herbal Medicine* 5(4): 190–198.

Odason, E.E. & Kolawole, J. 2007. Anti-diabetic properties and brine shrimp toxicity of the aqueous extract of the root of *Cassia siamea* Lam. (Ceasalpiniaceae). *Nigerian Journal of Pharmaceutical Research* 6: 66–69.

Odugbemi, T.O., Akinsulire, O.R., Aibinu, I.E. & Fabeku, P.O. 2007. Medicinal plants useful for malaria therapy in Okeigbo, Ondo state, southwest Nigeria. *African Journal of Traditional Complementary and Alternative Medicines* 4: 191–198.

Oladeji, O.S., Adelowo, F.E. & Odelade, K. 2016. Mass spectroscopic and phytochemical screening of phenolic compounds in the leaf extract of *Senna alata* (L.) Roxb. (Fabales: Fabaceae). *Brazilian Journal of Biological Sciences* 3(5): 209–219.

Oladeji, S.O., Adelowo, F.E., Oluyori, A.P. & Bankole, D.T. 2020. Ethnobotanical description and biological activities of *Senna alata. Evidence-Based Complementary and Alternative Medicine* 13: 1–12.

Oladeji, O.S., Odelade, K.A. & Oloke, J.K. 2019. Phytochemical screening and antimicrobial investigation of *Moringa oleifera* leaf extracts. *African Journal of Science, Technology, Innovation and Development*, 12(1): 79–84.

Owuor, B.O. & Kisangau, D.P. 2006. Kenyan medicinal plants used as antivenin: A comparison of plant usage. *Journal of Ethnobiology Ethnomedicine* 2: 7.

Rahman, S. & Husen, A. 2021. Potential role of medicinal plants in the cure of liver and kidney diseases. In: *Non-Timber Forest Products* (Eds. Husen, A., Bachheti, R.K. & Bachheti, A.). Springer International Publishing AG, Cham, pp. 229–254. https://doi.org/10.1007/978-3-030-73077-2_10

Rahman, S. & Husen, A. 2022. *Asparagus racemosus* (Shatavari) and *Dioscorea villosa* (Wild yams). In: *Herbs, Shrubs, and Trees of Potential Medicinal Benefits* (Ed. Husen, A.). CRC Press, Boca Raton, FL.

Rahman, S. & Husen, A. 2023a. *Trigonella foenum-graecum* L.: A useful medicinal plant for diabetes and other diseases. In: *Antidiabetic Medicinal Plants and Herbal Treatments* (Ed. Husen, A.). CRC Press, Boca Raton, FL.

Rahman, S. & Husen, A. 2023b. Role of Green Chiretta [*Andrographis paniculata* (Burm. F.)] Nees in diabetes and other ailments. In: *Antidiabetic Medicinal Plants and Herbal Treatments* (Ed. Husen, A.). CRC Press, Boca Raton, FL.

Rajendran, K., Balaji, P. & Basu, M.J. 2008. Medicinal plants and their utilization by villagers in Southern districts of Tamil Nadu. *Indian Journal of Traditional Knowledge* 7: 417–420.

Ramírez, O., Blair, S., Vaupés Medio, T.V., Cardona, F. & Pabón, A. 2017. Ethnobotany of medicinal plants used to treat malaria by traditional healers from ten indigenous Colombian communities located in Waupes Medio. *Biodiversity International Journal* 1(4): 151–167.

Ramstad, E. 1959. *Modern Pharmacognosy.* McGraw-Hill, New York.

Ratnam, K.V. & Raju, R.R.V. 2005. Folk medicine used for common women ailments by adivasis in the Eastern Ghats of Andhra Pradesh. *Indian Journal of Traditional Knowledge* 4: 267–270.

Ratnam, K.V. & Raju, R.R.V. 2008. Folk remedies for insect bites from Gundla brahmeswaram Wild Life Sanctuary, Andhra Pradesh. *Indian Journal of Traditional Knowledge* 7: 436–437.

Rieckmann, K.H. & Antuñano, F.J.L. 1971. Chloroquine resistance of *Plasmodium falciparum* in Brazil detected by a simple in vitro method. *Bulletin of the World Health Organization* 45(2): 157–167.

Ruth, W.M., John, M.M., Isabel, N.W. & Raposa, L.B. 2021. The medicinal properties of *Cassia fistula* L: A review. *Biomedicine & Pharmacotherapy* 144: 112240.

Saidu, K., Onah, J., Orisadipe, A., Olusola, A., Wambebe, C. & Gamaniel, K. 2000. Antiplasmodial, analgesic and anti-inflammatory activities of the aqueous extract of *Erythrina senegalensis*. *Journal of Ethnopharmacology* 71: 275–280.

Sanon, S., Ollivier, E., Azas, N.C., Mahiou, B.V., Gasquet, M., Ouattara, C.T., Nebie, D.I., Traoré, A.S., Esposito, F., Balansard, G., Timon-David, C.P. & Fumoux, F. 2003. Ethnobotanical survey and in vitro antiplasmodial activity of plants used in traditional medicine in Burkina Faso. *Journal of Ethnopharmacology* 86: 143–147.

Sansores-Perazaa, P., Rosado-Valladoa, M., Brito-Loezaa, W., Mena-Rejon, G.J. & Quijano, L. 2000. Cassine, an antimicrobial alkaloid from *Senna racemosa*. *Fitoterapia* 71: 690–692.

Schultes, R.E. & Raffauf, R.F. 1990. *The Healing Forest: Medicinal and Toxic Plants of the Northwest Amazonia*. UNESCO, París, Francia, vol. 12, pp. 236–237.

Selvi, V., Isaivani, V. & Karpagam, S. 2012. Studies on antimicrobial activities from flower extract of *Cassia alata* Linn. *International Journal of Current Science* 2: 299–303.

Silva, C.R., Monteiro, M.R., Rocha, H.M., Ribeiro, A.F., Caldeira-de-Araujo, A., Leita, A.C., Bezerra, R.J. & Pandula, M. 2008. Assessment of antimutagenic and genotoxic potential of *Senna* (*Cassia angustifolia* Vahl.) aqueous extract using in vitro assays. *Toxicology in Vitro* 22: 212–218.

Singh, S., Singh, S.K. & Yadav, A. 2013. A review on *Cassia* species: Pharmacological, traditional and medicinal aspects in various countries. *American Journal of Phytomedicine and Clinical Therapeutics* 1: 291–312.

Siva, R. & Krishnamurthy, K.V. 2005. Isozyme diversity in *Cassia auriculata* L. *African Journal of Biotechnology* 4: 772–775.

Sob, S.V.T., Wabo, H.K., Tane, P., Ngadjui, B.T. & Ma, D. 2008. A xanthone and a polyketide derivative from the leaves of *Cassia obtusifolia* (Leguminosae). *Tetrahedron* 64: 7999–8002.

Sung, B.K., Kim, M.K., Lee, W.H., Lee, D.H. & Lee, H.S. 2004. Growth responses of *Cassia obtusifolia* towards human intestinal bacteria. *Fitoterapia* 75: 505–509.

Ta, T.H., Hisam, S., Lanza, M., Jiram, A.I., Ismail, N. & Rubio, J.M. 2014. First case of a naturally acquired human infection with *Plasmodium cynomolgi*. *Malaria Journal* 13: 68.

Tang, W. & Eisenbrand, G. 1992. *Chinese Drugs of Plant Origin, Chemistry, Pharmacology, and Use in Traditional and Modern Medicine*. Springer, Verlag.

Thabrew, M.I., Munasinghe, T.M., Senarathm, S. & Yapa, R.M. 2004. Effects of *Cassiaauriculata* and *Cardiospermum halicacabum* teas on the steady, state blood levels of theophylline in rats. *Drug Metabolism and Drug Interactions* 20: 263–272.

Thongsaard, W., Deachapunya, C., Pongsakorn, S., Boyd, E.A., Bennett, G.W. & Marsden, C.A. 1996. Barakol: A potential anxiolytic extracted from *Cassia siamea*. *Pharmacology Biochemistry and Behavior* 53(3): 753–758.

Tripathi, Y.C. 1999. *Cassia angustifolia*, A versatile medicinal crop. *International Tree Crops Journal* 10(2): 121–129.

Tyler, V.E. 1982. *The Honest Herbal: A Sensible Guide to the Use of Herbs and RelatedRemedies*. George F. Stickley, Philadelphia.

Vadivel, V., Kunyanga, C.N. & Biesalski, H.K. 2012. Antioxidant potential and type II diabetes-related enzyme inhibition of *Cassia obtusifolia* L.: Effect of indigenous processing methods. *Food and Bioprocess Technology* 5: 2687–2696.

Veeresham, C. 2012. Natural products derived from plants as a source of drugs. *Journal of Advanced Pharmaceutical Technological Research* 3(4): 200–201.

Verma, N.K., Singh, A.K. & Chaurasiya, A.K. 2021. *Cassia tora* Linn. Importance and properties: A review. *International Journal of Pharmacological Research and Applications* 6(4): 631–634.

Verpoorte, R. 2015. Forward. In: *Evidence-Based Validation of Herbal Medicine* (Ed. Mukherjee, P.K.). Elsevier, Amsterdam, Netherlands, pp. 13–14.

Viegas, C., Bolzani, V.S., Furlan, M., Barreiro, E., Young, M.C., Tomazela, D. & Eberlin, M.N. 2004. Further bioactive piperidine alkaloids from the flowers and green fruits of *Cassia spectabilis*. *Journal of Natural Products* 67(5): 908–910.

WHO. 2015. *World Malaria Report*. World Health Organization, Geneva, Switzerland.

WHO. 2016. *Global Report on Diabetes*. World Health Organization, Geneva, Switzerland.

Wiwied, E., Aty, W. & Cholies, Z.N. 2009. Antimalarial activity of cassiarin A from the leaves of *Cassia siamea*. *Heterocycles A* 78: 1831–1836.

Woradulayapinij, W., Soonthornchareonnon, N. & Wiwat, C. 2005. In vitro HIV type 1 reverse transcriptase inhibitory activities of Thai medicinal plants and *Canna indica* L. rhizomes. *Journal of Ethnopharmacology* 101(1–3): 84–89.

Xu HY, Zhang YQ, Liu ZM, ... Huang LQ. 2019. ETCM: an encyclopaedia of traditional Chinese medicine. *Nucleic Acids Research.* 47(D1):D976-D982. DOI: 10.1093/nar/gky987.

Yadav, J.P., Arya, V., Yadav, S., Panghal, M., Kumar, S. & Dhankhar, S. 2010. *Cassiaoccidentalis* L.: A review on its ethnobotany, phytochemical and pharmacological profile. *Fitoterapia* 81(4): 223–230.

Yun-Choi, H.S., Kim, J.H. & Takido, M. 1990. Potential inhibitors of platelet aggregation from plant sources, V. Anthraquinones from seeds of *Cassia obtusifolia* and related compounds. *Journal of Natural Products* 53: 630–633.

Zhu, L., Zeng, X. & Zhao, M. 2008. Preparative separation and purification of five anthraquinones from *Cassia tora* L. by high speed counter-current chromatography. *Separation and Purification Technology* 63(3): 665–669.

17 Antimalarial Response, Traditional and Other Potential Uses of *Solanum* Genus

Shakeelur Rahman and Azamal Husen

17.1 INTRODUCTION

Solanum is one of the largest and most taxonomically challenging genera of plants. It is a diverse genus of flowering plants which include three food crops of high economic importance: the potato, the tomato, and the eggplant. It is the largest genus in the nightshade family Solanaceae, comprising around 1,500–2,000 species. *Solanum* species show a wide range of growth habits, such as annuals and perennials, vines, sub-shrubs, shrubs, and small trees (Knapp et al. 2004). *Solanum nigrum*, which is commonly known as nightshade, is one of the important herbs of the family Solanaceae. There are two varieties of *S. nigrum*, that is, black-coloured fruit and reddish-brown-coloured fruit. Both varieties have been used traditionally to treat various ailments, such as pain, inflammation, and fever and malaria (Acharya and Pokhrel 1970). The antioxidant and phytochemicals present in *S. nigrum*, such as alkaloids, flavonoids, glycosides, and steroidal lactones, play an effective role in the treatment of fever, cancer, microbial diseases, dermatitis, hydrophobia, and optical problems. The herb is also used as oral contraceptives, sexual hormones, and steroidal medicines (Kiran et al. 2009; Rajathi et al. 2015). *S. nigrum* is rich in calcium, phosphorus, iron, vitamin A and C, protein, fats, and fibres (Ikeda et al. 2000). Therefore, *S. nigrum* possesses several pharmacological activities, like antitumorigenic and antioxidant (Lee and Lim 2003), diuretic, antipyretic, anti-inflammatory (Zakaria et al. 2006), hepatoprotective (Raju et al. 2003), antibacterial (Rani and Khullar 2004), and antifungal (Eltayeb et al. 1997). Several studies have been conducted to investigate the nutritive value of the 'vegetable black nightshades'; these are summarized in the table. From this it is evident that these species constitute nutritious vegetables. The leaves can provide appreciable amounts of protein and amino acids; minerals, including calcium, iron, and phosphorus; vitamins A and C; fat and fibre; as well as appreciable amounts of methionine, an amino acid scarce in other vegetables. Moreover, the berries can apparently yield high amounts of iron, calcium, and vitamin B and appreciable amounts of vitamin C and carotene (Fortuin and Omta 1980). The seeds, too, contain vitamin C and carotene (Watt and Breyer 1962). The nutrient values may, however, vary with soil fertility, plantage, and type (i.e. variant or species). It was found that the leaf protein content of *S. nigrum* was dependent on the age of the plant. Moreover, the application of nitrogen increases the amount of ascorbic acid and protein while decreasing the calcium content in the leaves (Chweya 1997). Mathooko and Imungi observed that ascorbic acid content decreased with both an increase in the cooking time and in the volume of water used for cooking. The species accumulate nitrates and contain oxalates and phenolics (Mathooko and Imungi 1994).

17.2 BOTANICAL DESCRIPTION

Solanum spp. that comprise the nightshades are erect, multi-branching, herbaceous perennial shrubs able to grow up to 13 ft in height. Stems are covered with short, dense hairs that are often

DOI: 10.1201/9781003378396-17

armed with thorns. Leaves are alternate and vary in morphology, often ovate to elliptic, with an acute tip and rounded to oblique base. Leaf margins are shallowly and irregularly lobed. Flowers are borne in corymbs at intervals on the stems. Yellow globose fruit contain many seeds. Frugivorous birds eat the fruit and spread the seeds. *Solanum* spp. are prolific seed producers which may produce seed throughout the year. New plants can emerge from seed or roots. Root buds on existing plants are able to generate shoots, producing new plants. Root systems can be fairly extensive, reaching as far as 3 to 6 ft into the ground. In case of *S. nigrum* (Figure 17.1), which is most effective in malarial disease, it was observed as sub-glabrous to villous annuals up to 70–75 cm high, covered with simple multicellular hairs with glandular or eglandular heads. It shows taproots with few branches and numerous small lateral roots, pale brown, easily peeled off, exposing pale-yellow wood. It has erected glabrous or pubescent slightly woody green unbranched stem. Leaves are ovate, ovate-lanceolate, ovate-rhombic to lanceolate, 2.5–7.0cm long, 2.0 to 4.5(6.0) cm broad, margins entire to sinuatedentate. Inflorescences is simple, lax, and often extended cymes, (3) 5- to 10-flowered; peduncles (8) 14 to 28 mm fruiting when usually erecto-patent; pedicels much shorter, recurved in fruit. Calyces 1.2–2.5 mm long, slightly accrescent, deflexed, or adhering to base of mature berry, sepals usually ovate. Corollas stellate, white with translucent basal star (4) 5 to 7(9) mm radius, usually 1.5–3 times as long as calyx. Anthers are yellow, 1.5 to 2.5(2.8) mm long. Pollen is (26.6) 29.5 to 33.9(35.7) μm diameter. Styles are 2.8 to 3.5(4.5) mm long, not exerted beyond anthers. Berries are usually broadly ovoid, dull purple to blackish or yellowishgreen, 6–10 mm broad, remaining on plants or falling from calyces when ripe. Seeds are 1.7–2.4 mm long, (15) 26 to 60(96) per berry. Sclerotic granules are absent. Petiole and midrib of leaf show covering, uniseriate trichomes that are three- to five-celled, having pointed tips; they show arc-shaped bicollateral vascular bundle arrangement. Lamina of leaf shows anisocytic stomata scattered on both upper and lower surfaces but more abundant in lower surface. Stomatal index 15–17(on upper epidermis) and 22–33(on lower epidermis). Fruit shows thin, papery epicarp, pulpy mesocarp, and exile placentation; seeds at first remain attached to the placenta but afterwards separate from it and lie free in pulp of fruit (Annonymus 1999a; Edmonds and Chweya 1997).

17.3 DISTRIBUTION AND CULTIVATION

Solanum is one of the largest genera of flowering plants with about 1,500 species, which are distributed in all continents, except Antarctica (Vorontsova et al. 2013). Major centres of species diversity are the Andean and Atlantic forest regions of South America, but a secondary centre of diversity occurs in East Africa. Several major crops of global agricultural importance and of economic and medicinal value, such as *S. tuberosum*, *S. lycopersicum*, and *S. melongena* (Figure 17.1a, b, and c), belong to the family Solanaceae. In India, the genus *Solanum* is the most specious, with 49 species distributed throughout the country. *Solanum* species grow in various habitats and can be annuals and perennials, vines, sub-shrubs, shrubs, and small trees. Most of *Solanum* species are found in wastelands, roadsides, often on arable land, especially the edges of cultivated fields and plantations, in hedgerows, on railway cuttings, quaysides, and rubbish tips, in areas around buildings and houses, under trees, on forest and grassland margins, as garden weeds, on shingle beaches, riverbanks, and gullies (Karschon, and Horowitz 1985). Biophysical limit altitude is 3048m at sea level; optimal temperatures for the growth of these plants lie between 20 and 30°C (Rogers and Ogg 1981). Photoperiod is approximately 16 hr (daylight is usually supplemented by artificial light) (Henderson 1974). A seed moisture content of 35–40% is a necessary contributory factor for optimal germination (Wakhloo 1964). Annual rainfall of 500–1200 mm and/or frequent watering is required during dry periods in temperate areas (Burgert et al. 1973). The plants grow well in light (sandy), medium (loamy), and heavy (clay) fertile soil, especially those high in nitrogen or phosphorus and requires well-drained acid-neutral or alkaline soil, pH range of 6.0 to 6.5 (Fawusi 1983).

FIGURE 17.1 (a) *Solanum tuberosum*, (b) *Solanum lycopersicum*, (c) *Solanum melongena*, and (d) *Solanum nigrum*.

17.4 PHYTOCHEMICAL CONSTITUENTS

Phytochemical study on *Solanum* species led to the identification of steroidal saponins, steroidal alkaloids, terpenes, flavonoids, lignans, sterols, phenolic compounds, coumarins, amongst other compounds. More than 350 *Solanum* species have yielded more than 100 different forms of glycoalkaloids. In case of different species of *Solanum*, the green unripe fruits contain glycoalkaloids, and their eating is toxic to human beings as well as livestock, as they include solamargine, solasonine, solanine, α and β-solamagrine, solasodine, solanidine (0.09–0.65%). The former two are also found in leaves. Solanine is found in all parts of *S. nigrum* (Figure 17.1d), with the level increasing as the plant matures, though it is apparently modified by soil type and climate. The total alkaloid content of the fruits and leaves are 0.101 and 0.431, respectively (Cooper and Johnson 1984; Weller and Phipps 1978). Ripe fruit contains very little alkaloids and can be consumed without ill-effects. The little work done on glycoalkaloids, which are said to be responsible for anticancer activity, indicated that solasonine and solamargine, from leaves and unripe fruits, are the two most important. All authors reported that these compounds are present in the greatest concentrations in green (i.e. unripe) berries, and that the actual concentration can be very variable and can vary according to the stage of plant development, as well as being affected by genetic, seasonal, edaphic, and environmental factors. Solasonine concentrations are reported highest during fruiting, while solamargine is highest during flowering (Aslanov and Novruzov 1978). Also, the phytochemical studies revealed the presence of steroidal oligoglycosides called nigrumnin-I and nigrumnin-II (Ikeda et al. 1975). Earlier, it was reported on the identification of three known steroidal glycosides, namely, β-2-solamargine, solamagrine, α, β-solansodamine, and degalactotigonin. Spirostanol and furostanol glycosides, viz, uttroside A, uttroside B, uttrin A; flavonoids, viz, quercetine [3-O-β-glucosyl (1→6)]-O-[β-rhamnosyl(1→2)-O-β-galactoside], and 3-O-α-rhamnosyl (1→2)-O-β-galactoside; and other glycosides of quercetin are also present in *S. nigrum* (Annonymus 1999b). Two new pregnane saponins, solanigroside-A and solanigroside-B, were isolated from 60% ethanolic extract of the dried herb of *S. nigrum* (He et al. 2007). Chemical constituents of the non-saponins from *S. nigrum* were studied on 60% ethanol extract; separation was performed by HPLC and was identified by spectral methods as (I) 6-methoxyhydroxycoumarine, (II) syringaresinol-4-O-β-Dglucopyranoside, (III) pinoresinol-4-O-β-D-glucopyranoside, (IV) 3,4-dihydroxybenzoic acid, (V) adenosine, (VI) phydroxybenzoic acid, (VII) 3-methoxy-4-hydroxybenzoic acid. II, III, and V were isolated from this genus for the first time (Wang et al. 2007).

17.5 PHARMACOLOGICAL STUDIES

Many species belonging to this *Solanum* genus present a huge range of pharmacological activities, such as anticancer, hepatoprotective, antimalarial, anthelmintic, and other activities (Elizalde et al. 2021). *Solanum* species have been widely used in healthcare systems as a source for various phytochemicals, including steroidal alkaloids. *Solanum* is distinguished by the presence of the steroidal alkaloid solasodine, which is a potential starting material for the manufacture of steroid hormones. Because of the wide spectrum of biological activities such as antibacterial, anti-inflammatory, antioxidant, and anticancer, *Solanum* alkaloids have been a topic of interest in pharmacological and therapeutical investigations. Because of metabolites such glycoalkaloids, some of the *Solanums* species are poisonous. Several pharmacologically important lead compounds are found in *Solanum* species, including steroidal alkaloids, such as solasodine, solasonine, solamargine, and other medicinally important alkaloids; solasodine and its glycosylated derivatives, such as solamargine and solanine; and other chemicals with medicinal potential for developing new drugs against various human diseases. According to the recent literature, silver nanoparticles (AgNPs) synthesized from aqueous extracts of *Solanum* species have demonstrated adulticidal and insecticidal activity against *Ae. aegypti*, implying that AgNPs produced from plant extracts have higher levels of toxicity than the extracts alone (Rawani et al. 2013). Thus, these herbal-based AgNPs hold great promise as potent larvicides, but their environmental impact requires further investigation for controlling target vector mosquitoes.

Hepatoprotective activity. The ethanol extract of fruits of *Solanum nigrum* L at a dose of 250mg/kg orally was investigated in male albino rats for its hepatoprotective activity. It has provided significant protection against most of the biochemical alterations produced by carbon tetrachloride (CCl4). The activities of serum aspartate aminotransferase (AST), alanine aminotransferase (ALT), alkaline phosphatase, and total bilirubin were significantly increased with administration of CCL4. The degree of protection afforded by ethanol extract, when administered, significantly lowered the elevated serum enzyme and total bilirubin levels. This suggests the maintenance of structural integrity of hepatocytic cell membrane or regeneration of damaged liver cells by the extract, thus exhibiting potent hepatoprotective activity. The ethanol extract was also evaluated as a hepatoprotective agent by observing the liver for its histopathological changes. Histopathological examination of liver section of rat treated with toxicant (CCl4) showed intense centrilobular necrosis and fatty changes. The rat treated with extract along with toxicant (CCl4) to the considerable extent as evident from the formation of normal hepatic cards and absence of necrosis and vacuoles (Raju et al. 2003).

Antiulcerogenic and ulcer healing activity. Gastric ulcers have multiple etiopathogenesis. Ulcers caused by pyloric ligation (PL) are due to the increased presence of acid and pepsin in the stomach, and damages by indomethacin are due to decreased prostaglandin (PG) synthesis, which is essential for mucosa. Stresses are due to both physiological and psychological factors, which are crucial for gastrointestinal defence and increased accumulation of acid pepsin, leading to autodigestion of gastric mucosa. *Solanum nigrum* extract (SNE) showed significant dose-dependent ulcer-protective effect against cold restraint–induced ulcer (CRU), indomethacin (IND), and pyloric ligation (PL) induced gastric ulcer is mainly due to increase omeprazole. In CRU, the incidence of ulcer is mainly due to increased acid secretion and the generation of free radical, etc. The protective efficacy against CRU may be due to antioxidant activities of SNE and its control on acid secretion, thereby strengthening the animal's physiological capability to decrease stress ulcer. In this way, SNE was able to produce significant reduction of gastric mucosal damage induced by IND, indicating probable local increase in PG synthesis. Moreover, the activation of vagus-vagal reflux by stimulation of pressure receptors in the antral gastric mucosa in the hypersecretion model of pylorus ligature is believed to increase gastric acid secretion. The data clearly demonstrated that SNE dose-dependently decreased the gastric acid and pepsin secretion, which is concordant with earlier reports of antisecretory action of aerial parts of *Solanum nigrum*. Ethanol-treated rats showed significant increase in plasma concentration of gastric hormone gastrin and increase in gastric mucosal H+ K+ ATPase and suppression of gastrin release (Mallika and Chennam 2006).

Neuropharmacological activity. The ethanol extract of the fruit of *Solanum nigrum* Linn. was studied for its neuropharmacological properties on wistar rats. Different neuropharmacological tests were performed on experimental animals with ethanolic extract, such as pentobarbital-induced sleeping time, motility test, exploratory behaviour pattern (head dip test, Y-maze test, evasive test), test for motor incoordination (rotarod test, chimney test, traction test, inclined test), anticonvulsant activity, etc. The fruit extract showed dose-dependent activity against various aforementioned tests. The extract in particular at the dose of 255mg/kg caused decreased alertness and restlessness. There were neither tremors, twitches, convulsions, nor Straub tail response. No effects were noticed on alarm, reaction, body posture, limb position, gait, righting reflex, muscle tone, pinna, and corneal reflexes. From such observation, it is possible to conclude that the depressant effect of extract on locomotory activity was probably not due to peripheral neuromuscular blockage. The extract was well tolerated by mice, and there were no signs of acute (during the 2 hr observation period) or delayed (three days after extract administration) toxicity. It appears that the ethanolic extract of *Solanum nigrum* Linn., which can cause central nervous system depression, may be correlated with an increased parasympathetic tone (Perez et al. 1998).

Cytoprotective role (Vero cells). The 50% ethanol extract of the whole plant of *S. nigrum* was tested invitro for its cytoprotection against gentamycin-induced toxicity in Vero cells. The cytotoxicity study was assessed by the trypan blue exclusion assay, mitochondrial dehydrogenase activity (MIT) assay, and hydroxyl radical scavenging method. Incubation of Vero cells with the different concentration of dried 50% alcoholic extracts of the whole plant of *S. nigrum* significantly protected the cells from the effect of gentamycin in dose-dependent manner. The observed cytoprotection may be due to increase in activity of free radical scavenging enzyme or counteraction of the free radicals (Kumar et al. 2001).

Ethno-medical activity. The leaves extract of *S. nigrum* was found to be used in oral health care. Several field visits were made in Dharwad district of Karnataka to different parts of study area, and in the survey, 245 herbal healers were contacted during different seasons of the year. Information was collected about *S. nigrum* used in oral health care, the method of medicinal preparation, and its administration. The ethno-medicine survey showed that *S. nigrum* has its application in toothache. Leaves were ground and juice was extracted by filtering through cotton cloth, and it was found to be used as ear drop for relief from toothache (Hebbar et al. 2004).

Antitumour activity. The effect of crude polysaccharide isolated from *S. nigrum* Linn. (SNL-P) was examined on tumour growth. SNL-P had a significant growth inhibition effect on cervical cancer (U14) of tumour-bearing mice. Further, the analysis of tumour inhibition mechanism indicated that the number of apoptotic tumour cells increased significantly, the expression of Bax increased, and the expression of Bcl-2 and mutant P-53 decreased dramatically in cervical cancer section after oral administration of SNL-P for 12 days. Moreover, SNL-P treatment decreased the level of blood serum TNF-alpha. These results indicated that the tumour growth inhibition of SNL-P administration might correlate with the reduction if TNF-alpha level of blood serum, which resulted in massive necrosis in tumour tissue and upregulation of Bax and downregulation of Bcl-2 and mutant-53 gene expression, which triggered apoptosis in tumour cells. These findings demonstrated that SNL-P is a potential antitumour agent (Li et al. 2007). Also, in another experiment, the results showed that the total alkaloid isolated from *S. nigrum* interferes with the structure and function of tumour cell membrane. Disturbing the synthesis of DNA and RNA changed the cell cycle distribution so that total alkaloid could play in inhibition of tumour cells, while the glycoprotein isolated from *S. nigrum* might have shown anticancer abilities by blocking the antiapoptotic pathway of NF-kappaB, activating caspase cascades reaction and increasing the production of nitric oxide (An et al. 2006).

Antioxidant effect. Glycoprotein from *S. nigrum* (SNLGlycoprotein) was isolated and tested for antioxidative effects on oxygen free radicals using a 1,1-diphenyl-2-picrylhydrazyl (DPPH) assay. The free radical scavenging activities of the SNL-glycoprotein are optimal in acidic pH and up to 600 C. However, it has minimal activities in the presence of EDTA, although such activities are not dependent on M2+ ions (Ca2+ and Mg2+) in the presence of EDTA. Interestingly, when SNL-glycoprotein was treated with deactivating agents (Pronase E and NaIO4), the DPPH radical scavenging activity was decreased as compared with SNL-glycoprotein treatment alone. The antioxidative effect of SNL-glycoprotein on superoxide anion and hydroxyl radical under optimal conditions revealed that SNL-glycoprotein has remarkable scavenging effect on both radicals but exhibited slightly higher scavenging effect on superoxide anion generated by the enzymatic hypoxanthine/xanthine oxidase system than on hydroxyl radicals generated by the Fenton reaction. However, SNL-glycoprotein was more effective against hydroxyl radical in cell culture (NIH/3T3). Consequently, 20 mcg/mL SNL-glycoprotein has scavenging ability superoxide anion corresponding to that of ascorbic acid. On the other hand, its hydroxyl scavenging activity corresponds to 0.1mcg/mL catalase. From these results we can say that SNL-glycoprotein has potent anti-oxidative potential (Heo and Lim 2004).

Molluscicidal and larvicidal activities. The ethanolic extract preparation of *S. nigrum* Linn. leaves was made by soaking powder overnight in cold 70% ethanol, evaluated for molluscicidal and larvicidal effect. The extract showed the highest molluscicidal activity (LC-50 3.37mg/L within 24 hour) as well as larvicidal activity against larvae of two mosquito species, *Aedes caspius* and *Culex pipiens* (LC-50 51.29 and 125.89mg/L within 24 hr and 21.38 and 38.11mg/L within 48 hr, respectively). Sunlight, pH, and turbidity did not affect the activity of this extract, but molluscicidal activity seems to be correlated with the increase of temperature. The concentrated extract (1,000mg/L) can be stored at room temperature for six months without any change in its activity, but the diluted solution of this extract lost their activity after four weeks (Ahmed et al. 2001).

Antinociceptive, anti-inflammatory, and antipyretic effects. The chloroform extract was prepared by soaking (1:20; w/v) the air-dried powdered leaves (20 g) in chloroform for 72 hr, followed by evaporation (40°C) under reduced pressure to dryness (1.26 g), and then dissolved (1:50; w/v) in dimethylsulfoxide (DMSO). The supernatant, considered as the stock solution with dose of 200 mg/kg, was diluted using DMSO to 20 and 100 mg/kg, and all doses were administered (subcutaneous; 10 ml/kg) in mice/rats 30 min prior to tests. The extract exhibited significant antinociceptive activity when assessed using abdominal constriction, hot plate, and formalin tests. The extract also produced significant anti-inflammatory and antipyretic activities when assessed using the carrageenan-induced paw oedema and brewer's yeast–induced pyrexia tests. Overall, the activities occurred in a dose-independent manner. They demonstrated that the lipid-soluble extract of *S. nigrum* leaves possessed antinociceptive, anti-inflammatory, and antipyretic properties and confirmed the traditional claims (Zakaria et al. 2006).

17.6 ANTIMALARIAL RESPONSE

Solanum spp. contains a diversity of phytochemicals that can be culled for their insecticidal properties, particularly mosquito larvicide, adulticide, and repellent. Although crude extracts have higher insecticidal potency than pure components, probably due to synergy among their bioactive constituents, optimal utilization of crude extracts is limited by an inability to control their contents, which can vary depending on plant species, cultivation conditions, the season of harvest, and extraction methods and solvents used. Thus, an understanding of the mechanism of insecticidal activity of the major bioactive compounds can lead to consistent and optimal formulation and adjustment to match the target insect pests of interest. In addition to the claimed traditional use, the 80% hydromethanolic fruit extract of the plant has promising antimalarial activity, *in vitro* (IC50 =10.29 µg/ml) and in vivo, with a percentage parasitaemia inhibition of 60.68% ($P < 0.05$). Additionally, in the *in vitro* test, the 50% cytotoxic concentration (CC50) of the fruit extract of the plant was 66.3 ug/ml on the Raji cell line (B lymphocyte cell line), showing that the plant is potentially toxic at much higher doses (Haddad et al. 2017).

A huge number of phytochemicals show effective antimalarial activity belonging to the class of alkaloids, terpenes, flavonoids, sesquiterpenes, and other related compounds. Therefore, the demonstrated antimalarial activity of *S. nigrum* might be attributed to the phytochemicals present in the extract, which may work individually and/or synergistically with plenty of possible parasite-suppressive mechanisms. Furthermore, the antimalarial activities of the plant may not be only by directly attacking the infecting *Plasmodium* parasite but also by indirectly enhancing the defence mechanisms of the host. Overall, considering all the aforementioned points, the extract of *S. nigrum* is reasonably a potential antimalarial agent, justifying the claimed use of the plant for malaria control and also the *in vitro* report. Eugenol and (E)-6-hydroxy-4,6-dimethyl-3-heptene-2-one in *S. nigrum* crude extract were proposed as the main compounds responsible for mosquito-larvicidal activity (Kumar et al. 2018). Glucosisaustrin, a glucosinolate group of bioactive compounds from *S. nigrum*, was responsible for larval mortality. Glucosinolate is a plant-derived secondary metabolite and hydrophilic, having potent mosquito-larvicidal properties against *Cx. Quinquefasciatus*, and found to be safe for the environment (Silva et al. 2008).

17.7 TRADITIONAL AND OTHER POTENTIAL USES

Solanum genus has several species found in tropical and subtropical areas and is used in folk medicine and dietary supplements. Among them, *S. nigrum* has been considered ethnobotanically important due to its use in the traditional healthcare system to cure various ailments. The leaves and bitter berries with pungent smell have been traditionally used against severe ulcers, heart diseases, piles, dysentery, gastritis, and stomach ache (Jain et al. 2011). The leaves of *S. nigrum* are traditionally used to cure ringworm, stomach ulcer, rabies, and as a tonic for the liver (Muthu et al. 2006). Fruits are used to stop bedwetting of kids. It is also given to treat blindness, optical bacterial infection, and cataract (Moshi et al. 2009). The root of the herb is effective to increase feminine fertility, cough problems, and other respiratory infections (Sikdar and Dutta 2008). Whole plant parts are used as a remedy for dermal infections, snakebite, burns, and injury (Javed et al. 2019). The ethnic community of Rhodesia used the plant as one of their remedies for malaria, black water fever, and dysenteries (Kirtikar and Basu 1975). In Tanzania, boiled leaves are eaten as a vegetable, green fruits edible, roots eaten raw for stomach ache (Edmonds and Chweya 1978). The leaves of the herb are used to treat burns and peptic ulcers in Arabian countries (Latis 2002). The herb is very popular in India and used in several diseases. The young roots are given in chronic skin diseases and are used with great success in psoriasis. It is given internally for cardialgia and gripe and also a household treatment for anthrax pustules when it is applied locally. Berries apparently possess tonic, diuretic, and cathartic properties and are also useful in heart diseases and as a domestic treatment for fevers, diarrhoea, ulcers, and eye troubles (Kirtikar and Basu 1975; Edmonds and Chweya 1997). People of Mexico use the fruit of *S. nigrum* in traditional medicine as a remedy for the treatment of nervous condition (Perez et al. 1998); used an infusion made from boiled roots of this 'species' to administer to babies with worms, and crushed green leaves mixed with a grease to make poultices for sores; and used as weak infusion to cure insomnia (Vogel 1990). The indigenous community of Kenya applies the unripe fruits to aching teeth and squeezed onto babies' gums to ease pain during teething; roots are boiled in milk and given to children as a tonic (Edmonds and Chweya 1997). Traditionally, the Chinese people apply the juices of leaves to alleviate the pain, inflammation of the kidney, bladder, and in virulent gonorrhoea. The leaves are used as a febrifugal or detoxicant drug (Schilling et al. 1992)

17.8 SAFETY ISSUES

Most species associated with *Solanum* genus are reputedly poisonous to both humans and livestock. The widely reported toxicity of *S. nigrum* has been attributed to the alkaloid solanine, causing varying degrees of poisoning in humans, cattle, pigs, goats, ducks, and chickens, with death resulting in some cases. The effects of solanine poisoning in humans are reported to be nausea, vomiting, diarrhoea, colic, headache, dizziness, loss of speech, fever, sweating and tachycardia, reduced heartbeat, pupil dilation, blindness, mental confusion, convulsions, coma, and death. Such effects normally appear around eight hours after ingestion (Aslanov and Novruzov 1978). These plants also contain high levels of nitrate nitrogen (NO 3 -N) and are included in the group of plants which can cause N03-N toxicity in livestock. Acute nitrate toxicity can lead to death, with chronic toxicity resulting in a decrease in milk yield, abortion, impaired vitamin A and iodine nutrition, muscle tremors, staggering gait, rapid pulse, frequent urination, laboured breathing, followed by collapse and coma, with or without convulsions (Watt and Breyer 1962). Indeed, a chemical survey of various members of the section *S. nigrum* reported the presence of potentially toxic alkaloids only in unripe fruits, with ripe berries. Therefore, it is concluded that the plant is probably only poisonous to indiscriminate feeders such as livestock who might consume the whole plant (Grieve 1931).

17.9 CONCLUSION

Solanum genus–based botanical phytochemicals should not be dismissed as a potential future alternative to synthetic insecticides. The species of *Solanaceae* should be mined for their inexpensive, eco-friendly, safe, and effective alternatives to current chemical larvicides to control malaria and other insecticidal diseases. The present review suggests that *Solanum* genus–based phytochemical research should be continued as a potential future alternative to synthetic antimalarial and other insecticidal drugs. The plant-based drug should be less-expensive, environment-friendly, safe, and effective alternatives to the present chemical-based antimalarial medicines. Overall, this chapter attempts to compile recent and updated information on various aspects of *Solanum* genus, as the plants of the genus are used all over the world. Most of the species of the genus consist of dried whole plant shrub that grows wild and abundantly in open fields. The extensive survey of literature revealed that some of the plant species of genus *Solanum* are traditionally popular to use in malarial and insecticidal ailments. *S. nigrum* is one of the effective antimalarial plant species belonging to *Solanum*, commonly known as Makoya or kakamachi. *S. nigrum* is important medicinal plant with diverse pharmacological spectrum. The plant shows the presence of many chemical constituents, which are responsible for the various activities of the plant. *S. nigrum* has shown a tremendous potential that deserves special attention of the scientific fraternity to emerge as a milestone for medical science of this millennium due to its various medicinal uses. Further evaluation needs to be carried out on *S. nigrum* in order to explore the concealed areas and their practical clinical applications, which can be used for the welfare of mankind.

REFERENCES

Acharya, E. S. & B. Pokhrel. 1970. Ethno-medicinal plants used by bantar of bhaudaha, morang, Nepal. *Our Nature* 4(1):96–103.

Ahmed, A. H., I. H. Kamal & R. M. Ramzy. 2001. Studies on the molluscicidal and larvicidal properties of *Solanum nigrum* L. leaves ethanol extract. *Journal of the Egyptian Society of Parasitology* 31(3):843–852.

An, L., J. T. Tang & X. M. Liu. 2006. Review about mechanisms of anti-cancer of *Solanum nigrum*. *Materia Medica* 31(15):1225–1226.

Annonymus. 1999a. *The Ayurvedic Pharmacopoeia of India*. Government of India Ministry of Health and Family Welfare Department of Ayurveda, Yoga & Naturopathy, Unani, Siddha and Homeopathy, New Delhi, 1st ed., part 1, vol. 2, pp. 66–68.

Annonymus. 1999b. *Indian Herbal Pharmacopoeia*. Joint Publication of Indian Drug Manufacturers Association and Regional Research Laboratory, New Delhi, vol. 2, p. 141.

Aslanov, S. M. & E. Novruzov. 1978. Studies on glycoalkaloids of *Solanum nigrum* growing in Azerbaijan. *Biologicheskikh Nauk* 3:15–18.

Burgert, K. L., O. C. Bumside & C. R. Fenster. 1973. Black nightshade leaves its mark. *Farm, Ranch and Home Quarterly*:8–10.

Chweya, J. A. 1997. Genetic enhancement of indigenous vegetables in Kenya. *Traditional African Vegetables*:90–99.

Cooper, M. R. & A. M. Johnson. 1984. Black nightshade-*Solanum nigrum*. Poisonous plants in Britain and their effects on animals and man. *HMSO* 4:219–210.

Edmonds, J.M.; Chweya, J.A. 1997. Black Nightshades: Solanum Nigrum L. and Related Species; Bioversity International: Rome, Italy.

Elizalde-Romero, C. A., L. A. Montoya-Inzunza, L. A. Contreras-Angulo, J. B. Heredia & E. P. Gutierrez-Grijalva. 2021. *Solanum* fruits: Phytochemicals, bio-accessibility and bioavailability, and their relationship with their health-promoting effects. *Frontiers in Nutrition* 8:790582.

Eltayeb, E. A., A. S. Al-Ansari & J. G. Roddick. 1997. Changes in the steroidal alkaloid solasodine during development of *Solanum nigrum* and *Solanum incanum*. *Phytochemistry* 46(3):489–494.

Fawusi, M. O. 1983. Nitrogen fertilization and storage temperature effects on the nutritive value of *Solanum nigrum*. *Journal of Plant Foods* 5:161–167.

Fortuin, F. T. & S. W. Omta. 1980. Growth analysis and shade experiment with *Solanum nigrum* L., the black nightshade, a leaf and a fruit vegetable in West Java. *Netherlands Journal of Agricultural Science* 28(4):199–210.

Grieve, M. 1931. *Nightshade, Black. A Modem Herbal*. Jonathan Cape, Middlesex, England. Reprinted by Penguin Books, Middlesex, England, pp. 582–583.

Haddad, M. H. F., H. R. Mahbodfar, Z. Zamani & A. Ramazani. 2017. Antimalarial evaluation of selected medicinal plant extracts used in Iranian traditional medicine. *Iranian Journal of Basic Medical Science* 20(4):415–422.

He, X. J., G. X. Zhou & W. C. Ye. 2007. Pragnane glycosides from *Solanum nigrum*. *Journal of Asian Natural Products Research* 9(6):517–523.

Hebbar, S. S., V. H. Harsha & V. Shripathi. 2004. Ethnomedicine of Dharwad district in Karnataka, India plants used in oral health care. *Journal of Ethnopharmacology* 94:261–266.

Henderson, R J. F. 1974. *Solanum nigrum L. (Solanaceae) and related species in Australia*. Contributions from the Queensland Herbarium 16: 1–78.

Heo, K. S. & K. T. Lim. 2004. Antioxidative effects of glycoprotein isolated from *Solanum nigrum* L. *Journal of Medicinal Food* 7(3):349–357.

Ikeda, T., H. Tsumagari & T. Nohara. 1975. Steroidal oligoglycosides from *Solanum nigrum*. *Chemical and Pharmaceutical Bulletin* 48(7):1062–1064.

Ikeda, T., H. Tsumagari & T. Nohara. 2000. Steroidal oligoglycosides from *Solanum nigrum*. *Chemical and Pharmaceutical Bulletin* 48(1):1062–1064.

Jain, R., A. Sharma, S. Gupta, I. P. Sarethy & R. Gabrani. 2011. *Solanum* nigrum: Current perspectives on therapeutic properties. *Alternative Medicine Review: A Journal of Clinical Therapeutic* 16(1):78–85.

Javed, A., M. Usman, S. M. Haider, B. Zafar & K. Iftikhar. 2019. Potential of indigenous plants for skin healing and care. *American Scientific Research Journal for Engineering, Technology and Sciences* 51(1):192–211.

Karschon, R. & M. Horowitz. 1985. Intraspecific variation of *Solanum nigrum*. L. and *S. villosum* Miller, important crop weeds in Israel. *Phytoparas* 13(1):63–67.

Kiran, K. R., M. Rani & A. Pal. 2009. Reclaiming degraded land in India through the cultivation of medicinal plants. *Botany Research International* 2:174–181.

Kirtikar, K. R. & B. D. Basu. 1975. *Indian Medicinal Plants*. Lalit Mohan Basu Publication, Allahabad, 2nd ed., vol. 3, pp. 1748–1750.

Knapp, S., L. Bohs, M. Nee & D. M. Spooner. 2004. Solanaceae – a model for linking genomics with biodiversity. *Comparative and Functional Genomics* 5(3):285–291.

Kumar, P. M., A. Tiwari & S. V. Laxmi. 2018. Larvicidal activity of crude *Solanum nigrum* leaf and berries extract against dengue vector-*Aedes aegypti*. *International Journal of Current Research and Review* 10(14):16–21.

Kumar, V. P., S. Shashidhara & M. M. Kumar. 2001. Cytoprotective role of *Solanum nigrum* against gentamicin-induced kidney cell (Vero cells) damage in vitro. *Fitoterapia* 72:481–486.

Latis, K. M. 2002. Tumbuhan Ubatan Malaysia Universiti Kebangsaan Malaysia in collaboration with ministry of science. *Technology and Innovation* 589.

Lee, S. J. & K. T. Lim. 2003. Antioxidative effects of glycoprotein isolated from *Solanum nigrum* Linne on oxygen radicals and its cytotoxic effects on the MCF-7 cell. *Journal of Food Science* 68(2):466–470.

Li, Q., T. Feng & T. Zhang. 2007. Antitumor activity of crude polysaccharides isolated from *Solanum nigrum* Linn on U14 cervical carcinoma bearing mice. *Phytotherapy Research* 21(9):832–840.

Mallika, J. & S. S. Chennam. 2006. Antiulcerogenic and ulcer healing effects of *Solanum nigrum* (L.) on experimental ulcer models: Possible mechanism for the inhibition of acid formation. *Journal of Ethnopharmacology* 104:156–163.

Mathooko, F. M. & J. K. Imungi. 1994. Ascorbic acid changes in three indigenous Kenyan leafy vegetables during traditional cooking. *Ecology of Food and Nutrition* 32:239–245.

Moshi, M. J., D. F. Otieno, P. K. Mbabazi & A. Weisheit. 2009. The ethnomedicine of the Haya people of Bugabo ward, Kagera Region, north western Tanzania. *Journal of Ethnobiology and Ethnomedicine* 5:24.

Muthu, C., M. Ayyanar, N. Raja & S. Ignacimuthu. 2006. Medicinal plants used by traditional healers in Kancheepuram district of Tamil Nadu, India. *Journal of Ethnobiology and Ethnomedicine* 2:43.

Perez, R. M., J. A. Perez & L. M. Garcia. 1998. Neuropharmacological activity of *Solanum nigrum* fruit. *Journal of Ethnopharmacology* 62:43–48.

Rajathi, M., D. Modilal, R. Anandan, R. Sindhuand & M. N. Logeshwari. 2015. Screening of *Solanum nigrum* for its phytochemical and antimicrobial activity against respiratory tract pathogens. *International Journal of Pure and Applied Zoology* 3(3):210–215.

Raju, K., G. Anbuganapathi, V. Gokulakrishnan, B. Rajkapoor, B. Jayakar & S. Manian. 2003. Effect of dried fruits of *Solanum nigrum* against CCl4-induced hepatic damage in rats. *Biological and Pharmaceutical Bulletin* 26(11):1618–1619.

Rani, P. & N. Khullar. 2004. Antimicrobial evaluation of some medicinal plants for their anti-enteric potential against multidrug resistant *Salmonella typhi*. *Phytotherapy Research* 18(8):670–673.

Rawani, A., A. Ghosh & G. Chandra. 2013. Mosquito larvicidal and antimicrobial activity of synthesized nano-crystalline silver particles using leaves and green berry extract of *Solanum nigrum* L. (Solanaceae: solanales). *Acta Tropica* 128(3):613–622.

Rogers, B. S. & A. G. Ogg. 1981. Biology of weeds of the *Solanum nigrum* complex (Solanum Section Solanum) in North America. US Dept. of Agriculture. Science and education administration. *Agricultural Reviews and Manuals* 23:1–30.

Schilling, E. E., Q. S. Ma & R. N. Anderson. 1992. Common names and species identification in black nightshades, Solanum sect. Solanum (Solanaceae). *Economic Botany* 46(2):223–225.

Sikdar, M. & U. Dutta. 2008. Traditional phytotherapy among the Nath people of Assam. *Ethno-Medicine* 2:39–45.

Silva, T. M. S., C. A. Camara, K. R. L. Freire, T. G. Silva, M. F. Agra & J. Bhattacharyya. 2008. Steroidal glycoalkaloids and molluscicidal activity of *Solanum asperum* rich fruits. *Journal of the Brazilian Chemical Society* 19(5):1048–1052.

Vogel, V. J. 1990. *American Indian Medicine*. The Civilisation of America University of Oklahoma Press, Oklahoma, vol. 95, p. 342.

Vorontsova, M. S., Stern, S., Bohs, L. & Knapp, S. 2013. African spiny Solanum (Subgenus Leptostemonum, Solanaceae): A thorny phylogenetic tangle. *Botanical Journal of the Linnean Society* 173:176–193.

Wakhloo, J. L. 1964. Ecological and physiological studies on two species of Solanum: I. Germination and development of *S. xanthocarpum* S chrad. & Wendl. and *S. nigrum* L. *Flora* 155:237–249.

Wang, L. Y., N. L. Wang & X. S. Yao. 2007. Non-saponins from *Solanum nigrum*. *Zhong Yao Cai* 30(7):792–794.

Watt, J. M. & M. G. Breyer. 1962. *The Medicinal and Poisonous Plants of Southern and Eastern Africa*. E. and S. Livingstone Ltd., Edinburgh and London, pp. 996–1000.

Weller, R. F. & R. H. Phipps. 1978. A review of the black nightshade (*Solanum nigrum* L.). *Protection Ecology* 1:121–139.

Zakaria, Z. A., H. K. Gopalan & H. Zainal. 2006. Antinociceptive, anti-inflammatory and antipyretic effects of *Solanum nigrum* chloroform extract in animal models. *Yakugaku Zasshi* 126(11):1171–1178.

18 Antimalarial Response, Traditional and Other Potential Uses of *Stachytarpheta* Genera

Shuvadip Mondal, Manjula Rai, Tuyelee Das, Abhijit Dey, and Surjit Sen

18.1 INTRODUCTION

In today's world, malaria remains a leading concern of mortality and morbidity in many populous countries, especially from tropical and subtropical regions, having a persistent detrimental effect on healthcare operations (Laryea and Borquaye, 2021). In 2018, the World Health Organization stated that it remains endemic in 80 countries and adjoining regions (WHO, 2019; Laryea and Borquaye, 2021). The artemisinin-based combination therapy has long been used to curb this life-threatening disease in children and adults, but a disadvantage has appeared, that there is a development of drug resistance by the malaria parasite *Plasmodium* spp. (Eastman and Fidock, 2009; Dondorp et al., 2010; Laryea and Borquaye, 2021). These challenges put forth the concept to develop new potent plant-derived drug to control malaria.

In Africa and parts of Asia and America, traditional herbal preparations proved to be useful therapeutics against malaria and malaria-like symptoms (Laryea and Borquaye, 2021). One of such plants that have been used traditionally in various ailments, including malaria, is *Stachytarpheta* spp., belonging to the family Verbenaceae. The ethanolic extracts from the leaves of *Stachytarpheta angustifolia* are reported to act against plasmodial infections (Laryea and Borquaye, 2021). South-Eastern Nigerian people hold the traditional belief of repelling mosquitoes, the vector of malaria parasite, by using the leaf of *Stachytarpheta cayennensis* on the burning lantern (Ezenyi et al., 2016; Okonkwo and Onyeji, 2018). Investigation shows that 0.5 g of essential oil from *S. cayennensis* can pose 100% mortality to *Anopheles gambiae* (Okonkwo and Onyeji, 2018). Some studies also show the ability of this plant to act as antiplasmodic, anti-inflammatory, antimicrobial, antidiabetic, and antiulcer, antiallergic, bronchodilatory, digestion-stimulating, antacid, antidiarrhoea, anticonvulsant, and sedative agent (DeSouza et al., 2010; Okoye et al., 2014; Ezenyi et al., 2016). The isolated phytochemicals from *S. cayennensis* that raised it to provide pharmacological activities against malaria and other diseases are ipolamiide, verbascoside, betulinic acid, martinoside, hydroxyipolamide, isoverbascoside, leucosceptoside A, and Jionoside D (Froelich et al., 2008; Ezenyi et al., 2016). Further, it has been ascertained that the immunomodulatory potentiality of methanolic extracts from the leaves of this plant is as a standard antimalarial agent by providing cellular and humoral immune response (Okoye et al., 2014). Various phytochemical literatures have shown the effectiveness of the whole plant or leaves of the genus *Stachytarpheta* in various diseases, like ulcers, skin lesions, internal inflammations, asthma, fever, renal abnormalities, atherosclerosis, and venereal infections (Iroka et al., 2015; Yadav et al., 2021). The present book chapter provides concise information on antimalarial activity, phytochemistry, and other potential ethnomedicinal properties of *Stachytarpheta* spp.

 DOI: 10.1201/9781003378396-18

18.2 BOTANICAL DESCRIPTION

The genus *Stachytarpheta* Vahl. belongs to the family Verbenaceae and is popularly known as 'porter weed', 'gervao', or 'orgevao' (Yadav et al., 2021). The genus name is composed of two Greek words, 'stachy' and 'tarpheta', meaning 'spike' and 'thick', respectively (Yadav et al., 2021). *Stachytarpheta* is firstly placed under the tribe Verbeneae, subtribe Verbeneae, along with *Verbena*, *Bouchea*, and *Lippia*, but not *Lantana*, but later it is placed under the tribe Lantaneae (Schauer, 1847; Briquet, 1895; Atkins, 2005).

The stem of the plant is slightly angled and grows on erect, half-woody main branch (Yadav et al., 2021). Sometimes, the stem has hairs on opposite faces, where hairs can be simple or uniseriate, occasionally gland-tipped, and may be erect with multidirectional point (Atkins, 2005; Yadav et al., 2021). Leaves are elliptical or oblong to ovate in shape, with serrated margins, decurrent base, and 2 to 10 cm long (Yadav et al., 2021). Inflorescence is spike, and it is terminal in nature, slender in shape, and 10–30 cm long, 3–4 mm thick (Yadav et al., 2021). Flower calyx is small, oblique, with four-toothed margin, whereas corolla is dark blue to bluish purple in colour and 1 cm long (Yadav et al., 2021). Fruit remains in the enclosed calyx and somehow sinks in the smooth and oblong rachis (Olayiwola et al., 2013; Yadav et al., 2021). Further, morphological comparison between few pharmacologically important species of *Stachytarpheta* is presented in Table 18.1.

18.3 GEOGRAPHICAL DISTRIBUTION

It is estimated that the total number of species of *Stachytarpheta* is around 133 and a worldwide distribution (Atkins, 2005). It is a monophyletic genus distributed in countries of tropical and subtropical regions, like the Americas, Asia, Africa, and Oceania (Atkins, 2005; Yadav et al., 2021). The abundance of species distribution of this genus is mostly from American countries (Table 18.2). Therefore, a definite prediction about the origin of *Stachytarpheta* is exclusively from the New World, except *S. indica* having an origin from the Old World (Atkins, 2005). The most richness of species diversity is found in the tropical and subtropical countries of Americas, for example, Mexico and Brazil (Atkins, 2005).

TABLE 18.1
Morphological Comparison between Few Pharmacologically Important Species of *Stachytarpheta*

Species Name	Plant Height (m)	Leaf Nature	Flower Nature	Key Reference
S. jamaicensis	0.6–1.2	Ovate to elliptic and oblong in shape	5–7 mm long, mix of bluish and pinkish purple to deep blue in colour	Thangiah, 2019; Yadav et al., 2021
S. cayennensis	1.5	Ovate to elliptical in shape	4–5 mm long, dark blue–purple or violet-white in the centre	Froelich et al., 2008; Okoronkwo and Echeme, 2015; Yadav et al., 2021
S. indica	0.6–0.9	Lanceolate to oblong lanceolate or elliptic ovate in shape	13 mm wide, deep blue, with white-coloured centre	Joshi et al., 2010; Musa et al., 2019; Yadav et al., 2021
S. mutabilis	0.1–0.2	Elliptic-oblong, lance-shaped structure	13–18 mm long, with bright rose or red-pink in colour	Osorio et al., 2014; Yadav et al., 2021
S. urticifolia	0.5–1.5	Ovate to elliptic ovate in shape	Purple-blue or royal blue, with a striking white throat	Daniel and Rajendran, 1992; Yadav et al., 2021

TABLE 18.2
Numbers of Species Diversity and Numbers of Endemics of *Stachytarpheta* in South American Countries

Country Name	Total Number of Species	Number of Species as Endemics	Key Reference
Mexico	22	14	Moldenke, 1971; Atkins, 2005
Central America	24	1	Atkins, 2005
Colombia	10	0	
Venezuela	7	1	
The Guianas	8	1	
Ecuador	5	2	
Peru	7	4	
Bolivia	4	0	
Paraguay	7	2	
Argentina	5	2	
Brazil	79	73	Zappi et al., 2003; Atkins, 2005

18.4 PHYTOCHEMISTRY

The genus *Stachytarpheta* is rich in a variety of secondary metabolites commonly known as bioactive compounds or phytochemicals. Eventually, these bioactive compounds are responsible in exhibiting therapeutic activities. Preliminary phytochemical screening of *Stachytarpheta* exhibits the presence of several major groups of compounds, including alkaloids, flavonoids, tannins, saponins, glycosides, steroids, and phenols (Table 18.3, Figure 18.1). The bioactive components from the aqueous leaf extract of *S. jamaicensis* include a variety of iridoid glycosides and phenylethanoid glycoside (Udodeme et al., 2016). Studies also showed some important secondary metabolites in the leaves of *S. jamaicensis*, which include terpene glycoside and steroidal glycosides (Melita and Castro, 1996; Okwu and Ohenhen, 2009, 2010; Yuliana et al., 2019). The active phytoconstituents found in the entire plant of *S. indica* are different types of steroidal glycosides, phenolic acids, and phenylethanoid glycoside (Roengsumran et al., 2002; Princely et al., 2013; Yuvaraj et al., 2019). *S. cayennensis* is rich in iridoids glycosides, mainly ipolamiide; phenylpropanoid glycosides, such as jinoside-D, martynoside, acetoside, iso-acetoside, and leucosceptoside-A; and phenylpropanoid glycosides, such as verbascoside and isoverbascoside (Leitao et al., 2005; Froelich et al., 2008; Okoronkwo and Echeme, 2015). *S. mutabilis* showed the presence of iridoid glycoside, mainly ipolamiide and 6 β-hydroxy ipolamiide (Osorio et al., 2014). Phytochemical analysis of various extracts of root, stem, leaf, and inflorescence of *S. urticifolia* showed the presence of saponins, flavonoids, terpenoids, steroids, glycosides, quinines, alkaloids, tannins, and phenols. The methanolic extract of the leaves is rich in iridoid glycoside, whereas the n-hexane root extract showed the presence of steroidal glycosides-α-spinasterol (Chowdhury et al., 2003; Sreelatha et al., 2013).

18.5 ANTIMALARIAL ACTIVITY

Malaria continues to be a significant global health concern, complicated by the emergence of drug resistance. *Stachytarpheta cayennensis*, a Nigerian fever remedy, was tested for antiplasmodial activity. Methanol extract and solvent fractions were produced and analyzed. Glycosides, flavonoids, and phenolic substances were found in the methanol extract, which was nontoxic. The extract and fractions reduced parasitaemia in early, repository, and established infestations. The methanol extract reduced haematotoxicity and DPPH radicals. *S. cayennensis* has substantial antiplasmodial activity, presumably due to its antioxidant characteristics (Ezenyi

TABLE 18.3
Phytochemical Constituents of Pharmacologically Active Few Species of *Stachytarpheta*

Stachytarpheta spp.	Plant Part	Bioactive Compounds	Subtype	Key Reference
S. jamaicensis	Leaves, whole plant	Iridoid glycosides	Ipolamiide	Melita and Castro, 1996; Okwu and Ohenhen, 2010; Udodeme et al., 2016; Yuliana et al., 2019
		Phenylethanoid glycoside	Verbascoside	
		Flavonoid	Scutellarin, Hispidulin	
		Terpene glycosides	Lanostane triterpenoids, lanostane phenylacetate	
		Steroidal glycosides	16-β-(β-D-glucopyranosyl-3,8,22-trihydroxy)-cholestan-1-β-yl-6-O-(3,4,5-trimethoxybenzoyl)-β-D-glucopyranoside-1; 16-β-(β-D-glucopyranosyl-2)-3,8,22-tri-hydroxy cholest-5,14,16,23-tetraene-1-β-yl-6-O-(3,4,5-trimethoxybenzoyl)-β-D-glucopyranoside-2	
S. indica	Leaves, whole plant	Sterols	*β*-sitosterol, stigmasterol	Roengsumran et al., 2002; Princely et al., 2013; Yuvaraj et al., 2019
		Iridoid glycosides	Ipolamiide	
		Phenolic acid	Flavone apigenin	
		Terpene glycosides	Ursolic acid, lupeol	
		Phenylethanoid glycoside	Verbascoside	
S. cayennensis	Leaves and roots	Iridoid glycosides	Ipolamiide	Leitao et al., 2005; Froelich et al., 2008; Okoronkwo and Echeme, 2015
		Phenylpropanoid glycoside	Acetoside, iso-acetoside, jinoside-D, martynoside, leucosceptoside-A	
		Phenylethanoid glycoside	Verbascoside, isoverbascoside	
S. mutabilis	Whole plant	Iridoid glycosides	Ipolamiide, 6 *β*-hydroxy ipolamide	Osorio et al., 2014
		Terpene glycosides	Saponins	
S. urticifolia	Root, stem, leaf, and inflorescence	Iridoid glycosides	Ipolamiide	Chowdhury et al., 2003; Sreelatha et al., 2013
		Terpene glycosides	Saponins	
		Steroidal glycosides	*α*-spinasterol	

et al., 2016). In certain Latin American nations, the boiled juice of *S. cayennensis* whole plant material is utilized ethnomedicinally to cure malaria symptoms. Five phenylethanoid glycosides were found in ethyl acetate. Leucosceptoside A, martynoside, and jionoside D were discovered in *Stachytarpheta. S. cayennensis* ethanol leaf extract was tested for *in vivo* antiplasmodial efficacy in *Plasmodium berghei*–infected mice. In early and established infections, the extract was blood schizonticidal. Leaf extract showed antimalarial activity. The active chemicals must be identified and tested in people (Okokon et al., 2008). Laryea et al., 2021 studied the antiplasmodial activity of *Stachytarpheta angustifolia*, which has traditionally been used to treat malaria in Ghana. Plant ethanol extracts were tested *in vitro* against *Plasmodium falciparum*. However, *S. angustifolia* exhibited no action (IC50 > 50 g/mL).

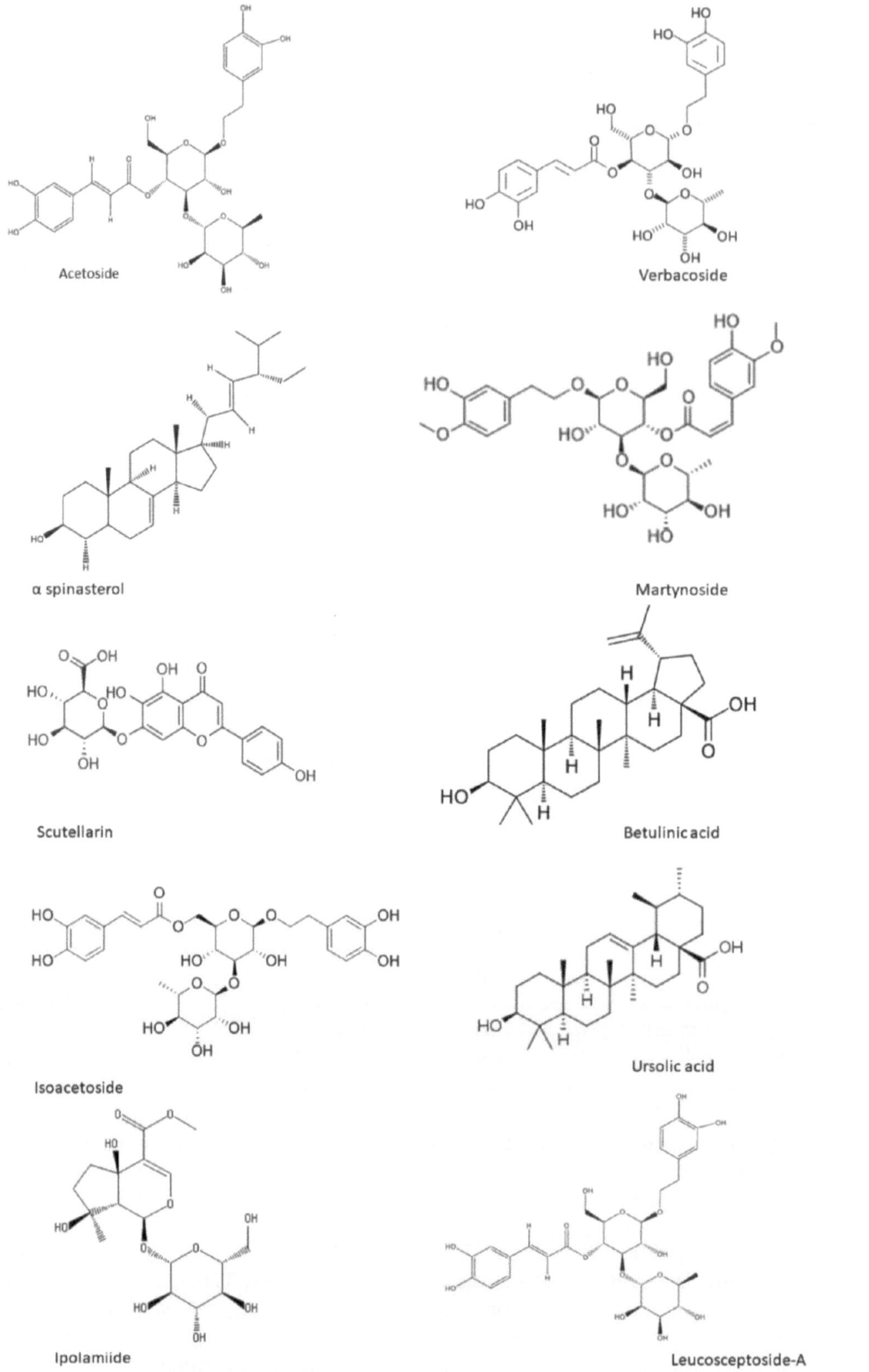

FIGURE 18.1 Structure of some important bioactive compounds of *Stachytarpheta*.

18.6 TRADITIONAL USES OF *STACHYTARPHETA*

Stachytarpheta is an important medicinal plant with great medicinal properties in traditional and folk medicinal systems. This plant has been used traditionally by the elderly for decades. Various parts of the plants are used in Indian traditional and folk medicine. It is used as antimicrobial, antispasmodic, antidiabetic, analgesic, anti-inflammatory, antipyretic, antifungal, anticancer, hepatoprotective, anxiolytic, haemolytic, laxative, larvicidal, anthelmintic, etc. This plant has been reported to possess medicinal properties due to the presence of many bioactive phytochemicals.

S. jamaicensis is considered as an important medicinal plant in South-Eastern Asia and West Tropical Africa and used as antiallergic, antacid, analgesic, anti-inflammatory, antiasthmatic, hypotensive, antihelminthic, diuretic, laxative, lactagogue, purgative, sedative, spasmogenic, vasodilator, vulnerary, and also as a cure for cough, cold, fever, and digestive complications (Idu et al., 2007; Jagadish and Gopalkrishna, 2008; Sulaiman et al., 2009; Joshi et al., 2010; Putera and Anis Shazura, 2010; Valdés et al., 2010; Sivaranjani et al., 2014). The plant is often used by the old as a stomach tonic and cooling agent. Leaf and stem extracts are often processed as teabags before use. Drinking this cooling tonic can help cure digestion problems, including indigestion, acid reflux, ulcers, constipation, dyspepsia, and poor digestion, by stimulating the function of the gastrointestinal tract (Idu et al., 2007). In Southern Nigeria, *S. jamaicensis* is practiced by women for birth control, abortion, to treat menstrual disorders, and female complaints (Melita and Castro, 1996; Idu et al., 2007; Liew and Yong, 2016). To restore the uterus, return to its original position after childbirth, the leaves of *S. jamaicensis* are cooked in water and provided to the mothers as a tea. It is also known to control hormones and enhance milk production in lactating mothers (Okwu and Ohenhen, 2010). It has been reported that its infusion has been used as a remedy for headache and to cure sore in children's ears (Liew and Yong, 2016).

S. cayennensis is used in traditional medicine to treat gastrointestinal diseases, manage mental illness, and as an anti-inflammatory, analgesic, antipyretic, hepatoprotective substance (Vela et al., 1997). The leaves are utilized in traditional medicine in Nigeria and other regions to treat anxiety and sleeplessness. In Brazil and other countries, crushed leaves and roots are used as folk medicine to cure inflammation, varicose ulcers, fever, skin blemishes, hepatic and renal problems, hypertension, and diabetes (Olayiwola et al., 2013). It is also reported that the plant is used ethnomedicinally as anti-allergic, bronchodilatory, digestion-stimulating, antacid and antidiarrhoea (Mesia-vela, 2004), pain-relieving, antispasmodic and anti-inflammatory (Lee, 2004; Penido, 2006), cellular-protective, antioxidant antimicrobial, and sedative agents (Lee, 2004; Park, 2004; Okoye et al., 2008; Okoye et al., 2010).

S. indica has been used as a folk remedy to treat allergies, respiratory conditions, cough, cold, fever, constipation, digestive problems, dysentery, and has even been reported to encourage menstruation (Yuvaraj et al., 2019). It is widely used in the Indian traditional system of medicine to treat diabetes and liver complaints. Traditional medicine considers stem bark as an analgesic, anti-inflammatory, and antipyretic agent. Folklore suggests that *S. urticifolia* can treat fever, rheumatic inflammations, venereal infections, dropsy, ulcers, and other stomach issues (Chowdhury et al., 2004). It is recommended to infuse the bark while treating diarrhoea and dysentery. In Bangladesh, the ethnic people utilize it as an abortifacient. The leaves of *S. mutabilis* are mashed with lime and then applied to sores and inflamed wounds. The plant is mostly grown in Amravati, India, and is beneficial for treating cancers in addition to being decorative (Ingole, 2011).

18.7 CONCLUSION

The whole plant and leaves of most species of *Stachytarpheta* have long been used in traditional medicine for the treatment of a number of ailments and have shown the potentiality to act against different diseases. Most promisingly, it includes various types of natural bioactive compounds which make a clear portfolio of this plant to make future drugs for malaria. As malaria is a life-threatening disease and still occurring in underdeveloped and populous countries of Asia, Africa, and South America, there is an urgent need to find immediate ways to eradicate this disease. Henceforth, our work will provide an innovative approach to researchers and investigators to make safe and active natural drug from this plant to curb malaria and other ailments. Moreover, collaborative efforts of ethnobotanists and pharmacologists could be a useful strategy for assessing and validating the use of *Stachytarpheta* species with advanced and innovative scientific techniques.

REFERENCES

Atkins, S. (2005). The genus Stachytarpheta (Verbenaceae) in Brazil. *Kew Bull.* 60, 161–272.

Briquet, J. (1895). Verbenaceae. In: H. G. A. Engler & K. Prantl (eds.), *Die natfirlichen Pflanzenfamilien IV*, pp. 132–182.

Chowdhury, R., Rashid, M. U., Khan, O. F. & Choudhury, M. H. (2004). Bioactivity of extractives from *Stachytarpheta urticaefolia. Pharm Biol.* 42, 262–267.

Chowdhury, R., Rashid, M. U., Khan, O. F. & Hasan, C. M. (2003). Ipolamiide and α-spinasterol from *Stachytarpheta urticaefolia. Biochem Syst Ecol.* 31, 1209–1211.

Daniel, P. & Rajendran, A. (1992). *Lectotypification of Stachytarpheta urticifolia Sims (Verbenaceae).* Taxon.

DeSouza, P. A., Silva, C. G., Machado, B. R. P., Lucas, N. D., Leitão, G., Eleutherio, E., Ortiz, G. M. D. & Benchetrit, L. (2010). Evaluation of antimicrobial, antioxidant and phototoxic activities of extracts and isolated compounds from *Stachytarphet acayennensis* (Rich.) Vahl, Verbenaceae. *Rev Bras Farmacogn.* 20(6), 922–928. https://doi.org/10.1590/S0102-695X2010005000042

Dondorp, A. M., Yeung, S., White, L., et al. (2010). Artemisinin resistance: Current status and scenarios for containment. *Nat Rev Microbiol.* 8(4), 272–280. https://doi.org/10.1038/nrmicro2331

Eastman, R. T. & Fidock, D. A. (2009). Artemisinin-based combination therapies: A vital tool in efforts to eliminate malaria. *Nat Rev Microbiol.* 7(12), 864–874. https://doi.org/10.1038/nrmicro2239

Ezenyi, I. C., Akah, P. A. & Okoli, C. O. (2016). Stachytarpheta cayennensis methanol leaf extract and solvent fractions display antiplasmodial activity against Plasmodium berghei berghei in vivo. *Int J Phytopharmacol.* 7(1), 1–9.

Froelich, S., Gupta, M. P., Siems, K. & Jenett-Siems, K. (2008). Phenylethanoid glycosides from *Stachytarpheta cayennensis* (Rich.) Vahl, *Verbenaceae*, a traditional antimalarial medicinal plant. *Rev Bras Farmacogn.* 18(4), 517–520. https://doi.org/10.1590/S0102-695X2008000400003

Idu, M., Omogbai, E. K. I., Aghimien, G. E., Amaechina, F., Timothy, O. & Omonigho, S. E. (2007). Preliminary phytochemistry, antimicrobial properties and acute toxicity of *Stachytarpheta jamaicensis* (L.) Vahl. Leaves. *Trends Medical Res.* 2(4), 193–198.

Ingole, S. N. (2011). Diversity and useful products in some verbenaceous member of Melghat and Amravati regions, Maharashtra, India. *Biodiversitas.* 12, 146–163.

Iroka, F. C., Okeke, C. U. & Okereke, C. N. (2015). Taxonomic significance of alkaloids and phenols in the species of *Stachytarpheta* found in Awka, *Nigeria. J Glob Biosci.* 4(4), 1961–1965.

Jagadish, N. R. N. & Gopalkrishna, B. (2008). Evaluation of analgesic activity of different extracts of *Stachytarpheta indica* L. (Vahl). *Biomed.* 3(4), 229–233.

Joshi, V. G., Sutar, P. S., Karigar, A. A., Patil, S. A., Gopalakrishna, B. & Sureban, R. R. (2010). Screening of ethanolic extract of *Stachytarpheta indica* L. (Vahl) leaves for hepatoprotective activity. *Int J Res Ayurveda Pharm.* 1(1), 174–179.

Laryea, M. K. & Borquaye, L. S. (2021). Antimalarial, antioxidant, and toxicological evaluation of extracts of *Celtis africana*, Grosseria vignei, physalis micrantha, and Stachytarpheta angustifolia. *Biochem Res Int.* https://doi.org/10.1155/2021/9971857

Lee, J. H. (2004). The effect of aceteoside on histamine release and arachidonic acid release in RBL-2H3 mast cells. *Arch Pharm Res.* 29(6), 508–513.

Leitao, G. G., de Souza, P. A., Moraes, A. A. & Brown, L. (2005). Step-gradient CCC separation of phenylpropanoid and iridoid glycosides from roots of *Stachytarpheta cayennensis* (Rich.) Vahl. *J Liq Chromatogr Relat Technol.* 8, 2053–2060.

Liew, P. M. & Yong, Y. K. (2016). *Stachytarpheta jamaicensis* (L.) Vahl: From traditional usage to pharmacological evidence. *Evid Based Complement Altern Med.* 16, 1–7.

Melita, R. S. & Castro, O. (1996). Pharmacological and chemical evaluation of *Stachytarpheta jamaicensis* (*Verbenaceae*). *Rev Biol Trop.* 44, 353–359.

Mesia-vela, S. (2004). Pharmacological study of Starphytarpheta cayennensis Val in rodents. *Phytomedicine.* 11(7), 616–624.

Moldenke, H. N. (1971). A fifth summary of the verbenaceae, avicenniaceae, stilbaceae, dicrastylidaceae, symphoremaceae, nyctanthaceae and eriocaulaceae of the world as to valid taxa, geographic distribution, andsynonymy. *Phytologia*. 34(3), 247–281.

Musa, A. D., Ogbiko, C., Dabai, M. U., Ali, I. J., Yelwa, A. S. & Buhari, H. B. (2019). *Stachytarpheta indica* leaf extract: Oral acute toxicity, *in vitro* phytochemical and antimicrobial potentials. *Earthline J Chem Sci*. 2(1), 163–173. https://doi.org/10.34198/ejcs.2119.163173

Okokon, J. E., Ettebong, E. & Antia, B. S. (2008). In vivo antimalarial activity of ethanolic leaf extract of Stachytarpheta cayennensis. *Indian J Pharmacol*. 40(3), 111–113. https://doi.org/10.4103/0253-7613.42303

Okonkwo, C. O. & Onyeji, C. M. (2018). Insecticidal potentials and chemical composition of essential oils from the leaves of phyllanthus amarus and Stachytarpheta cayennensis in Nigeria. *Int J Biochem Res Rev*. 22(3), 1–16.

Okoronkwo, N. E. & Echeme, J. O. (2015). Isolation and characterisation of compound from *Stachytarpheta cayennensis* (Rich.) Vahl leaves. *Chem J*. 1(3), 74–80.

Okoye, T. C., Aguwa, C. N., Okoli, C. O., Akah, P. A. & Nworu, C. S. (2008). Evaluation of the anticonvulsant and sedative effects of leaf extracts of Stachytarpheta cayennensis. *J Trop Med Plants*. 9(1), 17–22.

Okoye, T. C., Akah, P. A., Ezike, A. C. Uzor, P. F. Odoh, U. E., Igboeme, S. O., Onwuka, U. B. & Okafor, S. N. (2014). Immunomodulatory effects of Stachytarpheta cayennensis leaf extract and its synergistic effect with artesunate. *BMC Complement Altern Med*. https://doi.org/10.1186/1472-6882-14-376

Okoye, T. C., Akah, P. A., Okoli, C. O., Ezike, A. C. & Mbaoji, F. N. (2010). Antimicrobial and antispasmodic activity of leaf extract and fractions of Stachytarpheta cayennensis. *Asian Pacific J Trop Med*. 3(3), 189–192.

Okwu, D. E. & Ohenhen, O. N. (2009). Isolation, characterization and antibacterial activity of lanostane triterpenoid from the leaves of *Stachyterpheta jamaicensis* Linn Vahl. *Pharma Chem*. 1, 32–39.

Okwu, D. E. & Ohenhen, O. N. (2010). Isolation and characterization of steroidal glycosides from the leaves of *Stachytarpheta Jamaicensis* Linn Vahl. *Der Chem Sin*. 1(2), 6–14.

Olayiwola, G., Ukponmwan, O. & Olawode, D. (2013). Sedative and anxiolytic effects of the extracts of the leaves of *Stachytarpheta cayennensis* in mice. *Afr J Tradit Complement Altern Med*. 10(6), 568–579. https://doi.org/10.4314/ajtcam.v10i6.32

Osorio, D. V., González, I. R., Fermín, L. R., Silva, B. S., Arzola, J. C. & Meza, M. A. (2014). Composición del aceite esencial y caracterización físicoquímica de las hojas de *Stachytarpheta mutabilis* (Jacq.) *Vahl Avances Quim*. 15–19.

Park, J. C. (2004). Effects of methanol extract of Cirsium japonicum var ussuriense and its principle, hispidulin-7–0-neohesperidoside on hepatic alcohol metabolizing enzymes and lipid peroxidation in ethanol-treated rats. *Phytother Res*. 4, 19–24.

Penido, C. (2006). Anti-inflammatory and anti-ulcerogenic properties of Staphytarpheta cayennensis (L.C. Rich) *Vahl. J Ethnopharmacol*. 104(1), 225–233.

Princely, S., Basha, N. S., Kirubakaran, J. J. & Dhanaraju, M. D. (2013). Preliminary phytochemical screening and antimicrobial activity of aerial parts of *Stachytarpheta indica* L. (Vahl.). *Med Plants*. 5, 96–101.

Putera, I. & Anis Shazura, K. (2010). *Antimicrobial Activity and Cytotoxic Effects of Stachytarpheta Jamaicensis (L.) Vahl Crude Plant Extracts* [Master dissertation], Universiti Teknologi Malaysia.

Roengsumran, S., Sookkongwaree, K., Jaiboon, N., Chaichit, N. & Petsom, A. (2002). Crystal structure of ipolamiide monohydrate from *Stachytarpheta indica*. *Anal Sci*. 18, 1063–1064.

Schauer, J. C. (1847). Verbenaceae. In: A. P. de Candolle (ed.), *Prodromus*, vol. 11, pp. 522–570. https://doi.org/10.1007/978-3-642-18617-2_25

Sivaranjani, R., Ramakrishnan, K. & Bhuvaneswari, G. (2014). Pharmacognostic studies on root of *Stachytarpheta jamaicensis*. *Int Lett Nat Sci*. 8(2), 100–105.

Sreelatha, R., Challa, M., Kasturi, A., Edavana, S. K. & Sanivada, S. K. (2013). Phytochemical screening, total phenol content and antioxidant activity of different parts of *Stachyterpheta urticifolia* (Salisb) Sims. *JPR*. 1, 718–723.

Sulaiman, M. R., Zakaria, Z. A., Chiong, H. S., Lai, S. K., Israf, D. A. & Azam Shah, T. M. (2009). Antinociceptive and anti-inflammatory effects of *Stachytarpheta jamaicensis* (L.) Vahl (Verbenaceae) in experimental animal models. *Med Princ Pract*. 18(4), 272–279.

Thangiah, A. S. (2019). Phytochemical screening and antimicrobial evaluation of ethanolic-aqua extract of *Stachytarpheta jamaicensis* (L.) vahl leaves against some selected human pathogenic bacteria. *Rasayan J Chem*. 12(1), 300–307. http://doi.org/10.31788/RJC.2019.1215042

Udodeme, H. O., Odoh, U. E., Ugwu, P. N., Diovu, E. O., Okonta, E. O., Onyekere, P. F., et al. (2016). Pharmacognostic studies of the leaves of *Stachytarpheta jamaicensis* Linn. (Vahl) (*Verbenaceae*). *Int J Pharmacogn Phytochem Res*. 8, 1503–1508.

Valdés, A. F. C., Martínez, J. M., Lizama, R. S., Gaitén, Y. G., Rodríguez, D. A. & Payrol, J. A. (2010). In vitro antimalarial activity and cytotoxicity of some selected cuban medicinal plants. *Rev Inst Med Trop Sao Paulo*. 52(4), 197–201.

Vela, S. M., Souccar, C., Lima-Landman, M. T. & Lapa, A. J. (1997). Inhibition of gastric acid secretion by the aqueous extract and purified extracts of *Stachytarpheta cayennensis*. *Planta Med*. 63, 36–39.

World Health Organization. (2019). *World Malaria Report*. World Health Organization.

Yadav, P. D., Modi, K. P. & Shah, M. B. (2021). Phytochemistry, pharmacology, and botanical aspects of *Stachytarpheta* species – A review. *Int J Green Pharm*. 15(2). https://doi.org/10.22377/ijgp.v15i2.3078

Yuliana, A., Auwaliyah, F. & Fatmawati, S. (2019). 6β-hydroxyipolamiide of *Stachytarpheta jamaicensis* leaves. *J Technol Sci*. 30, 68–72.

Yuvaraj, R., Rao, M. R., Prabhu, K., Sundram, R. L., Shil, S., Kumar, M. S., et al. (2019). The gas chromatography-mass spectrometry study of one medicinal plant, *Stachytarpheta indica*. *Drug Invent Today*. 12, 1665–1669.

Zappi, D. C., Lucas, E., Stannard, B. L., Nic Lughadha, E., Pirani, J. P., de Queiroz, L. P., Atkins, S., Hind, D. J. N., Giulietti, A. M., Harley, R. M. & de Carvalho, A. M. (2003). Lista das plantas vasculares de Catolks, Chapada Diamantina, Bahia, Brasil. *Bol Bot Univ Sao Paulo*. 21(2), 345–398.

19 Antimalarial Response, Traditional and Other Potential Uses of *Tabebuia* Genera

Esmail Khosropour, Leila Hakimi, Ahmad Mirjalili, and Ayda Alavian

19.1 HERBAL MEDICINE

The study of pharmacognosy and the application of medicinal plants, which form the basis of conventional medicine, is known as herbal medicine. As a result of global pharmacology research, several herbal medications have been transformed into contemporary treatments, such as the artemisinin antimalarial drug class, which was isolated from the herb *Artemisia annua*, which was used to cure fever in traditional Chinese medicine (Luo et al., 2019; Khan and Ahmad, 2019). Herbal remedies provide benefits over modern medications, such as fewer side effects, cost-effectiveness, and affordability, which promote the development of herbal drug candidates. Plants have long been used for healing purposes (Husen, 2020, 2021, 2022a, 2022b, 2023; Rahman and Husen, 2021, 2022, 2023a, 2023b). Different kinds of metabolites are produced by plants. Plants produce phenolic compounds and alkaloids as secondary metabolites, and these substances can be used therapeutically to fight pathogens and other pathogen-related issues. Natural substances can also be helpful in determining the receptors' functional pathways (Chaudhari et al., 2019). For instance, acetylcholine receptors were divided into two groups based on how well they responded to the stimuli muscarine and nicotine, which were isolated from the fungi *Amanita muscaria* and *Nicotiana tabacum* L., respectively. Morphine was initially isolated from opium in 1803, and this marked the beginning of the isolation of substances having pharmacological properties from various portions of plants (Wicks et al., 2021). Digoxin and quinine were later produced using the genus digitalis and cinchona, respectively. What plant should we start with? Although it seems irrational, it is occasionally required to screen a big herbarium for a specific effect without any prior knowledge. Investigating some plant species that have previously shown promise as potential sources of important medicines is another strategy. Traditional remedies are also the greatest place to start your research; for instance, Egyptian traditional medical texts recommend using plant mixes in your prescriptions. Chinese herbal formulations, which have been used for thousands of years and have been studied for their therapeutic potential, have also been examined Chang and Luo et al., 2019; Wang et al., 2020).

19.2 MALARIA

Malaria is one of the serious parasite infections. A significant rate of death results from the increased prevalence of malaria brought on by treatment resistance (Talapko et al., 2019). One method for discovering new antimalarial medicines is the utilization of natural chemicals derived from plants. The drug artemisinin and its semi-synthetic derivatives, including artemether, arteether, artesunic acid, and dihydroartemisinin, are quite effective against strains of malaria that are resistant to a variety of antimalarial drugs. Artemisinin was first isolated from the plant *Artemisia annua* L. in 1972 (Patel et al., 2021). Several significant plants, including *Artemisia* spp., *Cinchona* spp., *Cryptolepis*

DOI: 10.1201/9781003378396-19

sanguinolenta (Lindl.) Schltr., *Handroanthus impetiginosus* (Mart. ex DC.), and *Tabebuia* spp., are regularly evaluated for their antimalarial activity in the literature (Mohammadi et al., 2020). Tropical locations have higher rates of illness and mortality due in large part to the widespread disease of malaria. The parasites that cause malaria in humans include *Plasmodium falciparum*, *P. vivax*, *P. ovale*, and *P. malariae* (Talapko et al., 2019; Mohammadi et al., 2020).

Primary diagnosis is crucial due to the varying treatment sensitivity of diverse species and the likelihood of mixed infections. Light microscopy is used to detect malaria infection in the bloodstream because it is suitable for different *Plasmodium* species diagnoses, highly accurate, and reasonably priced (World Health Organization [WHO], 2015). Depending on the pathogen, the appearance of infected red blood cells (RBCs) can be used to determine the type of parasite present. Worldwide instances of malaria infection decreased from 219 million in 2010 to 219 million in 2017, according to a WHO data (239 million infections). Moreover, 92% of infected cases in 2017 were reported from the African continent, where the majority of the patients carried *P. falciparum* parasites. In 2016, malaria claimed 451,000 lives worldwide, while in 2017, only 435,000 people perished from the disease. A potential therapy discovery may result from an understanding of the life cycle of the malaria parasite (Talapko et al., 2019; Dabaro et al., 2020, Figure 19.1).

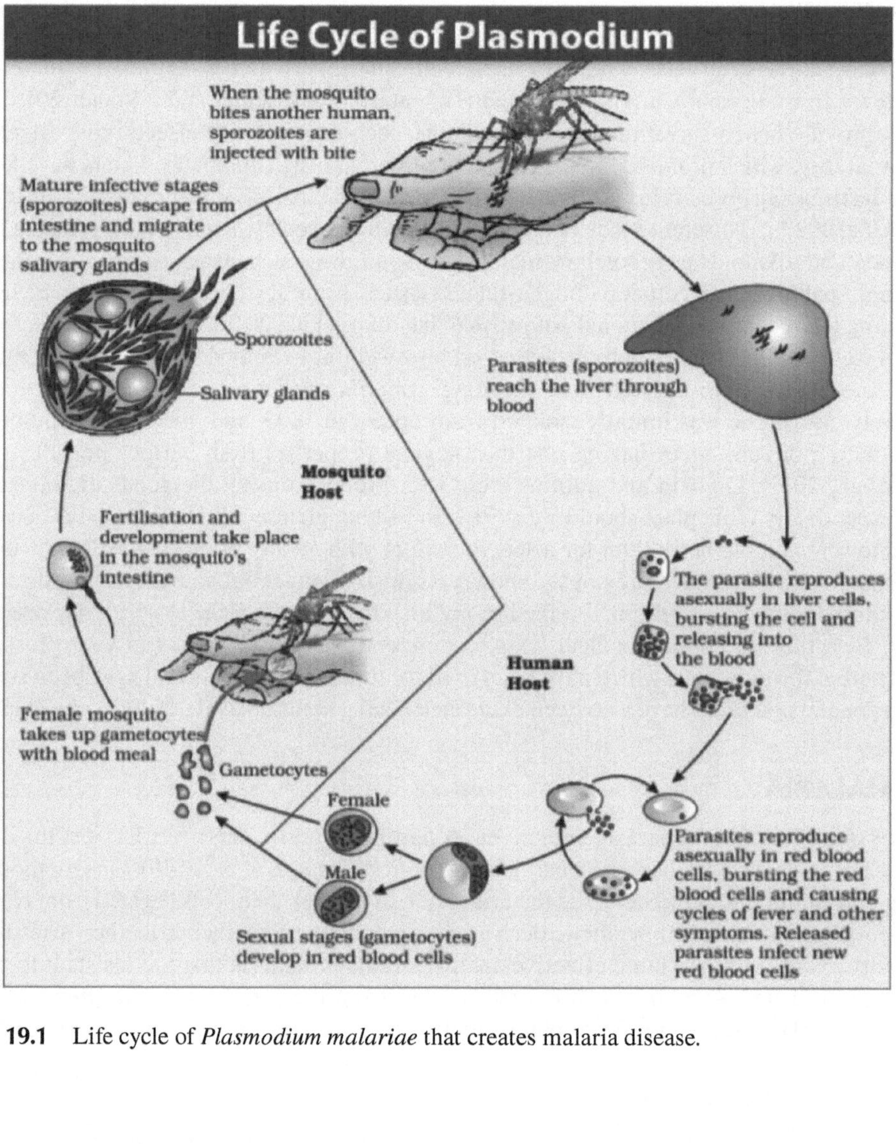

FIGURE 19.1 Life cycle of *Plasmodium malariae* that creates malaria disease.

19.3 MALARIA AND HERBAL MEDICINE

Chloroquine was the first-choice malaria treatment for many years; the quinine product from the *Cinchona* tree was effective up until chloroquine resistance developed as a result of improper use and monotherapy. Chloroquine is only currently indicated for the therapy of *Plasmodium vivax* malaria, and *Plasmodium falciparum* malaria is no longer treated with it (Price et al., 2020). Artemisinin was eventually created as a novel antimalarial agent as a result of comprehensive antipyretic herb screening for antimalarial activity (AMA) to combat medication resistance. The following aims for the creation of new antimalarial medications were suggested by Yadav et al. (2019). Artemisinin-based combination treatments (ACTs) were suggested as the first line of treatment for chloroquine-resistant malaria according to WHO recommendations. Plant-based antimalarial remedies have some benefits and drawbacks relative to totally synthetic medications. Therefore, the advantages are (Mohammadi et al., 2020):

1. The best medications for treating malaria, especially in cases where resistance has developed.
2. They have less severe adverse effects because they are compatible with the human body.
3. On the basis of natural compounds, new antimalarial pharmacophores can be developed.
4. Patient compliance is higher.
5. Since indigenous populations frequently use medicinal plants, by keeping track of these patients, additional clinical data can be gathered before it is actually employed in clinical practice.

And drawbacks of herbal medicines are:

1. Unlike chemical medicines, herbal sources may not be accessible in all countries.
2. The manufacture of synthetic endoperoxides is prioritized due to the increased expense of some herbal medicines, such as artemisinin, from herbal sources.
3. Due to limited drug half-lives and possible drug resistance, patients should take more than one dose of several herbal treatments each day, unless they are available in multiple formulations.
4. Insufficient high-quality clinical research and validation techniques for antimalarial plants.
5. Since the use of medicinal herbs is generally unrestricted, their safety and usage guidelines may be disregarded, which can lead to misuses and abuses.

19.4 *TABEBUIA* GENUS

Bignoniaceae has between 116 and 120 genera and 650 and 750 species. This family included seven tribes, most of which were found in tropical and subtropical regions of the globe (El-Hawary et al., 2021b). Bignoniaceae, often known as the trumpet vine or trumpet creeper family, receives its name from the genus *Bignonia*. Some of the family's well-known members include *Tecoma*, *Catalpa*, *Tabebuia*, and *Jacaranda*. Many nations, including Bangladesh and the United States, use a lot of members of this family in traditional medicine (Mohammadi et al., 2020). In the Bignoniaceae family, *Tabebuia* is the biggest and most significant genus. Antonio Gomes, a taxonomist, was the first to use '*Tabebuia*' in literature in 1803; later, in 1838, de Candolle adopted the word as a generic name (Hamed et al., 2020). The term '*Tabebuia*' is an abbreviation for the phrase 'tacyba bebuya', which means 'ant wood', and refers to ants that live in the hollow twigs of some *Tabebuia* species. A vast genus of blooming trees in the tropical and subtropical regions, *Tabebuia* contains over 100 species (El-Hawary et al., 2021b). Figures 19.2 and 19.3 show two main species of *Tabebuia*.

FIGURE 19.2 *Tabebuia Avellanedae* (pink trumpet tree).

19.5 MALARIA AND *TABEBUIA*

Traditional medicine routinely employs *Tabebuia* species to treat a variety of ailments, including syphilis, malaria, cutaneous infections, stomach disorders, cancer, inflammation, pain, bacterial and fungal infections, anxiety, poor memory, irritability, and depression, as well as diabetes, prostatitis, constipation, and allergies (El-Hawary et al., 2021b). The biological efficacy of secondary metabolites extracted from various members of this genus and employed in clinical investigations as adjuvants in cancer therapy, such as lapachol, has been reported in several publications. Bark extract from *Tabebuia* species is referred to as 'taheebo', 'lapacho', 'pau d'arco', or 'ipê', and its active ingredients include napthoquinones, quinines, furanonapthoquinones, benzoic acid, cyclopentene dialdehydes, and flavonoids (Zhang et al., 2020). Table 19.1 shows the antimalarial responses of different *Tabebuia* species. For example, Gómez-Estrada et al. (2012) reported that traditional chromatographic methods and bioassay-guided fractionation (*Artemia salina*) enabled the isolation of one naphthoquinone (lapachol) and four naphtho-furan-4,9-diones from the active fractions. As compared to chloroquine's performance in the same experiment, these compounds' results showed

FIGURE 19.3 *Tabebuia Argentea* (silver trumpet tree).

TABLE 19.1
Antimalarial Responses of *Tabebuia* Species

Species	Results	Key References
T. ochracea	The highest antimalarial activity against *Plasmodium berghei* was given by a mixture of two compounds which could not be separated, but the isomeric structures of 5- and 8-hydroxy-2-(1'-hydroxy)-ethyl-naphtho-[2,3-b]-furan-4,9-dione (1 and 2) were determined from spectroscopic data.	Pérez et al., 1997
T. billbergii	One naphthoquinone (lapachol) and four naphtho-furan-4,9-diones were able to be isolated from the active fractions using conventional chromatographic techniques and bioassay-guided fractionation (*Artemia salina*). These substances demonstrated a significant antiplasmodial impact and had extremely encouraging IC50 values, especially when compared to chloroquine's performance in the same assay. Moreover, the bark was used to extract the triterpenes b-sitosterol and stigmasterol.	Gómez-Estrada et al., 2012
T. incana	The trunk wood of *Tabebuia incana* afforded the lignans pawlownin and cycloolivil as the major constituents together with a cycloolivil acetonide, la.	de Oliveira et al., 1993

a significant antiplasmodial action and extremely encouraging IC50 values. B-sitosterol and stigmasterol, two triterpenes, were also extracted from the bark.

19.6 PHYTOCHEMICALS OF *TABEBUIA*

Naphthoquinones are thought to be the primary component among the 292 chemical compounds that have been identified from *Tabebuia* to date. Alkaloids, iridoids, tannins, and flavonoids are further groups of secondary metabolites that have been described (Tahara et al., 2020). Several studies recommend basic phytochemical screening as the initial step in identifying chemical classes. The leaves, bark, and wood of *Tabebuia* have been found to contain a variety of phytochemical groups. For instance, 19 glycosides, including four iridoid glycosides, two lignan glycosides, two isocoumarin glycosides, three phenylethanoid glycosides, and eight phenolic glycosides, were extracted using methanol from the bark of *T. impetiginosa*. Furanonaphthoquinones, naphthoquinones, anthraquinones (such as anthraquinone-2-carboxylic acid quinones), benzoic acid, flavonoids, cyclopentene dialdehydes, coumarins, iridoids, and phenolic glycosides are the main components of *T. impetiginosa* (Zhang et al., 2020). Scientists were interested in the presence of naphthoquinones; those in the medical field were particularly interested in lapachol and -lapachone. While -lapachone demonstrates significant toxicity in murine and human cells, lapachol prevents the development of malignant cells (Stepczyńska et al., 2021). In *Drosophila melanogaster* heterozygous for the tumour suppressor gene, lapacol has been demonstrated to lessen the frequency of tumours brought on by doxorubicin. Invasion of HeLa cells can be reduced by lapachol, which may be a useful building block for the creation of brand-new antimetastatic drugs. The bark of *Tabebuia* contains fatty acids, including oleic, palmitic, and linoleic acids. The presence of free sugars was also found in the bark; glucose was the most prevalent, followed by fructose and sucrose (Roy et al., 2022). There are other organic acids, including oxalic acid, and the fatty alcohols -tocopherol and -tocopherol. Cardiovascular disease and neurological illnesses are made less likely by -tocopherol. Moreover, *Tabebuia* has a few volatile components that have antioxidant properties. The main volatile components of *Tabebuia* include 1-methoxy-4-(1E)-1-propenylbenzene, 4-methoxybenzaldehyde, 4-methoxyphenol, 5-allyl-1,2,3-trimethoxybenzene, and 4-methoxybenzyl alcohol (Zhang et al., 2020). The main phytochemicals in *Tabebuia* follow.

19.6.1 NAPHTHOQUINONES

Naphthoquinone is a form of quinone that is present in nature and plays important biological roles. These substances have actions that are leishmanicidal, analgesic, anti-inflammatory, antimalarial, antifungal, antiviral, antitrypanosomal, antischistosomal, anticancer (antineoplastic), and antiulcerative (Pereyra et al., 2019; Rahman et al., 2022). The 1,4-naphthoquinone ring can be directly substituted, which changes the naphthoquinone's acid/base and oxidation/reduction properties as well as its activity. The scientific world has been very interested in naphthoquinones because of their pharmacological characteristics, such as anticancer activity and possible therapeutic application (Pereyra et al., 2019). It has been discovered that some substances with a quinone ring in malignant cells have antiproliferative properties. The lethal effects of naphthoquinones are linked to their ability to block electron transporters, decouple oxidative phosphorylation, produce ROS, and create protein adducts, particularly with -SH enzyme groups (de Almeida et al., 2021). The main components of *Tabebuia* are naphthoquinones. Lapachol, a naturally occurring 1,4-naphthoquinone that is widely dispersed in this genus, as well as -lapachone, the most prevalent naphthoquinone isolated from the genus, are currently in the clinical trial stage as plant-derived anticancer agents. Additionally, β-lapachone is a potential depigmentation agent for various hyperpigmentation disorders in skincare preparations. Naphthoquinones received a special consideration in *Tabebuia* species due to their pharmacological, anti-inflammatory, antimicrobial, antimalarial activity, and antileishmanial activities (El-Hawary et al., 2021b).

19.6.2 Lignans

Lignans are secondary metabolites found in plants that are produced when phenylpropanoids undergo oxidative coupling. These phenolic compounds have biological effects that range from anti-inflammatory, antitumour (terminaloside P, IC50 = 10 nM), anti-inflammatory, anti-neurodegenerative (schibitubin B, IC50 = 3.2 nM), antiviral (patentiflorin A, IC50 = 14–23 nM), to antibacterial (Zálešák et al., 2019). Also, it was shown that certain individuals in this group, specifically enterolactone and its biochemical precursors, which are also referred to as phytoestrogens, have significant protective effects. As most of these lignans are consumed in acceptable amounts, their ability to protect against cancers like breast and colon is all the more crucial to notice. The leaves of *T. argentea* have 5-hydroxysesamin 5-*O*-β-D-glucopyranosyl-(1–2)-[β-D-gluco-pyranosyl-(1–6)]-β-D-glucopyranoside. Among 26 lignans isolated and identified in *Tabebuia*, avallandae A has anti-inflammatory activity, icariside E4 shows antinociceptive activity, and lyoniresinol-3a-*O*-b-D-*gluco*-pyranoside has a potent antioxidant activity (El-Hawary et al., 2021a).

19.6.3 Flavonoids and Phenolic Compounds

Different parts of plants like fruits, grains, bark, roots, stems, and flowers all contain flavonoids – a class of organic compounds with varying phenolic structures (Dias et al., 2021). There are efforts being undertaken to extract the components known as 'flavonoids' because these natural compounds are well-known for their positive impacts on health. Nowadays, flavonoids are regarded as a crucial ingredient in a wide range of nutraceutical, pharmacological, therapeutic, and cosmetic applications. This is explained by their ability to modulate important cellular enzyme function in addition to their antioxidative, anti-inflammatory, antimutagenic, and anticarcinogenic qualities (Khosropour et al., 2021; Hernández-Rodríguez et al., 2019). The majority of the flavonoids in *Tabebuia* are flavonols, while the presence of other flavonoids appears to be sparse. The leaves and flowers of *T. argentea*, *T. pentaphylla*, *T. ochracea*, and *T. caraiba* were where the majority of the reported flavonoids were found. Antioxidant, anti-inflammatory, and neuroprotective properties are known for phenylethanoids and phenylpropanoids. *T. avellanedae* and *T. chrysotricha* were the sources of the majority of these compounds. The presence of flavonoids and phenolic compounds in *Tabebuia* extracts is responsible for the plant's antioxidant properties (El-Hawary et al., 2021a).

19.6.4 Coumarins

Natural substance known as coumarin, which is present throughout the entire plant kingdom, has aromatic and fragrant properties. It can be found in a variety of plant sources, including fruits, vegetables, spices, and medicinal plants, as well as in all the plant's parts, including the fruits, roots, stems, and leaves. One of the most common sources of coumarin exposure for people are certain types of cinnamon, which has significant amounts of the chemical. Since there are no formal measurements of the consumption of foods containing cinnamon, the precise amount of coumarin that humans are exposed to has not been established. Pure coumarin cannot be added to food because it can be toxic to the liver in large doses (Lončar et al., 2020). Coumarins have thrombolytic, inflammatory, vasodilator, and bactericidal properties (Kontogiorgis et al., 2015). The bark of *T. avellanedae* was the source of the latter two, 6-hydroxymellein and the novel coumarin 1-(5-(hydroxymethyl) furan-2-yl) isochroman-6,7-diol (El-Hawary et al., 2021b).

19.6.5 Carotenoids

Tetraterpene pigments known as carotenoids are found in a variety of organisms, including some types of algae, plants, mammals, and some species of archaea and fungi. Up to 2018, 850 naturally occurring carotenoids had been reported (Maoka, 2020). Carotenoids can be made by photosynthetic bacteria, fungus, algae, and plants. Together with chlorophylls, carotenoids are crucial pigments in photosynthetic organs (Zia-Ul-Haq, 2021). In non-photosynthetic organs of plants, carotenoids also function as photoprotectors, antioxidants, colour attractants, and precursors of plant hormones. Animals cannot naturally produce carotenoids; thus, those that are found in them are either directly ingested from food sources or partially altered via metabolic processes. Animal carotenoids exhibit structural variation as a result (Maoka, 2020). Animals utilize carotenoids for a variety of functions, including as precursors to vitamin A, photoprotectors, antioxidants, immune system boosters, and contributors to reproduction. The abundance of carotenoids in Tabebuia species may be the cause of the ornamental flower colours. The yellow blooms of *T. argentea* were used to isolate and identify four carotenoid components (El-Hawary et al., 2021b).

19.6.6 Aldehyde, Acids, and Esters

Several aldehydes, acids, and esters have been recorded for *Tabebuia* species. All aldehydic substances have been included in *T. avellanedae*. The identified acidic compounds were found in a variety of *Tabebuia* species, including *T. rosea*, *T. heptaphylla*, *T. aurea*, *T. avellanedae*, *T. palmeri*, and *T. roseo-alba* (El-Hawary et al., 2021b; Sirigeri et al., 2022). These compounds included benzoic acid, 4-hydroxycinnamic acid, and caffeic acid. The *Tabebuia* species isolated included cyclopentenyl esters from *T. avellanedae*, *T. rosea*, and *T. heptaphylla*, among other ester compounds. The stem bark of *T. aurea* was used to isolate the cinnamate derivative (El-Hawary et al., 2021a).

19.6.7 Hydrocarbons, Triterpenoids, and Sterols

Analysis of *T. palmeri* and *T. pentaphylla* extracts revealed three fatty alcohols and two fatty acids. Moreover, *T. pentaphylla* and *T. rosea* were the sources of four hydrocarbons. The trunk bark of *T. heptaphylla* was used to extract the linear triterpene squalene (Salatino et al., 2020;El-Hawary et al., 2021b). Squalene was found in several trials to stimulate the immune system against HIV, H1N1, leukaemia, and herpes, as well as to limit the growth of tumours in the colon, skin, lung, and breast. Plant sterols are renowned for their capacity to lower cholesterol levels, aid in the prevention of heart disease, and reduce the risk of heart attacks (Gylling et al., 2014).

19.7 TRADITIONAL USES OF SOME *TABEBUIA* SPECIES

Portuguese and Spanish population used the names of Pau d'arco and lapacho to identify about 26 species of shrubs and trees belong to *Tabebuia*. These species are indigenous to the American tropics, from Mexico to southern South America; majority of the species are found in Brazil and neighbouring countries (Bussmann, 2018). For curative purposes, native people preferred the inner bark, although the heartwood is more potent. Leaves and flowers are less commonly used. In the early 1980s, d'arco became known in North America and Europe (El-Hawary et al., 2021b). The infusion and decoction of the bark or wood was ingested regularly by at least one million people. In 1995, d'arco was listed among the top 25 selling herbs in the United States, representing 1.7% of herb sales in United States in 1996. Old native populations used *Tabebuia* extracts as an antidote for snakebites. *Tabebuia* has been widely used for treating a large number of diseases in human beings. Different parts like leaf and bark have been mainly used among the local people. Many cancers, diabetes, malignant tumours, leukaemia, and different types of inflammatory disorders have been treated by *Tabebuia* species. For instance, *T. avellanedae* has treated eczema, psoriasis, fungal infections, skin cancers, malaria, leishmaniosis, fevers, fungal, bacterial infection, cold, cough, and flu (de Miranda et al., 2001; Rodrigues, 2006; Yamashita et al., 2007). *T.* aurea has a good anti-inflammatory potential (Malange et al., 2019), and *T. heptaphylla* has been mainly used as anti-inflammatory, anticancer, and for treating wounds (Schmeda-Hirschmann and Papastergiou, 2003) (Table 19.2).

TABLE 19.2
Traditional Use of Different *Tabebuia* Species

Species	Region	Traditional Use	Key Reference
T. avellanedae Lorentz ex Griseb	USA and Brazil	To treat tumours and cancers in folk medicine	Ueda et al., 1994Zhang et al., 2015
		To treat eczema, psoriasis, fungal infections, and even skin cancers	Pereira et al., 2013; Suo and Yan, 2016
		To treat ulcers, bacterial and fungal infections	De Miranda et al., 2001; Twardowschy et al., 2008
		To treat malaria, leishmaniasis, fevers, fungal, bacterial infections, and syphilis	Duke, 1985; Duke and Vasquez, 1994
		For gastrointestinal disturbances, inflammation, and tropical diseases (Rodrigues, 2006); to treat colds, coughs, and flu	Rodrigues, 2006
		To treat colds, coughs, and flu	Grenand et al., 2018
		To treat uterine cancer and liver cirrhosis	
T. impetiginosa	Amazon rainforest, Argentina, Bolivia, Brazil, Colombia, Ecuador, French Guinea, Paraguay, Peru, Suriname, Trinidad, Tobago, and Venezuela	To treat diabetes, malignant tumours, leukaemia, other cancers, anaemia, and Parkinson's disease	Lewis et al., 2004
		Anti-inflammatory and for treatment of fungal infections	Castellanos et al., 2009
T. aurea (Manso) S. Moore	South America (from Venezuela to Argentina)	Anti-inflammatory	Nunes et al., 2003; Reis et al., 2014; Malange et al., 2019
T. argentea Britt	South America (from Venezuela to Argentina) and India	Anti-inflammatory and for treating influenza	Agra, 1996; De Abreu et al., 2014
T. chrysotricha	Brazil	Analgesic, antitumour agent, antidiabetic, and for treatment of peptic ulcer	Grazziotin et al., 1992
T. incana A. H. Gentry	Amazon	Anti-inflammatory, antimalarial, anticancer, and for the treatment of kidney and liver disorders	da Silva et al., 1977; de Oliveira et al., 1993
T. heptaphylla	Eastern Paraguay	Anti-inflammatory, anticancer, and for treating wounds	Gupta, 1995; Ortega Torres et al., 1989; Schmeda-Hirschmann and Papastergiou, 2003
T. ochracea ssp. neochrysantha	Tropical America, from El Salvador to north-west Venezuela	Antimalarial and for healing ulcers	Franco Ospina et al., 2013
T. rosea (Bertol.)	Guatemala, Costa Rica, Colombia	Antipyretic and for treating eye infections	

TABLE 19.3
Antioxidant Potential (Such as Anticancer an Anti-Inflammatory Properties) of *Tabebuia*

Species	Plant Tissue	Attribute	Results	Key References
T. pallida	Leaf	Anticancer	High phenolic and flavonoid content; high potential in antioxidant capacity by scavenging free radicals, which show high anti-ROS and anticancer potential.	Rahman et al., 2019
T. argentea, T. guayacan, T. pulcherrima, T. pallida, T. rosea	Stem and leaf	Anticancer	The highest activity observed for *T. rosea* (Bertol.) DC stem extract against HepG2 cell line (IC_{50} 4.7 μg/mL), *T. pallida* L. stem extract against MCF-7 cell line (IC_{50} 6.3 μg/mL), and *T. pulcherrima* stem extract against CACO2 cell line (IC_{50} 2.6 μg/mL).	El-Hawary et al., 2021a
T. guayacan	Leaf	Anticancer	Sixteen compounds were isolated, and their structures were elucidated using different spectroscopic techniques. The drug likeness of the isolated compounds, as well as their binding affinity with three anticancer drug target receptors – CDK-2/6, topoisomerase-1, and VEGFR-2 – were evaluated. Additionally, the most promising compounds were *in vitro* evaluated for inhibitory activities against CDK-2/6 and VEGFR-2 enzymes using kinase assays method. The results showed high anticancer activity by obtaining low IC50 value.	El-Hawary et al., 2022
T. avellanedae	Bark	Antidepressant-like action	The tail suspension test (TST) immobility duration was reduced in mice-administered ethanol extract (10 and 30 mg/kg) over the course of 14 days without affecting locomotor activity, suggesting an antidepressant-like effect. At dosages of 10–30 and 30 mg/kg, respectively, EET per se enhanced CREB (Ser133) and GSK-3 (Ser9) phosphorylation in sham-operated animals. Olfactory bulbectomy increased immobility time in the TST, CREB, and ERK1 phosphorylation, BDNF immunocontent, and hyperactivity. It also resulted in a lack of motivation and self-care behaviour.	Freitas et al., 2013
	Bark	Anti-inflammatory	From the water extract, five new compounds were extracted, and their structures were determined through study of NMR spectroscopy and mass spectrometry. In cultured human myeloma THP-1 cells co-stimulated with lipopolysaccharide, compounds 1–5 at 25 M demonstrated strong inhibitory activity on the production of inflammatory cytokines, tumour necrosis factor, and interleukin-1 without any significant cytotoxicity, and their anti-allergic and antioxidant activities were assessed.	Suo et al., 2012
	Inner bark	Anti-inflammatory	The ethnopharmacological action of *T. avellanedae* in treating inflammatory illnesses was based on its components, which have a notable anti-inflammatory effect on inflammatory reactions.	Zhang et al., 2016
	Bark	Anti-obesity	After 16 weeks of feeding, mice given a 0.5% n-BuOH extract of Taheebo showed significantly lower body weight compared to control mice, and the amount of fat mass also fell dramatically.	Iwamoto et al., 2016

T. aurea	Leaf and bark	Antimycobacterial	Aqueous and alcoholic extracts showed that the aqueous extracts of stem bark were more effective as compared to aqueous and leaf extracts and alcoholic stem bark and leaf extracts.	Agarwal et al., 2015
	Stem bark	Inflammatory, myotoxic, and haemorrhagic activities	Hydroethanolic extract prevented polymorphonuclear cell recruitment into the peritoneal cavity and paw oedema. When *Bothrops neuwiedi* (BnV) was neutralized with hydroethanolic extract, the haemorrhagic activity was diminished, and skeletal muscle tissue histopathological investigation revealed that the haemorrhagic area was smaller and multifocal. Less severe muscle necrosis and a leukocyte infiltration were present. BnV promoted the creation of hydrogen peroxide, and BnV neutralized and decreased this production.	Reis et al., 2014
T. impetiginosa	Inner bark	Antibacterial	At 0.1 mg/disc, anthraquinone-2-carboxylic acid, lapachol, and metronidazole showed only mild anti–*Helicobacter pylori* efficacy. The two substances with the strongest activity at 0.005 mg/disc were amoxicillin and tetracycline. At this dosage, 2-(Hydroxymethyl) anthraquinone showed average activity. At 0.001 mg/disc, tetracycline had considerable action, whereas amoxicillin exhibited negligible effect. Lapachol (4 g/mL), 2-(hydroxymethyl) anthraquinone (2 g/mL), and anthraquinone-2-carboxylic acid (8 g/mL), on the other hand, were less active than amoxicillin (0.063 g/mL) and tetracycline (0.5 g/mL) in the MIC bioassay.	Park et al., 2006
Tabebuia impetiginosa	Bark	Antiparkinsonian	The bark extract showed higher levels of dopamine. In the reserpine-induced Parkinsonian model, the extract decreased postural flexion, somewhat decreased tremor, muscle rigidity, and postural immobility scores, but did not significantly lower the score for hypokinesia. In neuroleptic-induced Parkinsonism, a high dose of the extract (500 mg/kg) is more effective than the conventional medication at reducing catatonia and catalepsy scores.	Roy et al., 2022

19.8 OTHER MEDICINAL POTENTIAL (LIKE ANTICANCER AND ANTI-INFLAMMATORY PROPERTIES) OF *TABEBUIA*

Tabebuia species showed higher potential in scavenging reactive oxygen species (ROS), which represents high antioxidant capacity against concerns and inflammatory. Park et al. (2006) used paper disc assay to evaluate anti–*Helicobacter pylori* activity of seven 1,4-naphthoquinone derivatives (structurally related to lapachol), 1,4-naphthoquinone, 5,8-dihydroxy-1,4-naphthoquinone (naphthazarin), 2-methyl-1,4-naphthoquinone (menadione), 2-hydroxy-1,4-naphthoquinone (lawsone), 5-hydroxy-2-methyl-1,4-naphthoquinone (plumbagin), 5-hydroxy-1,4-naphthoquinone (juglone), and 2,3-dichloro-1,4-naphthoquinone (dichlone) by inner bark of *T. impetiginosa.* Zhang et al. (2016) represented that ethnopharmacological action of *T. avellanedae* in treating inflammatory illnesses was based on its components, which have a notable anti-inflammatory effect on inflammatory reactions. Roy et al. (2022) showed that bark extract of *T. impetiginosa* decreased postural flexion, somewhat decreased tremor, muscle rigidity, and postural immobility scores, but did not significantly lower the score for hypokinesia. For lowering catatonia and catalepsy scores in neuroleptic-induced Parkinsonism, 500 mg/kg is superior to the standard drug (Table 19.3).

19.9 CONCLUSION

Medicinal woody plants play a vital role in human health because numerous parts, including leaves, fruits, and bark, can be employed in herbal remedies. *Tabebuia* is a genus of the medicinal trees that has a number of benefits for treating a wide range of illnesses. These trees are capable of treating cancer, inflammatory conditions, malaria, and microbial infections due to their main chemical constituents, which also include napthoquinones, quinines, furanonapthoquinones, benzoic acid, and flavonoids. Therefore, it is a promising practice to develop different species of *Tabebuia* according to how well they adapt to the climate to replace chemical drugs globally, especially in places where the availability of chemical drug is restricted.

REFERENCES

Agarwal, M., & Chauhan, S. (2015). Anti-Mycobacterial potential of *Tabebuia aurea* (Manso) Benth & Hook (Bignoniaceae). *Journal of Medicinal Plants Studies*, *3*, 63–68.

Agra, M. D. F. (1996). *Plantas da medicina popular dos cariris velhos, Paraíba, Brasil*. João Pessoa, Editora União.

Bussmann, R. W. (2018). Tabebuia avellanedae Lorentz ex Griseb. In *Medicinal and Aromatic Plants of South America: Brazil* (pp. 439–451). Springer, Dordrecht.

Castellanos, J. R. G., Prieto, J. M., & Heinrich, M. (2009). Red Lapacho (Tabebuia impetiginosa) – a global ethnopharmacological commodity? *Journal of Ethnopharmacology*, *121*(1), 1–13.

Chaudhari, S. K., Arshad, S., Amjad, M. S., & Akhtar, M. S. (2019). Natural compounds extracted from medicinal plants and their applications. In *Natural Bio-active Compounds: Volume 1: Production and Applications* (pp. 193–207). Springer, Singapore.

Dabaro, D., Birhanu, Z., & Yewhalaw, D. (2020). Analysis of trends of malaria from 2010 to 2017 in Boricha District, Southern Ethiopia. *Malaria Journal*, *19*, 1–10.

da Silva, M. F., Lisbôa, P. L. B., & Lisbôa, R. C. L. (1977). *Nomes Vulgares de Plantas Amazônicas*. INPA, Manaus.

De Abreu, M. B., Temraz, A., Vassallo, A., Braca, A., & De Tommasi, N. (2014). Phenolic glycosides from Tabebuia argentea and Catalpa bignonioides. *Phytochemistry Letters*, *7*, 85–88.

de Almeida, P. D., Jobim, G. D. S. B., dos Santos Ferreira, C. C., Bernardes, L. R., Dias, R. B., Sales, C. B. S., . . . & Lima, E. S. (2021). A new synthetic antitumor naphthoquinone induces ROS-mediated apoptosis with activation of the JNK and p38 signaling pathways. *Chemico-Biological Interactions*, *343*, 109444.

De Miranda, F. G. G., Vilar, J. C., Alves, I. A. N., Cavalcanti, S. C. D. H., & Antoniolli, Â. R. (2001). Antinociceptive and antiedematogenic properties and acute toxicity of Tabebuia avellanedae Lor. ex Griseb. inner bark aqueous extract. *BMC Pharmacology*, *1*(1), 1–5.

de Oliveira, A. B., Raslan, D. S., de Oliveira, G. G., & Maia, J. G. S. (1993). Lignans and naphthoquinones from Tabebuia incana. *Phytochemistry*, *34*(5), 1409–1412.

Dias, M. C., Pinto, D. C., & Silva, A. M. (2021). Plant flavonoids: Chemical characteristics and biological activity. *Molecules*, *26*(17), 5377.

Duke, J. A. (1985). *CRC Handbook of Medicinal Herbs*. Boca Raton.

Duke, J. A., & Vasquez, R. (1994). *Amazonian Ethnobotanical Dictionary*. CRC Press.

El-Hawary, S. S., Mohammed, R., Taher, M. A., AbouZid, S. F., Mansour, M. A., Almahmoud, S. A., . . . & Amin, E. (2022). Characterization of promising cytotoxic metabolites from tabebuia guayacan hemsl.: Computational prediction and in vitro testing. *Plants*, *11*(7), 888.

El-Hawary, S. S., Mohammed, R., Tawfike, A. F., AbouZid, S. F., Taher, M. A., Abdelmohsen, U. R., & Amin, E. (2021). Metabolic profiling of cytotoxic metabolites from five Tabebuia species supported by molecular correlation analysis. *Scientific Reports*, *11*(1), 8405.

El-Hawary, S. S., Taher, M. A., Amin, E., AbouZid, S. F., & Mohammed, R. (2021). Genus Tabebuia: A comprehensive review journey from past achievements to future perspectives. *Arabian Journal of Chemistry*, *14*(4), 103046.

Franco Ospina, L. A., Castro Guerrero, J. P., Ocampo Buendía, Y. C., Pájaro Bolívar, I. B., & Díaz Castillo, F. (2013). Actividad antiinflamatoria, antioxidante y antibacteriana de dos especies del género Tabebuia. *Revista Cubana de Plantas Medicinales*, *18*(1), 34–46.

Freitas, A. E., Machado, D. G., Budni, J., Neis, V. B., Balen, G. O., Lopes, M. W., . . . & Rodrigues, A. L. S. (2013). Antidepressant-like action of the bark ethanolic extract from Tabebuia avellanedae in the olfactory bulbectomized mice. *Journal of Ethnopharmacology*, *145*(3), 737–745.

Gómez-Estrada, H., Gaitán-Ibarra, R., Díaz-Castillo, F., Pérez, H. A., & Medina, J. D. (2012). In vitro antimalarial activity of fractions and constituents isolated from Tabebuia billbergii. *Revista Cubana de Plantas Medicinales*, *17*(2), 172–180.

Grazziotin, J. D., Schapoval, E. E. S., Chaves, C. G., Gleye, J., & Henriques, A. T. (1992). Phytochemical and analgesic investigation of Tabebuia chrysotricha. *Journal of Ethnopharmacology*, *36*(3), 249–251.

Grenand, P., Moretti, C., Jacquemin, H., & Prévost, M. F. (Eds.). (2018). *Pharmacopées traditionnelles en Guyane: créoles, wayãpi, palikur*. IRD éditions.

Gupta, M. P. (1995). 270 plantas medicinales iberoamericanas. CYTED-SECAB. *Presencia, Bogotá*, *617*.

Gylling, H., Plat, J., Turley, S., Ginsberg, H. N., Ellegård, L., Jessup, W., . . . & Chapman, M. J. (2014). Plant sterols and plant stanols in the management of dyslipidaemia and prevention of cardiovascular disease. *Atherosclerosis*, *232*(2), 346–360.

Hamed, A. N. E., Mahmoud, B. K., Samy, M. N., & Kamel, M. S. (2020). An extensive review on genus "Tabebuia", family bignoniaceae: Phytochemistry and biological activities (1967 to 2018). *Journal of Herbal Medicine*, *24*, 100410.

Hernández-Rodríguez, P., Baquero, L. P., & Larrota, H. R. (2019). Flavonoids: Potential therapeutic agents by their antioxidant capacity. In *Bioactive Compounds* (pp. 265–288). Woodhead Publishing.

Husen, A. (2020). *Exploring Medicinal Plants* (Book Series). CRC Press.

Husen, A. (2021). *Traditional Herbal Therapy for the Human Immune System*. CRC Press. https://doi.org/10.1201/9781003137955

Husen, A. (2022a). *Herbs, Shrubs and Trees of Potential Medicinal Benefits*. CRC Press. https://doi.org/10.1201/9781003205067

Husen, A. (2022b). *Exploring Poisonous Plants: Medicinal Values, Toxicity Response and Other Potential Uses*. CRC Press. https://doi.org/10.1201/b23017

Husen, A. (2023). *Antidiabetic Medicinal Plants and Herbal Treatments*. CRC Press. https://doi.org/10.1201/b23347

Iwamoto, K., Fukuda, Y., Tokikura, C., Noda, M., Yamamoto, A., Yamamoto, M., . . . & Moriyama, T. (2016). The anti-obesity effect of Taheebo (Tabebuia avellanedae Lorentz ex Griseb) extract in ovariectomized mice and the identification of a potential anti-obesity compound. *Biochemical and Biophysical Research Communications*, *478*(3), 1136–1140.

Khan, M. S. A., & Ahmad, I. (2019). Herbal medicine: Current trends and future prospects. In *New Look to Phytomedicine* (pp. 3–13). Academic Press.

Khosropour, E., Weisany, W., Tahir, N. A. R., & Hakimi, L. (2021). Vermicompost and biochar can alleviate cadmium stress through minimizing its uptake and optimizing biochemical properties in Berberis integerrima bunge. *Environmental Science and Pollution Research*, 1–11.

Kontogiorgis, C., Nicolotti, O., Mangiatordi, G. F., Tognolini, M., Karalaki, F., Giorgio, C., . . . & Barocelli, E. (2015). Studies on the antiplatelet and antithrombotic profile of anti-inflammatory coumarin derivatives. *Journal of Enzyme Inhibition and Medicinal Chemistry*, *30*(6), 925–933.

Lewis, W. H., Okunade, A. L., & Elvin-Lewis, M. P. F. (2004). Pau d'Arco or Lapacho (Tabebuia). In *Encyclopedia of Dietary Supplements* (pp. 527–535). CRC Press. https://doi.org/10.1081/E-EDS-120022122

Lončar, M., Jakovljević, M., Šubarić, D., Pavlić, M., Buzjak Služek, V., Cindrić, I., & Molnar, M. (2020). Coumarins in food and methods of their determination. *Foods*, *9*(5), 645.

Luo, H., Vong, C. T., Chen, H., Gao, Y., Lyu, P., Qiu, L., . . . & Wang, Y. (2019). Naturally occurring anti-cancer compounds: Shining from Chinese herbal medicine. *Chinese Medicine*, *14*(1), 48.

Malange, K. F., Dos Santos, G. G., Kato, N. N., Toffoli-Kadri, M. C., Carollo, C. A., Silva, D. B., . . . & Rondon, E. S. (2019). Tabebuia aurea decreases hyperalgesia and neuronal injury induced by snake venom. *Journal of Ethnopharmacology*, *233*, 131–140.

Maoka, T. (2020). Carotenoids as natural functional pigments. *Journal of Natural Medicines*, *7*.

Mohammadi, S., Jafari, B., Asgharian, P., Martorell, M., & Sharifi-Rad, J. (2020). Medicinal plants used in the treatment of Malaria: A key emphasis to Artemisia, Cinchona, Cryptolepis, and Tabebuia genera. *Phytotherapy Research*, *34*(7), 1556–1569.

Nunes, G. P., Da Silva, M. F., Resende, U. D., & De Siqueira, J. M. (2003). Plantas medicinais comercializadas por raizeiros no Centro de Campo Grande, Mato Grosso do Sul. *Revista Brasileira de Farmacognosia*, *13*, 83–92.

Ortega Torres, E., Status de Ortega, L., & Spichiger, R. (1989). *Noventa Especies Forestales Del Paraguay*. Consertatoire et jardin botaniques de la Ville de Gèneve, Missouri Botanical Garden.

Park, B. S., Lee, H. K., Lee, S. E., Piao, X. L., Takeoka, G. R., Wong, R. Y., . . . & Kim, J. H. (2006). Antibacterial activity of Tabebuia impetiginosa Martius ex DC (Taheebo) against Helicobacter pylori. *Journal of Ethnopharmacology*, *105*(1–2), 255–262.

Patel, O. P., Beteck, R. M., & Legoabe, L. J. (2021). Exploration of artemisinin derivatives and synthetic peroxides in antimalarial drug discovery research. *European Journal of Medicinal Chemistry*, *213*, 113193.

Pereira, I. T., Burci, L. M., da Silva, L. M., Baggio, C. H., Heller, M., Micke, G. A., . . . & de Paula Werner, M. F. (2013). Antiulcer effect of bark extract of Tabebuia avellanedae: Activation of cell proliferation in gastric mucosa during the healing process. *Phytotherapy Research*, *27*(7), 1067–1073.

Pereyra, C. E., Dantas, R. F., Ferreira, S. B., Gomes, L. P., & Silva-Jr, F. P. (2019). The diverse mechanisms and anticancer potential of naphthoquinones. *Cancer Cell International*, *19*(1), 1–20.

Pérez, H., Diaz, F., & Medina, J. D. (1997). Chemical investigation and in vitro antimalarial activity of Tabebuia ochracea ssp. neochrysantha. *International Journal of Pharmacognosy*, *35*(4), 227–231.

Price, R. N., Commons, R. J., Battle, K. E., Thriemer, K., & Mendis, K. (2020). Plasmodium vivax in the era of the shrinking P. falciparum map. *Trends in Parasitology*, *36*(6), 560–570.

Rahman, M. M., Hossain, A. S., Mostofa, M. G., Khan, M. A., Ali, R., Mosaddik, A., . . . & Alam, A. K. (2019). Evaluation of anti-ROS and anticancer properties of Tabebuia pallida L. Leaves. *Clinical Phytoscience*, *5*, 1–12.

Rahman, M. M., Islam, M. R., Akash, S., Shohag, S., Ahmed, L., Supti, F. A., . . . & Thiruvengadam, M. (2022). Naphthoquinones and derivatives as potential anticancer agents: An updated review. *Chemico-Biological Interactions*, 110198.

Rahman, S., & Husen, A. (2021). Potential role of medicinal plants in the cure of liver and kidney diseases. In *Non-Timber Forest Products* (Husen, A., Bachheti, R. K., & Bachheti, A., pp. 229–254). Springer International Publishing AG. https://doi.org/10.1007/978-3-030-73077-2_10

Rahman, S., & Husen, A. (2022). *Asparagus racemosus* (Shatavari) and *Dioscorea villosa* (Wild yams). In *Herbs, Shrubs, and Trees of Potential Medicinal Benefits* (Ed. Husen, A.). CRC Press.

Rahman, S., & Husen, A. (2023a). *Trigonella foenum-graecum* L.: A useful medicinal plant for diabetes and other diseases. In: *Antidiabetic Medicinal Plants and Herbal Treatments* (Ed. Husen, A.). CRC Press.

Rahman, S., & Husen, A. (2023b). Role of Green Chiretta [*Andaphis paniculata* (Burm. F.)] nees in diabetes and other ailments. In *Antidiabetic Medicinal Plants and Herbal Treatments* (Ed. Husen, A.). CRC Press.

Reis, F. P., Bonfa, I. M. S., Cavalcante, R. B., Okoba, D., de Souza Vasconcelos, S. B., Candeloro, L., . . . & Toffoli-Kadri, M. C. (2014). *Tabebuia aurea* decreases inflammatory, myotoxic and hemorrhagic activities induced by the venom of *Bothrops neuwiedi*. *Journal of Ethnopharmacology*, *158*, 352–357.

Rodrigues, E. (2006). Plants and animals utilized as medicines in the Jaú National Park (JNP), Brazilian Amazon. *Phytotherapy Research: An International Journal Devoted to Pharmacological and Toxicological Evaluation of Natural Product Derivatives*, *20*(5), 378–391.

Roy, S. P., Kadiri, S. K., Karkar, V. V., & Konijeti, S. R. (2022, November). Antiparkinsonian activity of *Tabebuia impetiginosa* bark and biochemical analysis of dopamine in rat brain homogenates. In *Annales Pharmaceutiques Françaises* (Vol. 80, No. 6, pp. 853–863). Elsevier Masson.

Salatino, A., Ferreira, M. J., Udulutsch, R. G., Palacios, C. E., & Salatino, M. L. F. (2020). Fingerprinting of cerrado species based on cork lipophilic constituents. *Biochemical Systematics and Ecology*, *88*, 103989.

Schmeda-Hirschmann, G., & Papastergiou, F. (2003). Naphthoquinone derivatives and lignans from the Paraguayan crude drug "tayï pytá" (Tabebuia heptaphylla, Bignoniaceae). *Zeitschrift für Naturforschung C*, *58*(7–8), 495–501.

Sirigeri, S., Vadiraj, K. T., & Belagali, S. L. (2022). Tabebuia rosea: A prospective non-edible biodiesel feedstock. *Biofuels*, *13*(1), 17–19.

Stepczyńska, M., Pawłowska, A., Moraczewski, K., Rytlewski, P., Trafarski, A., Olkiewicz, D., & Walczak, M. (2021). Evaluation of the mechanical and biocidal properties of lapacho from Tabebuia plant as a biocomposite material. *Materials*, *14*(15), 4241.

Suo, M. R., Isao, H., Kato, H., Takano, F., & Ohta, T. (2012). Anti-inflammatory constituents from Tabebuia avellanedae. *Fitoterapia*, *83*(8), 1484–1488.

Suo, M. R., & Yan, S. Y. (2016). Iridoid glycosides from Tabebuia avellanedae. *Chemistry & Biodiversity*, *13*(12), 1611–1616.

Tahara, T., Watanabe, A., Yutani, M., Yamano, Y., Sagara, M., Nagai, S., . . . & Iida, A. (2020). STAT3 inhibitory activity of naphthoquinones isolated from Tabebuia avellanedae. *Bioorganic & Medicinal Chemistry*, *28*(6), 115347.

Talapko, J., Škrlec, I., Alebić, T., Jukić, M., & Včev, A. (2019). Malaria: The past and the present. *Microorganisms*, *7*(6), 179.

Twardowschy, A., Freitas, C. S., Baggio, C. H., Mayer, B., dos Santos, A. C., Pizzolatti, M. G., . . . & Marques, M. C. A. (2008). Antiulcerogenic activity of bark extract of Tabebuia avellanedae, Lorentz ex Griseb. *Journal of Ethnopharmacology*, *118*(3), 455–459.

Ueda, S., Umemura, T., Dohguchi, K., Matsuzaki, T., Tokuda, H., Nishino, H., & Iwashima, A. (1994). Production of anti-tumour-promoting furano-naphthoquinones in Tabebuia avellanedae cell cultures. *Phytochemistry*, *36*(2), 323–325.

Wang, S., Long, S., Deng, Z., & Wu, W. (2020). Positive role of Chinese herbal medicine in cancer immune regulation. *The American Journal of Chinese Medicine*, *48*(7), 1577–1592.

Wicks, C., Hudlicky, T., & Rinner, U. (2021). Morphine alkaloids: History, biology, and synthesis. *The Alkaloids: Chemistry and Biology*, *86*, 145–342.

World Health Organization. (2015). *Global Technical Strategy for Malaria 2016–2030*. World Health Organization.

Yadav, D. K., Kumar, S., Teli, M. K., Yadav, R., & Chaudhary, S. (2019). Molecular targets for malarial chemotherapy: A review. *Current Topics in Medicinal Chemistry*, *19*(10), 861–873.

Yamashita, M., Kaneko, M., Iida, A., Tokuda, H., & Nishimura, K. (2007). Stereoselective synthesis and cytotoxicity of a cancer chemopreventive naphthoquinone from Tabebuia avellanedae. *Bioorganic & Medicinal Chemistry Letters*, *17*(23), 6417–6420.

Zálešák, F., Bon, D. J. Y. D., & Pospíšil, J. (2019). Lignans and neolignans: Plant secondary metabolites as a reservoir of biologically active substances. *Pharmacological Research*, *146*, 104284.

Zhang, J., Hunto, S. T., Yang, Y., Lee, J., & Cho, J. Y. (2020). Tabebuia impetiginosa: A comprehensive review on traditional uses, phytochemistry, and immunopharmacological properties. *Molecules*, *25*(18), 4294.

Zhang, L., Hasegawa, I., & Ohta, T. (2016). Anti-inflammatory cyclopentene derivatives from the inner bark of Tabebuia avellanedae. *Fitoterapia*, *109*, 217–223.

Zhang, L., Tatsuno, T., Hasegawa, I., Tadano, T., & Ohta, T. (2015). Furanonaphthoquinones from Tabebuia avellanedae induce cell cycle arrest and apoptosis in the human non-small cell lung cancer cell line A549. *Phytochemistry Letters*, *11*, 9–17.

Zia-Ul-Haq, M. (2021). Historical and introductory aspects of carotenoids. In *Carotenoids: Structure and Function in the Human Body* (pp. 1–42). Springer, Cham.

20 Antimalarial Response, Traditional and Other Potential Uses of *Swertia* Genus

Shakeelur Rahman and Azamal Husen

20.1 INTRODUCTION

For centuries, medicinal plants and their traditional knowledge systems have been applied to prevent and treat different ailments of human beings as well as veterinaries (Husen 2020, 2021, 2022a, 2022b, 2023; Rahman and Husen 2021, 2022, 2023a, 2023b; Bachheti et al. 2023). *Swertia* (family Gentianaceae) is a diverse genus comprising about 170 species of herbs mostly found in the mountainous areas of tropical regions at an altitude of 1,250–3,550 m. In Bhutan, China, India, Japan, and Nepal, some *Swertia* species are known for uses in traditional medicines to treat different diseases and illnesses. All underground and aerial parts of *Swertia* species, including the leaf, stem, root, and seed, are used in the preparations of mixtures and decoctionsto treat various illnesses. Many *Swertia* species, for instance, *Swertia corymbosa, S. hookeri* C.B. Clarke, *S. macrosperma* (C.B. Clarke) C.B. Clarke, *S. lawii* Burkill, *S. punctata* Baumg., *S. calycina* Franch., *S. petiolata*, *S. purpurascens* Boiss., *S. bimaculata* (Siebold and Zucc.) Hook. f. and Thomson ex C.B. Clarke, *S. japonica* Makino., and *S. frachetiana* Harry Sm. are associated with traditional folk medicine. They are also used as substitutes for *S. chirayita* (Roxb. ex Fleming) H. Karstin India, China, Pakistan, Japan, and other Asian countries in the treatment of liver disorders, fevers, dysentery, diarrhoea, stomach problems, and other ailments (Brahmachari et al. 2004). Moreover, about 30 different *Swertia* species have been noticed in China, and they are used for medicinal purpose for treating hepatitis, cholecystitis, pneumonia, osteomyelitis, dysentery, and scabies (Geng and Chen 2018). *S. chirayita* is the important medicinal species of the genera used in different ailments (Anonymous 1976). It is a therapeutic herb recognized in the Indian pharmaceutical codex as well as the British and American pharmacopoeias. The healing importance of the herb has been documented and described by the ancient Indian herbal medicine system Ayurveda and other conventional medical systems (Joshi 2000). *S. chirayita* is an annual herb having medicinal properties such as antipyretic, antidiabetic, antispasmodic, anti-inflammatory, hypoglycaemic, hepatoprotective, antibacterial, and wound healing (Laxmi et al. 2011; Mahmood et al. 2014). The herb is mostly found in sub-temperate region of Himalayas (Kumar and Van Staden 2016). Some important medicinal compounds have been extracted from the herbs, which are effective against cancer, tumour, and AIDS (Sultana and Ahmed 2013). *S. chirayita* is also a highly effective medicinal herb in Unani and Ayurveda and has been used in the cure of malarial fevers (Bhat and Surolia 2001). About 200 phytochemical compounds with different structural patterns have been investigated; these are xanthone glycosides, xanthones, iridoids, or secoiridoids, and triterpenoids comprise major classes of compounds reported from genus *Swertia* (Brahmachari et al. 2004).

20.2 BOTANICAL DESCRIPTION

S. chirayita (Figure 20.1) is an herbaceous plant of up to 1 m in length and 6 mm in diameter. The whole plant contains a specific type of bright yellowish colour in fresh sample. The stem of the herb

 DOI: 10.1201/9781003378396-20

is porous, purple-brown, smooth, woody, thin bark; basal part is cylindrical, while quadrangular towards ascending branches. The herb contains large, continuous, and readily detachable yellow pith. The leaves are opposite sessile, broad at base, smoothed surface, entire ovate lanceolate; five-nerved with rounded base, acuminate apex and entire margin, narrow oval shape tapering to a point at each end, acuminate, generally with five to seven easily recognizable lateral veins. The flowers are 2–3 mm wide, tetramerous, and the shapes are oval, consist of calyx and corolla each four-lobed, with two glandular depressions near the base of each of corolla lobes, stamens 4, perigynous. The ovary is oval-shaped and pointed, having two carpels, unilocular, with two parietal placentae; style slender with recurred stigmas, fruit an ovoid yellowish brown unilocular many-seeded capsule; odour is indistinct, and taste is very bitter. The fruit has several tiny reticulated seed around 0.25 mm in length and 0.16 mm to 0.45 mm in breadth with irregular oval shape. It bears simple, slenderical, with conical roots (Kirtiker 1996).

Microscopically, the inner tissue or parenchyma of leaves surround several chloroplasts; mesophyll tissue has little differentiation. The epidermis is single-layered and covered outwardly, with a thick-lined cuticle that is much distinctly developed on the upper surface than the lower one. The naked eye observation shows the upper epidermis cells have straight walls and are larger than the lower epidermis cells, which show sinus outline. Stomata are of cruciferous type and found only on the lower surface. The transect stem illustrates single-layered epidermis, externally covered with a thick-lined cuticle present in the stem in its early age; the older epidermis remains intact, but cells become flattened and tangentially elongated. Four ribs also consist of an epidermis and parenchymatous cortical cells; endodermis distinct, showing anticlinal or periclinal walls. Very small needle-shaped crystals are also present in large quantity; cells are cortical, and in some cortical cells, resin along with small drops of oil are present as dark-brown mass. The roots transaction

FIGURE 20.1 *Swertia chirayita.*

shows two to four layers of cork; secondary cortex represented by 4–12 layers of parenchymataous cells which are thick-walled. Some of them show radial wall, tangentially elongated with sinuous walls. The secondary phloem constituted of companion cells, thin-walled strands of sieve tubes and phloem parenchyma; secondary xylem composed of tracheids cells, vessels, and xylem fibres (Anonymous 1992).

20.3 DISTRIBUTION AND CULTIVATION

S. chirayita is indigenous to the shaded area of the hilly regions of northern India. It is well distributed in the temperate Himalayas at altitudes between 1,200 and 3,000 m from Kashmir to Bhutan, and in the Khasi hills in Meghalaya at 1,200 and 1,500m (Joshi and Dhawan 2005; Anonymous 2007). *S. chirayita* is an annual herb that generally grows up to a height of 3 ft or 1 m. The blooming season of the plant is between September and October. The plant grows well in loamy, sandy, and clay soils of acidic, neutral, and alkaline nature. The plant flourishes well in humid and humus-rich soils along the streams or in marshlands. Cool summer region is the most suitable place for proper development of the plant. *S. chirayita* is able to survive temperatures as low as -15°C. The plant is propagated through its seeds. Spring season is the most suitable for the seed sowing when the temperature is low. Seedlings are transplanted in the early part of summer. The dried plants are harvested in the late stages of flowering and tied up in flattish bundles to sell in the market (Bentley and Trimen 1880).

20.4 PHYTOCHEMICAL CONSTITUENTS

Pharmacologically active phytochemicals from different *Swertia* spp. have been studied that show that *Swertia* plants yielded structurally diverse compounds, including xanthonoids, iridoids/secoiridoids, flavonoid, alkaloids, and terpenoids (Pant et al. 2000). Recent pharmacological studies have confirmed that the extracts or monomeric compounds of the genus *Swertia* possess anticancer, antioxidant, antibacterial, hepatoprotective, and antidiabetic properties. These wide applications of *Swertia* spp. indicate that the presence of several bioactive molecules in this plant enhances its utility for extraction of drugs of diverse constituents for health benefits. *S. chirayita* constitutes a large number of phytochemicals which have pharmaceutical values, including swertiamarin, ursolic acid, xanthones, flavonoids, terpenoids, iridoids, amarogentin, glycosides, and secoiridoids (Table 20.1). The most important constituents that develop better test in the herb are chirantin and ophelic acid. Chirantin or amarogentin develop a light yellow colour soluble in warm water, and ophelic acid is yellow in colour, dissolved easily in water (Kumar et al. 2010). The extracts of *S. chirayita* constitute transparent gentianine, which is a monoterpene alkaloid. It is soluble in aqueous solution and used as an anaesthetic, antihistamine, anticonvulsant, and anti-inflammatory (Kumar and Van Staden 2015). Swertiamarin is measured as secoiridoid glycoside, which is effective in arthritis, malaria, diabetes, hypertension, atherosclerosis, and abdominal ulcers. Lignan are extracted from the *Swertia* spp. that is a kind of fibre-linked polyphenolic substances used as hepatoprotective. Swerchirin is extracted from the species of *Swertia*, which is a member of xanthone and tetraoxygenated xanthones, characterized as antimalarial, antihepatotoxic, and hypoglycaemic (Leong et al. 2016; Ghosal et al. 1973). *S. chirayita* contains mangiferine, a phytochemical constituent having antimicrobial, immunomodulatory, anticancer, antiallergic, and antioxidant properties (Du et al. 2018; Imran et al. 2017). Apart from the aforementioned phytochemical constituents, *S. chiraiyta* also constitutes different types of triterpenoids, including pichierenol, swertanone, gammacer-16-en-3β-ol, 21-a-H-hop-22(29)-en-3β-ol, taraxerol, swertenol, oleanolic acid, ursolic acid, episwertinol, swerta-7, 9(11)-dien-3β-ol, pentacyclic triterpenoids like β-amyrin, friedlin, chiratenol, kairatenol (Tabassum et al. 2012). It was found in the current research that naturally produced triterpenoids are effective against various human ailments, including cancers. Similarly, Taraxerol and oleanolic have analgesic properties, while ursolic acid is used as anti-inflammatory, chemoprotective, and antimicrobial (Rastogi et al. 1991). Further details are presented in Table 20.1.

TABLE 20.1
***Swertia chirayita* Phytochemical Constituents and Their Known Biological Activities**

Phytochemical Constituents	Biological Activities	Key References
Mangiferin	Antiviral, immunomodulatory, anti-inflammatory, antioxidant, antidiabetic, antitumour, anti-HIV, chemopreventive, hypoglycaemic, antiatherosclerotic, antiparkinson	Kumar and Van Staden 2015
Swertiamarin	CNS depressant, anticholinergic, antibacterial, anticancer, antihepatitis, antiatherosclerotic, cardioprotective, antidiabetic, antiarthritic	Kumar 2017
Amaroswerin	Gastroprotective	Kumar and Van Staden 2015
Amarogentin	Antileishmanial, topoisomerase inhibitor, anticancer, antidiabetic, gastroprotective, anthelmintic	Kumar and Van Staden 2015
Swerchirin	Hepatoprotective, hypoglycaemic, pro-hematopoietic, chemopreventive, blood glucose–lowering activity	Kumar and Van Staden 2015
Ursolic acid	Antitumour, antimicrobial	Kumar 2017
Sweroside	Hepatoprotective, antibacterial, hyperpigmentation, osteoporosis, anthelmintic	Sankar et al. 2017
Swertanone	Anti-inflammatory	Kumar and Van Staden 2015
Gentianine	Antimalarial, anti–hepatitis B virus, antipsychotic	Kumar and Van Staden 2015
Bellidifolin	Hypoglycaemic	Kumar and Van Staden 2015
Oleanolic acid 3,25,56,57	Antimicrobial, antitumour, anti-inflammatory, antioxidant activity	Kumar 2017
Syringaresinol	Hepatoprotective	Kumar 2017
Isobellidifolin	Hypoglycaemic	Kumar 2017
1-Hydroxy-3,7,8-Trimethoxyxanthone	Antiulcerogenic, spasmogenic agent	Kumar 2017
1-Hydroxy-3,5,8-Trimethoxyxanthone	Antimalarial	Kumar 2017
1,5,8-trihydroxy-3-Methoxyxanthone	Blood sugar lowering	Kumar 2017
Alkaloids	Antipyretic	Sankar et al. 2017
Chiratol	Anti-inflammatory	Kumar 2017
Flavonoids	Antipyretic	Sankar et al. 2017
β-Amyrin	Anti-inflammatory, antimicrobial, antifungal	Sankar et al. 2017
Xanthones	Anti-inflammatory, CNS depressant	Sankar et al. 2017

20.5 PHARMACOLOGICAL STUDIES

The genus *Swertia* show a wide range of biological and pharmacological activities, including anticancer, antidiabetic, antibacterial, antioxidant, antihepatitis, anthelmintic, antileishmanial, and neuroprotective. It has also been used as medicinal herbs for treating hepatitis, cholecystitis, pneumonia, osteomyelitis, dysentery, and scabies (Cao et al. 2013). *S. chirayita* is one of the well-known species for its efficacy in antimalarial, anthelmintic, anti-inflammatory, hepatoprotective, antiviral, and anticarcinogenic. The entire plant is used in traditional medicine; however, the root is mentioned to be the most powerful part. The antioxidant activity of *S. chirayita* against CCl4-induced toxicity mice was studied by applying different standard invitro methods to examine the properties of malondialdehyde (MDA) and antioxidant enzymes, which include superoxide dismutase (SOD), glutathione (GSH), and catalase (CAT). Powder generated by evaporation and lyophilization of ethanolic extract of *S. chirayita* was diagnosed with potent antioxidant ability (Zhao et al. 2009). The antioxidant characteristic of the herb was studied by Sharma et al. (2013) by analyzing its free radical hunting potential. It was clearly indicated that methanolic extract of all parts of the herb have antioxidant property. It was assumed that the presence of secoiridoids and flavonoids mostly determines the antioxidant property of methanolic extract of the herb (Suryawanshi et al. 2009). A group of scientists agreed that the presence of tannins and glycosides in the extract of *S. chirayita* is responsible for the antimicrobial activities (Rehman et al. 2012). The aqueous and ethanolic extracts of the herb were observed effective against some pathogenic bacteria. The plant species is useful against some gram-negative and gram-positive bacteria, such as *E. coli*, *K. pneumoniae*, and *P. vulgaris* (Rehman et al. 2012; Khalid et al. 2011). The analgesic property of ethanolic root, leaf, and stem extract of the *S. chirayita* was screened over rat, where it was observed that the extract was effective as an analgesic (Das et al. 2012; Alam et al. 2010). An experiment was conducted over albino rat to find out the anti-inflammatory effect of the herb through xanthone derivative of the plant. It was found in the experiment that the plant extract is effective against anti-inflammation (Banerjee et al. 2000). The herb also has anticarcinogenic and cytotoxic effect, as the inhibitory property of amarogentin was observed over cell growth and induced apoptosis (Saha et al. 2004). Antiviral properties of the herb extract showed effective against herpes simplex viruses (Gopalkrishna et al. 2008). The hepatoprotective nature of the herb was analyzed by executing paracetamol and galactosamine models (Nagalekshmi et al. 2011). Nagalekshmi et al. analyzed how the plant extracts develop safety to defend acute hepatotoxicity, imposed by paracetamol dosages of 150 (mg/kg) on Swiss albino mice. Significant protection against paracetamol-induced hepatotoxicity was developed with oral feeding of specific doses (100–200mg/kg) of *S. chirayita* extract. It was identified by some researchers that the extracts of the herb are effective against paracetamol-induced toxicity in initial single-layer cultures of rat hepatocytes. The lethal cell injury against CCl4 and high development of lipid peroxidation compounds due to deficiency of antioxidant enzyme activities can be prevented by xanthones, isolated from *S. chirayita* plant extract. Invivo experiments using different animal models were conducted in this study to confirm antiradical properties of xanthones-rich extract specifically to prevent toxic effects of free radicals (Balasundari 2005). Twenty-six types of phytochemical constituents have been isolated from the ethanolic extract of *S. chirayita*, and each of the compounds was analyzed invitro condition to develop protection against hepatitis B 17 virus using Hep G 2.2.15 cell lines (Zhou et al. 2015). It was found in the aqueous and methanolic extracts of *S. chirayita* that the plant has good anthelminthic properties (Iqbal et al. 2006).

20.6 ANTIMALARIAL RESPONSE

Malaria is known as one of the most prevalent and deadly infectious diseases in Asia, Africa, and the America. It has been reported that malaria causes around one million deaths every year in the world. The United Nations included the disease as one of the Millennium Development Goals, with the aim of combating malaria (WHO 2015). Malaria is an intermittent fever with irregular onset and remission, excessive thirst, heaviness in the body, generalized body pains, headache, rigors, nausea, vomiting, etc. There is a large number of plant species used in the treatment of various kinds of fevers, including malaria. *S. chirayita* is one of the popular herbs traditionally used as antimalarial drug. There are some active phytochemical constituents of *S. chirayita* reported to be effective in malarial fever; these are gentianine, swerchirin, and swertiamarin (Kwak et al. 2005; Natarajan et al. 1974). Gentianine is an effective antimalarial bioactive compound obtained from several plant species of family Gentianaceae. It is a translucent monoterpene alkaloid which is also found in *S. chirayita* (Natarajan et al. 1974). It was found that there is analgesic activity of ethanolic and methanolic extracts of various plant parts of *S. chirayita*, such as leaf, stem, and roots, applying acetic acid–induced mice model. Results of the study clearly suggest that ethanolic extract from the leaf and stem part of the plant showed inhibition into parasite proteins. The result found in their study with [35S] methionine absorption suggests antimalarial activity of the plant extract (Alam et al. 2010). Swerchirin is a medicinally foremost xanthone, obtained from *S. chirayita*, having good antimalarial properties (Saxena 1996). Swertiamarin is a secoiridoid glycoside obtained from *S. chirayita* having analgesic property (Lei et al. 1982).

20.7 TRADITIONAL AND OTHER POTENTIAL USES

Traditionally, herbs of *Swertia* genus are used as a bitter tonic in stimulating appetite and as a febrifuge in India, whereas in Japan, most of the herbs are used for treating gastrointestinal diseases, including nausea, gastroparesis, and gastric atony. *S. chirayita* is one of the important medicinal herbs traditionally used in malarial fever, blood purification, laxative, anthelmintic, jaundice, skin diseases, and eye care. Whole parts of the herb are effective in curing the ailments (Nadkarni 1989; Baitar 2000; Negi et al. 2011). The herb is an important indigenous knowledge that belongs to different tribal and indigenous communities. They are used as home remedy for treatment of malaria, chronic fever, hypertension, asthma, hepatitis, vomiting, constipation, inflammation, tuberculosis, and indigestion (Chen et al. 2011; Saha et al. 2004). Traditionally, the plant species is used in hypoglycaemic, antibacterial, antiaging, antifatigue, and antifungal diseases, which are formulated on different concentrations of the herb extract by some popular herbal companies. These formulations are Melicon V ointment, Ayush-64, Diabecon, and Mensturyl syrup. Research findings clearly reported that such plant formulations can act as active protectant of heart and can regulate blood pressure effectively (Kumar and Staden 2015).

20.8 SAFETY ISSUES

There is no adverse effect of *Swertia* spp. reported till now by any of the experts. Some of the researchers have observed mild toxicity of the herb. The herb should be avoided by those patients who are suffering from gastric or duodenal ulcers. This herb is considered safe when taken as prescribed. However, further studies are required to evaluate its biological activities in vivo and the toxicological and mutagenic properties, in order to better validate the safety of these different plant-derived compounds.

20.9 FUTURE REMARKS

Swertia spp. have several promising prospects for both traditional and modern medicines. There is a need of time to conduct comprehensive research on different antimalarial plant species, including *S. chirayita*, to compare and analyze the efficacy of phytoconstituents for its scientific authentication. The chemical extraction of these plant contents can be established as antimalarial drug. There is no report on any serious side effects or toxicity of *S. chirayita*, but further toxicological studies are still needed to confirm the safety of the herb in humans. Clinical trials of the plant constituents must be conducted to establish its efficacy. *S. chirayita* is one of the popular herbs used all over the world for various diseases. The unsustainable harvesting of the herb and destruction of habitat of the plant species are diminishing it fast. Hence, the International Union for Conservation of Nature and Natural Resources (IUCN) has also categorized the plant species as critically endangered. The National Medicinal Plant Board, government of India, has been prioritized for conservation and cultivation of the herb by developing agro-technology packages. Further research is needed in wider context of conservation and commercialization of this critically endangered herb. Some innovative biotechnological tools, including micropropagation, cryopreservation, and bioreactors, are required for conservation of the herb. Moreover, in the near future, hairy root technology can be used as a model system to improve the valuable phytochemicals of the herb. More research and innovative findings may improve the therapeutic importance of *S. chirayita* and promote their future use in modern medicine.

20.10 CONCLUSION

Swertia genus contains important herbal medicinal plants and belongs to family Gentianaceae. Traditionally, the herb is used in Ayurveda, Unani, and other folk system of medicines in the treatment of fevers, jaundice, diabetes, skin diseases, microbial, and the diseases caused by protozoan. The pharmacological importance of the herb is growing day by day because of its valuable phytochemical constituents. Different phytoconstituents of the herb, such as amarogentin, chiratin, gentiopicrin, and gentianine, are considered effective in malarial fever and other diseases. However, toxicological and histopathological studies are required to understand its mechanism and standardize the plant product against various diseases. Conservation of *S. chirayita* is an important concern because of its increasing demand and overharvesting. Applying advanced biotechnological techniques can help in the production of bioactive constituents of the herb and its conservation. This review summarized the existing botanical description, phytochemical, pharmacological properties as well as safety and conservation of the *S. chirayita*.

REFERENCES

Alam, K.D., M.S. Ali, S. Mahjabeen, S. Parvin, M.A. Akbar & R. Ahamed. 2010. In vitro antimicrobial activities of different fractions of Swertia chirata ethanolic extract. *Pakistan Journal of Pharmaceutical Sciences* 23:455.

Anonymous. 1976. *The Wealth of India*. CSIR, New Delhi, pp. 79–80.

Anonymous. 1992. *The Unani Pharmacopoeia of India, CCRUM, Department of AYUSH Ministry of Health & Family Welfare*. Government of India, Seema Offset Press, New Delhi, vol. 2, pp. 117–125.

Anonymous. 2007. *The Unani Pharmacopoeia of India, CCRUM, Department of AYUSH Ministry of Health & Family Welfare*. Government of India, Rakmo Press, New Delhi.

Bachheti, A., B.R.K. Deepti, A. Singh, M. Zebeaman, Y. Hunde & A. Husen. 2023. Bioactive constituents and health promoting compounds of underutilized fruits of the northern Himalayas of India: A review. *Food Production, Processing and Nutrition*. https://doi.org/10.1186/s43014-023-00140-5

Baitar, I.E. 2000. *Al Jame- Al- Mufrada- Al AdviaWaAlAghzia (Urdu translation)*. CCRUM, New Delhi, vol. II.

Balasundari, P., S.K. Singh & S. Kavimani. 2005. Free radical scavenging of xanthones from Swertia chirata Buch Ham and tumor cell growth inhibition. *Main Group Chemistry* 4:177–185.

Banerjee, S., S.T. Kumar, S. Mandal, P. Das & S. Sikdar. 2000. Assessment of the anti-inflammatory effects of Swertia chirata in acute and chronic experimental models in male albino rats. *Indian Journal of Pharmacology* 32:21.

Bentley, R. & H. Trimen. 1880. *Medicinal Plants* (eds. Bentley, R. and Trimen, H.). J and A Churchill, London, p. 183.

Bhat, G.P. & N. Surolia. 2001. In vitro antimalarial activity of extracts of three plants used in the traditional medicine of India. *The American Journal Society of Tropical Medicine and Hygiene* 65:304–308.

Brahmachari, G., S. Mondal, A. Gangopadhyay, D. Gorai, B. Mukhopadhyay, S. Saha& A.K. Brahmachari. 2004. *Swertia* (Gentianaceae): Chemical and pharmacological aspects. *Chemistry & Biodiversity* 1:1627–1651.

Cao, T., C. Geng, Y. Ma, K. He, H. Wang, N. Zhou, X. Zhang, Y. Tao & J. Chen. 2013. Xanthones with anti-hepatitis B Virus activity from *Swertia mussotii*. *Planta Medica* 79:697–700.

Chen, Y., B. Huang, J. He, L. Han, Y. Zhan & Y. Wang. 2011. In vitro and in vivo antioxidant effects of the ethanolic extract of Swertia chiravita. *Journal of Ethnopharmacology* 136:309.

Geng, C.A. & J.J. Chen. 2018. The progress of anti-HBV constituents from medicinal plants in China. *Natural Products and Bioprospecting* 8:227–244.

Ghosal, S., P.V. Sharma, R.K. Chaudhuri & S.K. Bhattacharya. 1973. Chemical constituents of the gentianaceae V: Tetraoxygenated xanthones of Swertia chirata Buch.-Ham. *Journal of Pharmaceutical Sciences* 62:926.

Gopalkrishna, V., H. Verma, P. Patil & R. Kolhapure. 2008. Antiviral activity of the Indian medicinal plant extract Swertia chirata against herpes simplex viruses: A study by in vitro and molecular approach. *Indian Journal of Medical Microbiology* 26:322.

Husen, A. 2020. *Exploring Medicinal Plants* (Book Series). CRC Press, Boca Raton, FL.

Husen, A. 2021. *Traditional Herbal Therapy for the Human Immune System*. CRC Press, Boca Raton, FL. https://doi.org/10.1201/9781003137955

Husen, A. 2022a. *Herbs, Shrubs and Trees of Potential Medicinal Benefits*. CRC Press, Boca Raton, FL. https://doi.org/10.1201/9781003205067

Husen, A. 2022b. *Exploring Poisonous Plants: Medicinal Values, Toxicity Response and Other Potential Uses*. CRC Press, Boca Raton, FL. https://doi.org/10.1201/b23017

Husen, A. 2023. *Antidiabetic Medicinal Plants and Herbal Treatments*. CRC Press, Boca Raton, FL. https://doi.org/10.1201/b23347

Imran, M., M.S. Arshad, M.S. Butt, J.H. Kwon, M.U. Arshad & M.T. Sultan. 2017. Mangiferin: A natural miracle bioactive compound against lifestyle related disorder. *Lipids in Health and Disease* 16:84.

Iqbal, Z., M. Lateef, M.N. Khan, A. Jabbar & M.S. Akhtar. 2006. Anthelmintic activity of Swertia chirata against gastrointestinal nematodes of sheep. *Fitoterapia* 77:463.

Joshi, P. & V. Dhawan. 2005. *Swertia chirayita*: An overview. *Current Science*: 635–640.

Joshi, S.G. 2000. *Medicinal Plants*. Oxford and IBH, New Delhi, p. 216.

Kirtiker, K.B. 1996. *Indian Medicinal Plants*. International Book Distributors, Rajpur Road, Dehradun, India, vol. 2:3, pp. 1664–1666.

Kumar, K.P.S., D. Bhowmik, C. Biswajit & M. Chandira. 2010. Swertia chirata: A traditional herb and its medicinal uses *Journal of Chemical and Pharmacological Research* 2:262.

Kumar, N. 2017. Review on *Swertia chirata* Buch.-Ham. Ex Wall: A bitter herb W.S.R. to its phytochemistry and biological activity. *International Ayurvedic Medical Journal* 5:9.

Kumar, V. & J. Van Staden. 2016. A review of *Swertia chirayita* (Gentianaceae) as a traditional medicinal plant. *Frontiers in Pharmacology* 6:308.

Laxmi, A., S. Siddhartha & M. Archana. 2011. Antimicrobial screening of methanol and aqueous extracts of *Swertia chirata. International Journal of Pharmacy and Pharmaceutical Sciences* 3(4):142–146.

Lei, Wei Ya. 1982. Swertiamarin central inhibitory effects. *Journal of Chinese Materia Medica* 13:368.

Leong, X.Y., P.V. Thanikachalam, M. Pandey & S. Ramamurthy. 2016. A systematic review of the protective role of swertiamarin in cardiac and metabolic diseases. *Biomedicine and Pharmacotherapy* 84:1051.

Mahmood, S., S. Hussain, S. Tabassum, F. Malik & H. Riaz. 2014. Comparative phytochemical, hepatoprotective and antioxidant activities of various samples of *Swertia chirayita* collected from various cities of Pakistan. *Pakistan Journal of Pharmaceutical Science* 27(6):1975–1983.

Nadkarni, A.K. 1989. *Indian Materia Medica*. Bombay Popular Prakashan, pp. 1184–1185.

Nagalekshmi, R., A. Menon, D.K. Chandrasekharan & C.K.K. Nair. 2011. Hepatoprotective activity of Andrographis paniculata and Swertia chirayita. *Food and Chemical Toxicology* 49:3367.

Natarajan, P.N., A.S. Wan & V. Zsaman. 1974. Antimalarial, antiamobeic and toxicity tests on gentianine. *Planta Medica* 25:258–260.

Negi, J.S., P. Singh & B. Rawat. 2011. Chemical constituents and biological importance of *Swertia*: A review. *Current Research in Chemistry* 3(1):1–15.

Pant, N., D.C. Jain & R.S. Bhakuni. 2000. Phytochemicals from genus *Swertia* and their biological activities. *Indian Journal of Chemistry* 39B:565–586.

Rahman, S. & A. Husen. 2021. Potential role of medicinal plants in the cure of liver and kidney diseases. In: *Non-Timber Forest Products* (eds. Husen, A., Bachheti, R.K. & Bachheti A.). Springer International Publishing AG, Cham, pp. 229–254. https://doi.org/10.1007/978-3-030-73077-2_10

Rahman, S. & A. Husen. 2022. *Asparagus racemosus* (Shatavari) and *Dioscorea villosa* (Wild yams). In: *Herbs, Shrubs, and Trees of Potential Medicinal Benefits* (ed. Husen, A.). CRC Press, Boca Raton, FL.

Rahman, S. & A. Husen. 2023a. *Trigonella foenum-graecum* L.: A useful medicinal plant for diabetes and other diseases. In: *Antidiabetic Medicinal Plants and Herbal Treatments* (ed. Husen, A.). CRC Press, Boca Raton, FL.

Rastogi, R.P., B.N. Mehrotra, S. Sinha, P. Pant & R. Seth. 1991. *Central Drug Research Institute (India), and Council of Scientific& Industrial Research (India)*. Publications & Information Directorate, Compendium of Indian Medicinal Plants, CDRI, Lucknow and National Institute of Science Communication, New Delhi: India (Central Drug Research Institute and Publications & Information Directorate, New Delhi.

Rehman, S., A. Latif, S. Ahmad & A.U. Khan. 2012. Antibacterial screening of Swetia chirayta Linn. against some Gram negative pathogenic strains. *International Journal of Appllied Biology and Pharmaceutical Technology* 4:188.

Saha, P., S. Mandal, A. Das, P.C. Das & S. Das. 2004. Evaluation of the anticarcinogenic activity of Swertia chirata Buch. Ham, an Indian medicinal plant, on DMBA-induced mouse skin carcinogenesis model. *Phytotherapy Research* 18:373.

Sankar, A.S.S., N.V. Sugathan & M. Mohan. 2017. A review on synergistic and hermetic action of *Swertia chirayita* from homoeopathic perspective. *International Journal of AYUSH Medicine & Research*:2456–6209.

Saxena, A.M. 1996. Mode of action of three structurally different hypoglycemic agents: A comparative study. *Indian Journal of Experimental Biology* 34:351–355.

Sharma, N., V. Varshney, R.P. Kala, B. Bisht & M. Sharma. 2013. Antioxidant capacity and total phenolic content of Swertia chirayita (Roxb.ex Fleming). *International Journal of Pharmaceutical Sciences Review and Research* 23:259.

Sultana, M.J. & F.R.S. Ahmed. 2013. Phytochemical investigations of the medicinal plant *Swertia chirata* Ham. *Biochemistry and Analytical Biochemistry* 2(145):2161–1009.

Suryawanshi, S., R.K. Asthana & R.C. Gupta. 2009. Assessment of systemic interaction between Swertia chirata extract and its bioactive constituents in rabbits. *Phytotherapy Research* 23:1036.

Tabassum, S., S. Mahmood, J. Hanif, M. Hina & B. Uzair. 2012. An overview of medicinal importance of Swertia chirayita. *International Journal of Applied Science and Technology* 2:298.

World Health Organization. 2015. *World Malaria Report*. WHO, Geneva, p. 201.

Zhao, H., C.L. Jones, G.A. Baker, S. Xia, O. Olubajo & V.N. Person. 2009. Regenerating cellulose from ionic liquids for an accelerated enzymatic hydrolysis. *Journal of Biotechnology* 139:47.

Zhou, N.J., C.A. Geng, X.Y. Huang, Y.B. Ma, X.M. Zhang, J.L. Wang & J.J. Chen. 2015. Anti-hepatitis B virus active constituents from Swertia chirayita. *Fitoterapia* 100:27.

Index

Note: Page numbers in *italics* indicate a figure and page numbers in **bold** indicate a table on the corresponding page. Page numbers followed by "n" with numbers refer to notes.

D

E

F

G

H

I

K

L

M

Q

R

S

T